Telecommunications in the Pacific Basin

COMMUNICATION AND SOCIETY
edited by George Gerbner and Marsha Siefert

IMAGE ETHICS
The Moral Rights of Subjects in Photographs, Film, and Television
Edited by Larry Gross, John Stuart Katz, and Jay Ruby

CENSORSHIP
The Knot That Binds Power and Knowledge
By Sue Curry Jansen

THE GLOBAL VILLAGE
Transformations in World Life and Media in the 21st Century
By Marshall McLuhan and Bruce R. Powers

SPLIT SIGNALS
Television and Politics in the Soviet Union
By Ellen Mickiewicz

TARGET: PRIME TIME
Advocacy Groups and the Struggle over Entertainment Television
By Kathryn C. Montgomery

TELEVISION AND AMERICA'S CHILDREN
A Crisis of Neglect
By Edward L. Palmer

PLAYING DOCTOR
Television: Storytelling, and Medical Power
By Joseph Turow

THE VOX DEI
Communication in the Middle Ages
By Sophia Menache

THE EXPORT OF MEANING
Cross-Cultural Readings of DALLAS
By Tamar Liebes and Elihu Katz

TELEVISION IN EUROPE
By Eli Noam

TELECOMMUNICATIONS IN EUROPE
By Eli Noam

TELECOMMUNICATIONS IN THE PACIFIC BASIN
An Evolutionary Approach
By Eli Noam, Seisuke Komatsuzaki, and Douglas A. Conn

Telecommunications in the Pacific Basin

An Evolutionary Approach

EDITED BY

ELI NOAM

SEISUKE KOMATSUZAKI

DOUGLAS A. CONN

New York Oxford

OXFORD UNIVERSITY PRESS

1994

Oxford University Press

Oxford New York Toronto
Delhi Bombay Calcutta Madras Karachi
Kuala Lumpur Singapore Hong Kong Tokyo
Nairobi Dar es Salaam Cape Town
Melbourne Auckland Madrid

and associated companies in
Berlin Ibadan

Published by Oxford University Press, Inc.
200 Madison Avenue, New York, New York 10016

Library of Congress Cataloging-in-Publication Data
Telecommunications in the Pacific Basin : an evolutionary approach /
edited by Eli Noam, Seisuke Komatsuzaki, Douglas A. Conn.
p. cm.—(Communication and society)
Includes bibliographical references and index.
ISBN 0-19-508421-7
1. Telecommunication—Pacific Area. 2. Telecommunication—Pacific Area—Deregulation. 3. Telecommunication policy—Pacific Area.
I. Noam, Eli M. II. Komatsuzaki, Seisuke, 1927– . III. Conn, Douglas A. IV. Series: Communication and society (New York, N.Y.)
HE8620.7.T45 1994
384′.099—dc20 92-46617

9 8 7 6 5 4 3 2 1

Printed in the United States of America
on acid-free paper

Preface

Although telecommunications are in flux throughout the world, nowhere has the transformation advanced further, faster and in more interesting ways than in the Pacific region. This change is closely related to the more general progress of the electronic sector, in which the countries bordering the Pacific are leaders in design and manufacture.

It is also based on the region's voracious appetite for telecommunications services. Geographically distant from the rest of the world and from each other by vast stretches of ocean, Pacific countries' economies are increasingly tied together by electronic information flows. Given an environment of enormous economic and technological change, it would be surprising if the traditional institutions of telecommunications were left unaffected. Indeed, they are being transformed almost everywhere.

The subject of this book is the telecommunications systems of the Pacific region—their past, present, and future. Their evolution provides an abundance of rich material, yet it had not been previously comprehensively presented and analyzed. This absence was unfortunate in several respects.

First, the region is comprised of economies at various stages of development, permitting an evolutionary perspective for propositions regarding the relationship between economic development and the nature of telecommunications regulation. National telecommunications systems range from public networks that have barely begun to reach the broader population to systems where a variety of economic and political forces have broken down the traditional monopoly system and prompt users and new entrants to fashion alternative types of networks. One question is whether the traditional structure of telecommunications—based on natural monopoly characteristics of networks and the use of telecommunications for social and development policy—has been only a passing phase, dependent on the state of economic and technological development, and is now changing significantly with the growth of an information economy.

Second, there is much interaction among countries. Each Pacific Basin country has been affected by reforms in telecommunications policies of the United States and Japan, and by industrial development policies countries. Interrelated policies raise the question of the dynamics and stability of an international environment comprised of different telecommunications regimes.

Third, the electronics industry in many Pacific Basin countries is growing explosively and figures prominently in national industrial policy. The role of this industry in the telecommunications structure of various countries and its interaction with institutional change in telecommunications is important, because it sheds light on broader trends of transformation.

Fourth, the Pacific Basin includes some of the world's most vigorously developing economies of our time. The telecommunications policies in these countries shape the nature of international collaboration and trade arrangements in the information sectors generally and is therefore important to understand.

This book deals with these questions. It represents a multinational effort, built upon a bilateral partnership between the United States and Japan. On the American side, the partner was Columbia University, through two of its research centers at the Graduate School of Business—the Center on Japanese Economy and Business (CJEB) and the Columbia Institute for Tele-Information (CITI) directed respectively by Professors Hugh Patrick and Eli Noam, who undertook the selection of the authors. Douglas Conn, associate director of CITI, was chiefly responsible for the management of the project and the book's editing process. In Japan, the partner was the Research Institute of Telecom-Policies and Economics (RITE), directed at the time by Professor Seisuke Komatsuzaki. Joining him as part of the organizational committee in Tokyo were Professors Tsurihiko Nambu, Susumu Nagai, Yoichi Ito, and Takehiko Musashi.

Funding for the entire project was shared by RITE in Tokyo and the Pacific Basin Studies Project of Columbia University. At Columbia, Professor Patrick played an instrumental role in conceiving the concept and setting it in motion. Professors Noam and Komatsuzaki provided the intellectual leadership, structure of the themes discussed in the chapters and identified the authors, based, in Noam's case, on travel to and interviews in fourteen countries. Special thanks also goes to Mr. Hideaki Toda, senior vice president, International Affairs, NTT, for his interest and assistance.

These bilateral partners recruited thirty-one authors covering fourteen countries and the Pacific Island nations into one of the most outstanding group of experts on Pacific Basin telecommunications brought together. An authors' conference was held at Hosei University, Tokyo, in October 1988, a larger public event took place at Gakushuin University, Tokyo, shortly thereafter. This volume presents the project's studies.

Many people beyond the authors and editors participated in the creation of this volume. In New York, at Columbia Business School, the conference organizers were assisted by Justin Doebele, Christopher Dorman, and Richard Kramer at the Columbia Institute for Tele-Information and Francis Rosenbluth, Sherry Ranis, Robert Uriu, Charles Curtis, and Jonathon McHale at the Center on Japanese Economy and Business. Additional research and editorial assistance at Columbia was offered by Mark Lofstrum, Robert Baker, and Peggy Dannemann. Larry Meissner completed the complex final editorial stages and was helpful in updating portions of the book.

In Tokyo, RITE staff, including Junichi Ikejima, Haruko Yamashita and Tadashi Himeno, handled much of the conference logistics.

New York E. N.
Tokyo S. K.
February 1993 D. C.

Contents

III Advanced Networks in Transition

IV Beyond Universal Service

Contributors

Thomas G. Aquino is head of the Business Economics School of the Center for Research and Communication in the Philippines. He earned an A.B. degree in economics from the University of the Philippines, an M.S. in industrial economics from the Center for Research and Communication, and a D.B.A. from the University of Navarre in Spain. His fields of specialization include industrial policy and planning, corporate strategy, and technology development.

Trevor Barr is associate professor, Swinburne Institute of Technology, and a director of the Australian Electronics Development Centre in Melbourne, a joint public-private sector educational consortium. A regular media commentator on public policy issues affecting Australian media and telecommunications, he was senior consultant to the Commission For The Future in 1987, an Australian Commonwealth government advisory agency within the Department of Industry, Technology and Commerce. His publications include *The Electronic Estate: New Communications Technology and Australia* (Penguin, 1985) and (as general editor) *Challenges and Change: Australia's Information Society* (Oxford University Press, 1987).

Srisakdi Charmonman is distinguished professor of computer engineering at King Mongkut's Institute of Technology Ladkrabang and chairman of the Board of the Graduate School of Computer Information Systems at Assumption Business Administration College, both in Bangkok. He earned his Ph.D. from Georgia Institute of Technology in 1964. He was director of graduate studies in computer science at the University of Missouri and a professor of a computer science at the State University of New York before returning to Thailand in 1974. He has directed more than twelve research projects and published more than 250 articles on various aspects of computers and communications in local and international journals.

Kwang Yung Choo is professor of communication at Seoul National University. He was a visiting scholar at the department of journalism, University of Texas, Austin during 1992. His major areas of interest are international communications and telecommunications. He is author of *The Challenge of the Information Society*. He received a B.A. from Seoul National University, an M.A. in journalism from University of North Texas, and a Ph.D. in communications from the University of Texas, Austin.

Douglas A. Conn is associate director of the Columbia Institute for Tele-Information. He has published widely on U.S. and international telecommunications issues. His publications include contributions to *U.S.–Japan Trade In Telecommunications: Conflict and Compromise* (Greenwood, 1993) and *Telephone Company and Cable Television Competition* (Artech House, 1990) and he has been a contributing columnist to the *Hong Kong Economic Journal.* He holds a B.S. from Northwestern University, an M.A. from The Annenberg School of Communications at the University of Pennsylvania, and an M.B.A. from Columbia University.

Peter F. Cowhey is a professor in the Graduate School of International Relations and Pacific Studies and the political science department at the University of California, San Diego. His books on international telecommunications and trade include *Managing the World's Economy* (1992) and *When Countries Talk* (1988).

Jerome J. Day, Jr., is director of the Centre for Computing Applications and Services at Hong Kong Baptist College. In Hong Kong since 1972, he previously taught in the M.B.A. programs of the Chinese University of Hong Kong and developed their Computer Services Centre. In addition to information systems management and educational computing, his specializations include telecommunications, and he has served as chairman of the Hong Kong Management Association Telecommunications User Group and as member of the Hong Kong Government Telecommunications Board. He earned a B.S. degree in physics from Holy Cross College and an M.B.A. from the Wharton School of Business, University of Pennsylvania.

Herbert S. Dordick is professor of communications and chair of the Department of Radio, Television, and Film at Temple University in Philadelphia. He is the author or co-author of eleven books including *The Emerging Network Marketplace, Understanding Modern Telecommunications, The Executive's Guide to Information Technology,* and *Innovative Management Using Telecommunications,* and has published numerous scholarly articles dealing with both international and domestic communications policy and regulation. In 1985 he was a Fulbright Scholar at Victoria University in New Zealand. His degrees are in electrical engineering from Swarthmore College and the University of Pennsylvania.

Fong Chan Onn is deputy minister of education in Malaysia. He was professor of applied economics and chairman of the Division of Applied Economics in the Faculty of Economics and Administration, University of Malaya, Kuala Lumpur. He received a BEng (Electrical) from the University of Canterbury, and an M.B.A. and Ph.D. from the University of Rochester. He has served as a consultant to the Malaysian government, World Bank, ILO, ESCAP, and UNFPA, and has published extensively in international journals, including *Management Science, Operations Research, Transportation Science, Journal of Developing Areas,* and *The Developing Economies*. His books include *Technological Leap: Malaysian Industry in Transition* (Oxford University Press,

1986) and *New Economic Dynamo: Structures and Investment Opportunities in the Malaysia Economy* (Allen & Unwin, 1986).

Youichi Ito is professor in the faculty of policy management at Keio University (Fujisawa) and vice director of the Institute for Communications Research, Keio University (Tokyo). A member of the International Council of the International Association for Mass Communication Research, he serves as a member of the editorial boards of *Communication Theory, Journal of Development Communication, Media, Culture & Society,* and *The Journal of Development Communication,* and as editor of the *Keio Communication Review.* He has been active in numerous advisory committees for Japanese government agencies, especially the Ministry of Post and Telecommunications. He was educated at Keio, the School of Public Communications at Boston University, and the Fletcher School of Law and Diplomacy at Tufts. Among his publications in English are "Johoka [Information] as a Driving Force of Social Change" (*Keio Communication Revw* issue 12, pp. 33–58) and "Telecommunications and Industrial Policy in Japan" (in *Telecommunications Regulation and Deregulation in Industrial Democracies* edited by Marcellus Snow, Longman/Annenberg, 1986).

Hudson Janisch is professor, faculty of law, at the University of Toronto. He has been a visiting professor at the Department of Communications at Simon Fraser University. He teaches administrative law and communications law and policy. He has consulted extensively for business and governments, and in 1992 was counsel to the Canadian Senate Standing Committee on Communications. He is chair of the telecommunications section, editorial board, *Media and Communications Law Review,* and was the recipient of the 1991 Canadian Business Telecommunications Alliance Honorary Award. He holds a B.A. from Rhodes University, an M.A. LL.B. Cambridge University, and an M.C.L. LL.M. and J.S.D. from the University of Chicago. He is a member of the Bar of Ontario, Canada.

Meheroo Jussawalla is a research associate economist at the East–West Center, Honolulu, and affiliate faculty member in the Department of Economics at the University of Hawaii. Before migrating to the United States in 1975, she was Professor of Economics and Dean of the Faculty of Social Sciences at Osmania University in India. She has published extensively on telecommunications economics and development, including *The Calculus of International Economics: A Study in the Political Economy of Transborder Data Flows, The Cost of Thinking: Information Economies of Ten Pacific Countries,* and *Information Economics: New Perspectives and Information Technology and Global Interdependence,* and serves on the editorial board of *Information Economics and Policy* and *Transborder Data Report.* She serves on the board of trustees of the Pacific Telecommunications Council, on the International Advisory Board of the Atwater Institute, Montreal, Canada, and is a director of the Hawaii Information Network System.

Myung Koo Kang is associate professor in the Department of Communications at Seoul National University. He has published numerous articles in the field and is author of *A History of Professionalization Among Korean Journalists.* He earned his M.A. in journalism and Ph.D. in mass communications from the University of Iowa.

Seisuke Komatsuzaki is professor at Tokyo University for Information Science (TUIS) and is advisor to the Research Institute of Telecom Policies and Economics (RITE), Japan. He also worked in a variety of roles at NTT for more than 20 years, primarily in the area of terminal development.

Eddie C.Y. Kuo is professor and head of the department of mass communication at the National University of Singapore and a former director of the University's Centre for Advanced Studies. A University of Minnesota Ph.D., his fields of specialization include information society, language and communication, and pluralism and social change. Major publications include *Communication Policy and Planning in Singapore* (London: Kegan Paul International, 1983) and *Language and Society in Singapore* (Singapore University Press, 1980). He also has articles in a number of journals, including *Telecommunications Policy, Journal of Communication, International Journal of Sociology of Language, Anthropological Linguistics, Columbia Journal of World Business,* and *Asian Survey.*

Liang Xiong-Jian is professor and dean of the management engineering department in Beijing University of Posts and Telecommunications. He teaches courses in the management and organization of telecommunications, and the economics and organization of posts and telecommunications. He graduated from the Department of Telecommunication Engineering in Shanghai Jiaotong University in 1955 and from the Department of Engineering Economics of the graduate school of Beijing University Post and Telecommunications in 1958. He is co-author of *Management & Organization of Telecommunications* (in Chinese), and has published numerous scholarly articles.

Yeo Lin is director of the Industrial Economics Research Center at the Industrial Technology Research Institute in Hsinchu, Taiwan. She previously taught at the University of San Diego and National Cheng-Chi University in Taiwan. She received her Ph.D. in economics from Northwestern University. Her fields of specialization include industrial organization and applied microeconomics.

Vincent Lowe is professor of communications and dean of the Institute of Post Graduate Studies at the University of Sains Malaysia. He has written books and published articles in journals worldwide on issues relating to communications policies. He received a Ph.D. in political science from the Massachusetts Institute of Technology, his M.Sc. in journalism from Columbia University, and his B.A. Hons. in economics from the University of Malaya.

Mao Chi-Kuo is professor of management science at the National Chiao-Tung University, Taiwan, and director general of the Provisional Engineering Office of High Speed Rail. He participated in the 1988 Telecommunication Systems Regulatory Reform Project as a consultant on changing DGT's management organization. Other positions he has held include director general of the Tourism Bureau and secretary general of MOC. He received a B.S. in civil engineering from National Cheng-Kung University, an M.E. in system engineering from Asia Institute of Technology, Bangkok, and a Ph.D. in management science from the Massachusetts Institute of Technology.

Patrick G. McCabe is senior advisor for telecommunications and postal policy in the New Zealand Ministry of Commerce. He has been employed in the New Zealand government since 1977. His previous positions include three years as private secretary to the minister of agriculture. He has a B.A. and M.P.P. from Victoria University of Wellington, and a postgraduate diploma of agricultural science, in economics, from Lincoln University.

Susumu Nagai is professor of economics at Hosei University in Tokyo. He received an M.A. and completed doctoral courses in 1972 at Hitotsubashi University. A member of Japan Society of Information and Communication (JSIC), he is the author of *Evaluation and Analysis of Problems about Deregulated Industry, The Case of Telecommunications* (1987, in Japanese), part of the Fair Trade Commission Monograph Series, and many articles.

Tsuruhiko Nambu is professor of economics on the faculty of economics at Gakushuin University. He earned his B.A. and M.A. degrees in economics from the University of Tokyo. He has been an associate editor of the *Journal of Industrial Economics, Economics of Innovation and New Technology, Managerial and Decision Economics* and the *International Journal of Industrial Organization.* He has published many articles in such journals as *The Economic Journal, Economic Studies Quarterly,* and *Business Review* and is a contributor to *Changing the Rules* (ed. by Robert B Crandall and Kenneth Flamm, Brookings, 1989).

Eli M. Noam is professor of finance and economics at the Columbia University Graduate School of Business. He has served as public service commissioner for New York State and is currently director of the Columbia Institute for Tele-Information. He has also taught at Columbia Law School and Princeton University. He was a member of the Advisory Board for the U.S. federal government's FTS-2000 telephone network, of the National Academy of Science's review of the US Internal Revenue Service computer system, and of the National Computer Laboratory. He has published widely on telecommunications and is the author, editor, or co-editor of, among others: *Telecommunications Regulation: Today and Tomorrow* (Harcourt, 1984); *Video Media Competition* (Columbia, 1985); *The Impact of Information Technology on the Service Sector* (Ballinger, 1986); *The Law of International Telecommunications in the United States* (Nomos, 1988); *Television in Europe* (Oxford, 1992) and *Telecommunications in Europe* (Oxford, 1992). He received an A.B., a Ph.D. in economics, and J.D. law degree from Harvard University.

Jonathan L. Parapak is secretary general of the Ministry of Tourism, Posts and Telecommunications of the government of Indonesia. Previously, he was the president of Indosat, the state-owned company for international telecommunications. He has been the chairman of the board of governors of Intelsat and a member of the ITU High Level Committee. He has written and published many papers on telecommunications, information technology and management. He earned a M.Eng Sc. from the University of Tasmania, Australia.

Bohdan (Don) Romaniuk is director, regulatory matters, at Alberta Government Telephone Limited (AGT). He is also national chairperson of the Telecommunications Committee of the Media and Communications Law Section of the Canadian Bar Association. He has previously been employed at Bell Canada as a regulatory economist and was engaged in private practice law in Toronto. He has authored and co-authored a number of scholarly articles in telecommunications regulation and competition law. He received his B.A. from the University of Alberta, an M.A. in economics from Queen's University, Kingston, Ontario, and an L.L.B. from the University of Toronto. He has been admitted to the Law Society of Upper Canada (Ontario).

Chintay Shih is executive vice president of the Industrial Technology Research Institute in Hsinchu, Taiwan. He received his Ph.D. in electrical engineering from Princeton University, his MS in management from Stanford University, and his undergraduate degree from National Taiwan University. His fields of specialization include integrated circuits, semiconductor devices, and management of R&D.

Keuk Je Sung is a research fellow at the Korea Institute for International Economic Policy (KIEP). He was previously a research fellow at the Korean Information Society Development Institute (KISDI). His fields of interest include Korea–U.S. bilateral trade relations and multilateral trade negotiations. He is a member of the Korean delegation to the UR/GNS. He has written extensively on the Korean Telecommunications Industry and trilateral comparisons of business transactions in Korea, the United States, and Japan. He earned a B.A. in economics from Seoul National University and a Ph.D. degree in managerial economics and decision sciences at Northwestern University.

Tseng Fan-Tung is commissioner of the Research and Planning Committee of the Directorate General of Telecommunications (DGT), Taiwan, a researcher on tele-information industries for the Science and Technology Advisory Group (STAG) to the Executive Yuan, and deputy executive secretary of the Committee on Industry Automation within the Executive Yuan. He has previously served as deputy managing director of the Telecommunications Training Institute; and held positions at the DGT; the Taipei Satellite Communications Center; and COMSAT. He also served as executive secretary of the ad hoc working group of the Executive Yuan on cable television. He was secretary general of the China Interdisciplinary Association (CIDA) and led a study on the modernization of the legal framework on telecommunications and information. He received an M.S. in communications engineering from George Washington University and a Ph.D. in electrical engineering from National Taiwan University.

Zhu You-Nong is a professor at Beijing University of Posts and Telecommunications. Since the early 1960s she has taught courses in economics and management of posts and telecommunications. She is the co-author of *Economics of Posts and Telecommunications* and author of *Economic Accounting in Posts and Telecommunications,* as well as numerous scholarly articles. She graduated from the Department of Telecommunications Engineering in Shanghai Jiaotong University in 1952.

Ken Zita is principal at Network Dynamics in New York, a consultancy in New York. He is formerly senior market planner at AT&T Network Systems responsible for China, Taiwan, and Hong Kong. He has written numerous papers on telecommunications development in China for the U.S. Joint Economic Committee, the U.S. China Business Council and the Pacific Telecommunications Council and is author of *Modernising China's Telecommunications.* He is a member of the APEC/PECC working committees on telecommunications. He holds a B.A. from Wesleyan University.

Telecommunications in the Pacific Basin

Introduction

This book is divided into four parts. The first contains three chapters that provide a supranational and overarching perspective. The other parts, which look at fourteen countries plus the Pacific Island Nations, are grouped by the level of development within an evolutionary framework that focuses on the goals and nature of the telecom service providers. This grouping scheme, described later, often parallels, but is distinct from, an ordering by quality or technology of their telecommunications networks.

The focus of the studies presented in this volume is not on technology but on the social, economic, and political processes of change. They examine the regulatory, economic, and social change caused by technological evolution, marketplace developments, and institutional reorganization.

The chapters of the first section focus on general themes. First, Eli Noam, professor at Columbia University and former commissioner on New York State's Public Service Commission, sets the stage with a multistage evolutionary model of public telecommunications networks.

Noam argues that the evolution of networks is driven by the dynamics of group formation and their transformation into politically based redistribution; not by technology alone or by an ideological preference for competition. Inevitably, networks evolve beyond the point of stability and cohesion, and new coalitions of network users emerge. Multiple networks emerge because of its very success in establishing a universal network, not because the publicly shared network has failed to provide adequate service. Ultimately, networks grow beyond its ''tipping point'' and their cohesion breaks down as telecommunications penetration and service quality grows.

Initially, the network evolves as a means of sharing costs; growth occurs through political transfers from some users to others. It then moves to a stage that is unstable economically and politically, where the interests of participants become sufficiently divergent to generate centrifugal forces that break apart the shared system. Network expansion in the first phase makes economic and technical sense; later, expansion is politically motivated. In the third phase, the success of full service provision actually undermines the network's exclusivity. A final stage is the consolidation of various national subnetworks into global carriers.

Several of the most developed countries in the region have reached the third stage. Their policy instabilities also tend to transmit themselves to other countries in the second and first stages of network evolution; thus they have a multiplier effect. This dynamic process leads to a new type of a telecommunications environment—that of a network federation—and to a new type of policy agenda for the future.

Professor Tsurihiko Nambu of Gakushuin University, Tokyo, presents a comparison of telecommunications policies from a deregulatory perspective, encompassing evolution in Australia, Canada, South Korea, the United States, and Japan. Nambu isolates three common policy issues: the introduction of competition, the importance of universal service, and the effect of value added network (VAN) market growth in the United States and Japan on international trade in services with less-developed countries.

Professor Peter Cowhey of the University of California-San Diego discusses the role of private firms in an increasingly international environment. Cowhey posits that international institutions reinforce domestic ones, forcing them to rethink strategic positions. He devises a two-dimensional model to explain their globalization. The first dimension relies on the open or closed nature of economies, while the second is the international reach of firms. Cowhey observes that the current movements of firms is from domestic to multinational to global, transforming most information and telecommunications firms, as well as the standards-setting process, to a global orientation.

The second section of this book contains country-specific chapters that were broken down, where possible, into two parts for each country that were sometimes authored by different contributors: (1) the institutional and historical foundations of telecommunications in a country, and (2) on the domestic policy process.

The fourteen countries covered in this volume are organized along stages of the evolution of public networks described by Noam:

1. *Cost-sharing stage networks* whose growth are based on the sharing of costs and increasing the value of interconnectivity. China, Indonesia, Malaysia, the Pacific Islands, the Philippines, and Thailand are presently in this stage.
2. *Redistributory stage networks,* which grow through politically directed expansion and through transfers from some users to others. In this stage, although at various levels of development and moving toward the third stage, are Australia, Hong Kong, Singapore, South Korea, and Taiwan.
3. *Pluralistic stage networks* are not necessarily more advanced technologically than redistributory stage networks, but they have progressed institutionally. They are not a uniform system but a federation of subnetworks. The United States, Japan, New Zealand, and Canada are in transition in this stage.

The analysis of the *cost-sharing* stage countries begins with an institutional analysis of *China* by Professors Liang Xiong-Jian and Zhu You-Nong, Beijing University of Posts and Telecommunications, and is followed in Chapter 5 with an analysis of government policy in China by Mr. Ken Zita, a private consultant, formerly with AT&T International. Throughout much of its ancient history, China enjoyed relatively advanced postal communications systems. By the mid-nineteenth century, many of these postal systems, as well as the new telecommunications systems, were foreign owned. After 1949, the Ministry of Posts and Telecommunications (MPT) was faced with the task of refurbishing an antiquated and disabled network in the postrevolution era.

The MPT initiated a series of five-year plans with specific goals for network upgrades, investment and growth. However, through the 1970s, these plans failed to keep pace with other infrastructure services such as electric power and water. By 1980, there was a marked switch in policy as the central government made telecommunications a priority. Investment funds were allocated, localized pricing and procurement policies were allowed, and tax rates were lowered to regional telecommunications departments. In the manufacturing sector, the Ministry of Electronics Industry was created to coordinate domestic production, but synergies between research and development (R&D), manufacturing, and marketing has remained poor.

While China has seen fast progress in basic service since 1980, development is still too low to meet national demand. China's long-term goals include the addition of 37.6 million telephones by the year 2000, thereby increasing urban density to 25 percent. The primary obstacles to development are funding and investment administration exacerbated by the lack of convertible currency. The choices for the government are to either remain fairly centralized, or increase the telecommunications governance of the provinces, or to permit more foreign involvement. Within the Chinese MPT the decision-making process has stalled, but autonomous decisions are already being made in the provinces. However, the ultimate viability of choices has been challenged by the Post-Tiananmen economic conservatism.

Indonesia is discussed by Jonathan Parapak, secretary general of the Department of Tourism, Posts, and Telecommunications and former president and director, PT Indosat. Indonesia, a country of 187 million people, has a state-owned telecommunications system comprised of Perumtel (domestic service) and Indosat (international). This structure can be attributed to the large investment capital needed to improve the network, such as government plans to more than double installed lines between 1988 and 1994.

However, incentives for private investment have been strengthened. A 1989 law allows for private value added services, paging, and citizen-band communications. Central exchanges have been licensed to the private sector, usually in association with new residential housing development. In the electronics sector, the Indonesian government has offered reduced import duties and other incentives to attract foreign and domestic manufacturers to establish both assembly and research facilities locally.

Indonesia, together with *Malaysia,* is also the subject for Professor Vincent Lowe of the University of Sains Malaysia. Lowe points out that primary determinants of growth in Indonesian and Malaysian telecommunications are well-established relations between local telecommunications firms and the ruling parties. These relations have an overwhelming impact on procurement and contract bidding.

Malaysia, however, has had a different response to this development. There, privatization of telecommunications, unlike in Indonesia, has been sought for growth and the redistribution of wealth, which was part of a larger economic policy to have ethnic Malays own 30 percent of businesses by 1990, compared with only 1 percent in 1970. Syrarikat Telekom Malaysia (STM) was changed from a public to a private monopoly in 1987. This privatization in the telecommunications market has also promoted new value-added service ventures.

Lowe's analysis is complemented by Fong Chan Onn, deputy minister of education, Malaysia, and professor at the University of Malaya, in Chapter 7. Fong analyzes how the Malaysian telecommunications industry had developed rapidly since the 1957 Malaysian independence. In particular, with intensified export-oriented industrialization after 1970, the industry has surged in its value-added services offerings and manufacturing capability. The electronics sector, which took shape in the 1960s, has achieved tremendous annual growth rates, due in part to a readily available workforce and low-wage conditions.

Despite these developments, demand continued to outstrip supply. This remains a crucial issue even after the privatization of some telecommunications services in 1987. Fong points out three areas to stimulate development: further liberalization, including increased competition in VANs; the encouragement of foreign investments; and upgrade.

In the *Pacific Islands,* telephone penetration and sectoral investments are modest. As a result, a high percentage of the population is without basic services. In her analysis of telecommunications in the Pacific Islands, Dr. Meheroo Jussawalla from the East–West Center in Honolulu, Hawaii, highlights the importance of satellite, High Frequency (HF) single side-band radio, and microwave technologies in ameliorating these problems. The range of services and quality across the Pacific Island Nations and Papua New Guinea, among others, is quite different. The problems faced by their PTTs stem from a lack of financial and administrative autonomy while government ownership has generally prevented tariffs from reflecting true costs. While all Pacific Island Nations have Postal Telephone and Telegraph (PTT) monopolies for domestic services, some islands, such as Papua New Guinea, are slowly modifying their political ideologies and considering privatization.

Professor Thomas Aquino, Center for Research and Communications—Philippine Business Economics School, covers the *Philippines*. Here, economic and political crises in the 1980s have resulted in overall public sector investments in telecommunications and electronics that lagged far behind more basic public benefits, such as agriculture, power, social services, and transportation. The telecommunications sector is comprised of both public and private companies, which can have as much as 40 percent foreign ownership. The Philippines Long Distance Telephone Company (PLDT), a privately held company, is the largest telephone operator in the country, accounting for 93 percent of total telephone lines. The remainder is served by forty-seven private operators.

The government encouraged PLDT to be more flexible in allowing competitors into its markets; a national telephone plan aims at promoting private competition, while improving regulatory oversight of entry. PLDT does have direct competition in the domestic record carrier market and in leased line services, but the government intends to lessen PLDT's exclusivity, especially in newly developed markets.

The low level of service penetration is perhaps the greatest dilemma confronting the Philippines: Density is only 1.31 per 100 persons, and there are less than 1 million main stations in a country of 56 million. 1989 legislation provided for the installation of public telecommunications service in every town and municipality through direct government subsidies. Government policies to open the sector, however, have sent mixed signals to PLDT, which is inclined to concentrate its new efforts on high traffic areas such as Metro Manila; in turn, this may have a deleterious effect on governmental plans to encourage investments in more remote areas through its subsidies.

Professor Srisakdi Charmonman of King Mongkut Institute of Technology and Assumption Business Administration College provides the historical background and the institutional development in *Thailand* in Chapter 11. The role of the private sector in Thailand is limited to that of equipment manufacturers, suppliers, and users. Many of these are subsidiaries of large multinational corporations, which also have established manufacturing and assembly facilities geared for exportation.

By law, government agencies have a monopoly over the provision and regulation of all telecommunications. Charmonman indicates the weaknesses of Thailand's tripartite form of government regulation of telecommunications—poor quality control, slow serviceability, high rates, low penetration in rural areas, and insufficient capital investment—and suggests movement toward deregulation and competition policy as a means to correct the problems.

The countries in the cost-sharing stage rely heavily upon government intervention, through subsidies and long-term planning, to develop their telecommunications infrastructure. However, lack of capital and sometimes political instability have resulted in inadequate investment (often necessitating foreign investment) and slow expansion of service.

The countries in the second and *redistributory* stage have gone well beyond the early developmental stages and have begun to expand their networks through a redistributory process, though most operations are still closely controlled by a monopoly carrier. *Australia's* telecommunications system has been driven by the historic national goal of overcoming distance and geographic isolation. However, Australian governments have generally lacked an integrated frame of reference for decision making and policy implementation in the communications field. These characteristics of Australian telecommunications are examined by Trevor Barr, of Ericsson's Australia Electronics Development Center and the Swinburne Institute of Technology.

In the past, Australia was often considered the counterexample to open markets and liberalization, despite a fairly high state of development. But this is changing. In 1989, the government initiated cautious steps with the creation of Austel, an independent regulatory body, while reaffirming Telecom Australia's monopoly over public switched services. In 1991, AUSSAT, the domestic satellite carrier, was sold to an international consortium in order to compete with the newly merged Telecom Australia and Overseas Telecommunications Commission (OTC), which had provided all international connections.

Prior to this change, Telecom Australia had struggled to maintain its prime market power and position, though it was aware that the political climate demanded change and compromise. On one hand, organized business interests wanted more participation in the growth of the industry, and they argued for a diminution of the monopoly's power and a more laissez-faire regime. On the other hand, trade union interests wanted to ensure that an industrial status quo would be maintained. Also, traditional equipment suppliers, with a guaranteed market to the PTT, were eager to see no major structural changes.

In 1987, the Minister for Industry, Technology, and Commerce declared that Australia's information industries trade gap would be the single largest contributor to Australia's trade deficit in the 1990s. This induced a sense of urgency that prompted the government to foster a more entrepreneurial and export-oriented high-tech sector.

In *Hong Kong,* telecommunications regulation has been in a state of flux, although there has been a noticeable migration toward competition. Professor Jerome Day, Hong Kong Baptist College, presents the unique characteristics of telecommunications in Hong Kong in the next chapter. Prior to 1990, Cable & Wireless (C&W), a British-owned company, offered all international services and domestic telex, while Hong Kong Telephone, a subsidiary of C&W, provided nearly all domestic services. In 1988 C&W's local operations were consolidated into Hong Kong Telecommunications (HKT) with 10 percent government ownership and 10 percent local participation. The colony's telecommunications system historically has been operated by foreign corporations; this is likely to continue into the future, beyond the 1997 return of Hong Kong to China. Hong Kong's substantial consumer electronics industry also has historically produced goods for overseas markets, yet is often prohibited by its exporters from selling products locally.

Indeed, the regulatory power of the Hong Kong colonial government over C&W has been tenuous in the past, due to C&W's status as an instrument of the British government. In 1989, the government granted an exclusive cable television franchise, which was also to compete in the local telecommunications market, to a consortium comprised of local and foreign interests. The license was eventually vacated, but the government remains committed to competition in the local market.

Singapore's controlled evolution toward liberalization is described by Professor Eddie Kuo, National University of Singapore. This evolution has been aided by a strong government employing centralized planning and intervention in markets. Liberalization has been limited to terminal equipment and to maintenance services. There is, however, a heavy dependence on international trade in services, and this has manifested itself in the government's industrial strategic plans, which rely on the telecommunications and the computer industries. This plan, which was created and implemented by the National Computer Board (NCB) and Singapore Telecom, attempts to develop a strong export-led information sector and improve competitiveness and productivity in the domestic economy through information technology. In 1986, a plan for a more integrated national communication environment to include teleshopping, telebanking, and telecommuting was adopted for the domestic market.

One chapter on *South Korea* was written by Professors Kwang Yung Choo and Myung Koo Kang, Seoul National University. A contribution by Keuk Je Sung of the Korean Information Society Development Institute (KISDI) was added to this. The Korean Telecommunications Authority (KTA) is a government-owned corporation that was corporatized in 1982; the Ministry of Communications has exclusive regulatory authority. KTA has introduced competition since 1982 in the data communications and international long distance markets in the form of DACOM, a specialized private carrier. DACOM however, is 34 percent owned by KTA. KTA was renamed Korea Telecom in 1990.

After DACOM's success in enhanced services, other private information service providers, in areas such as data processing and information retrieval, were permitted into the South Korean market. By 1991 some 130 firms were participating in the VANS market. While politicians remained wary of surrendering Korea Telecom's power, corporations, especially in the strong computer and electronics sectors, lobbied to privatize some parts of the monopoly. These same reforms are also raising the level of R&D expenditures by South Korean manufacturing firms and government development funds, resulting in greater exports of telecommunications and computer equipment.

Foreign interests, including of the United States, have pressured for further privatization and competition through domestic joint ventures and tougher trade negotiations. However, Korea Telecom's privatization is likely to be selectively administered by the government due to Korea's low telephone penetration rate, which is still about one third that of other advanced countries.

Professor Mao Chi-Kuo, National Chiao-Tung University, and Dr. Tseng Fan-Tung, Research and Planning Committee of the DGT and deputy executive secretary, Committee on Industry Automation within the Executive Yuan, present an overview of the development of telecommunications in *Taiwan* that is followed by a chapter by Chintay Shih and Yeo Lin from the Industrial Technology Research Institute of Taiwan, who analyze the changes in the policy environment. The directorate general of telecommunications (DGT) operates as a monopoly and annual profits over total sales average 37 percent. It has long suffered from some bottlenecks in the procurement procedure and difficulty in recruiting top personnel. These problems have had long-term adverse effects on construction schedules and growth, especially in data communications services.

The DGT and its operations had not undergone any significant review until the 1980s, when a new opposition party and a strong labor constituency began to demand change. As a response to these pressures, various governments agencies, including a 1988 ad hoc committee established by the Ministry of Communications, studied the needs and feasibility to implement institutional change. The ad hoc committee's primary proposal involved separating the DGT's administration body and operating entity, and following the Japanese model of Type I and Type II carrier regulation.

On a much smaller scale, by 1987 these reviews had led to a more liberalized terminal equipment market, followed in 1990 by deregulation in the leased lines and VAN markets. In the electronic sector, however, Taiwan has been a dominant force since the 1970s. In 1984, electronic products became Taiwan's largest export sector as it moved aggressively into the microelectronic and computer industries.

The networks in the five countries analyzed in the second stage are still essentially centralized, still expanding toward more universal coverage. This is not the case for networks in third-stage countries discussed in the final chapters of the book. Technologically, they are of high standards; however, they are still largely intact institutionally. In the *pluralistic* stage, their monopoly networks begin to break apart.

Canada has a mixture of public and private ownership, discrete territorial monopolies, significant vertical integration in the private sector, and both provincial and national regulatory schemes. Telecommunications in Canada is characterized by a tangled and complex patchwork of foreign, domestic, and mixed ownership of its infrastructure. The regulatory regime consists of three different levels of government, each acting independent of each other yet with no single level of authority responsible for the market as a whole. These unique characteristics are analyzed in Chapter 19 by Professor Hudson Janisch, University of Toronto and Bohdan Romaniuk, a communications lawyer currently with Alberta Government Telephone (AGT).

Proximity to the United States has affected the business community in particular, where the U.S. model of liberalization has been long resisted but not ignored. Conversely, this proximity engenders a selective, "wait-and-see" attitude toward U.S. moves, but the long history of U.S. business involvement in Canadian Telecommunications often leads to a Canadian reaction. One example is the potential for bypass of Canadian long-distance providers through the more competitive U.S. long-distance market. However, Canada is more than a follower or stepchild of the United States. In the equipment market, for instance, Northern Telecom has captured 9 percent of the world telecommunications equipment market and over 40 percent of the U.S. switch market, and it continues to be a technological leader in several other areas.

The Canadian telecommunications services industry has been dominated for nearly sixty years by a cartel whose members are both vertically and horizontally integrated. During the 1980s, the federal government became more willing to assert itself in matters of national telecommunications policy as the procompetition business lobby grow considerably in strength. This manifested itself two ways. First, in strong pressures for competition in the long distance market, which was permitted in 1992. Second, in the challenges to the role of the provincial form of regulation such as the 1989 Court ruling that brought more companies across the country under the purview of the CRTC, the national telecommunications regulator.

New Zealand is an especially far-reaching example for deregulation in the field of telecommunications. Patrick McCabe of the New Zealand Ministry of Commerce discusses the background of the extensive deregulation which has occurred in New Zealand and Professor Herbert Dordick, Temple University, examines policy changes. New Zealand's economy has been characterized since the 1930s by long term policies of protectionist import licensing. Prior to 1984, the New Zealand economy had been a heavily regulated one. This contributed to the lack of incentive for manufacturers to innovate, slower investment in crucial public infrastructures, and ultimately, in the 1970s and 1980s, an economy heavily burdened by debt.

However, economic liberalization was surprisingly introduced by the Labour Party after its electoral victory in 1984. In 1988, the details for telecommunications deregulation were announced. The government chose to avoid what it perceived as the mistakes made in the United States, United Kingdom, and Japan. Its aim was to remove barriers to entry and allow the threat of competition to elicit the "desired response" from the incumbent, New Zealand Telecom (NZ Telecom). This response includes the rebalancing of the prices and the improvement of services.

In 1989, NZ Telecom lost its monopoly rights. The market for domestic services was opened to competition, and international telecommunications was deregulated; there are no restrictions placed on carriers which offer telecommunications services. To compete on a more equal footing, NZ Telecom was restructured into a holding company with five regional operating companies, a long-distance company, and a number of new venture subsidiaries operating on an arm's length basis. In 1990, NZ Telecom was privatized and sold to a consortium of U.S. and New Zealand companies. While New Zealand's small population size hinders widespread competition, these reforms put competitive pressures on NZ Telecom and have reshaped the domestic market. Clear Communications, also comprised of U.S. and New Zealand firms, was formed in 1991 to offer competitive long-distance service.

Japan has the largest annual gross national product of all Asian countries, which is reflected in its annual investment in telecommunications of some U.S.$12 billion. Nowhere is the dynamic of change in telecommunications more evident than in Japan, which has moved toward competition in several sectors of its telecommunications industry. In Chapter 20, Professors Yoichi Ito and Atsushi Iwata, Keio University, analyze the history of domestic Japanese telecommunications until 1978 and the development of international telecommunications service after that. The authors conclude that technological innovation in Japanese telecommunications has influenced the development of policy goals and their implementation, adding that these innovations have been of global importance, making it impossible for any nation to monopolize them.

Professor Susumu Nagai, Hosei University, then offers an analysis of institutional developments in Japanese telecommunications. By 1953, both NTT and KDD, the two dominant carriers were formed as public corporations regulated by the newly created Ministry of Posts and Telecommunications (MPT). Upon recommendation of the Ministry of International Trade and Industry (MITI) in the 1960s and 1970s, national policies were implemented to strengthen the communications and information industries.

The Japanese equipment industry, which includes such giants as NEC, Fujitsu, Hitachi, and Oki, has consistently had annual growth rates of between 12 and 15 percent. Their rise in the microelectronics and consumer electronics provided a strong foundation for their entry into the computer and telecommunications industries, where they have garnered substantial shares in such overseas markets as PBXs and underseas cable.

The telecommunications reform of 1984 made it possible for private companies to enter both international and domestic telecommunications markets in Japan. In both cases, these reforms also permitted limited foreign participation. Nagai notes that the process of deregulating data communications occurred in response to strong pressure for market entry from the data services industry and others in the electronics and telecommunications sectors. International pressure also had a considerable influence on reforms in telecommunications. Negotiations over U.S.–Japan trade imbalances have resulted in the reorganization of the procurement practices of NTT, a limited opening of the VANs market to foreign competitors, and entry by U.S. firms into the satellite and mobile communications market.

Nagai also discusses the regulations and subdivisions of so-called Type I and Type II carriers in Japan. This scheme, which separates leased or software defined carriers from facilities based carriers, has been adopted by South Korea and Taiwan. Three reasons are offered for the success of the introduction of competition both domestically and internationally in Japan: cheaper rates, efficient self-selection of carriers, and easy interconnection of networks. The relative success of competition in Japan has heightened high-level policy discussions about further deregulation and the breakup of NTT, which by 1989 had become valued as the world's largest corporation.

The next chapter, dealing with the *United States of America* is authored by Eli Noam. Perhaps the most prominent change in U.S. telecommunications has been the divestiture by AT&T of its local operating companies in 1984. The U.S.'s well-ordered system dominated by one large vertically integrated telephone carrier/manufacturer was transformed by divestiture and two decades of liberalization into a complex system with a growing number of players and institutions.

Virtually all U.S. telecommunications carriers are privately owned. Since the AT&T Divestiture, the operations of various types of networks has become highly decentralized; there are numerous, local, long-distance and international carriers. Competition has emerged in local transmission, mostly for business customers, in the form of local area networks and alternative local telecommunications services (ALTs), and soon by cable television companies and mobile radio carriers.

Deregulation and divestiture did not result in the inadequate service and rising rates that many had predicted. Long-distance service rates have declined significantly while local rates, after an initial spurt, rose at about the rate of inflation (CPI); they even declined for a while. Prices for network and customer equipment are both substantially lower.

However, as new networks and markets developed, the framework of government regulation grew more complex. On the federal level, the three branches of the U.S. government play significant roles in the regulatory process while the state public service commissions, operating both in concert and in conflict with Washington, oversee intrastate operations of regulated carriers.

The twin developments of competition and deregulation were important in dismantling the telecommunications monopoly; at first, they led to a change in number but not necessarily in kind. The trend toward private networking, both facilities-based and "virtual," though largely outside the public view, has been gathering momentum. For example, while virtually 100 percent of network investments were made by public network carriers in 1980, this figure had already dropped to 66 percent by 1986; the remainder was accounted for by large users and private networks. A fundamental question is the extent to which private networks should fall within the scope of traditional regulatory principles, such as common carriage, nondiscrimination, and universal service.

Finally, Douglas Conn, Columbia University, provides a concluding chapter on the transformations ocurring in domestic and international telecommunications in the Pacific Basin. It describes the regional collaborations occuring in Pacific Basin telecommunications and discusses ten common themes shared by most countries in this volume. Collaboration is not only tied to regional organizations that facilitate trade and set standards, but also to public and private consortiums providing satellite and underseas fiberoptics links. On one hand, the interregional changes have stimulated change among those countries that had not evolved away from a monopoloy structure. On the other hand, the domestic changes have spurred companies and government entities to seek international growth through more global cooperation.

The evolution of U.S., and other countries', telecommunications toward a diverse set of networks and service providers allowed an illustration that the monopoly system was not the only feasible institutional structure. Deregulation and liberalization expanded to others in the Pacific Basin, sometimes for internal reasons, and in other instances, partly as a result of increased demand from international customers and political pressures of trade negotiations. Yet, to believe that each step taken by other industrialized economies of the Pacific Basin was merely an adjustment to trade pressure by Washington, London, or Brussels trivializes the underlying change.

There are larger forces at play. The widespread changes in policy and industry are part of a broad transition caused by diverse forces, identified in the first chapter of the book. This transition is leading toward a normalization of telecommunications sectors so that they begin to resemble much of the rest of national economies. In time, the network environment will form a pluralistic network of user associations, or a network of networks.

The liberalized telecommunications sectors, together with dynamic electronics manufacturing, have driven as well as benefitted by the development of flourishing service economies. In Singapore, for example, the expansion of the national economy is in many ways linked to service sector growth. Given the correlation between a strong service sector and economic growth, policies governing service sector productivity will remain important public issues.

As Pacific Basin economies continue to expand, so too will their prominence in world affairs. In the telecommunications field, their policies will affect international trade of goods and services and international investment behavior. The interconnection of these nations' networks to each other and to the world is but one example. Tokyo, Hong Kong, and Singapore are competing with each other to become international communications hubs for multinational corporations as information flow to and from the region. The movement toward a more global economy will not occur without this region's participation. We should expect the Pacific Basin's role in international telecommunications to grow in importance and leadership.

I

THE EVOLUTION OF TELECOMMUNICATION NETWORKS

1

The Three Stages of Network Evolution

ELI M. NOAM

Because several of the changes in international telecommunications policy originated in the United States, they are often viewed as the product of particularly U.S. business interests, wrapped in a Chicago economic ideology. Beginning in the mid-1980s, however, several other industrialized countries began adopting similar policies, or at least discussed previously unthinkable changes.

The question then arose whether the changes reflected something more fundamental, beyond the governments in power. I posit that these policy changes are indeed part of a broad transition, one in which a multiplicity of centrifugal forces transforms the traditional network into a loosely interconnected federation of subnetworks, a network of networks.[1]

Three stages of evolution in networks can be distinguished and will be discussed further:

1. *The cost-sharing network.* Expansion is based on the logic of spreading fixed costs across many participants, and increasing the value of telephone interconnectivity.
2. *The redistributory network.* The network grows through politically mandated transfers among users.
3. *The pluralistic network.* The uniformity of the network breaks apart because the interests of its numerous participants cannot be reconciled, and a federation of subnetworks emerges.
4. *The global network.* Various domestic subnetworks stratify internationally and form networks that transcend territorial constraints.

There is a logical progression to these trends. The network first expands because of economic and technical considerations; later, it expands due to political imperatives. As the network provider succeeds in offering full service to every household, however, it also undermines the foundation of its exclusivity.

Most countries are still engaged in the cost-sharing and redistributory network. A few have reached substantial penetration and have begun moving toward the pluralistic stage. Economic growth and telephone penetration are strongly correlated—historically, roughly an additional telephone per $50,000 of GNP.

Thus a progression through the various stages is reaching the high-growth nations of the Pacific Basin and Europe.

This chapter is primarily concerned with the forces leading to the emergence of the pluralistic network. The first section looks at the nature of networks. Their dynamics and evolutionary stages are taken up. The second section looks at the trends and fares leading to network centrifugalism. In the final sections the new network coalitions and their international implications are discussed.

1.1 A Theory of Network Evolution

Almost all analysis of telecommunications concentrates on the *suppliers* of services. Issues are inevitably posed in terms of AT&T versus MCI, NTT versus the NCCs (new common carriers), Intelsat versus Cable & Wireless, VANs versus basic carriers, and so on.

It is more useful, however, to examine a "*demand*-side" telecommunications analysis. Telecommunications should not be considered primarily as a service produced by carriers, but as an interaction of societal groups, with the interaction facilitated by service vendors called *carriers*. Left to its own devices, supply structure reflects the underlying interactions of users, whether in an all-encompassing "user coalition" or in smaller groupings. A universal public network interconnecting everybody with anybody under a single organizational roof is technically and financially merely one arrangement out of several.

Thus, deregulation should be seen as far more than a policy liberalizing the *entry* of suppliers. It is also the liberalization of *exit* of users from a sharing coalition that has become confining.

Integration and centrifugalism are two basic types of forces common to many social processes. In telecommunications, their current purest expressions are the moves toward the integrated services digital network (ISDN) as the "superpipe," and the establishment of modularized interconnection arrangements such as open network architecture (ONA) that introduces segmentation into the very core of the network.

Telecommunications is but one instance of the widespread ascendancy of centrifugalism within previously shared arrangements. Wherever one looks, people are breaking up social networks of interaction to form new ones. Examples abound in the United States, including public education, mass transit, public safety, dispute resolution, pension systems, health provision, electrical power and gas distribution, stock exchanges, and so on.

One way to look at a network is as a cost-sharing arrangement between several users. Fixed costs are high, marginal costs low, and a new participant helps existing ones lower their cost. In that way it is similar to a swimming pool or national defense (i.e., to a "public good"). While there is only one national defense system, however, there are many types of arrangements for swimming pools. We may want to share the pool with a few dozen families, but not with thousands. A pure public good admits everyone, but a pure private good, only one. There is a wide spectrum between (Buchanan 1965).

A telecommunications network is one intermediate example. It is not a private good, nor does it meet the two main conditions for a public good, namely nonrival consumptions and nonexcludability. In fact, nonexcludability had to be established as a legal requirement, and we call it the *universal service obligation.*

We will now develop, in a stepwise fashion, a model for network evolution and diversification.

1.1.1 The Basic Model[2]

Let the total cost of a network serving *n* subscribers be given by a function of fixed costs and variable costs. We assume that users are homogenous. (Of course, some network participants are much larger than others, but that poses no problem if we define a large organization to consist of multiple members of type *n,* such as telephone lines or terminals rather than accounts.) We assume positive network externalities to exist though at a declining rate (i.e. a subscriber is better off the more other members there are on the network, *ceteris paribus*) (including network performance and price).

We assume that the network membership is priced at average cost (i.e., that users share costs equally). (This assumption will be dropped later.) This can be shown schematically in Figure 1.1, where utility *u(n)* is steadily increasing, though at a declining rate, and price = average cost is declining, at least at first. At this stage, the network is in its cost-sharing phase.

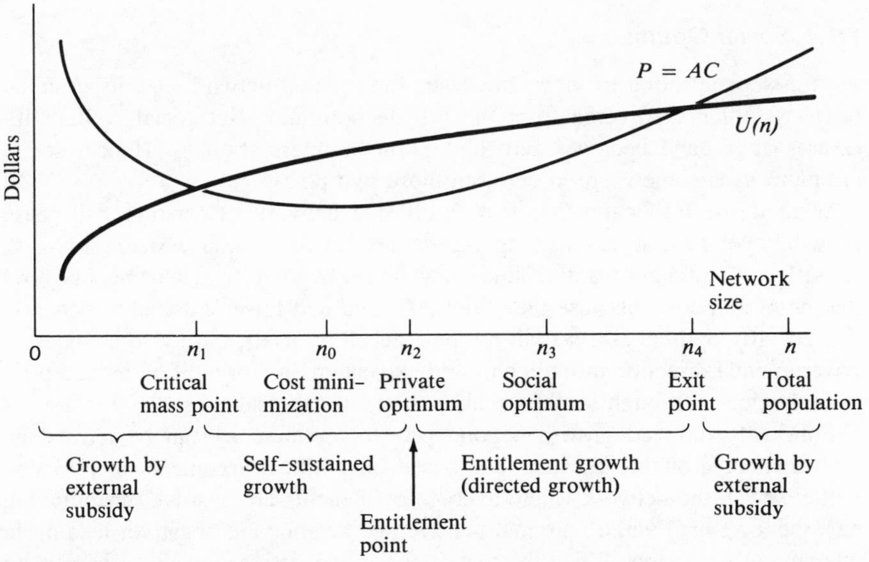

Figure 1.1

1.1.2 Critical Mass

Subscribers will find it attractive to join a well-sized network because total costs are shared by many, making average costs low, while the number of subscribers n adds to utility. This can be seen in Figure 1.1, where the utility of joining a network rises at first. Conversely, where the network is small, average cost is high and externalities small. In that range, below a "critical mass" point n_1, a network will not be feasible, unless supported by external sources.

To reach n_1 requires a subsidy of sorts, either by government or by the network operator's willingness to accept losses in the early growth phases of operations. The strategic problem is to identify in advance a situation in which such a break-even point n_1 can be reached. It is possible that such a point does not exist, and subsidies would have to be permanent in order to keep the network from imploding. We will return to the critical mass issue later.

1.1.3 Private Optimum

Through the cost-sharing phases of network growth, the earlier network users can lower their cost by adding members. However, at some point average cost AC increases.

Some further expansion would be accepted by the network members since newcomers beyond the low cost point would still add to utility. This will be up to the point n_2. Left to themselves, the existing subscribers of the network would not accept members beyond n_2, the private optimum.

1.1.4 Social Optimum

From a societal point of view, however, the optimal network size in an equal price system may diverge from the private optimum. Net social welfare increases at n_2, and becomes zero at a point of intersection n_4. Hence, social optimum n_3 is somewhere in between those two points.

What is the implication? Left to itself, the network association will cease growth beyond n_2, at least as long as costs are equally shared. Existing network subscribers would not want to admit newcomers beyond n_2. Latecomers beyond that point add cost, because they raise AC, and add fewer externality benefits. The socially optimal size n_3 will not be reached by itself, but by some external governmental direction through required expansion, and/or a differentiated pricing scheme, or through some internal politics of expansion.

Politically directed growth beyond private optimum n_2 can be termed an "entitlement growth" because it is based on political arguments of *rights* to participate in the network where average net benefits are positive (encouraging attempts of entry) while marginal net average benefits are negative, leading to attempts at exclusion. When the marginal net benefits are positive, there is no need to resort to the language of entitlements, since growth is self-sustaining and sought by network insiders. It is only beyond that point that entitlements, rights, and universal service rights (i.e., obligations by the network) become an issue.

1.1.5 Exit from the Network

If $n_2 < N$, with N being the total population, some people would be left out of the network. As discussed previously, a government would require the network to be open to other users. Yet there may well be a point where the network is expanded to an extent that, given its increasing price, a user is better off by not participating. We define n_4 as the "exit point" (i.e., the largest n such that the indifference exists between dropping off the network and sharing in the cost of supporting the expanded network).

It is possible that this exit point lies beyond the total population, $n_4 > N$. But this seems not likely under an average-pricing scheme, because the last subscribers may impose a heavy burden on the rest of subscribers, and the subsequent departure of some subscribers would lead to further reduction in the utility of the remaining members and may induce a secondary exodus. Thus, assuming $n_4 < N$, a government's aim to establish a truly universal service, without resorting to a subsidy mechanism or price discrimination, is likely to be infeasible. In other words, a universal service policy is dependent on a redistributive policy. This is the second stage of evolution, that of the redistributory network.

1.1.6 Political Price Setting and Redistribution

We have so far assumed that universal service is something imposed externally by government. In this section, however, it will be shown that the *internal* dynamics of network members can take the network toward universal service, and towards its own disintegration.

As has been shown, a network will cease to grow on its own after private optimum n_2. This conclusion, however, was based on a pricing scheme of equal cost shares. Yet there is no reason why such equality of cost shares would persist if they are allocated through a decision mechanism that permits the majority of network users to impose higher cost shares on the minority. (This assumes that no arbitrage is possible.)

Suppose for purposes of the model that decisions are made through voting by all network members.[3] Let us assume at this stage that all users are of equal size (or that voting takes place according to the number of lines a subscriber uses) and that early network users have lower demand elasticity for network use. The determinative vote is provided by the median voter located at $n/2$. A majority would not wish to have its benefits diluted by a number of beneficiaries larger than necessary. This is the principle of the "minimal winning coalition." Its size would be $n/2+1$.

A majority will establish itself such that it will benefit maximally from the minority. The minority that can be maximally burdened are the users with less elastic demand for telephone service, which are the early subscribers. But there is a limit to the burden, given by utility curve *u(n)*. If price gets pushed above *u(n)*, subscribers will drop off. Hence, the majority $n_2/2+1$) will burden the minority ($n_2/2-1$) with a price up to positive utility, and they will bear the rest of the cost.

This then is the redistributory outcome, assuming no discrimination within majority and minority.

1.1.7 Monopoly and Expansion

Such redistribution, however, is not a stable equilibrium. Before, network size n_2 was reached (once the critical mass threshold was crossed) by voluntary association. Further members were not admitted because they lowered utility to the incumbents. With internal redistribution, however, several things happen. There are now incentives for the minority network members to exit the network and form a new one in which they would not bear the redistributory burden. This would be possible if the minority were of a size larger than critical mass, $n_2/2>n_1$. Even where that were not the case, the minority could band together with those beyond network size n_2 who desire telephone service but were previously excluded.

This exit would deprive the majority of the source of its subsidy and is therefore held undesirable. The only way for the majority to prevent this "cream-skimming" or "cherry-picking" is to prohibit the establishment of another network, both by those wanting to leave the original network and similarly by those not admitted to it by being beyond n_2.

Thus, a monopoly system and the prevention of arbitrage become essential to the stability of the system.

At the same time the model predicts that the network must expand beyond n_2. For the majority, there is added utility from added network members, while most of its cost is borne by the minority. They will therefore seek expansion. The cost to the majority is only that the subsidy by the minority must be shared with more network participants. Therefore, the majority would admit new members up to the point n_5 where marginal utility to its members is equal to the marginal price due to the diluted subsidy.

This is not the end of the story, however. With expansion to n_5, the majority is now $n_5/2$ rather than $n_2/2$ (i.e. larger than before), and it can also tax a larger minority ($n_5/2$) than before. Hence, the expansion process would take place again. This process would continue, until an equilibrium, n_5* would be reached at the point where $du/dn=o$ for a minority member.

n_5* is the point up to which the network will grow under the internal dynamics described earlier. It will be larger, the greater the marginal utility from added network members is, the smaller marginal cost, and the greater fixed cost are.

1.1.8 Network Tipping

As this process of expansion takes place, the minority is growing, too. The likelihood that its size increases beyond the point of critical mass n_1 is increased, and the utility of its members, given the burden of subsidy, may well be below that of membership in a smaller but nonsubsidizing alternative network. We have so far assumed that there is only one network, and that a user's choice is whether to join or not. Suppose there are no legal barriers to the formation of a new network. In that case, a user's choice menu is to stay, to drop off altogether, or to join a new network association. Assume that the new network would have the same cost characteristics as the does traditional network. (In fact, it may well have a lower cost function for each given size if there has been accumulated monopolistic inefficiency in the existing network and rent-seeking behavior by various associated groups.)

Then, minority coalition members would find themselves to be better off in a new network *B*, and they would consider such a network, abandoning the old one. The only problem is that of transition discontinuity. A new network, in its early phases, would be a money-losing proposition up to its critical mass point n'_1. The minority will strive to exit the redistributory network once the latter's size is more than twice the size of critical mass. The network has entered its pluralistic stage.

1.1.9 Subsidies for Reaching Critical Mass

We have mentioned before that waiting for demand to materialize prior to the introduction of a network or network service may not be the optimal private or public network policy. Demand is a function of price and benefits, both of which are in turn functions of the size of the network. Hence, early development of a network may require internal or external support in order to reach critical mass.

This suggests the need, in some circumstances, to subsidize the early stages of the network—up to the critical mass point n_1—when the user externalities are still low but cost shares high. These subsidies could come either from the network provider or its membership as a start-up investment, or from an external source such as a government as an investment in "infrastructure," a concept centered around positive externalities. The question is how the internal support is affected by the emergence of a system of multiple networks.

The private start-up investment in a new form of network is predicated on an expectation of eventual break-even and subsequent positive net benefits to members. If one can expect the establishment of additional networks, however, which would keep network size close to n_1, there would be only small (or no) net benefits realized by the initial entrants to offset their earlier investment. This would be further aggravated by interconnection rights because a new network could make immediate use of the positive network externalities of the membership of the existing network that were achieved by the latter's investment. Hence, it is less likely that the initial risk would be undertaken if a loss were entirely borne by the initial network participants while the benefits would be shared with other entrants who would be able to interconnect and thus immediately gain the externality benefits of the existing network users, but without contributing to their cost-sharing. The implication is that in an environment of multiple networks that can interconnect, less start-up investment would be undertaken. It pays to be the second entrant rather than the first. A situation of market failure exists.

In such a situation, there may be a role for direct outside support, such as by a government subsidy. At first this may seem paradoxical. Should a competitive system of multiple networks not be *less* in need of government involvement than a monopoly? On second thought, however, there is some economic logic to this. Just as the subsidies to individual network users that were previously *internally* generated by other network users will have to be raised *externally* (through the normal mechanism of taxation and allocation) if at least some users are still to be supported, so might subsidies to the start-up of a network as a whole have to be provided externally. This will also be done through taxation and allocation, where network externalities as well as start-up costs are high enough to make the establishment of a network desirable.

1.1.10 Social Welfare and Multiple Networks

If network associations can control their memberships, stratification is inevitable. They will seek those members who will provide them with the greatest externality benefits—those that have many actual or potential contacts with. Furthermore, they will want to admit low-cost, high-volume, good-risk customers as members. Thus, different affinity-group networks and different average costs will emerge.

What, then, about social welfare in such a differentiated system? The traditional fear is that the loss of some cost-sharing and externalities brought by a second network would reduce social welfare. However, where the network was at n_3 or substantially larger than the socially optimal size n_4, the fracture of the network could increase social welfare, depending on the cost and utility functions, if cost closer to n_0 is reached. Where mutual interconnection is assured, one can keep the externalities benefits (and even increase them) while moving down the cost curve toward a lower AC. Furthermore, the cost curves themselves are likely to come down with the ensuing competition.

The welfare implications of the formation of collective consumption and production arrangements is something analyzed by theorists of clubs, among whom are Schelling, Buchanan, Tullock, Rothenberg, Tiebout, and McGuire. The club analysis, applied to networks, can show:

1. Given mobility of choice, different user groups will cluster together in associations according to quality, size, price, interactive density, and ease of internal decision making. When "voting with one's telecommunications node," the economically optimal association size need not encompass the entire population.
2. Service quality and optimal group size are interdependent. Thus, the optimizing size of a network's membership will vary according to quality levels sought by different user classes.
3. Optimal group size depends on the ratio of marginal utilities for different

services, set equal to the ratio of transformation in production. Thus, if different network services operate on different layers of the physical network, they will have different optimal sizes.

4. Most importantly, it is rarely Pareto-efficient to attempt income transfer by integrating diverse groups and imposing varying cost shares according to some equity criteria. Allowing homogenous groups to form their own associations and then redistributing income by imposing charges on some groups can be more efficient. Politically, however, the former may be easier to accomplish because the subsidy is not transparent (i.e., is not explicit).

The theoretically based analysis of the model described earlier means that a network coalition, left to itself under majority-rule principles, would expand beyond the size that would hold under rules of equal treatment of each subscriber. Such an arrangement can be stable only as long as arbitrage is prevented, as long as the minority cannot exercise political power in other ways, and, most importantly, as long as it has no choice but to stay within the burdensome network arrangement.

Beyond that point, however, the proexpansion policy creates incentives to form alternative networks. The more successful the network policy is in terms of achieving universal service and "affordable rates," the greater the pressures for fracture of the network. Hence, the very success of network expansion bears the seed of its own demise. This is what I call the "tragedy of the common network," in the Greek drama sense of unavoidable doom, and borrowing from the title of J. Hardin's classic article "The Tragedy of the Commons"[4] on the depletion of environmental resources. In the case of telecommunications the tragedy is that the breakdown of the common network grew from its very success—the spread of service across society and the transformation of a convenience into a necessity.

1.2 Forces of Change

Let us now become more concrete. Several broad trends contribute to new network coalitions becoming an increasingly realistic proposition. These are discussed in the following.

1.2.1 Saturation of Basic Service

For decades a primary policy goal has been to establish a network reaching every household, achievement of universal penetration in advanced industrial economies is a fairly recent phenomenon. In the United States household penetration rates peaked (at 41 percent) with the stock market in 1929 and then fell ten percentage points to a 1933 low. In 1946 household penetration passed 51 percent, reached 75 percent in 1957, 80 percent in 1962, and 90 percent by 1970 (Census 1975, p. 783). West German penetration was only 12 percent in

1960, and 75 percent by 1980. In France, it was 6 percent in 1967 and 54 percent in 1983. Universal Service was achieved by substantial redistribution.

1.2.2 Increasing Cost of Incremental Subscribers

Network characteristics typically include high costs for first users, and declining costs of subsequent users. Eventually, as in most economic processes, the marginal costs of additional users increase again, causing increased average costs. This trend is also evident in the entire Bell System, where average capital investment cost per new telephone steadily increased (in 1982–1983 dollars, *Telecom Factbook* 1986.)

1945	1955	1965	1975	1985
$1928	2050	2580	3960	4624[5]

1.2.3 An Activist Role by the Equipment Industry

Once universal penetration is reached, the supplying industry must reorient itself or face a dramatic drop in its level of activity. Having spread telephony, the supplying industry becomes a victim of its own success in saturating the market. This leaves several strategies.

Upgrade. The equipment industry advocates upgrading the network. This means an accelerated supply push rather than demand pull, and may include videotex, ISDN, broadband networks, and cable television projects.

Export. Increased attention to international activities can substitute for a shrinking domestic market. However, many markets in industrial and industrializing countries are protected by governments that use the network as a way to promote a domestic electronic industry. This results in trade frictions, but in most cases eventually leads to partial opening to achieve reciprocity.

Targeting Users as Equipment Buyers. Manufacturers turn to large users as a stable long-term market. In the United States virtually all capital investment in equipment in 1975 was by the carriers, but in 1986 the figure fell to only two-thirds. Noncarriers bought PBXs, multiplexers, concentrators, network management systems, satellite, and microwave facilities.

Users have increasingly assumed control over the network segments closest to them—first over equipment on their premises, then over the wiring segments in their buildings. Another type of user control has been through local area networks (LANs), which are privately established high-volume links serving the data flows within an organization, and among its equipment. In some organizations LAN traffic reaches 60 percent of the total communication flow. Here, too, expansion is inevitable; LANs often grow geographically into wide area networks (WANs), which may cover several continents.

The equipment industry, once a protector of the old order, has increasingly aided the creation of creating alternatives to the traditional shared network.

1.2.4 Reductions in Equipment Costs and Increases in Productivity

Another factor leading to greater subnetworking is the considerable downward shift in the economics underlying transport and switching. Network costs drop further as equipment becomes cheaper, more powerful, and lower in operating costs. Switch prices in North America fell from $230 per line in 1983 to less than $100 in 1992, while manpower requirements declined considerably and productivity increased.

Similarly, the price of fiber and of LEDs has radically dropped while their transmission capacity increased enormously. By the mid-1990s, fiber may be cheaper to install than copper.

For local distribution—in the past the segment with the greatest characteristics of "natural" monopoly—several types of new carriers have been emerging, based on radio, fiberoptic, and coaxial transmission. As the economic incentives to share in one large "network club" decline, alternative arrangements become more viable.

1.2.5 Increases in User Size

As the traffic volume of large users rises, it takes fewer users to travel down the cost curve and benefit from economies of scale. Average use per line increases annually, on average, by about 4–7 percent, as society and economy exhibit more information-intensive activities. This transition to a white-collar service sector as an area of major activity is observable around the world in economically advanced societies.

The purchase of communications capability at advantageous prices has become more important, and this has led to the emergence of a new breed—private telecommunications managers—whose function is to reduce costs for their firms and establish expertise outside the postal–industrial coalition. These managers aggressively seek low-cost transmission and customized equipment systems in the form of private networks of power and scope far beyond previous efforts.

The growth of large users means it takes a smaller number of them to reach any given volume. This reduces transaction costs of organizing and coordinating a new network, and makes it possible for a smaller number of users to enjoy economies of scale.

1.2.6 Upward Drift of the Old Network's Cost Curve

Costs and efficiencies of networks are a function of market structure as well as engineering. The traditional network operating as an exclusive arrangement tends to drift upward in cost terms. This can be exacerbated by regulatory arrangements that lead to wrong incentives, such as rate-of-return regulation with its overcapitalization. The cost reductions achieved when U.S. and Japanese companies experience competitive pressures are good indicators of such trends. This implies a new network, unencumbered by the accumulated high-cost attributes of the old one, could operate on a lower cost curve even in the presence of economies of scale.

1.3 The New Network Coalitions

The success of communalism engenders forces for particularism because the level of use and technical requirements of the some users are increasingly differentiated from those of average users. Consequently, where legally permitted, new coalitions of users are emerging. Examples are private intraorganizational networks, shared tenant services, LANs, WANs, and other specialized services.

Such private networks have begun to carve out slices from the public network. It does not take a large number of private networks to have an impact, as the operation and administration of some of them may require hundreds of skilled technicians and managers. The largest 3 percent of users typically account for 50 percent of all telephone revenues. While these activities are spearheaded by private firms, they are not exclusive to them; nonprofit institutions such as hospitals and universities, and public organizations such as state and local governments, have actively pursued similar strategies.

While most entities might participate in several networks, the pluralist networks would not require separate transmission links. Transporting the traffic of several low-volume users in "virtual" private networks over the general networks will make sense, and provide functions for traditional carriers and new forms of systems integrators. The economies of sharing are not abolished, but they must prove superior rather than being imposed by a legal requirement.

Many advocates of the traditional shared network system believe the demands of pluralism could be met by software options limited to the exclusive physical network of the traditional monopolist. This is wishful thinking. Granted, permitting software networks on a transmission monopoly is a correct first response to the emerging pressures. However, it is unlikely to be adequate in the long run. At some point users will also want to supplement transmission offerings in ways that satisfy their preferences in terms of technology, control, and economics. An exclusive network cannot be the superior solution in each instance, particularly if it has to follow political mandates or cannot bargain individually on prices.

Is it possible for the public network provider to supply each user grouping with whatever it needs, without requiring new network arrangements? Theoretically, the answer is *yes*. Indeed, some of the change is taking place on private networks supplied by the monopolists. However, it is unlikely to be institutionally adequate, if only because it requires heroic willingness by the traditional network to collaborate with schemes designed to reduce its revenue. In addition, it needs a substantial lowering of a cost-structure that has crept up over time, as suppliers and employees have shared in monopoly profits. Finally, an enormous upgrading of innovativeness and responsiveness would be required, and traditional large firms often cannot match upstart organizations.

1.4 Global Networking and the Interaction of Liberalization

The pluralist network groupings need not be territorial. The idea of telecommunications as consisting of interconnected national systems is likely to be transcended in many instances as specialized and general transnational networks emerge, spurred by the drop in cost of international circuits, and consolidating various national subnetworks.

For satellite transmission the marginal cost with respect to distance is close to zero. Fiberoptic links also have lower distance-sensitive costs. This implies communication flows can be routed in ways to exit previously shared arrangements, or to join new and more congenial ones. Opportunities for arbitrage arise, creating incentives for a country to liberalize its regulatory regime, becoming a communications "haven." This undermines administrative attempts to set rules for prices and service conditions.

Eventually, specialized global networks will emerge for a variety of groups requiring intensive communication with each other. Their relationships are functional rather than territorial, and communication links will relate participants, making traditional physical clustering of related activities less necessary.

There are unique domestic elements that affect telecommunications in each country. There are also pressures such as technological change common to all, and interactions among countries. The more interrelated countries and their economic activities become, the less likely stable solutions to domestic policy issues exist, and instability in one country affects everyone else in the system at least to some extent.

Hence, politically optimal regulation in an interrelated world may be different than for single activities in isolated jurisdictions. In most instances one would encounter a positive cross-elasticity of regulation, where liberalization in *A* leads to greater liberalization in *B*, and vice versa. In some instances, however, the cross-elasticity would be negative. One example is transborder data flow protection laws, which may lead to unstable equilibriums. The less protected data is in one country, the tighter another may become in response.

Consider the response of regulators to each other's level of regulatory strictness. One possibility is a gradual convergence that does not require coordination—an equilibrium can be reached by unilateral actions and reactions. It is also possible, however, that regulatory strictness in each country either moves successively higher or lower in response to the other countries, causing excessive deregulation or regulation (corner solutions) or cyclical change.

The alternative is coordinated "supraregulation," but it may not be stable either. One jurisdiction's adherence to an agreement provides the other with an opportunity for gain by seeking a noncooperative policy. In each jurisdiction there are pressures to seek one's own ideal regulatory level, which is likely to be different from the agreed-on level or from the interactive equilibrium. Going-it-alone can be due to short-sightedness or lack of understanding of the interaction involved, or it could be based on the rational desire to gain advantages over others by breaking joint policy, at least in the short run.

Domestic instability in the regulation of telecommunications is therefore linked to international instability in such regulation; if a country's domestic system unravels, then the resulting repercussions of adjustment in turn affect other countries with stable domestic situations.

These changes lead to unstable situations affecting the entire system, where a single inconsistency with multiple secondary effects can lead to further inconsistencies. At the same time, collaborative regulatory adjustments become more difficult, because they cannot be confined to subsectors.

In telecommunications one might therefore expect the trend toward liberalization to be spreading, though accompanied by efforts of stabilization. This process, however, is not entirely one-sided. For example, due in part to the resistance of other countries, the United States has liberalized domestic services far more than such international ones as satellite carriers. Hence, a greater internationalization of communications will make it more difficult for any one country to go its own way.

Oscillations may occur as the matrix of interrelations steadily becomes more cross-elastic, but the long-term tendency should lead to reduced international protection of the traditional network system. In this manner, network pluralism is an expansionary process. It is less an ideological choice than a response to an internal inability to structure a stable equilibrium serving multiple domestic interests and goals. One may predict that similar inconsistencies will spread throughout the system.

In the past, international interactions have often been used to stabilize domestic arrangements. Now, however, a symmetrical scenario is being played out in the opposite direction, as an international trend toward liberalization undermines domestic stability.

Notes

1. Attempts at a broad interpretation of the transformation of networks are rare. One is the Huber Report, a study of the post-divestiture American network by the U.S. Justice Department (1987) based on the relative cost of transmission and switching. Another approach is that of NTT's Hayashi (1988), who discusses the economics of networks.

2. I will follow the network analysis as developed in Noam, Eli, "The Next Stage in Telecommunications Evolution: The Pluralistic Network," paper presented at the Pacific Telecommunications Conference, Japan, October 1988; section 1.2 contains parts of the methodology of my Columbia colleague Geoffrey Heal, "The Economics of Networks," Columbia University, unpublished paper, 1989.

3. This analysis should not suggest that a voting mechanism is governing in reality (although it exists for telephone cooperatives in Finland and the United States) but rather

to understand the pressures and dynamics that are transmitted to the governmental institutions that embody the different user interests.

4. Hardin, Garrett, "The Tragedy of the Commons," *Science,* vol. 162, Dec. 13, 1968. *Tragedy* is used in the sense of Alfred North Whitehead: "The essence of traumatic tragedy is not unhappiness. It resides in the solemnity of the remorseless working of things."

5. Encompasses all U.S. carriers; translated for new telephones from data on access lines using 1975 ratio. *Sources:* Telecom Factbook, 1986; FCC Statistics of Communications Common Carriers, 1945, 1955, 1965, 1975, and 1985.

Bibliography

Baumol, William, John Panzar, and Robert Willig. 1982. *Contestable Markets and The Theory of Industry Structure.* New York: Harcourt Brace Jovanovich.

Buchanan, James M. 1965. "An Economic Theory of Clubs." *Economica* 35(125): 1–14.

Hayashi, Koichiro. 1988. "The Economies of Networking—Implications for Telecommunications Liberalization." International Institute for Communications Conference, Washington, D.C.

McGuire, Martin. 1972. "Private Good Clubs and Public Good Clubs." *Swedish Journal of Economics* 74(1): 84–99.

Noam, Eli M. 1991. "Network Tipping and the Tragedy of the Common Network: A Theory for the Formation and Breakdown of Public Telecommunications Systems." *Communications and Strategies,* pp. 43–72 (Spring).

Telecom Factbook. 1986. Washington, D.C.: Television Digest.

Tullock, Gordon. 1971. "Public Decisions as Public Goods." *Journal of Political Economy* 179(4): 913–18 (Jul.–Aug.).

U.S. Bureau of the Census. 1975. *Historical Statistics of the United States, Colonial Times to 1970,* Part 2. Washington, D.C.: U.S. Government Printing Office.

U.S. Department of Justice. 1987. *The Geodesic Network.* Washington, D.C.: U.S. Government Printing Office. The "Huber Report".

U.S. Federal Communications Commission. 1945, 1955, 1965, 1975, 1985, *Statistics of Communications Common Carriers.* Washington, D.C.: U.S. Government Printing Office.

2

A Comparison of Deregulation Policies

TSURUHIKO NAMBU

Many Pacific Basin countries will have to face issues related to deregulation or privatization even if their governments opt not to adopt such policies. Because technological configurations and demand for telecom services are as diverse as the levels of development found in various countries, it is impossible to draw general conclusions applicable to each specific situation. Rather, by stressing differences among the countries that have already embarked on the process, policy alternatives for countries in intermediate stages can be made clearer. Special emphasis is thus placed on the United States and Japan because of several contrasts between their policy orientations and regulatory schemes. Australia, Canada, and South Korea are also discussed.

Developing countries must face the problems associated with constructing basic telecommunications infrastructure—that is, they need to achieve universal service. This is generally considered to require cross subsidy, in contrast to developed nations where unbundling competition and abolition of cross subsidy are hailed. However, the style of introducing competition is instructive for developing countries because they can contemplate ways of structuring their industry from a long-term perspective. There is a saying, "Learn from other's mistakes, you don't have time to make them all yourselves." By comparing the experiences others have already had with regulation, developing countries can better plot their way.

The first section looks at the patterns of competition triggered by deregulation. It is very important to recognize that competition does not mean the same thing to everyone. Rate rebalancing is the concern of the second section. The third section takes up universal service—again, something that means different things to different people. The focus of the fourth section is the problems of market dominance by incumbents, a situation every country faces soon after the emergence of competition. Policy implications, particularly cross subsidies and value added network (VAN) services, are considered in the fifth section. The chapter concludes with implications for policy analysis from a comparative perspective.

2.1 Patterns of Competition

It is important to stress that, from a policy perspective, *competition* has a number of meanings, so one cannot adhere to a strictly uniform definition. Rather, I will make clear what is meant by each group using the term. Competition in an economic sense has two aspects. First, it is a situation where all firms are price takers—the balance of supply and demand determines the price, it is not set by any one seller or buyer, or any group. Second, in a business context the term often simply means rivalry among firms even though one or more of them may in fact have some pricing power. Monopoly has historically been the standard in telecommunications. It is therefore impossible to realize competition in the economic sense all at once; moreover, this is often undesirable because the monopoly situation has been essential to achieve the cross subsidization needed to provide universal service, an overriding public interest.

One can distinguish two practical approaches to introducing competition into a telecommunications market. The first is to accept newcomers without any explicit restriction on the number of firms. The second way is more gradual, admitting competitors in limited numbers. Under each approach one presumes some kind of (money-losing) universal service must be provided through income redistribution within the industry.

In addition to this distinction, one must distinguish local and long-distance markets. The traditional institutional framework has been characterized by distortions of tariffs in local and long-distance services that enabled regulators to subsidize local services by pricing long-distance service well above costs. If a country wishes to have income redistribution from long-distance to local service, certain interventions are inevitable.

For purposes of comparative analysis, I have chosen five Basin countries—Australia, Canada, South Korea, the United States, and Japan—and summarized the pattern of competition being discussed by government or already brought about by deregulation, as shown in Table 2.1.

VAN services are provided in competitive markets in all five countries, reflecting a general policy attitude toward VAN services that favors economic competition. Until the end of the 1980s Australia was more cautious about this than the others. Local networks are monopolies in all but Australia—where a

Table 2.1. Competition in Telecommunications in Five Countries, 1992

Australia	Canada	Japan	South Korea	United States	Market
D[a]	M	U[b]	M	M>L	Local
D[a]	M>D	U	D	U	Long Distance
D	L	D>L	D	D	Mobile Phones
U	U	U	L>U	U	VANs

D: duopoly; L: limited number of competitors; M: monopoly; U: unlimited competitors, > indicates transition from one to another is underway.

second network is to be built in the 1990s—and Japan—where actual competition remains limited.

Government policies toward the telephone network show marked differences; the dichotomy between the local and long-distance markets in the United States is most pronounced. The long-distance market is free to competition and any number of firms can enter and exit. By contrast, local services are provided by regulated companies that still function as monopolists. Of course, the threat of bypass also makes the local market potentially competitive and political forces will no doubt reshape it in this direction.

2.1.1 Country Approaches

The U.S. approach to deregulation is understandable in the sense that the differences in demand characteristics and technology were taken into consideration to demarcate markets. Demand in the interexchange market is more price elastic and diversified, and technology affords many opportunities for new firms to emerge. The local market, on the other hand, generally must meet basic demand and historically has had fewer technological alternatives (see Crandall and Flamm 1989).

When governments put a great weight on the public interest, particularly universal service—securing access to telecommunications networks—they become more cautious about permitting new entrants into their long-distance markets. To preserve this market for funding deficit services, regulators prefer to control the rate of entry and to restrict the number of newcomers. This is the approach adopted by Canada and Australia.

Canada's 1987 and 1988 proposals introduced a new framework for regulation in which Type I (facilities-owning) carriers continue to provide universal service based on cross subsidization. New entrants into Type I business are obliged to prove their entry enhances social welfare. Carriers providing local services are regulated. The situation is almost identical in Australia, where, in a 1988 announcement, the government confirmed that one of its regulatory objectives is to assure universal access to the telephone network, which means cross subsidization will be retained, albeit (as later announced) in the context of a duopoly.

In Canada and Australia, the importance of economies of scale and scope in telecommunications networks is referred to more often than they are in U.S. deregulation arguments. At the same time, structured cross subsidization in telephone pricing is still, in the early 1990s, taken for granted. Of course, these governments fully recognize that the trend of international restructuring in telecommunications has, and will continue to have, a big influence on their domestic telecom sector, and that some policy measures are required to cope with technological innovation. This cautious attitude is worth noting because each country has different conditions—such as market size, level of technological development, business and residential needs, geographical conditions, and so on.

The particular feature of deregulation in Japan is that competition has been introduced to both local and long-distance markets. The Telecommunications Business Law of 1985 divided carriers into Type I and Type II. Type I carriers are defined as businesses, large or small, that own telecommunications facilities. They require authorization for entry from the Ministry of Post and Telecommunications (MPT). The number of Type I carriers in each market is not predetermined, although MPT can regulate entry through the act's demand and supply adjustment clause. Nippon Telegraph and Telephone Corporation (NTT) provides both local and long-distance services. It also continues to provide "universal and equitable" service, as it did when it was a monopoly.

Economists find it difficult to rationalize the Japanese regulatory framework. Because local as well as long-distance markets face competition from newcomers, NTT must at some point encounter difficulties in funding universal service. Newcomers necessarily concentrate on the most profitable areas, so the old cross subsidization structure will eventually collapse. If one assumes MPT is clever enough to effectively regulate the rate of entry and to manipulate the number and geographical dispersion of entrants, what does "competition" mean? It could be that in Japan it is simply a restructuring of the old public monopoly and the substance of economic competition is not desired. It is not clear that MPT is so omnipotent as to deal successfully with changes in the telecommunications industry that will result from continuing technological advances—including changes in the very definition of universal service.

South Korea's science and technology policies are based on "The Long-Term Science and Technology Development Plan Toward 2000" which was released in 1986. Information sectors were also promoted under the Sixth Economic and Social Development Five-Year Plan (1987–1991). South Korea seems to be taking a unique approach to reshaping its telecom sector. In its long-term plan, the government announced that it will adopt nationwide flat-rate pricing for telephony while deregulating the industry. The approach is very ambitious because new information technology will be made available to the general public at very low prices. It is not certain, however, that competition can coexist if economies of scale and scope are large. This question has been raised in countries like Canada and Australia, and doubts have been expressed.

2.2 Rate Rebalancing

In every country, telecommunications regulation historically has been structured to establish a system of cross subsidization among services provided by a monopoly. The reasons have included: (1) external economies associated with communications, (2) income redistribution, (3) economies of scale, and (4) economies of scope (including vertical integration). Measures to affect cross subsidization usually consist of rate discrimination between local and long-distance calls and between business and residential customers. As a result, great discrepancies between rates and costs have existed in each service, especially between local and long distance.

Rate rebalancing is inevitable after the emergence of competition. This is because when there is competition—that is, there are no entry barriers and prices are determined by the market—companies have an incentive to increase their level of business by cutting prices—which means reducing the margin between rates and cost. Overall revenue loss is not a foregone conclusion—demand for services inevitably increases when prices are lowered for services that had been overpriced.

In the United States rebalancing took place quickly; long-distance rates fell and local rates generally rose—although comparisons regarding effective local rates are complicated by extensive unbundling, particularly the ability to own rather than lease equipment. In addition, an "FCC subscriber line charge" accruing to the local telco was added to the cost of each access line as a way of making the contribution long-distance service made to the cost of maintaining the local network more explicit. One cannot opt out of paying the fee by foregoing use of long distance, so it is technically not an access charge, although that is what it was called when it was imposed in 1984 after extensive, often angry, debate and delay.

The charge is based on the premise that there is essentially no measurable marginal cost to making a call on the local loop—and it is immaterial whether the call is local or long distance. While all long-distance callers, however, historically had paid for use of the local loop by the minute, most local calls paid nothing per minute. Even where there were local per-minute charges, they were generally (substantially) less than what was paid for a long-distance call. This is, of course, the explanation of how long distance subsidizes local service. In opting for a line charge, federal regulators recognized that although the marginal cost of using the local network is negligible, the fixed costs involved in having the network at all are large, and chose to use a fixed charge to contribute to covering these costs. No other country has adopted this drastic approach to rebalancing.

Per call and per minute charges for local calls—called measured service or message units—are used in the United States, particularly in major cities. Increased use of such a method is certainly a logical part of rebalancing, but there has been opposition. Thus, in 1986 voters in the state of Oregon passed a prohibition against mandatory measured service. During the campaign measured service as such was painted as generally bad. Ironically, most organized support for passage came from self-styled consumer advocacy groups whose claimed constituents would benefit most from measured service. Only in 1991 did Oregon telcos begin actively tariffing and promoting measured-service options to unlimited calling.

In the United Kingdom, where competition is now duopolist, rebalancing has also been occurring. Local rates rose and long-distance rates fell sharply after deregulation. In Canada, rebalancing is regarded as a precondition for introducing competition. During 1987 and 1988, long-distance rates fell about 30 percent in Canada.

By contrast, the Japanese experience is very peculiar. Since privatization of the public monopoly, Japan has had minimal rebalancing. Rebalancing, in the usual sense, seems "impossible" for two reasons. First, the benefits of rebalancing are not seen as being equally distributed. Thus, changing the status quo by raising local rates meets strong political and social resistance. There is a naïve sentiment that there should be no price increases as long as NTT remains profitable. Second, and more importantly, in the long run MPT intends to let entrants—called new common carriers (NCCs)—grow by protecting them from price competition.

Differences between rates and costs in the long-distance market serve as a kind of subsidy to the NCCs. The difference also provides an investment incentive for the NCCs. MPT wishes to avoid discouraging the new competitors until they are big enough, although nobody knows just how big that is. Differences between rates and costs in the long-distance market are still remarkable when compared with rates in other countries. Table 2.2 shows the disparity between rates in Japan and in the United States, the United Kingdom, West Germany, and France.

It is striking that the ratio of closest to farthest band in Japan is 12.0 to 1 compared to 1.23 to 1 in the United States for the same distances. Germany, France, and the United Kingdom all have much larger ratios than the United States, but not as large as Japan's. It should be noted that the low U.S. ratio is partly the result of near-band calls that are relatively expensive compared to the four other countries, plus far-band calls that are much less than in Japan, Germany and France, and about the same as in the United Kingdom. In short, calls in the United States are only somewhat distance-sensitive. (In fact, where

Table 2.2. Comparison of Long-Distance Rates, 1991*

Band (*km*)	Japan	United States	United Kingdom	West Germany	France
−20	20	70	34	20	16
20−	50	74	79	61	48
30−	60				
40−	90	78	113	182	64
60−	120				
80−	140				
100−	180	86		202	177
160−	240				
320−	330				
469−		95			
1,482		100			
3,058−[a]		101			
exchange rate		137	257	87.7	26.2

*Rates are given in yen. They are for three-minute dialed calls during weekday busines hours (peak). Japanese rates are NTT; U.S. rates are ATT&T. Japanese bands are used through 320+ km, then U.S. bands. Actual break points for other countries differ somewhat.

[a]To 4,800 km. There are two additional bands (which cover Alaska and Hawaii from the mainland) with the highest rate being 132 yen.

[b]Yen per unit of local currency.

Source: Unpublished NTT study and ATT&T published rates.

intra-local access and transport area (LATA) calls remain a monopoly, it is not uncommon for them to be as expensive as far-band interLATA calls.)

Since 1985, NTT long-distance rates have been reduced only for the closest and farthest bands. Its three competitors have lower rates in most bands, particularly between Tokyo and Osaka (covered by the farthest band, which kicks in at a mere 160 km). Initially the three NCCs did not have identical rates, but they have since 1988—15 percent or more below NTT. Despite the sizable discounts, the NCCs succeeded in gaining profits quickly. Daini Denden was in the black by 1988, after starting its telephone business in September 1987, while Japan Telecommunications was profitable in 1989, and Teleway Japan was in 1990. In contrast, in the United States, MCI took more than three years to become profitable when it started out in the mid-1970s. Sprint became profitable on an ongoing basis in 1988.

Japan's newcomers concentrate on the central business districts in Tokyo and Osaka, where their primary customers are large businesses. Customers are provided automated routing devices permitting them to shift easily from NTT to NCCs when it is worthwhile. The result is that in 1988 NTT had high-attrition rates among customers in downtown Tokyo areas like Marunouchi and Kanda, and in downtown Osaka areas like Kitahama and Honmachi.

What has happened is exactly what was expected. NCC services are perfect substitutes for NTT's. It is natural that customers shift from NTT to the NCCs when they find price differentials in available services. The expansion of NCC business depends on its supply capacity and the ability to connect to NTT's local network. NTT cannot refuse connections without good reason, so the NCCs are generally assured access to NTT's network. (This issue is discussed at greater length in Chapter 23.)

2.3 Universal Service

Good reasons exist to preserve *universal service* because access to phone service is considered essential to daily life. In this sense, externalities and income redistribution do matter. There is a question as to what *universal service* means.

2.3.1 United States

In the United States, beyond a dial tone, the nature of universal service has never been uniform. For example, basic monthly charges are a function of the size of the calling area in both Japan and the United States. While the differentials are the same across Japan, however, they are not in the United States. This is because the pricing of available services—and what service must be offered—depends in the United States on decisions by each state's public utility commission. As a result, subscribers do not have identical access rights to the network in the sense that charges for dial tones (the basic monthly rate) and for local calling (whether unlimited or measured) vary from place to place. This is not unreasonable, given that some costs vary according to such things as population density and terrain. In rural areas ''party lines'' (two or more subscribers sharing a line) were common into the 1980s, and they still exist in some areas as a way to keep costs, and thus rates, down.

Quite simply, in the United States there is no assumption that *universal* means *uniform* service. Mark S. Fowler, chair of the U.S. Federal Communications Commission in the early 1980s, ambiguously defined universal service as ''service for all at reasonable prices'' in an editorial page piece for the *Wall Street Journal* (Oct 4, 1983).

2.3.2 Japan

In Japan, the concept of universal service generally implies rates that are uniform for subscribers throughout the country. It also implied the goal of catching up with demand for residential and business lines until into the late 1970s. While this backlog was being worked off, the cost of pay phones was kept quite low and small businesses (retail shops and eating places in particular) were encouraged to have pay phones.

In the days of public monopoly, NTT used profits from long distance, especially between Tokyo and Osaka, to fund universal service. As the NCCs increase market share, NTT will face problems in financing universal service. The NCCs have been paying the local rate (10 yen) at each end of a long-distance call for the use of NTT's local network. There is a controversy over whether this is enough to cover local network costs. NTT of course asserts that it is not, and that the NCCs should be paying more. Unfortunately, the data needed to discuss this issue objectively are not available and disclosure of such information, including well-defined costs and revenues, is necessary to reach any meaningful conclusions. If the local network is found to be in deficit, a natural and conventional remedy is to levy access charges on the NCCs on a per line basis.

Some have suggested NTT should continue to provide universal service without a change in tariff structure as long as the company is profitable. This is a very dangerous suggestion from a national standpoint. As long as the difference between rates and costs is very large, new investment by the NCCs will be profitable. The difference, however, is an artificial creation, not a reliable criterion for efficient investment or dynamic economic efficiency. Moreover, it may lead to overinvestment because of the unrealistically high profitability it engenders.

At the same time, increased usage of NCC service may bring about income redistribution by making NTT customers into NCC customers, primarily ones that frequently use long distance. If NTT is correct that NCCs have not paid the costs of the local network, then NTT customers are subsidizing NCC customers. More specifically, because businesses are generally bigger users of long distance than are individuals, the subsidy would be flowing from individuals to corporations. This is a situation that cannot be justified.

Even after privatization of telecommunication markets, NTT remains the dominant firm in part because it maintains a nationwide network. This leads to the argument that the NCCs must be protected from NTT. To attain a competitive structure in the industry, it seems obvious that new entrants need to grow to an "appropriate" size. MPT is trying to protect the NCCs by setting their tariffs about 20 percent below NTT's and restricting NTT price reductions.

Of course it is impossible for MPT to keep the present divergence between rates and costs in the face of competition; timing of the cessation of protection for the NCCs thus becomes an essential issue. In this regard, the situation in the U.S. interstate market is a good reference point. By 1990 AT&T's two major competitors, MCI and US Sprint, had gained substantial shares and both were considered strong competitors. That year AT&T received 65 percent of total toll service revenues from the interLATA market, MCI got 14 percent, and Sprint, 10 percent. (Including intraLATA toll calls, AT&T's share was 51 percent, LECs had 37 percent, MCI, 11 percent, and Sprint, 8 percent.) In this context, NCCs could effectively compete with NTT if their collective share reached 20–30 percent on the more profitable routes. Their shares in the long distance market were estimated at about 9 percent in 1986 and 40 percent in 1990.

2.4 Market Dominance and New Competition

The dominant carriers in the United States and Japan have faced a difficult problem directly related to their dominance. Provision of networks poses an entry barrier to competitors providing long-distance and other services. (Since divestiture AT&T is no longer the local network provider for its competitors in long distance, but the Baby Bells face the dominance issue in their geographical areas.)

It is generally agreed that to introduce competition successfully, dominant carriers should provide newcomers equal access to the local network; however, this is not a simple matter. Perhaps the ultimate problem is determining reasonable rates for the new entrants to access the local network. Entrants naturally argue that there should be no discrimination toward them and that they should not pay any extra tariff. However, it is generally understood there is no rational rule by which to economically divide common costs into separate items. Discussions always occur over the fairness of cost allocation among services.

In the United States, the Computer III decision introduced the concept of open network architecture (ONA). In Japan no definite rules have yet been established, but this kind of discussion is anticipated. ONA seems reasonable from the perspective of entrants. There is another view, however, that is frequently taken by incumbents—usually the dominant carriers that are in advantageous positions because of their ownership of the local network. The argument is that a local network is a huge capital investment that cannot be built by newcomers. This makes economic sense only if one additionally assumes duplicating the network is a waste of resources. In the process of competition, such duplication is not necessarily an economic loss; determination depends on the growth rate of demand created by competition. It may justify duplication.

Conversely, one should note that new entrants are not confined to the existing local network. Newcomers can choose to construct or configure alternative networks. Doing so includes a complex game between the incumbent and new entrants; the former is in a superior position in the short run, but the opposite may be the case in the long run. Since ultimate purpose of telecommunications policy is to foster competition, it is important to make balanced judgments in attempting to resolve the arguments concerning access.

Here one faces another problem—monopolization of customer information by the incumbent. This is closely related to the problem of privacy, where a policy to foster competition can only go so far. Disclosure of customer information can be achieved with the consent of customers. It is not at all clear whether partial information is or can be useful to competitors. In Japan the situation is more complicated because new entrants in the local network are often subsidiaries of electric utility companies that, as local monopolies, have exclusive information about their customers for electrical services.

The difficulties encountered in the United States and Japan might suggest several challenges that should be resolved before other countries attempt to introduce competition. Into the early 1990s local networks have been, ultimately, the source of most of the policy problems encountered in deregulation. Technologically speaking, however, bypass of the local network is possible, especially where competition is profitable. One needs to compare the short- and long-run consequences. If newcomers are protected too much and given all the conveniences they require, it may inhibit incentives to realize technological innovation. Moreover, the dependence of entrants on the incumbent might be the richest opportunity for collaboration or peaceful coexistence between them.

Another problem between an incumbent and new entrants occurs in customer-premises equipment (CPE) and electronics industries in general. In the United States, AT&T used to be integrated with equipment manufacturing through Western Electric. Since divestiture, this sector has faced fierce competition from outside the United States.

By contrast, the manufacturing sector was never fully integrated with telecom carriers through ownership in Japan and other Pacific Basin countries. In Japan, NTT maintained a very close relationship with a select group of domestic manufacturers. NTT was both the biggest buyer of CPE and other electronic devices as well as a collaborator with these companies in R&D activities. In this way, NTT subsidized Japanese electronics producers, helping them to become giants in the world market. The so-called denden family consists of NEC, Hitachi, Toshiba, Fujitsu, Oki, and Mitsubishi Denki. They no longer depend as much on NTT, but they still collaborate in R&D efforts.

NTT has been criticized for exerting its purchasing power in the equipment market because it is practically a domestic monopsonist for these products. Cost-effective, comparable if not superior, foreign (i.e., mostly German, Canadian, and American) equipment was routinely frozen out by narrow specifications that had little if anything to do with actual performance—such as the color of housings. The other side of the argument is that NTT seems to have used its power primarily to force its domestic suppliers to become very cost-efficient producers of high-quality equipment, which at least in the longer run has led to marketwide cost savings captured by NTT. In any case, NTT's domestic buying helped make Japan's electronics industry extremely competitive in the world market, and Japanese companies and policymakers feel no need to apologize for having achieved this result.

2.5 Policy Implications

Two types of competition are found in telecommunications. The United States and Japan have adopted policies basically aimed at creating a competitive market structure. Canada and Australia prefer very gradual entry by a limited number of entrants. A number of reasons account for the two policies. Some are economic, but others are political and technological. One economic reason is very simple: Market size determines the degree of competition that is possible, especially when economics of scale and scope are present. In the United States and Japan, business demands for telecom services are great and will grow rapidly. This enables firms to enter markets with bullish expectations. If expectations play a positive role, the significance of sunk cost will diminish because firms will be more confident about selling off their facilities if they are required to exit the market at some point in the future.

Another reason, partially connected to the first, corresponds to the stage of development of technology in telecommunications and computers. Again, the United States and Japan have been experiencing a merger and coevolution of the two fields for some time. Lack of an internationally competitive electronics industry can be considered an obstacle to development of a telecom equipment industry.

A third economic reason relates to the behavior of business customers and technological alternatives in telecommunications. It is evident in the United States that large business users are ready to bypass traditional carriers when bypass prices are competitive. A similar phenomena exists in Japan, where the major stockholders of the NCCs are large financial and industrial firms. China also is experiencing bypass by some of its biggest users—government agencies motivated as much by a desire to control their own communications network as by any special needs the public network could not offer given the chance (and the investment going into the private networks).

For these reasons, NDCs may prefer gradual approaches to introducing competition. Generally speaking, economies of scale and scope are important, and the extent to which they are available depends on the level of demand. Domestic demand will be a determinant of an appropriate pattern for the introduction of competition. In telecommunications, attainment of scale economies may be domestically possible with fewer fears about foreign competition since telecom services are usually nontradeable goods.

2.5.1 Cross Subsidization

The stress placed on universal service shows the difference between the United States and other countries. In Japan and other countries universal service is generally perceived as some level of basic service provided nationwide with uniform quality and rates. By contrast, in the United States, the cost and quality of service differs geographically because of a regulatory structure that allows state public utility commissions to be policy makers. However, it must be noted that if access to universal service is taken simply as a phone line passing (almost) every household, it has been a fact in the US since at least the 1950s—with installation available within days most of the time in most places.

Using the U.S. approach, an interpretation is possible that the vertical integration of the network is no more efficient than a decentralized system. But there is, of course, little empirical background for this belief. In fact, it is conceivable that among less developed countries the telecommunications network will show economies of scope achieved through vertical integration. It remains true that a vertically integrated monopolistic network can provide services efficiently.

In the United States, a mechanism to adjust the nonuniform provision of universal service exists, although this is usually not a priority of state or national policies. In part this reflects a commitment to cost-based pricing. In any case, each regulatory jurisdiction has implemented some form of "lifeline" (subsidized) rate, in some cases using a specific line item charge on regular subscriber's phone bills to fund it.

The U.S. situation stands in great contrast to Japan or other countries where uniformly priced universal service is a first priority behind development of the local network. These differences must be taken into consideration before drawing conclusions from comparative analyses.

2.5.2 VAN Services

In each country, a consensus on liberalization of VAN services has been achieved. In most countries a competitive structure has already been established. This is a natural result of cost and demand conditions in this segment of the industry. A possible threat to local providers (but not their customers) in developing countries might be the comparative advantage of developed countries in this area. Because VAN services can be traded, unlike a telecommunications network, pioneers who have accumulated knowledge and experience have a tremendous advantage over latecomers to the international market. The most advanced VAN services will be geared to the most sophisticated networks. In this case, pioneers can easily override latecomers and consequently international trade friction will arise in this area, creating and fueling both domestic and international pressures for policy reforms that allow competition. (An account of one of the earliest and so far most successful international VANs—Vitel—is in the *Far Eastern Economic Review,* Aug 2, 1990, p. 41.)

The Computer III decision freed AT&T from the requirement of structurally separating its R&D activities for enhanced services and network provision, whereas in Japan a division of NTT was spun off as NTT Data Communications Systems Corp. The new company is a wholly owned subsidiary of NTT, but it will eventually become fully independent. These contrasting policies suggest that Japan has no concrete idea about R&D activities in telecommunications on a national policy basis. In the early stages of telecommunications development, cross subsidization of R&D in enhanced services by network services can usually be justified under monopoly conditions. Such subsidization, however, may be harmful for the development of enhanced services in a more competitive marketplace. Japan will eventually come to a crossroads necessitating R&D policy choices.

2.6 Implications for Pacific Basin Countries

Telecommunications historically has been provided by a public monopoly, with the United States and Canada being exceptions. When one talks about deregulation or privatization, initial differences in supply structure do matter. Countries other than the United States can hardly adopt the approach of dividing the long distance and local market because universal service is deemed essential and cross subsidy is inevitably required. Among countries where the network is underdeveloped, realizing economies of scale and reaching minimum efficient scale are of primary importance. In this respect, monopolistic supply will be preferred, but one must note that this need not mean public monopoly.

However, because technological innovations in telecommunications networks are so frequent, area-specific monopoly and the old multilayer network may no longer be a requisite for developing nationwide networks. This is a great advantage to latecomers, who can make the best use of advances in electronics. For purposes of developing efficient telecommunications networks, flexible institutional settings might be the key to exploiting the situation we find in the 1990s.

Bibliography

Crandall, Robert B., and Kenneth Flamm, eds. 1989. *Changing the Rules: Technological Change, International Competition and Regulation in Communication.* Washington, D.C.: The Brookings Institution.

3

Public and Private Cooperation in International Informatics

PETER F. COWHEY

The melding of communications and information technologies forms the core of the world's electronics and advanced services sectors. Producers and service providers utilizing these merged technologies are commonly collectively called the informatics industries. Informatics faces the market traditions of its precursor components, and these traditions are incompatibly different. Thus, domestic communications markets have been routinely closed to foreign competition—particularly in services, but also even in goods. At the same time, significant, although far from complete, openness in information (computing) has been allowed. This means more firms have been able to trade internationally and invest freely in other countries (and their own) in computing than has been the case in communications.

More broadly, in part because historically they have been otherwise more easily tradeable (transportable), there has been more openness to trade in goods than in services. Services have surpassed goods as a share of national product and new job creation in advanced economies, and competitive advantages in goods and services are becoming more mutually dependent. There has thus been a fundamental change in the nature of the environment in which existing trade and regulatory relations were created. In a sense, therefore, informatics becomes a microcosm of the broad issues of international economic relations: Conflicting traditions have to be reconciled or superseded.

This chapter addresses the question of what sort of regime will prevail for informatics. The answers lie in the interaction of corporate strategies and government choices about the openness of markets.

The regime for a market is created by governments and by firms. Government rules, principles, and decision-making methods are obviously a major part of the environment in which the world economy works. Governments set the rules for competition, but they carefully observe what their firms want. Firms in turn calculate the economic and political parameters of the market and launch new strategies based on their assessments. These corporate strategies alter the nature of international interdependence. The resulting regime is a product of both the public rules of governments and the private governance of firms. All this is true at both the domestic and international level. The concern here is with the international market and environment.

A country can have a fairly competitive domestic market but, directly or through "structural impediments," restrict competition by foreigners in terms of trade (barriers to imports) or investment. Japan usually surfaces as an example. What I have termed "openness" (which is closely related to what is talked about as contestability or market access) is the basic factor in this chapter with regard to government impact on market environment. Market access is more important than any individual principle of free trade in the minds of some U.S. trade negotiators (Cowhey and Aronson 1992; *GATT Focus* 1991 Nov/Dec, p. 2).

Regimes can also change even if the degree of formal openness remains the same—the changes arise from the internationalization of firms. The degree to which dominant firms internationalize their business to become global firms is thus the second basic factor analyzed for its effect on market environment.

The question facing governments and corporations is: Where is the informatics sector heading? The answer depends on the extent to which internationalization and the degree of openness of the regime are changing. This chapter argues that there is a profound shift in both dimensions. During the 1980s the action is shifted away from a restrictive international regime. A traditional free-trade regime grew (and will grow) in importance because it is the simplest solution. Many of the important innovations in the 1990s, however, will occur in what I call "market access regimes" and "internationalization of regulation regimes." The concluding sections of this chapter show how.

To lay the groundwork for this, the next section outlines the determinants of change in the two basic factors—contestability and firm internationalization—and looks at the international regime for communications before 1970. The second section looks at the regime we are coming from, while the third examines evidence concerning economic and political change in the environment. A minimal-change scenario is developed in the fourth section. The last section reviews evidence that a more dramatic change is occurring in the environment for global informatics.

3.1 The Determinants of Change

Domestic politics set the initial agenda for what states seek from international regimes and interact with considerations about the distribution of international power and transaction costs to determine how open states will allow their markets to be. Countries seek regimes that reinforce domestic political bargains concerning markets—bargains reflecting domestic political demands that already incorporate expectations about international factors. In other words, firms have expectations about the international competitive and political environment, and they factor these into their claims with regard to both domestic and foreign economic policies. The same is true of government agencies.

Three broad changes in the global economy have been shifting the political preferences of many countries regarding the informatics sector. First, the firms involved are becoming more international as measured by the percentage of their total sales in other countries. This has created support among firms and their work forces for international openness. AT&T and British Telecom plan to have over 50 percent and 25 percent, respectively, of their revenues from international markets by the year 2000. As recently as 1983 neither had over 10 percent. This mirrors a general shift (Millner 1988).

Second, users of communications systems are becoming more sensitive to supply and pricing. This has boosted political support for competition. Third, many key users are global firms and government agencies that want both domestic and international reforms. (Regimes are never a simple extension of the preferences of particular industries and their government overseers because they also influence the welfare of other voters, industries, and government agencies.)

At the same time, many informatics technologies now have features such as high fixed costs or specialized and competing technical standards, which add the risk of oligopolistic strategies that negate some of the advantages of relying on free trade and openness (Krugman 1986). In short, there is pressure for more openness at the domestic and international levels, but also suspicion that classic free trade alone is not enough.

3.1.1 Transaction Costs

The current mix of incentives makes transaction costs—the costs of information, bargaining, monitoring, and enforcement in coordinating action in any group—a particularly important factor in determining the form of openness to pursue. Higher transaction costs make it easier for countries to renege on international bargains. (Economists describe this as "opportunistic behavior"; see, for example, Williamson 1985.) The creation of industrial policies to bolster international competitive positions is one example.

One purpose of international regimes is to reduce transaction costs. Successful international regimes often have inertia because there are transaction costs to governments if they change the regime significantly.

Transaction costs are also tied to power because the greater the concentration of international power, the easier the establishment of an open international regime. Open regimes are fragile because they require closer adherence to nondiscrimination than do closed regimes. They also impose concentrated short-term costs on many political constituents in return for diffuse long-term gains in efficiency. A clearly dominant power has a sufficiently large stake in the international system as a whole to offer rewards and apply pressures to get other countries to agree to an open world economy.

A dominant power can assist an open regime, but how can other nations guarantee the leading power will act in good faith, especially because powerful countries with less dependence on trade have to cope with the problem that there is always a constituency for protectionism. (In comparison, small states with greater dependence on trade usually create political institutions to restrain interests opposed to openness; Rogowski 1987). One way is for other countries to give the great power's domestic interest groups special incentives to support the international agreements, and this is often done. Thus, many features of the current international regime may now have to change to fit the idiosyncracies of the domestic Japanese political economy because of its importance in the global order.

As hegemony declines, a small club of countries can serve as the focal point for liberalization. However, member nations will emphasize careful monitoring of agreements and more elaborate systems of side payments among themselves (Yarbrough and Yarbrough 1987). The diffusion of power leads to a greater emphasis on minilateralism as a supplement to more universal international arrangements. *Minilateralism,* as the term implies, is narrower than *multilateralism.* That is, it is generally focused, ad hoc if you will, in the sense that it relies on selective, perhaps overlapping, groupings of like-minded countries to organize international bargaining rather than relying on more inclusive bargaining forums.

In the case of informatics, U.S. power was sufficient to break the old regime of closed markets and minimally internationalized firms. However, the emerging regime bears the stamp of growing minilateralism and careful safeguarding against opportunistic behavior.

3.1.2 Before 1970

World communications operated in an essentially restrictive international regime before 1970. Major global commercial strategies reflected relatively low degrees of internationalization. Moreover, the only major example of internationalization involved an essentially closed market; Intelsat was a single global organization created to provide a jointly owned global monopoly on satellite service. Let us examine the individual components.

International services were jointly provided in closed markets with complementary needs. The ITU provided the framework. It held that all international services were a bilateral monopoly characterized by joint investment from PTTs. However, because the United States was a major player, the ITU accepted private ownership of U.S. international communications carriers, which made it easier for them to operate outside the United States.

In theory each country supplied its half of an international circuit and exchanged traffic at midpoint. All revenues were split equally, based on a negotiated accounting rate, regardless of the country originating the telephone call or telex. Customers usually paid much more than the accounting rate because each monopolist was free to charge what it wished for traffic originating in its home market. This approach assumed national monopolies had complementary needs in sharing their monopolies. Markets were closed, but international exchanges were harmonious mutual accommodations among national monopolies.

Telecom services and companies were essentially domestically oriented, and governments vigorously attempted to structure competition within the domestic market (Cowhey 1990).

The ITU system was somewhat less restrictive for equipment than it was for services. Most equipment markets were controlled by national telephone companies. European countries and Japan actively kept national production in local hands as much as possible, and AT&T owned its own equipment manufacturer. Firms with desirable technology often accepted licensing as a way of increasing returns on their research. Technology licensing and consulting thus reflected complementary needs, and government policies in this area tacitly supported largely closed international markets.

Only after the United States began to open its own market unilaterally did tension arise over the communications equipment market. Protection at home combined with ambitions to sell abroad—mostly to the United States and some developing countries—helped create in several countries consortia of domestic firms that aimed to develop collective capabilities. It should be stressed that these export-oriented alliances rested in part on a refusal to allow foreign firms access to the home market. This was the classic strategy of Japan. In contrast, there was vigorous and fairly free trade in computers in large parts of the world market, even though Japan and other countries actively promoted local producers.

3.2 The Regime from Where We Come

The market for communications equipment is either network or customer premises. Network equipment, about two-thirds of the total, includes items such as central office switches and fiber optic cables; customer-premises equipment (CPE) covers private branch exchanges (PBX), telephone key systems, facsimiles, handsets, and the like. The total market for services far surpasses the market for equipment. Each of these three elements of the current regime are discussed in this section.

3.2.1 Network Equipment

Network equipment is fairly concentrated, with each country traditionally having had one to three market leaders and a public monopoly buyer. In the predivestiture United States, the service revenue of AT&T was bigger than any Post, Telephone, and Telegraph (PTT). In addition, unlike PTTs, AT&T had its own manufacturing arm. The economies of scale, in particular for switches and satellites, are further narrowing the market; however, fiberoptics and satellites would be only modestly internationalized except for the fact the customer base is strongly clustered among a handful of public authorities.

Central office switches in the 1990s will cost \$2–3 billion to develop; the previous generation cost from \$0.5–1 billion. Most analysts believe three or four independent technology platforms will exist, and all remaining switch makers will cluster around them (much as many makers of mainframe computers basically use one of the seven principal architectures available from the leading companies).

The market for switches is becoming more diverse because private users are now also buying them, but it is still highly concentrated and requires constant interactive support for software development. These conditions suggest that internationalization of the technology will be high even though market leaders are usually reluctant to enter corporate alliances. (This analysis omits the possibility that smaller switches may play a role like that of minicomputers. Minicomputers did not eliminate mainframes, but they changed the market.)

3.2.2 Customer Premises Equipment

The CPE market has several niches, which contributes to its high fragmentation. Technological capabilities are disseminated rapidly and economies of scale for individual products vary radically—for example, they are high for key telephone sets and low for data transmission devices. Market leaders are tapping economies of scope across a family of products, including some kinds of network equipment. This broadening is often done by buying out other companies in their own countries. For example, computer makers purchased producers of T1 switches and specialized communications technologies.

Although it is something of an oversimplification, into the early 1990s certain skills have been geographically concentrated. U.S. makers excelled at network management CPE that allows individual users of networks to manage or duplicate many central network functions. They also had the most sophisticated links between computers and communications. Although Japanese firms originally focused on lower-end technology, they moved in the late 1980s to CPE featuring specialized semiconductors that add value to network inputs, a move similar to what they did with television and fax equipment. European firms specialized in equipment that efficiently interconnects with network services. These differences reflected divergences in expectations regarding evolution of networks.

From the late 1970s consultants and companies were predicting and developing products for a CPE market intermeshed with the computer market, only to be largely disappointed in the success of their offerings. At last, as the 1980s ended, intermeshing had really begun to happen. There are predictions that integrated business systems—packages combining a digital PBX, LAN, and WAN—will capture a majority of the combined network and CPE market by the mid-1990s (e.g., Roobeek 1988).

There was a consolidation of major suppliers of central network equipment and mainframe computers in several countries in the late 1980s, such as the creation of GEC Plessey Telecom (GPT) and Ericsson's buy out of its British partner (Thorn) in Great Britain and the Alcatel purchase of ITT operations for $1.3 billion in France. Siemens purchased GTE's non–U.S. switch and domestic transmission businesses and entered a joint venture with GPT in the United

Kingdom—a move matched by Northern Telecom's joint venture with Britain's STL.

There has also been horizontal expansion for firms already in either computer or CPE markets. Hewlett Packard launched its own X.25 packet switched network products and bought a minority share in a U.S. electronic mail and voice processing firm (Octel), intending to market these services globally. Unisys bought Timeplex (a T1 vendor), IBM deepened its ties with Network Equipment Technologies, and Tandem Computers bought a leader in LANs to build better bridges between local and wide area networks.

Demand for particular classes of products is highly unstable because their capabilities overlap with other product lines, leading to battles over which approach should dominate. Moreover, the cutting edge of the customer base is still relatively concentrated among large global users who demand interactive customization of technology and interconnectivity among all of their equipment suppliers. Interconnectivity and customization make simple standardization hard. The result is hybrid forms through common alliances among firms promoting technological infrastructures (such as a particular architecture) and product networks (such as sharing technology to allow compatibility across production lines of participating companies).

3.2.3 Services

Services are becoming difficult to categorize. This is not just because modems and facsimile machines are routinely attached to "voice" (standard telephone) circuits. Value added networks (VANs) and information services are also called *enhanced services*. The older value added services include protocol conversion, packet switching, and remote computer services. This is where the meshing of computers and communications is most palpable—in application-specific integrated circuits (ASICs) and software that controls the electronics that make many services both needed and possible.

Traditional service markets will not prove as contestable as equipment markets. Nearly half of all service markets are local telephone services, which may never be opened to foreign firms, or opened by allowing partnerships of foreign and domestic firms to buy privatized national telephone companies. Politics and technology; however, have made competition in a number of areas more likely. These are international voice, data transmission (ranging from dedicated leased circuits to value added services), information and computer services, and cellular telephone franchises.

The underlying architecture and degree of competition in the public communications network strongly influence the feasibility and competitive advantage of different types of information services and computer equipment. Indeed, hardware- and software-defined services are often exchangeable within a network. Thus, Centrex and the PBX are rivals; however, services and equipment can also be complements. Huber (1987, p. 13-3) noted that the $900 million per year market in the United States for alarm monitoring was made possible by additional services and equipment for the monitored premise, at a cost of over $2 billion per year.

3.2.4 Data Transmission

The physical infrastructure of a network is a specialized asset with a limited range of uses. Once installed, there are strong incentives to exploit its limited range. Competition among new international facilities for traffic opens the prospect of highly volatile behaviors ranging from predatory pricing to dropping regulatory restraints on the use of facilities to lure customers. It may also lead to innovations in relationships between users and producers seeking to assure traffic loads. Many countries are permitting selective competition in specialized infrastructure facilities, such as mobile satellite systems and cellular phones. Companies are therefore bidding for franchises in a number of different countries with an eye to building interconnected global networks.

Consistent, current estimates of the world data communications market are difficult to come by. The 1987 market was estimated at $223 billion, 22 percent services and 78 percent products, with just under half of each in North America (Aronson and Cowhey 1988, p. 66). This excludes all voice services, although some of these will become subject to competition (particularly those for private corporate networks). While international voice services are the most profitable segment of the market, data-related services are seen as having more growth and profit potential. Hence the computing and information side of this market, including equipment and services, is the heart of most company strategies. It also represents the market segment where the political pressures for change in countries outside the United States are greatest.

3.3 The Regime in Transition

Demands from suppliers and users for openness and competition, problems with transaction costs due to the nature of the technologies, and moderate diffusion of global power combine to lead one to expect innovation in global regimes.

Historically, competitive assets often have been concentrated in one country early in a product cycle. For example, U.S. software firms and Japanese manufacturing firms may have carved out some unique advantages by the late 1970s (Mowery 1988, pp. 1–22). Neither of these edges will be as true in the 1990s as they once were. Openness encourages international diffusion of competitive capabilities, and that promotes internationalization of complementary assets. As research, manufacturing, and marketing ("key competitive assets") become less concentrated in a single country, companies will rely more on international operations to secure complementary assets.

Costs for research are rising while product cycles are shortening. Firms are responding in a variety of ways. They are entering alliances to share R&D costs. They are emphasizing quick global positioning of products to capture the margins necessary to recover those costs. They are seeking to enhance economies of scale and scope in manufacturing by combining global export platforms and local plants to customize products to national market niches and deploy them quickly (Thomas 1988).

3.3.1 The Politics of Openness

In the first thirty to forty years after World War II, under U.S. leadership, the world moved toward greater openness, but this was not in informatics. The reason was domestic politics.

The world today is more fragmented because the United States has become less of a hegemonic international power. However, there are now many players besides the United States who see the benefits of openness, and they wield enough market and political power to make further liberalization conceivable so long as they work together in arrangements emphasizing more careful checks and balances concerning their conduct. One symptom of the rise of minilateralism and elaborate safeguards to monitor burden sharing. U.S. bilateral trade agreements with Canada and Israel are good examples, as are the aggressive reciprocity provisions in the 1988 U.S. trade law.

Transaction costs in international bargaining depend significantly on technology and the way international institutions cope with costs. The most important characteristic of informatics is the difficulty of monitoring good faith agreements for services. For example, what constitutes a fair cost for a leased circuit?

Another problem is assuring interconnectivity of all communications facilities, services, network equipment, and CPE. For example, CPE and enhanced services suppliers benefit from a deeper knowledge of the technology and pricing of the basic services network. This poses transactions cost problems among the different segments supplying the informatics network. (The problem is analogous to the worries IBM's rivals used to have about their knowledge of the system's network architecture when IBM's dominance of the mainframe computer market mattered.)

When is the design of a public network discriminatory against foreign competitors? These issues are very difficult to resolve domestically, and even more so at the international level. The old regime for informatics solved these problems by simply providing a few common standards among divergent systems of national standards. Joint monopoly for international services removed temptations to cheat in services because each side could only deliver its monopoly service with the help of the other. The new technologies require more common designs for national networks, and politics now require curbs on monopoly solutions. Finding institutional solutions for monitoring and enforcement is very difficult and may drive companies to novel forms of global integration and alliances with other firms to overcome the problem. These experiments are especially important when it comes to the creating of new international communications carriers, as discussed later.

3.3.2 The Impact of Technology

Neither international power nor bargaining issues would matter if a revolution had not occurred in the interest group politics of informatics. Most countries long ago granted authority over communications to a single monopolist and then merged the telephone company and the government ministry that regulated it. In addition, countries often mixed their postal and telephone services under one operation, which became known as the PTT. As a rule, long-distance services subsidized both local telephone and postal services. In addition, the telco subsidized local equipment makers and their workers. Telcos generally did not own equipment makers, the U.S. and Canada being exceptions.

Regulatory reform in telecommunications always invokes issues concerning those who benefits from change. Many important technological breakthroughs in information use have been clustered among the relatively few corporations and government agencies that are the largest users. Typically, 5 percent of all users constitute over half of long-distance traffic (Saunders 1989). Members of this group have both the motivation and the resources needed to influence communications regulation. They stopped treating communications solely as a domestic issue long ago. They now have active interest groups promoting global regulatory change.

Similarly, the microchip revolution allowed many important new companies to enter the electronics industry, particularly in the US and Japan. Traditionally, a handful of older firms dominated the production of equipment in conjunction with the dominant national phone company. Newcomers wanted to break this cozy relationship because the dominant companies received too many benefits that spilled over to their operations in related markets, giving the established players an ''unfair'' advantage. Now that the newer firms are succeeding, they are split on how to balance goals of interconnectivity and proprietary technical advantages.

Finally, many service and equipment producers, as well as large users, wanted to produce new information- and telecommunications-based services. There are instances of large users moving into the long-distance and information-network businesses by offering their own networks to others—Westinghouse and Sears in the United States are examples. Electric utilities in both the United States and Japan also have gotten into the telecommunications business.

3.3.3 Reforming the Monopolies

Countries vary in how they sort out who wins and loses from reform because every sectoral regime is part of a broader effort by political leaders to cement general electoral or political support. Every country maintains some cross subsidies for households in order to avoid political trouble. There also is concern about how to offer smaller businesses the same services as those available to larger firms. Moreover, every country except the United States still has some form of support for its preferred providers of network equipment.

Still most countries are modernizing their monopolies by no longer treating the PTT as a cash cow for the government or other stakeholders in the telecommunications system. Profits are being reinvested to modernize services. Reform is also changing policymaking by separating post and telephone services, introducing independent regulatory commissions instead of letting the PTT set policy, and sometimes privatizing PTTs to make financial planning and labor policies less subject to the whims of government policy.

The reform movement is also curtailing the monopoly power of national telcos over the sale of CPE and liberalizing the provision of installation and repair services. This opens the way to independent systems integrators such as Computer Sciences Corp. and Electronic Data Services (EDS), which provide services, software, equipment, and enhanced communications services. Just as strikingly, the major telcos are privatizing their own services to become systems integrators and customers of systems for major customers.

PTTs argued either there were economies of scale in the production of the equipment or that control over the equipment was necessary for maintaining consistent technical standards and high-quality service. As a result, although few countries ever sanctioned a monopoly over supplies, most bought from a handful of national companies or a few national companies plus one foreign firm that was obligated to undertake extensive local production, as in the case of ITT in Europe. The FCC's abortive attempts in 1986–1987 to monitor purchases of foreign equipment by BOCs is a reminder that the United States is not exempt from this practice.

The provision of network equipment is also being modified. The new policy ostensibly liberalizes procurement practices, but only to the extent of permitting a greater array of foreign suppliers to achieve larger minority market shares. There are strong indications that local content requirements will prevail even in industrialized countries. Thus, the EEC proposed that open procurement of network equipment will apply only to firms with 50 percent or more local content after 1992. It would waive the rule if EEC firms have equal access to a foreign supplier's home market.

3.3.4 Openness

Trade negotiators have started to win acceptance of three principles: (1) technical requirements must be transparent to all buyers and sellers, (2) the process of setting standards should be subject to contributions by interested foreign participants, and (3) countries should recognize technical testing and certification of equipment done in other countries.

There is general agreement that virtually all information services should be competitive and that the telephone company should be subject to regulatory controls to curb unfair competitive advantages arising from its monopoly over the basic network. That is where agreement stops. Many countries want value added services—for example, protocol conversion and packet switching—to remain domestic monopolies, while allowing some choice in international value added carriage. The Latin American activities of IVANs such as BT, GE Information Services, Infonet, and AT&T EasyLink Services illustrate this. Other countries allow foreign firms to provide value added services to private corporate networks and to supplement limited public offerings.

Many countries restrict foreign provision of information services in order to boost local industry while others are open. Japan is relatively open. Latin American countries have reached agreement with PanAmSat, a new competitor to Intelsat, for private data and video networks.

PanAmSat is part of another important phenomenon. Countries in the Pacific Basin openly compete to become international traffic hubs. Hubbing brings revenue to support modernization and helps spread costs. It also brings ancillary business as firms move operations to the hub. Many Pacific Basin countries have reached agreements on voice services with all the major U.S. carriers in return for pledges to build traffic flows. One reason Australia converted to competition on basic services is to compete more effectively with Hong Kong, Singapore, and Japan for hub operations. Nations like Indonesia and China have experimented with satellite systems in order to start playing this game.

In short, the domestic politics of services and equipment suggest limited openness for local basic services, some chance for openness in domestic long-distance and value added voice, and greater opportunity for all international services. There is strong potential for international openness in CPE, value added, and information services. Prospects for network equipment are mixed. Meanwhile, the distribution of international power remains concentrated enough to favor liberalization, but there will be a tendency towards very elaborate safeguards and minilateralism.

3.4 Alternative Regimes

Internationalization of firms and a shift to greater openness are likely. The real question is how far the changes will go. This section explores the issue in two steps. It begins with a brief review of the most prominent proposal for minimum change in the international regime. To see if minimal change will suffice, I will then review the changes in the principal users and providers of telecom services and equipment—the United States, Japan, and the United Kingdom, which are collectively referred to here as the Triad.

3.4.1 Minimum Reform: The WATTC Process

The smallest degree of change that will occur is what I call "dynamic centralization." It is embodied in the proposed draft regulations submitted to the World Administrative Telephone and Telegraph Conference (WATTC) in 1989. Standardization will bridge the existing gaps among communications and computer systems around the world; *interconnectivity* is the watchword. In general, the European authorities have favored this approach.

Dynamic centralization expects the traditional phone system to become a powerful public network for completely integrated voice, data, and video services available to everyone over broadband, high-speed networks of fiberoptic cables. Competition will play an increased role in two senses.

First, there will be enough reform for policymakers to enhance the credibility of threats of competitive entry if the PTTs do not improve efficiency. One can interpret reforms in West Germany in the mid-1980s along this line. However, many analysts think privatization may be a necessary part of this policy mix (see, e.g., *Economist*).

Second, specialized information services will be offered over the network by new entrants. Some content will be competitive, but the pipe will be a public utility intent on capturing all economies of scale and scope as a way of raising capital. Pricing for network services will be more flexible for large users, but it will still permit cross subsidies supporting expanded definitions of universal service to include some enhanced services to households.

The trickiness of this is suggested by what is happening in Europe. The EEC (1987, pp. 34–35) notes that members disagree on whether packet-switched networks, circuit-switched data networks, teletex, electronic mail, and videotex are basic services.

The 1989 WATTC draft regulations proposed extended international regulation over all new types of value added and information services, and potentially over all entities that use the communications network. This would significantly extend the scope of regulation. For example, some of these services could be declared to be universal public services (UPS) and thereby subject to licensed entry, public tariff approval, and obligations to serve all interested parties. The regulations could also treat the providers like traditional carriers.

The WATTC draft allows some competition for some information services. GATT would presumably provide a framework for competition. (On the GATT initiative for services, see Aronson and Cowhey 1988.)

An examination of the WATTC formula suggests that it rests on a redefinition of jointly provided services, which significantly moves the regime from restrictive regulation toward the internationalization of regulation. It acknowledges that PTTs must rely much more heavily on global customers who demand global services than was true in the past.

The old system of universal service cannot reliably provide much more integrated and flexible services; therefore, it must change. The best example of a successor to jointly provided services is Global Information Movement and Management (GIMM), an ambitious undertaking of AT&T, KDD, and British Telecom, together with several closed users groups such as SITA for airlines.

GIMM assumes underlying basic services will remain relatively closed to competition, although less so than in the past. Nonetheless, the internationalization of customers requires each major national carrier to internationalize by collaborating on the development of common service offerings (involving similar technical architectures, more flexible pricing, and a single account manager who can arrange for services in all three countries). In short, AT&T, KDD, and British Telecom are creating international commercial alliances through GIMM to provide competitive global assets.

Closed users groups are viewed by many regulators as the best way to satisfy the demands of some very sophisticated and influential groups of users. Organizations like SITA for airlines and SWIFT for banks are recognized under ITU rules as deserving exceptions to rules that discourage the independent supply of enhanced services, such as the prohibition of shared use and resale of circuits. Firms serviced by these closed user groups have common needs that exceed the capabilities of public networks, so it is natural that they should seek to invest in a commonly owned specialized network to serve their particular needs. This limits the proliferation of independent networks, satisfies their strongest demands, and assures every firm in the banking or airline industries that it will have similar international capabilities.

It is unlikely the WATTC approach can prove to be a stable solution for several reasons. The political forces in key financial centers are pushing for more openness than WATTC favors. Customers demand more open competition and globally customized responses than efforts like GIMM can meet. Efforts by PTTs to field ''one-stop shopping'' for global services to large multinationals stumble over high coordination costs among the partners and corporate cultures built on more protection and less internationalization. One sign of this frustration was AT&T's consideration of buying a minority share of Mercury, Britain's second largest telephone company, in the hope that common ownership might permit more decisive strategies. Closed users groups are somewhat better suited to meeting customer demand, but they are at a disadvantage to fully meet the individual needs of major customers.

There are significant transaction costs problems among the different informatics segments. Information service suppliers want flexibility in the underlying facilities and basic services. CPE suppliers seek to weaken the market for most services because they want their specialized equipment to provide the added functionality, not the phone company. Therefore, users and CPE suppliers will not ultimately be satisfied with the limited coverage provided by GATT agreements intended to liberalize international enhanced services. Finally, as limited competition comes to international facilities, suppliers have fewer incentives to maintain restrictions on use of their facilities than did the old monopoly PTTs.

3.4.2 Alternative Models for Change

Reform means increased international competition is possible. At a minimum, trade negotiations could progress on information services and equipment issues. Many users and providers are not waiting.

Nearly round-the-clock trading has emerged in some types of financial markets—foreign exchange in particular—based primarily in London, New York, and Tokyo. This is both a result of and a force for further rationalization of communications and information systems. A more competitive and flexible information infrastructure is a natural concomitant of globalization of capital.

''Flexible decentralization'' is the U.S. vision of this change. Technological innovation will occur on more competitive terms and with a flexible architecture for future services and equipment because informatics is too important to allow a central network to dominate it. Every country needs one or more public networks with economies of scope in providing flexible interconnection of services. Moreover, individual public networks need a common understanding about how standards will make them interoperable. However, the public network should not exclude the flexible specialization of other networks. Indeed, finding the optimal mix of services even on a wide-band network depends on the ability of buyers and sellers to redesign the pipe, including shifting the point where command over the network resides, as well as to compete on its content. Competition in the underlying infrastructure of network facilities further encourages innovation. If dynamic centralization sees the public network as the highway of the future, the flexible centralization vision argues that there must always be competing highways and alternate modes of communication—just as waterways, roads, railroads, and airports form competing yet complementary networks within the physical transport infrastructure.

The Japanese approach is a hybrid. It embraces competition for facilities and services along the lines of flexible decentralization. It also uses government subsidies and industrial policies to promote a core set of information and video services for every home, which leads to experiments with pricing and network design that are somewhat akin to dynamic centralization. Table 3.1 is a simplified comparison of countries with regard to competition in services and facilities.

Most of Europe has embraced dynamic centralization. The European model probably does not go far enough, but it does not follow that the world market will fit the U.S. model: A Japanese–European understanding could also shape the future. However, the United States is so much a pivot point of trade, investment, and communications for Japan and Europe that sidestepping the United States will be difficult.

Table 3.1. Competition in Services and Facilities, 1991

Europe	Japan	U.S.	Service
No[a]	In theory	Bypass	Local Basic
No[a]	Yes	Yes	Domestic Long Distance
No[a]	Yes	Yes	International Long Distance
Some[b]	Yes	Yes	International Enhanced
Some[b]	Yes	Yes	Domestic VAN
			In the 1990s:
Broad[c]	Broad[c]	Voice	Universal Service defined to include.
High	Mixed	Low	Extent advanced processing functions are concentrated in the central public network.
Low	High	High	Level of facilities competition

[a]The United Kingdom and Sweden permit competition.

[b]Limited on data in some countries until mid-1990s. Many restrictions on enhanced voice services.

[c]Many enhanced services will be considered part of universal service.

3.4.3 The United States–Japan–United Kingdom Triad

The United States, Japan, and United Kingdom had 177 million of the world's 341 million access lines for communications in 1985. They constitute the majority of the world's equipment and domestic services markets. They are only a minority of the international basic services market, but they dominate traffic over the major transoceanic routes (as calculated by the author from data in Kitchen, Lewin, and Schoof 1987).

Countries seek international rules for markets that reinforce their domestic regulatory arrangements. Reform means that most industrial countries want to tinker with international rules, but divergence in their approaches to reform sends them in somewhat different directions.

The Triad has had a sometimes tumultuous, sometimes cooperative relationship when attempting to forge common understandings about the international communications market. U.S.–Japanese bilateral trade talks have been particularly tough and some matters remain unresolved. Still, the three appear headed toward several common positions.

U.S. firms hold varying positions on the best form of change. AT&T is most comfortable with stressing reform of ITU regulations. Both AT&T and the BOCs give higher priority to bilateral talks than they give to GATT. However, they have supported the GATT services initiative except for a strong insistence that the United States not grant unilateral access to foreign carriers to serve the U.S. voice services market. The computer industry strongly supports GATT, but sometimes splits because of feuds between IBM and its rivals on computer communications architectures. Large users and specialized service providers are the most outspoken critics of the ITU. They are also the most sympathetic toward GATT, which they would like to see supplemented by bilateral negotiations.

The Triad have accepted competition in the provision of international basic services and facilities. The United States has three big international carriers and many smaller ones; Japan has three; the United Kingdom has two. Each has accepted "private" international fiberoptic cables that can compete only for the traffic of specialized private networks. The United States and United Kingdom accept private international satellites for specialized services. Japan has accepted a significant ownership role by a foreign carrier, Cable & Wireless (C&W), in one of its international carriers. The United Kingdom planned to allow U.S. carriers ownership and control privileges over basic carriers as long as its firms could do the same in the United States.

The three countries have agreed to curtail traditional restraints on the shared use and resale of international circuits. This makes it much easier for providers of enhanced services to operate between countries. These I-VAN agreements open the way to much freer and flexible networks that tie together the three countries and other interested nations. Foreign ownership of domestic enhanced services is allowed in each country, although the United States accepted more administrative controls over licensing I-VANs than it prefers.

There is also agreement that equipment markets should be open, although implementation may fall short of the goal. The U.S.–Japanese talks established some very important precedents on the crucial issue of technical standards. The United States has convinced several non-Triad countries to certify equipment as long as it meets the relatively unrestrictive standard of "no harm to the network." It has further advanced the principle of "self-certification" by the supplier rather than relying on a few national testing establishments. It has also obtained a pledge from some countries to set tariffs for leased circuits on terms compatible with the needs of US users and enhanced service suppliers (Aronson and Cowhey 1987).

For all these agreements, there are important differences among members of the Triad. The U.S. exercises fewer controls over enhanced services—for example, allows more licensing—than the other two, and is reluctant to use communications policies as a form of limited industrial policy. Perhaps most crucially, the Japanese Ministry of Post and Telecommunications (MPT) was able to retain control over telecommunications policy after a major challenge by the Ministry of International Trade and Industry (MITI), which represents the computer and general electronics industries. The contest between MITI and MPT is not over, and MPT must write rules that take the MITI position into account or else face another serious challenge.

Many analysts believe Japan's MPT is using international deliberations to strengthen its hand. The MPT insists that all service categories be covered by explicit administrative rules. For example, it wants rules to govern emerging fax networks while the United States insists no rules apply and none are needed. In 1988 Japan declared that all I-VANs had to use the X.75 standard to maximize connectivity and all I-VANs had to offer interconnection to other VANs. The United States protested, saying the regulation prevents suppliers of specialized services from designing proprietary technical standards and choosing their own business partners. MPT subsequently offered to have the United States and Japan jointly design a supplementary standard to X.75 to satisfy both parties. U.S. firms responded that the problem was Japan's insisting on *any* single interconnect protocol; they wanted flexibility. MPT then offered a new formula for protocols and interconnection, which the industry found more interesting, although the U.S. government still found it less than fully satisfactory.

Japan uses its administrative powers over services not only to prevent competition in services but also (and more importantly as a reason) to strengthen its competitive position in the international equipment market. The government spurs economies of scale in equipment tied to selected new service features, particularly the use of optical imagery in conjunction with voice and data systems. However, the MPT strategy continues to raise serious questions about government oversight of the computer industry, and some Japanese electronics firms want more freedom.

3.5 The Emerging Regime

There are three prominent trends in the strategies being employed by companies in the Triad.

1. Service firms are becoming multinationals, and even multinational manufacturers rely on even more integrated global strategies for both goods and services. The strategies, however, often rest on international corporate alliances to supply many of the key goods and products.
2. Firms rely more on joint efforts to create a common technical infrastructure and to do product networking.
3. Firms form international alliances to create important products or to provide a single global effort in markets being transferred from closed to more open environments, such as transoceanic cables or satellite systems.

These trends all involve a shift toward more openness and continued internationalization. At the same time, governments are accepting more international competition and continue with vigorous industrial policies. This in turn suggests that neither traditional approach—a restrictive regime like the ITU or a free trade system such as GATT—can handle the problem. We are in an era when a market-access regime is becoming more central to world order, which poses the troubling problems to be discussed at the close of this chapter.

3.5.1 Creating Global Networks

There are two principal approaches to the creation of global service networks. One is the creation of a network whose primary asset (technical planning) and competitive strategy are anchored on the global strategy of a single firm. The other is that these networks consist of a single firm linking together a series of commercial ventures that are jointly owned and developed with other firms.

Many companies are primarily working on multination domestic strategies, which invest in telecommunications franchises in key nations. For example, Motorola is taking on joint ventures in countries such as Japan and Argentina to supply cellular telephone networks. It has proposed the Iridium low-orbit satellite system as a partnership vehicle for telcos around the world. Each network then becomes a buyer of Motorola equipment.

The two British carriers, British Telecom (BT) and Cable & Wireless (C&W), are excellent examples of the multinationalization of services. BT has focused primarily on enhanced voice and data services with selected CPE, while C&W has covered all services.

BT's initial strategy has specifically led to two apparent results. First, BT is becoming a multinational firm through CPE, particularly items such as PBXs, modems, multiplexers, and high-speed local area networks (LANs). It is doing so primarily by entering local joint ventures in North America and Japan. Second, it is building a global network in value added and information services that complement the CPE market. It purchased Tymnet, a global VAN, to accelerate its provision of integrated global private networks. It also emphasized cellular through moves such as its 22 percent ownership share of the United States' largest cellular telephone system (McCaw, for which it paid $1.5 billion in 1989), paging, air call services, voice messaging, and voice response systems. Its initial target was the United States, but it is also involved in joint ventures to penetrate Europe and Japan. It experimented with agreements with AT&T and KDD to provide one-stop shopping for corporate networks, but it may prefer to go its own way.

C&W is building a global digital pipe. It is the leading partner in joint ventures providing private fiberoptic cable systems being installed in the Atlantic and the Pacific. C&W directly owns 30 percent of the Japanese end of the Pacific cable as well as 16.83 percent of the new Japanese international long-distance carrier that owns the other 70 percent. It also owns 20 percent of Pacific Telecommunications Corporation, the majority owner of the U.S. side of the Pacific cable. It holds a 50 percent share of PTAT, the trans-Atlantic fiberoptic cable; US Sprint owns the other half.

C&W owns most of Hong Kong Telecom, and has entered joint ventures for a teleport in Jamaica (with AT&T), cellular phone services in the Caribbean, and a VAN in Australia. It has a share of a long distance resale network, C&W Communications, in the United States.

British firms are not the only ones playing this game. France's Cables et Radio owns slightly less than 15 percent of a U.S. long-distance carrier, FTCC, and a 2 percent interest in International Telecommunication Japan. It also owns a 15 percent share of a new global joint venture for VANs based on the existing network of Computer Sciences Corporation.

Some firms are more concerned with products than are C&W or BT. Both IBM and NEC represent a slightly different approach to creating new multinational strategies for goods and services. Each remains primarily in the equipment business and basically an independent multinational, but both are building global service networks to supplement their equipment sales. Each is also doing more product networking and cooperating on common technical infrastructures. IBM is doing so through a series of national joint ventures, the largest of which is with NTT and Mitsubishi in Japan for VANs. Others are in Italy and France. At the same time, it has been building its own global VAN to interconnect these ventures.

In Japan NEC has the NEC Network, which is a joint venture with companies in the Sumitomo group and two major insurance firms, and its PC VAN. NEC relies largely on its joint venture with GEISCO, a General Electric subsidiary, for its global needs.

Perhaps the most ambitious example of an effort to build intercorporate alliances to yield a family of products is AT&T. AT&T and Philips formed APT, a venture that swapped AT&T digital switches for Philips network transmission products. Frustrations in cracking the European market led AT&T to assume majority (60 percent) control, and then sell part of Philips' share to Italy's Italtel in order to win major equipment orders in Italy. APT also added Telefonica de España as a partner. Each new member would contribute products to the venture.

Meanwhile, AT&T has tried to find a winning combination of ventures for CPE and computers. Its 22 percent equity holding in Olivetti was the most prominent move, but this relationship proved unsuccessful and will be sold off at some future date. It bought NCR in 1991 to fill this gap.

AT&T has joint ventures with Ricoh for digital key systems and Toshiba for

PBX that could contribute to a global product family. Its 20 percent equity in Sun Microsystems (discussed later) was intended to indirectly help build yet another product family. At the same time, it finally got its joint venture in Japanese VANs, JENs, into proper working order to provide the global prototype for other VANs that will complement its CPE offerings. In 1989 it bought Instel, a large British VAN, as a vehicle for creating a European VAN for very high-speed data. The AT&T experience, however, shows that putting together a coherent global family of common products through partnerships is a particularly difficult task because coordination covering a wide range of products is extremely hard to maintain.

Joint technical infrastructures and product networking for sharing know how, developing standards, and cross-selling products are becoming more important because the traditional international process for setting technical standards is no longer sufficient. The intersection of technologies and the integration of the global operations of customers have far outpaced the traditional forums for setting standards. Moreover, many companies want to establish selective arrangements for setting standards in order to establish a dominant market position for their specialized technologies.

Joint ventures to support a common technical infrastructure include efforts to establish a common set of standards among firms and common technical capabilities, but not specific commercial products. Japanese firms have long done this within their own market; now Europe, through the Common Market's Esprit and R&D in Advanced Telecommunications Technology in Europe (RACE) programs, and the United States, through industrial consortia receiving antitrust exemptions and financial support from the U.S. government, are following the Japanese example. U.S. examples are Sematech and MCC. The forces leading to minilateralism should also promote more domestic alliances as safeguards against possible moves by firms from other countries. The growth of domestic alliances in the United States represents a major shift away from classic free trade and toward something more akin to industrial policy.

At the same time, the growing pressure for domestic consortia for technical infrastructure will open up entry to foreign firms. Foreign governments will demand access and foreign firms are becoming increasingly important to the success of any individual country's project. Foreign firms may have unique technological capabilities or detailed knowledge of foreign market conditions that are vital for developing universal technologies. Access could range from rapid dissemination of findings before formal publication and licensing of findings to foreign firms at a reasonable price to actual foreign participation in such consortia.

Foreign firms participate in several of the major research projects sponsored by the EEC. Foreign membership has been secured at last for one of Japan's major projects, Galaxy, for the next generation of computers. Both Univac Japan and IBM are members of the AIST standards committee that is to bring OSI to Japan. Similarly, companies from the United States, Japan, and Europe have joined X/OPEN, a London-based consortium for the development of common software building blocks. Companies from the United States, the United

Kingdom, and Canada joined to promote a preferred set of network management protocols.

3.5.2 Product Networking

Product networking—the creation of families of products, often from a variety of original vendors, but offered as an optional package—has become ubiquitous. Firms require domestic and international partners to create selective interconnectivity among products. They do not attempt product networking on a universal basis either because the task requires sharing valuable proprietary technology or because competitive tensions are too strong to make it worthwhile to attempt to reach a consensus.

A classic example is the Sun Microsystems campaign to standardize future high-speed minicomputers and workstations around its SPARC architecture for semiconductors and Unix software in much the same way IBM made its microcomputer architecture the standard in the early 1980s. It offered equity to AT&T (the original developer of Unix) to consolidate the Unix deal in January 1988. The deal was for up to 20 percent of Sun (15 percent of the company in newly issued shares at a 25 percent premium to the market, and 5 percent AT&T could buy on the open market). AT&T never acquired the full 20 percent. Sun also aggressively licensed SPARC technology to selected U.S. and Japanese firms (Fujitsu and Matsushita). This prompted DEC, IBM, Nixdorf, Siemens, Groupe Bull, and Philips to join a rival consortium for development of Unix. Meanwhile, DEC was busily striking deals with German and Canadian firms to support its new Enterprise Management Architecture for network management. Sun and AT&T ultimately could not agree on how to cooperate, and Sun had bought all its stock back from AT&T by 1990.

When competitors bring out a new technology more or less at the same time, the products will inevitably be incompatible, if not outright noninterchangeable. An excellent example is videocassette recorders in the 1970s, where Sony's first (and in some ways superior quality) Betamax format was soon joined by JVC's VHS. The latter's longer-playing tapes helped make it the dominant standard, and Sony ultimately capitulated. In nonconsumer products, emergence of a single standard is rarely as necessary—as can be seen from the ongoing debate over operating systems and programming languages. Even for commercial products, however, proprietary systems are becoming anachronisms.

Fearing being shut-in by their advances, firms often see open architectures and product networking as important when defining the technologies of follow-on or next-generation products. Even firms that can take almost discontinuous steps in technology find it worthwhile encouraging smaller-scale innovators to follow their bandwagon, even if it means some become competitors in the basic product. Technological pioneers in the past were routinely rewarded with arrows in their backs; the risk now is being trampled in the rush of complementary product producers through a breach in the technological frontier.

Alliances to provide a major product or create a single global competitor are subject to growing competition from rival alliances. For example; in 1988 U.S.-based Computer Services Corp sold 70 percent of its global network for value added services to PTTs in Europe and Japan. (CSC sold its remaining stake to MCI in 1990.) A vivid glimpse into the logic of PTTs is available in an interview with Gerard Simonet of France's Transpac, the public packet switching network, given to Heywood (1988). Simonet believes this network will provide a common vehicle to implement a single international data network for these countries.

However, other global networks have been mounted by joint ventures involving IBM, GEISCO, and EDS (a subsidiary of General Motors). These compete with the GIMM venture of AT&T, BT, and KDD. For this reason, BT may not join in the purchase of the CSC network. EDS and GEISCO signed an agreement for GEISCO to provide VAN services for EDS in Europe. EDS then announced a new venture with PanAmSat for other new services to Europe.

Telecommunications infrastructure facilities in the global arena, like satellite systems, were traditionally joint ventures operating in closed markets. This is changing rapidly. The international satellite equipment market has rapidly turned into a set of competing global teams led by Ford with Mitsubishi Electric and Alcatel Espace; Matra with TRW, NEC, and Spar; GE with Aerospatiale and Messerschmitt; and Huges. Note that all but one of these is U.S. led. Insistence on domestic satellite industries was the principal cause of joint venturing.

The satellite services market will become similarly subject to competition. Intelsat is subject to competition from PanAmSat (a U.S. firm) and Orion (a joint venture involving British Aerospace). Others will probably follow. Intelsat will respond by selling more capacity for domestic services in developing countries, but new competitors will emerge there as well. For example, AsiaSat, a consortia of a U.S.–Canadian firms, C&W, firms from Hong Kong, and China will provide domestic satellite services for the Asian area, including China. This will provide competition to any expansion of Palapa, the Indonesian system.

Restrictions on competition between satellites and cables for international traffic are declining. There is a new generation of joint ventures for transoceanic fiberoptic cables on the major global routes.

3.5.3 The New Regime

In its initiatives on trade in services, intellectual property, technical standards, and rules governing government procurement practices, GATT and corporate strategies are moving closer to what can be called market access regimes. Existing international institutions are scrambling to respond. The ITU is seeking to move toward internationalization of regulation, and is trying to steer corporate strategies for all but information services in that direction. No international framework easily fits a regime that primarily focuses on internationalization of domestic regulation and market access issues.

What will happen? WATTC will modestly modernize and internationalize traditional regulations, but it will fail to substitute international rules for domestic ones. ITU will certainly play a major role in setting technical standards and continue to facilitate coordination of commerce among countries sticking to the monopoly model. It will not halt the broader movement toward more open markets. GATT will succeed in greatly extending its nominal jurisdiction on equipment and enhanced services simply because no major government will tolerate collapse of the major free-trade arrangement. Moreover, these GATT provisions allow many services and goods to become routine free-trade commodities. GATT, however, will not provide a very cohesive regime because it provides no framework for basic services and facilities, and its enforcement rules may not suffice in a market with such high transaction costs.

The most likely development is a high degree of internationalization of firms in the context of various national industrial policies and regulatory regimes—some more committed to competition and free markets than others. This will force more bilateral and minilateral bargains to provide specialized regulatory frameworks among like-minded countries. For example, ITU regulations for international services will likely include an article permitting countries to allow more competition by mutual consent. There is every indication that the United States will demand more parity with its major trading partners in communications services and equipment. Indeed, the 1988 Trade Act mandates this approach; after its passage the U.S. government quickly began considering special restraints on foreign service suppliers in the US if US firms do not get equal access. This may lead countries to bargain over the rights of their international service carriers along long-followed lines in the international airline industry.

It is reasonable to speculate that the United States will get less openness than it demands, but it will force significant liberalization. This may force more active involvement on the U.S. government and more openness on Japan, and there is evidence each country is settling on "open industrial policies" for informatics—open because they are market-oriented and suggestive, rather than cartel-oriented or indicative. It also means the additional opening and integration of the world economy will be accompanied by the decline of universal rules for the informatics sector.

Will Pacific Basin countries be further divided or become more integrated through minilateral negotiations? This depends on whether or not they start to create a major set of international institutions capable of fostering creative economic and political bargains among countries in the region. So far, such regional institutions are weak.

While there are cautious calls for the creation of Pacific-oriented economic organizations comparable to the Organization for Economic Cooperation and Development, all such proposals run into a common problem. The heart of the world market still lies in the Triad of the United States, Japan, and Europe. A more likely outcome, therefore, is minilateral bargaining featuring the major industrial countries and newly industrializing countries that are willing to redefine their role in the world informatics market.

Acknowledgments

The author thanks Jonathan D. Aronson, Douglas Conn, Seisuke Komatsusaki, and Eli Noam for their comments. This research was supported by the Council on Foreign Relations and the Rockefeller Foundation.

Bibliography

Auster, Ellen R. 1987. "International Corporate Linkages: Dynamic Forms in Changing Environments." *Columbia Journal of World Business* 22: 3 (Summer).

Aronson, Jonathan D., and Peter F. Cowhey. 1987. "Bilateral Telecommunications Negotiations." Paper prepared for the International Institute of Communications.

——— and Peter F. Cowhey. 1988. *When Countries Talk, International Trade in Telecommunications Services*. Cambridge: Ballinger.

Caves, R. E., H. Crookell, and J. P. Killing. 1983. "The Imperfect Market for Technology Licenses." *Oxford Bulletin of Economics and Statistics* 45: 249–68.

Christelow, Dorothy B. 1987. "International Joint Ventures: How Important Are They?" *Columbia Journal of World Business* 22: 7 (Summer).

Cohen, Stephen, and John Zysman. 1987. *Manufacturing Matters*. New York: Basic Books.

Cooney, Stephen. 1989. *EC-92 and US Industry*. Washington: National Association of Manufacturers.

Cowhey, Peter, and Jonathan D. Aronson. 1992. *Managing the World's Economy: The Implications of Corporate Alliances*. New York: Council on Foreign Relations.

Deigan, Morris and Michael Hergert. 1987. "Trends in International Collaborative Agreements." *Columbia Journal of World Business* 22:15–22 (Summer).

Economist. 1991. "A Survey of Telecommunications: The New Boys." Supplement bound in the Oct 5 issue.

Commission of the European Economic Community (EEC). 1987 Jun. *Towards a Dynamic European Economy: Green Paper on the Development of the Common Market for Telecommunications Services and Equipment,* Com (87) 290 final.

Franko, Lawrence G. 1987. "New Forms of Investment in Developing Countries by US Companies: A Five Industry Comparison." *Columbia Journal of World Business* 22: 39–55 (Summer).

Ghemawat, P., Michael E. Porter, and R. A. Rawlinson. 1986. "Patterns of International Coalition Activity." In Michael E. Porter, ed., *Competition in Global Industries*. Cambridge: Harvard Business School Press.

Harrigan, Kathryn. 1986. *Managing for Joint Venture Success*. Lexington, MA: Lexington.

Heywood, Peter. 1988. "PTT-Infonet Deal Offers Glimpse of Post-1992 VAN Reality in Europe." *Data Communications,* pp. 60–66 (Sept.).

Hladik, Karen J. 1985. *International Joint Ventures*. Lexington MA: Lexington.

Huber, Peter W. 1987. The Geodesic Network: 1987 Report on Competition in the Telephone Industry. U.S. Department of Justice Department, Antitrust Division.

Kitchen, Malcom, David Lewin, and Hans Schoof. 1987. *Telecommunications: The Opportunities of Competition*. Princeton: Ovum Ltd.

Krugman, Paul R., ed. 1986. *Strategic Trade Policy and the New International Economics*. Cambridge, MA: MIT Press.

Lake, David. 1988. "The State and American Trade Strategy in the Prehegemonic Era." *International Organization* 42: 33–58. (Winter).

Lipson, Charles. 1985. *Standing Guard.* Berkeley: University of California Press.

Millner, Helen V. 1988. *Resisting Protectionism.* Princeton: Princeton University Press.

Mowery, David C. 1988. "Collaborative Ventures Between U.S. and Foreign Manufacturing Firms, An Overview." and "Joint Ventures in the U.S. Commercial Aircraft Industry." In David C Mowery, ed., *International Collaborative Ventures in US Manufacturing.* Cambridge, MA: Ballinger.

——— and Nathan Rosenberg. 1985. "Competition and Cooperation: The U.S. and Japanese Commercial Aircraft Industries." *California Management Review* 27: 70–92.

Pisano, Gary P., Michael V. Russo, and David J. Teece. 1988. "Joint Ventures and Collaborative Arrangements in the Telecommunications Equipment Industry." In David Mowery, ed., *International Collaborative Ventures in US Manufacturing.* Cambridge, MA: Ballinger, 1988.

Promothee. 1988. "Deregulation in the 1990s." *Project Promethee Perspectives* 5 (Mar.).

Rogowski, Ronald. 1987. "Trade and the Variety of Democratic Institutions." *International Organization* 41: 203–24 (Spring).

Roobeek, Annemieke J. 1988. "Telecommunications: An Industry in Transition." In H. W. De Jong, ed., *The Structure of European Industry,* 2nd revised edition. Dordrecht: Kluwer.

Saunders, Anthony. 1989. "The Influence of the New Communications Technologies on Banking and Finance." In Paula Newberg, ed., *New Directions in Communications Policy,* vol 2. Durham, NC: Duke University Press.

Thomas, Lacy Glenn. 1988. "Multifirm Strategies in the U.S. Pharmaceutical Industry." In David Mowery, editor, *International Collaborative Ventures in US Manufacturing.* Cambridge, MA: Ballinger, 1988.

Williamson, Oliver. 1985. *The Economic Institutions of Capitalism.* New York: Free Press.

Yarbrough, Beth V., and Robert M. Yarbrough. 1987. "Cooperation in the Liberalization of International Trade: After Hegemony, What?" *International Organization* 41: 1–26 (Winter).

II
NETWORK FORMATION

4
China

LIANG XIONG-JIAN AND ZHU YOU-NONG

Confucius noted, "The propagation of morals is a faster means of ruling than to have the post to send commands." Still, as befits a cradle of civilization, communication systems developed very early in China. During the Shang Dynasty (sixteenth to eleventh century B.C.) there were codes of drum beats to give alarms. In the Zhou (eleventh century to 771 B.C.) beacon fires and smoke signaling were used. As the state grew in size and complexity, organized communication systems developed to fulfill the requirements of the ruling class, as was the case in other ancient empires such as Persia and Rome.

Development of courier post systems depend on the relative power of the state and the prosperity of society. By Confucius' time in the sixth century B.C. a courier post was well known, quite speedy, and was used by the state to transmit commands and pass on orders. During the Han Dynasty (206 B.C. to 446 A.D.), courier service reached distant countries, including Persia. Indeed, the first Han emperor had once been a courier station chief.

The Tang Dynasty (618–907) was a period of power, culture, and prosperity when the courier service reached a high state of development. There was a well-organized system of 1,639 stations along land and water routes. Stations specifically built for the service handled official documents and farmlands were allocated for their upkeep. Couriers were exclusively engaged as carriers under a system requiring certification or credentials to transfer documents.

The Song (960–1279) further developed procedures for document transmittal by finding ways to eliminate delays at stations where couriers and horses were changed. Messengers would travel day and night, covering as much as 250 km in a day.

By the Ming (1368–1644) as a result of the development of commerce, civilian letter carriers emerged serving private citizens. The first such carriers appeared in the important seaport of Ningbo in Zhejiang province. Half the fee for carrying a letter was collected from the sender and the other half from the receiver. By 1840, several thousand unofficial postal bureaus had been set up to handle letters, parcels, and remittances to and from overseas Chinese. After 1840, a number of foreign countries operated postal services in China. These were initially mostly British. However, by the early twentieth century the Japanese and Russians had the most offices. Their offices were primarily in Manchuria, and to a lesser extent in other parts of northern China.

Telegraph and telephone were successively introduced to China after the 1870s, with facilities largely provided by foreign countries for their own use. The first submarine cable was laid in 1871 by the Danish Great Northern Co. connecting Hong Kong to Shanghai. The first city telephone system was opened in Shanghai by the British in 1881. The system expanded for the next half century. By 1936 there were over 18,500 post and telecommunication (PT) offices (including branches), 193,000 urban telephone lines, and 180,000 km of long-distance telegraph and telephone lines.

4.1 Structure

Public telecommunications in China are a government monopoly, primarily under the Ministry of Posts and Telecommunications (MPT), which was established November 1, 1949. The minister is responsible to the premier of the State Council. Other ministries—including the railways, the petroleum industry, water resources and electric power, and public security—construct and operate telecommunications systems dedicated to their own use.

As an executive agency of the government, MPT directly controls the operations of all PT enterprises, exercising both administrative and managerial functions, including supervision of their production. It is responsible for development plans, technical standards, service policies and regulations, and tariffs. In addition, it researches international and domestic PT markets. In addition to administrative departments and special operations that are directly under MPT, the sector has a extremely complex hierarchial structure. MPT forms the first level.

The second level, directly controlled by the MPT, consists of thirty Post and Telecommunications Administrations (PTAs). These geographically coincide with the highest level of local government. PTAs function as the medium managerial level and play an important role overseeing operations in the twenty-one provinces, five autonomous regions, and four special municipalities (the metropolitan areas of Beijing, Shanghai, Guangzhou, and Tianjin are not in provinces).

At the third level, operations are separated into activities that are performed by about 2,500 enterprises. These enterprises include equipment production and research facilities. Also at the third level is the network of main offices where the public can conduct post and telecommunications business—buy stamps, place telephone calls, and so on. (Beijing is an exception to this, its Postal Bureau is directly under the MPT, skipping the PTA level, although there is still a PTA for Beijing to oversee the enterprises). These offices are generally in each prefectural capital—seats of the second level of local government administration. The branches and subbranches of prefectural offices could be considered fourth and fifth levels.

Local governments also have some say in PT activities within their borders.

Thus provincial administration (PPT) falls under the jurisdiction of the MPT and the provincial government. PT enterprises are also responsible to the city or county governments where they are located. Within this three-way control, the MPT can override the province, which can override the city or county. Under a 1990 reform local control was expanded by the state council to make county-level PT departments responsible for township and village telephones, with administration vested in the county government.

Education and training are important tasks of the MPT. In 1955, the first specialized institute, the Beijing University of Posts and Telecommunications, opened. Four such specialized institutes had been established by 1980. Course offerings are in telecommunications technology and PT economics and management. Most of the 40,000 graduates have been assigned to positions in the various PT departments. In addition to these highly specialized institutes, provincial MPT administrations also run vocational and technical schools.

4.2 Development Since 1949

By 1949 the telecommunications network had been ravaged by years of war. During 1950–1952 urgent attention was devoted to rehabilitation and resumption of services—annual investment for telecommunications was 3.5, 1.7, and 0.6 percent, respectively, of total state investment. The effort both restarted facilities and insured unimpeded communication between Beijing and all provincial capitals and important cities as well as afforded some additions and improvements to the network.

4.2.1 First Five-Year Plan (1953–1957)

China's first five-year plan was one of steady economic growth, and telecommunications expanded rapidly. Investment in telecommunications averaged 0.55 percent of total state investment annually. This was relatively less than it was during the rehabilitation years, but the actual amounts were much larger because total investments were greater.

During the plan, a national trunk network was completed with Beijing as the center. All the important industrial cities were linked by direct circuits to Beijing for telegraphic and telephone communication. Wire photo service was also provided between a number of big cities in 1955. The networks primarily were wireline; remote areas such as Xizang (Tibet) and Xinjiang were linked by radio. Long-distance telegraph and telephone facilities increased only a little because a great deal of refurbishing was carried out; much of the "make-do" equipment from the rehabilitation period was replaced by better quality facilities. Table 4.1 summarizes the period's main additions to facilities and equipment.

National telephone density was only 0.05 in 1949, and there were no rural telephones. In 1952, only 84 cooperative farms had telephones, but the number

Table 4.1. Telecommunication Capacity Growth During the First Five-Year Plan (1953–1957)

Capacity in 1957	Growth during Plan (%)	Facility
4,946	11	Telegraph circuits
4,684	24	Long-distance telephone circuits
647,000	64	Urban telephone exchange capacity (lines)
326,000	327	Rural telephone exchange capacity (lines)

Source: MPT *Statistical Yearbook*

Growth is over the entire five-year period of the plan.

increased to 16,412 cooperatives by 1956. As population increased from 540 million in 1949 to 650 million in 1957, telephone density rose to 0.13.

4.2.2 Comparative Sectoral Growth, 1953–1980

During the first five-year plan the development of telecommunications generally matched the growth of the national economy and the demand for service. Taking 1949 as a base, the growth rate of the PT industry approximated that of manufacturing and agriculture taken together, but lagged behind the rate for growth of manufacturing alone.

Between 1957 and 1970 telecommunications did not grow steadily because of various political and economic reasons, and the gap between service availability and demand grew larger. However, improvement of the telecommunications system was an important aspect of the fourth five-year plan (1971–1975). Expanding and upgrading the system was given high priority in the allocation of scarce resources. Wideband trunk lines—both cable and microwave—were added. International connection was made by coaxial cable to Hong Kong and satellite ground stations. The ground stations were originally set up in 1972 so that the visits of Richard Nixon and Kakuei Tanaka could be covered live.

Still, taking the city of Beijing as an example, in the thirty years 1949–1979, the capacity of the city's telephone exchanges grew only 3.6 times—while manufacturing production increased by a factor of 104, water supply by 46, and electric power by 32.

4.2.3 Achievements, 1980–1985

China recognizes telecommunications as an advance agent that must take precedence in economic development. Accordingly, during the 1980s it became a priority in China's construction efforts. To raise the capital needed for this, emphasis has been given to the principle that funds ought to be raised in a variety of ways and from a number of sources, including state authorities, individuals, and investment by collective or private institutions. Installation fees for urban phones are allowed with the agreement of local government. Together with operating profits, these may be used for infrastructure investments.

State and local authorities furnish the necessary foreign exchange to support the importation of equipment and technology. Although basic tariffs are uniform nationally, local governments may permit PT enterprises to collect surcharges from users. For example, in addition to the unified rate users pay, there is a surcharge of 0.1 yuan per minute on trunk calls.

4.2.4 Enhancements During the 1980s

With ongoing adoption of advanced technologies, the capabilities of the national public telecommunication networks were significantly enhanced during the 1980s. Stored program controlled (SPC) exchanges were introduced in 1984 and by 1988 represented 25 percent of switch capacity. Fiberoptic cables, 960- and 600-channel microwave relay systems, and balanced and coaxial cables also became widely used. A domestic satellite network, first operational in 1986, links Beijing, Lhasa, Urumqi (in Xinjiang), Hohhot (Inner Mongolia), Guangzhou, and other cities. The number of long-distance telephone circuits doubled from 1980 to 1986 (to 44,000), and it exceeded 66,000 in 1988. Telegraph circuits have also increased. Table 4.2 provides additional data.

Since the sixth five-year (1980–1985) plan, a number of major construction projects have been completed to extend communication capacities and enhance the level of operation. These include the Beijing–Wuhan–Guangzhou 1,800-channel medium coaxial cable; a number of SPC exchanges in Beijing, provincial capitals, and coastal open cities; international gateway exchanges in Beijing and Shanghai; and the domestic satellite network.

A variety of means were used to raise funds for telecommunications development during the sixth five-year plan, as shown in Table 4.3. Enterprises have become less dependent on the state than was formerly the case. State funding accounted for only 31 percent, while capital raised by individual enterprises made up more than 50 percent of investment. This trend resulted from reforms

Table 4.2. Telephone Exchange Capacity and Telephone Density, 1980–1987*

		Urban Exchanges			
Year	Density	Lines	Automatic (%)	SPC (%)	Rural Exchanges Lines
1980	0.43	2.00	66	0	2.43
1983	0.49	2.62	75	0	2.54
1986	0.67	3.80	84	7	2.92
1987	0.75	4.64	87	19	3.09

Source: MPT *Statistical Yearbook*

*Million lines and percentages

During the 1980s about 95 percent of townships (*xiang*, the third level of local rural government) had telephone service.

Table 4.3. Investment Sources During the Sixth Five-Year Plan, 1981–1985*

31.05		Central government
4.52		Local governments
=	35.57	State investments
3.01		Localities
51.13		Enterprises
=	56.14	Individually raised
	2.87	Users
	3.69	Domestic loans
	0.29	Foreign capital
	1.44	Other

Source: MPT *Statistical Yearbook*
Shown in percentages.

to heighten the ability of enterprises to develop autonomously. Domestic loans and funds raised from users began to provide a small part of the total investment, also a new precedent. Foreign capital has remained negligible.

In 1985 investment equalled a whopping 56 percent of gross revenue generated by telecom services. In 1988 MPT said funds retained from operations allowed construction of only about 200,000 local lines a year. Private parties have received approval to set up switchboards with their own funds in order to improve rural communications. Some of these involve large numbers of people each contributing small amounts.

4.3 Networks

In many areas, particularly rural ones, the local network is not connected to the long-distance network. Instead, there are one or more phones at the town or county post office that are connected. The wait to use one is often hours. In 1988 MPT estimated about half of long-distance calls did not get through.

The long-distance telephone network is hierarchically structured, with up to four levels of manual and automatic transit centers depending on administrative districts and traffic conditions. Together with terminal exchanges connecting to subscribers these make up a five-level system. At present, the manual and automatic long-distance networks are used in combination, and the structures of these networks are identical. The four levels of transit centers for the network are described in detail later.

The highest level is interprovincial centers, the basic skeleton of the national network. These are the points where traffic between several provinces of a major region interconnect. There are six such centers—Beijing, Shanghai, Wuhan, Shenyang, Xi'an, and Chengdu. As the national capital, Beijing also has direct circuits to all the provincial capitals. Auxiliary centers are located in Tianjin, Nanjing, Lanzhou, and Chongqing. The network among them forms a mesh, while lower level networks form star configurations. Direct circuits are also provided in accordance with the demand for service, traffic volume and economic reasons to form an integrated and unified network.

The second level, usually located in provincial capitals, handles transit connections for intraprovince communication. At the third level, intercounty centers connect circuits between the counties; they are generally in cities that are seats of a subprovincial region or prefecture. County *(xian)* centers are the fourth level; they connect circuits within a district and are generally located in the seats of county government. A call from one county center to a distant one may have to go through seven circuits.

4.3.1 Private Networks

Because of the special requirements of various departments and limited public network capacity, there are a number of private networks, including ones for railways, electric power transmission, oil production, military departments, and broadcasting and television.

Local private networks handle calls within a long-distance numbering area of the public network. These networks are usually simply structured, using analog exchange equipment and urban telephone cables for transmission, and usually follow star configurations, though some have more complicated configurations.

Long-distance private networks traverse a number of areas of the public network. There are currently scores of departments using or planning to establish such networks. These may be categorized according to their usage or requirements. Two- and three-tier star-shaped networks connect a ministry or national commission with large-scale industrial or mining enterprises. Networks of mixed configuration incorporate a mesh type network between centers of higher order and a star from centers to lower-order stations. There are also international private networks which connect to international public networks via marine communications or satellite systems.

Many private networks have adopted digital technology. Digital microwave, fiber optic cables, satellite circuits, and SPC exchanges are increasingly used for transmission and switching. As these networks are usually smaller than the public networks and their members are in a position to make substantial investments, digitalization will probably occur faster than it does in the public networks.

An example is China National Petroleum & Chemical Co. GTE Spacenet built a satellite-based, interactive voice, data, and fax network for it under a $10 million contract (announced in 1992, after the project was operating). By 1997 the network was expected to cover 2,000 sites.

4.4 Technological Developments

China is implementing most of the technological advances that have been made in telecommunications, albeit on a limited scale.

Most digital SPC switches are imported, and the variety of systems occasions frequent difficulties in maintenance (see Chapter 5 for more on foreign suppliers). During the sixth five-year plan the MPT's First Research Institute in Shanghai developed a prototype of a switching system, dubbed the DS 2000 SPC. The Shanghai Bell Telephone Company produces S-1240 SPC switches with imported technology, and JD 1024 long-distance SPC switches have been developed domestically.

Domestic satellite communications were introduced in 1977 using Intelsat and ground stations in Beijing and Shanghai. A domestically produced experimental satellite was launched in April 1984, and a fully operational telecommunications satellite followed in February 1986. These are part of a public network linking Lhasa, Urumqi, Hohhot, Shanghai, and Guangzhou, with Beijing as the center. The 1984 satellite also has extensive television and radio broadcasting capability. In 1990 installation plans were announced to increase the number of earth stations to fourteen, including one more each in Beijing and Shanghai to link with Intelsat. The petroleum sector has its own network and other government departments are planning them.

MPT's First Research Institute has done much work on satellite communications. In 1980 it formulated the overall plan and technical specifications for earth stations for a domestic system to use Intelsat transponders. In 1982 it conducted tests and trials for leasing Intelsat facilities over the Indian Ocean and completed the installation and implementation of earth stations in Xinjing Nei in Mongolia, Xizhang, and Guangdong in 1985 and 1986. The Institute and Xi'an Communication Equipment Factory are positioned to provide and implement a complete system of earth stations.

China paid early attention to optical communications. A number of institutes conduct research in fiberoptics, including the Wuhan Post and Telecommunications Research Institute, Beijing University of Posts and Telecommunications, Jiaotong University in Shanghai, 46th Institute of the Ministry of Electronics Industry, and the University of Science and Technology in Shanghai. Fiberoptic transmission was in use by late 1986. It is used mainly for junction cables for urban telephone service in major cities such as Beijing, Wuhan, and Shanghai, although long-haul fiberoptic cables, such as one between Nanjing, Wuhan, and Chongqing, are under construction.

4.4.1 Public Data Telecommunications

China provides public data services through telex and low-speed data networks, open data services carried by the telephone networks, and the public packet switched data communication network.

4.4.1.1 Telex and Low Speed Data

More than sixty cities are equipped with telex switches, and over twenty have concentrators. A three-level telex network system has been established using Beijing and Shanghai as international traffic gateways and the provincial capital cities as subcenters. Beijing, Shanghai, and Guangzhou have direct connections with more than forty countries and regions. Provincial capitals are the centers of the domestic network.

Because mechanical electrical switching systems comprise a large part of the network, with resultant inherent poor transmission quality and low connection rates, some parts of China cannot even operate at 300 baud, although most of the system is capable of at least that and some routes have allowed 600 baud since the early 1980s.

A 300-baud network is being developed that will provide an effective means of communication for subscribers with light traffic and wide area communication needs. To develop domestic and international low-speed data and Chinese character telex services, advanced time division multiplexing (TDM) telegraphic equipment and the corresponding switching modules required for 300-baud service are being adopted to update the existing transmission network.

4.4.1.2 Open Data Service by Telephone Network

To meet subscriber demand for data transmission over the telephone network, MPT has worked out a ''technical system of opening data services on the telephone network'' that has been in use since January 1988. Since the late 1980s provincial capitals and medium coastal open cities have installed imported SPC switches. SPC systems for both international and domestic trunk services have been installed in Beijing, Shanghai, Guangzhou, and Tianjin; SPC for local telephone services account for almost half of the total access lines. These systems provide 2,400-baud data transmissions—and 9,600 for G3 facsimile services.

4.4.1.3 Public Packet Switching Data Communication Network

The packet switched network (CNPAC) consists of primary node switches in Beijing, Shanghai, and Guangzhou, and eight concentrators. The dual-system network management center is located in Beijing. Outgoing and incoming gateways are also located in Beijing, interconnecting with public packet switched networks around the world.

The system was put into operation in 1988 as a trial functioning network. Both the primary node equipment and concentrators are SESA DPS25s and use such CCITT (In English): (The International Telegraph and Telephone Consultive Committee) protocols as X.25, X.75, X.3, X.28, and X.29. The system also executes IBM's SNA/SDLC protocol and connects with an IBM mainframe. There are almost 500 ports, linked to the mainframe either via private line or through the public and telex networks. The system provides switched and permanent virtual circuits with some added services such as closed user group, reverse charge, and call transfer, as well as videotex. The public packet switched network can provide ports for various data bank connections, enabling information suppliers to serve their designated or public subscribers.

4.4.2 Mobile Communications

MPT has issued technical specifications for land-based mobile telephone networks and for public paging services. Depending on the particular area and network interface, a mobile telephone network must adopt TACS using 900 MHz (mainly in metropolitan areas such as Beijing, Tianjin, Shanghai, and Guangzhou) or NMT-450 using 450 MHz (used mainly for remote interior regions). Paging systems must conform to CCITT No. 1 code (POCSAG Code) and special service station modes with a frequency of 150 MHz.

Mobile communication services became available in Guangzhou, Shanghai, and Qinhuangdao in the late 1980s, and in 1990 in Beijing. Systems are under construction in Chongqing, Shenzhen, and Zhuhai. Other cities, including Tianjin and Dalian, are also making preparations. Furthermore, there are some private systems, such as the network set up by Liaohe Oil Field and the Beijing Tourism Administration, but they are prohibited from offering their services to the public. Some estimates put cellular subscribers in 1991 at 39,100.

There are still many private networks using other systems, including a dozen in Beijing, ranging from simple walkie-talkie and single-frequency radio-band telephone dispatching systems to automatic frequency selection systems with several high-frequency channels. Paging was introduced in 1984, and was available in forty-one cities at the end of 1987; there were 30,897 subscribers. The service operates twenty-four hours a day.

4.5 Equipment Manufacturing

Manufacturing telecom equipment falls mainly under MPT and the Ministry for Machine Building and Electronics Industry (MMBEI). The capacity of the plants under MMBEI is greater than that of plants under MPT. In 1980 there were twenty-nine such industrial enterprises directly under the ministry. The factories are capable of producing a wide range of items, including cable, microwave systems, telephone exchanges, and satellite communication equipment. In 1988 the factories employed about 40,000 and production amounted to 340 million yuan (about U.S.$91 million). Among the main types of equipment produced that year were:

47	Telephone systems (12-channel)
430	Telephone systems (60-channel)
3,512	Teletype machines
170	Facsimile machines
561	Shortwave radios
511	Microwave radios
244,900	lines of urban telephone exchanges
147,400	Telephone sets
11,343	km of communication cable

In addition to industrial plants directly under MPT, provincial administrations also run equipment factories. There were 120 such factories in 1980 with annual production valued at about 200 million yuan. They have been an important factor in the development of PT services and provide the basic facilities needed.

To increase development of equipment manufacturing, a number of production systems have been imported from abroad for MPT factories. These include production lines from Italy for making PCM (pulse code modulation) equipment at Factory 515 in Chongqing, from the United States for making plastic sheathed telephone cables at Factory 514 in Chendu, from Japan for making multifunction telephone sets at Factory 512 in Tianjin, from France for making telex equipment at Factory 524 in Guangzhou, from NV Philips for making PCM equipment by Factory 519 in Shanghai, and from Japan for making 140 Mbps digital microwave systems at Factory 506 in Beijing, as well as equipment for making SPC exchanges (Shanghai Bell and a joint-venture company with Belgium).

4.6 Urban Services

There are two categories of urban telephone service subscriber: those with telephones installed in their private residences and all others, such as business subscribers. There are three methods of billing: a flat monthly rental, a simple message rate, and a complex message rate system.

Under the first system, subscribers are charged a fixed monthly fee according to their service-fee class and subscriber category. For example, a category *A* subscriber in an area served by exchanges with more than 10,000 lines is charged 12 yuan; a category B subscriber in the same area pays 20.

According to the simple message rate system, subscribers are charged a fixed monthly rental according to their service-fee class and subscriber category in the same manner as the monthly rental system, with allowances for a certain number of free calls. If the number of calls made does not exceed the free allowance, only the monthly rental is charged. For example, the monthly rental for a *B* category subscriber in an area served by exchanges with over 10,000 lines is 16 yuan, with an allowance of 100 free calls. Each additional call is 0.04 yuan.

Under the complex message rate system, charges are based on the duration and distance of calls. The time unit for calls made to subscribers in the immediate urban area is three minutes. Calls made to subscribers outside the area are charged in the same way as long-distance calls. No free calls are allowed but the monthly rental fee is lower.

Letter service using fax machines at post offices is available between Beijing, Shanghai, Guangzhou, Shenzhen, and Zhuhai, and with Hong Kong, Macao, Japan, the United States, Canada, Germany, and Singapore. The service is expected to be expanded to more cities.

4.6.1 Services Rates

There are four categories for domestic rate setting—long-distance telephone, telegrams, leased circuits, and rental or maintenance fees for leased equipment—and an international category. The rates given in this section were in effect in 1988 unless noted otherwise; the official exchange rate for 1 yuan was U.S.$0.27.

Domestic long-distance charges vary with distance and time of day. There are twelve levels of per minute charge, the highest being 1.2 yuan for calls over 2,000 km. A call between Shanghai and Guangzhou is 1 yuan; it is 1.1 yuan between Guangzhou and Beijing. The minimum billing time for operator-connected or semiautomatic calls is three minutes, with additional time charged at the regular per minute rate. Automatic calls are charged in minute increments. Peak-rate hours are 7 A.M. to 9 P.M. except holidays and Sundays. The off-peak rate is half the peak rate.

Domestic telegrams are charged by the word and purpose. Regular rates range from 0.02 to 0.07 yuan per word; express or urgent telegrams are charged double.

The network access fee for 50-baud telex is 60 yuan per month; it is 100 yuan for 300-baud data. The call charge for 50-baud telex between domestic cities is 1 yuan per minute; it is 1.5 yuan for 300-baud data.

Telex rates in Beijing have come down significantly since early 1989, as has the time it takes for a connection. The wait was two years and installation was 30,000 yuan (including the machine) in early 1989. By the fall of 1990 the wait was two months and installation was under 6,000 yuan. However, per minute transmission rates to Hong Kong and New York increased—from 8 to 10 yuan and from 14.4 to 18, respectively. On the other hand, Tokyo rates dropped from 14.4 to 10.1. (At the 4.72 yuan/U.S.$ rate in 1990, these rates are high—$2.12 to Hong Kong, $3.05 to Tokyo, and $3.81 to New York.)

4.7 International Cooperation

China's international relations have progressed rapidly since implementation of its open door policy. In 1972 the Universal Postal Union and the International Telecommunications Union (ITU) accepted China. The ITU elected China a member of its Administrative Council. China joined Intelsat in August 1977. The country has participated in the activities of PT organizations in the Asian-Pacific Region.

International communication has increased phenomenally since implementation of reforms and the open door policy. In February 1985 Fuzhou started international direct dialing (IDD) service with Japan and the United States. In 1986, SPC exchange facilities for IDD were put into service in Beijing and Shanghai. In 1990 a third international gateway was opened in Guandong province feeding into facilities in Hong Kong. By the end of 1987, there were about 50,000 subscribers for IDD service in more than twenty cities, with access to fifty-one countries and regions. Foreign countries may direct dial subscribers in 310 cities. A 7,560-circuit cable linking Shanghai to Japan, to be built by KDD and AT&T, was announced in mid-1990. Completion is scheduled for 1993.

With regard to data retrieval, China is able to access information sources in the United States, Germany, the United Kingdom, France, and Luxembourg via Rome, and to Switzerland via Vienna. The service is available in Beijing, Shanghai, and Guangzhou. China is indeed working to be in touch with the rest of the world.

4.8 Conclusion

The growth rate of telephone density since 1980 has generally been slower, indeed, increasingly so, than the growth rate of GNP per capita. The number of telephone sets per million yuan of GNP has decreased, too. Still, telecommunications in China has developed since 1980 both in overall capacity and in operational efficiency. Both developments were required to meet the rising demand for services brought by China's opening and reforms. Of course, compared to most countries the level of development of telecommunications and density are still very low. China is the most populous country in the world; raising density one percentage point means the addition of 10 million phone lines, and total population continues to grow.

Two vignettes show how telecommunications is affecting China. As result of the spread of telephone access, telephone numbers now appear on product packaging. Before the mid-1980s even if the manufacturer had a telephone, it would have been at best difficult to call. In the late 1980s, in a mixture of an old custom with modern technology, a telephone had become the bride price in some areas of Fujian.

Bibliography

Ministry of Post and Telecommunications (MPT), *China Posts and Telecommunications,* various years, Beijing.

Ministry of Post and Telecommunications (MPT), *China Telecommunications Construction,* various years, Beijing.

Ministry of Post and Telecommunications (MPT), *Statistical Yearbook,* various years, Beijing.

5

China: Steps Toward Political and Financial Reform

KEN ZITA

China is ready to embark on its next revolution: the Information Age. The nation's economic reforms and the emergence of a market economy demand an efficient real-time communications network. Archaic switching and transmission equipment needs to be replaced, and a reliable long-distance network established. The entire telecommunications industry—including component manufacturing and system engineering and design, as well as network management and finance—needs to be brought to life. The scale of the effort is staggering, but China has high hopes for its telecommunications expansion.

When China ratified its seventh five-year economic plan in 1985, the policy blueprint formalizing its opening to the West, it sanctioned telecommunications as a national strategic priority. The following year, the first phase of the fifteen-year *China to the Year 2000,* telecommunications development study was made public. It stipulates that a minimum of $22 billion will be spent to quadruple the number of local telephone circuits and to unify the national network. The broad goals and specific recommendations for industrial self-sufficiency are ambitious.

Telecommunications in China is crippled by three strategic weaknesses. First, there is limited vertical integration of local and toll services, which skews economies of scale in capital investments and revenue collection and leads to technical inconsistencies among regional networks. Second, R&D and telecommunications manufacturing are split hodgepodge between the Ministries of Posts and Telecommunications (MPT) and Machine Building and Electronics Industry (MMBEI), and the rivalry between them splinters already scant resources. Third, state funds, and especially foreign exchange for imports and joint ventures, are extremely limited.

Impending financial reform in the telecom sector, paralleling broader central government economic policy changes, will sharpen the crisis. Massive cuts in direct appropriations to industrial and service ministries will dramatically reduce capital commitments from Beijing to local MPT bureaus. In response, municipal planners can be expected to manage phone networks as operating companies that issue bonds, raise rates, and, for the first time, come to terms with managing debt. The viability of this policy approach, which had gained both momentum and credibility in the late 1980s, has been called into question by the post-Tiananmen economic policies of the government of Li Peng. Politically, central government planners face the stiffest test yet to "socialist development with Chinese characteristics."

5.1 Modernization Goals

The shortcomings of China's telecommunications are widely recognized. Nationwide telephone density in 1990 was only 0.75. In the countryside density falls to 0.17, the equivalent of one telephone for every 500 persons. Moreover, most of China's telephone stations are in offices.

Public access is generally provided by private leased lines managed by small cooperative or entrepreneurial enterprises. For example, a typical community telephone "company," possibly a cigarette shop or a community high-rise bicycle garage, may manage two telephone lines for an entire neighborhood. Messages are posted on chalkboards or, luck permitting, runners search out the called party. In most areas only officials with deputy director status are entitled to private phone service; however, those with special political connections may possibly obtain a private line.

Public telecom services fall under the MPT. The MPT has exclusive responsibility for all international and domestic long-distance (interprovincial) calling. Regional and local networks are planned by local bureaus of the MPT; the exceptions are politically important minority regions, extremely poor areas, and the nation's borders, which are monitored by Beijing.

Strategies for the national network are devised in MPT's Department of Planning. The Long-Range Planning Division determines the national course; the Planning Institute crafts workable directives from long-range studies; and Import Planning sets line item priorities for major negotiations with foreigners. The Department of Policy and Regulation is responsible for policy analysis and overall strategic recommendations to MPT's minister. Via its academically oriented Research Academy, MPT also supports an important policy and econometric think tank known as the Research Center for Technical Economics, which provides quasi-independent reports and advice on timely issues.

China hopes to raise the number of telephones to 33.6 million by the year 2000 by adding at least 10–15 million virtual circuits or line equivalents to the approximately 11 million in place in 1991. Analog service is expected to be extended to the smallest towns. Digital switching and transmission corridors are slated to link provincial capitols and other big cities; fiberoptic and sophisticated switching systems will upgrade urban centers. Because China's own manufacturing capabilities are limited, equipment imports play an important role in network expansion. Still, about 1.4 million central exchange ("main office'' or ''public switch'') lines were manufactured in China in 1988, although an estimated 90 percent or more of them were older analog technology.

To date China has been unwilling to import analog switching systems, largely in an effort to protect its own manufacturers and to save foreign exchange for digital machines. Budgetary constraints may force at least partial reassessment of this approach. Since 1985 donations of some 500,000 lines of older generation crossbar and step-by-step switching technology have been accepted from Japan's NTT, Singapore Telecoms, and Hong Kong Telephone, apparently with success. The market for sales of used equipment, however, is soft due to Chinese demands for operating system source codes and extensive training on product lines that are being discontinued (Chen 1988; Wen 1988).

The success of China's development program to modernize the network is contingent on a number of factors. MPT needs to assume a more balanced leadership role. The Ministry's principal charter is to engineer and operate the national toll network. MPT has oversight of provincial and local telecom bureaus, but not in operations management and local network growth decisions. MPT, as an organ of the central government, is generally reluctant to relinquish authority to provincial planners, as this would sap its ability to direct the national toll network. While MPT provides important technical guidance to provincial bureaus, its discretionary control over import duties, investment credits for capital plant expansion underwritten by the state, and more broadly, its power of veto on major network projects, often overrides the will of local planners.

A coherent policy for managing domestic telecommunications manufacturing must be adopted. China's industrial base is fragmented by regionalism, interministry rivalries, inefficient labor and factory management, and an absence of reliable financing.

Local bureaus of MPT will have to boost internal generation of funds through more rational local tariffs and encourage municipal governments (and indeed the central government) to allow placement of long-term bonds. China's shortage of domestic capital and foreign exchange is a serious obstacle to growth, and reappraisal of the cost-pricing system deserves close scrutiny.

5.1.1 Services

Though MPT in Beijing is China's supreme telecommunications authority and the agent responsible for the nation's public trunking network, its role is more to lead a confederation of thirty separate provincial and municipal bureaus, each with virtually complete autonomy.

Local telephone service throughout China is a monopoly. A provincial authority usually has chief responsibility for local network development and financing throughout a region, characteristically defined by provincial boundaries. Local operators (hereafter referred to as *local PTTs* or ''bureaus'' for post telegraph, and telephones), which are organizationally a part of the MPT in Beijing, expect administrative and planning freedom from the central government according to long-standing bureaucratic relations.

Bureaus are often more closely aligned with the municipal government than with the ministry itself. Depending on the size of the town or city, a bureau may be an independent entity, an office within the city government, or an affiliate of the provincial bureau. The nearer the city is to Beijing, the more likely it will ascribe to ministry planning sentiments, and the less clout local authorities will have. Informal social ties, true to Chinese bureaucratic tradition, contribute greatly to the distribution of political influence between provincial authorities and the ministry in Beijing.

Big cities characteristically have their own telecom administration, separate from the province. Guangzhou, for instance, maintains a bureau to manage local plant, operations, and billing. Guangzhou municipality, Beijing, Shanghai, the five special economic zones (SEZs), and, to varying degrees, the fourteen largest coastal "open" cities are permitted to spend foreign exchange much as they see fit, so they have considerable discretion in selecting equipment imports. Other regions must secure central government approval. (SEZs and "open city" status are intended to encourage foreign trade and investment; see, e.g., Reardon 1991.)

Provincial authorities hold ultimate responsibility for big equipment contracts, network planning studies, tariff structures, and other long-range issues. Poor or conspicuously rural regions may look to Beijing for assistance if the organization and resources of the provincial authority are not suited to assist local matters.

MPT is the only agency authorized to carry toll traffic, a situation that makes for extremely high call charges and the sluggishness associated with a monopoly. No competitors to MPT are expected to be approved by the State Council at any time before 2000, and probably not soon after that either. A new carrier, Hong Kong-based AsiaSat, was established in 1990 to provide nationwide telephony and television. Jointly owned by Hutchinson Whampoa, the China International Trust and Investment Corporation (CITIC), and Britain's Cable & Wireless, AsiaSat will lease capacity to the domestic toll industry, but only MPT will be authorized to collect revenues.

5.1.2 Network Topology

The topology of network expansion reveals Chinese planning ideology and intentions for information resource distribution. The network is developing in two dimensions, each with unique technical characteristics, development plans, and political and economic implications. The public network, sponsored largely by MPT and local bureaus, constitutes the interprovincial trunking routes, international dialing services, and local telephony. The public network, significantly, is in turn subdivided into urban versus rural areas. "Private networks," national in scope but catering only to the internal administrative needs of five industrial and service ministries, are financed, planned, and operated independently of MPT.

5.1.2.1 The Public Network

The public network is managed as a utility; it is expensive, bureaucratic, and thus by definition hampered by insufficient funds. Affluent and relatively sophisticated cities are erecting modern networks based on the latest foreign technology, while poorer regions are making do with the basics. With reliable direct-dialing services urban users are beginning to regard the telephone as a vehicle of daily communication, at least in the work environment.

As usage rises, according to the prevailing logic, economic activity in the service area increases. The government hopes to spearhead advances in selected regions, which will in turn raise the overall capabilities of the network, thereby catalyzing cultural and commercial development. In essence, it is encouraging a technology and economic ''trickle down'' with Chinese characteristics.

The result is the formation of a two-tier network. Suburbs of cities with big construction budgets will benefit; most areas will have to wait. Until recently, telephone service in China was consistent: It was poor everywhere. The coastal cities are presently installing sophisticated imported systems while the nation's agricultural interior lags far behind.

Network stratification poses important social and economic questions. Any municipality that can afford imported equipment can proceed immediately with modernization. Other regions must compete for an ever-shrinking share of central government spending, attempt a ''middle-road'' course for network development based on domestic analog technology if production is available, or simply postpone development of a local telephone infrastructure. The last implies economic and cultural isolation. Rural network development, for instance, has none of the fanfare and little of the potential associated with bustling digital expansion in the cities. Even though 80 percent of the Chinese population lives in rural areas, rural network growth is expected to be modest. Before 1949, telephony in the hinterlands was nonexistent.

There is also the question of who gets access to the added lines. The waiting list for local service is estimated at 850,000, with 100,000 or more backlogged requests in Beijing alone (Chen 1988); the numbers would be higher if the public believed it was realistically possible to obtain service. Furthermore, distinction can be made among different ''classes of service'' actually received. With various grades of equipment being installed, priority calling status on quality equipment can be assigned only to those customers sufficiently well-connected politically. As a result, some users get instantaneous local dial tone and direct long-distance dialing over digital circuits, while others wait for lines on crackling and decrepit systems.

Information technology presents special difficulties for the Chinese socialist concept of public resource distribution. Although the state has had some success in providing public transportation, basic health care, and shelter, China simply cannot afford to provide every household with a telephone. The implication is disturbing. With access to effective telecommunications comes access to prosperity, social mobility, and virtually limitless horizontal communication within society.

Consider the discrepancy: A small packaged-goods enterprise in Nanjing is granted a clean local trunk with which it can price raw materials all over Jiangsu province. A competing enterprise, perhaps on the wrong side of a new central office exchange, can scarcely call across town over a single faltering or perennially blocked line. One firm can meet the dynamic demands of the emerging market economy, the other lags sadly behind. On the individual level the discrepancies of network access can be more poignant still: One person learns to perceive time and space, and indeed social access, as variables subject to that person's manipulation, and to regard the telephone as a tool for broader communication and personal growth. A cousin, with no access to or context for a telephone, lives a social and economic reality bound by the confines of the village, an outsider to many of the changes sweeping the country.

The great thrust of telecommunications expansion in China is to upgrade the business and administrative environment; most new local exchange lines are installed in government agencies, institutions, or businesses. The Western goal of universal service—that every citizen is entitled to affordable telephone service—is not a publicly stated goal of the MPT, nor is it likely to be for another decade or more.

5.1.2.2 Private Networks

The second dimension of Chinese telecommunications is the development of private, or overlay, networks operating independently of the MPT. Beginning in 1976 the central government granted permission to four ministries—coal, petroleum, railways, and water and power—to build their own nationwide systems to accommodate internal communications. It was widely recognized that the public network was a liability to effective communications, a critical consideration in the wake of the cultural revolution. At the time, the Ministry of Railways and the People's Liberation Army (PLA) already had systems in place. Interconnection between the various private networks and the public network is limited; only Railways and the PLA have formal interfaces, and these are few.

A number of additional systems emerged during the 1980s: the Ministry of Broadcasting operates an extensive microwave and satellite network to carry television signals, and the Bank of China is attempting to establish real-time links among primary banking centers in all thirty provinces. Smaller networks are employed by a number of others.

The private network development strategy is vital to the full-scale modernization of the economy. The intention is to streamline the state's ability to manage its industrial and commercial interests in a manner similar to a Western business conglomerate. Each ministry can be viewed as a separate business line, headquartered in Beijing, that contributes to the government's centrally planned bottom line. With economic decentralization racing to divest operational control from Beijing, ministries need national management information systems (MIS) to keep track of geographically disperse business concerns.

Without MIS to keep planning policies in line, ministries and, indeed, the central government itself, could lose more control over profit and loss centers (provincial or outlying offices, factories, independent enterprises, and cooperatives) than intended by the reforms. Beijing's economic initiatives are meant to stimulate production and to introduce responsible business management at the local level, not to disintegrate the power of the party or the state. The private networks are insurance that all roads will continue to lead to Beijing.

By leaving private network development to the individual ministries, the government is not obligated to underwrite network capital spending. In some cases, however, it will provide some funding, and it can concentrate instead on the national plan. By dedicating networks to unique business concerns, managerial efficiency is stimulated and accountability of regional enterprises to Beijing is improved. Like corporate networks elsewhere, China's private systems will probably remain relatively independent of the public network, both because MPT cannot itself shoulder the cost of development, and because the ministries are keen to preserve autonomy. The Ministry of Railways, which controls some 5–8 percent of all operating circuits in China, is estimated to have the largest of the private networks.

While not yet national networks, several major on-line processing systems have been inaugurated. The State Economic Commission will spend an estimated $1 billion by the end of the century on a wide area network (WAN) and office automation system linking major commercial centers. Xinhua, the national news agency and political tool of the Party, is putting a vast distributed processing network in place. MPT is installing an X.25 packet switched network with primary nodes in Beijing, Guangzhou, Shanghai, and several additional centers. The system will create a high-density traffic corridor; services will be marketed by the Beijing Telecommunications Authority to end users and government agencies.

5.2 Organization and Political Control

Decision making in China is complicated. A number of groups and individuals are inevitably involved, and it is rare that any one single group has definitive power to champion and approve an initiative. According to central government officials, decision makers often only have the power to negatively influence decisions; few, even at the highest levels, have power independently to approve projects.

Telecom equipment manufacturing is a fragmented, sometimes bitter, competition between MPT, manufacturer of selected products, and the MMBEI, the State Council's favorite child and highly subsidized research center for components and software. The Ministry of Railroads manufacturers almost all of its own equipment and runs several of its own telecommunications colleges. Despite formidable engineering and labor resources, China can report few economies of scale and only poor or insufficient synergy between R&D, product definition, and manufacturing (see, e.g., Zita 1988b, pp. 2–15).

The current strategic framework for electronics was formulated by the Group for the Revitalization of the Electronics Industry (also known as the Electronics Leading Group, ELG), a council of high-level technocrats within the State Council, during the group's 1985–1988 tenure. Under the leadership of Li Peng, who was subsequently premier, the ELG set the strategic path and development priorities for telecommunications, computers, software, integrated circuits, and electronic sensors. Although it has been functionally disbanded and its members reassigned to previous work units (only a skeleton staff remains), the ELG has left an indelible mark on the industry's future.

The ELG's conceptual recommendations, such as limiting the number of foreign electronics suppliers and targeting specific technologies for exploration and growth, are given tangible form by the State Planning Commission (SPC) and the Science and Technology Commission (SSTC). SSTC recommends how R&D funding should be spent, while SPC actually controls budgets. In late 1984 SPC sought to ease the rivalry between MPT and MMBEI by parceling specific R&D tasks to each. The compromise was for MPT to be the primary user of equipment and MMBEI the primary manufacturer.

Nonetheless, MPT will continue to manufacture a great deal of equipment. It makes most CO items and PABX, and it is a major supplier of optical electronics and line multiplexer equipment. MPT manages China's only operational digital switching facility, the Shanghai Bell Telephone Company. This is a joint venture set up in 1983 with ITT's Belgium subsidiary, which is now part of Alcatel. MMBEI is the country's largest producer of wire and cable. It also makes approximately 90 percent of all telecommunications components—ranging from mechanical relays and printed circuit boards to capacitors, transistors, and integrated circuits—and is an important joint venture partner for many foreign suppliers. Thus, it is slated to boost central office production with facilities being developed with Siemens and NEC. MMBEI's finished equipment is sold chiefly to the military and private networks; in contrast, MPT's systems are installed almost exclusively in the public network.

Under SPC–ELG guidance, a program has evolved to support "leading" research institutes and factories that pursue key development projects. MPT's Research Institute 1 in Shanghai, for instance, is slated to become China's foremost domestic PABX design center; MMBEI's Factory 738 in Beijing is destined to be the core for new research in large switches. Similar assignments—in some instances more than one—have been made for all strategic technologies: lightguide fiber (MPT in Wuhan and Shanghai), satellite earth stations (MMBEI in Nanjing), PCM (MPT in Chongqing); application-specific integrated circuits (MMBEI in Beijing and Nanjing), and so on. Leading research and manufacturing sites are reported to all have ample budgets, access to foreign exchange, highly qualified staff, preferential taxes, and, frequently, permission to license technology from abroad (see Zita 1988a).

Assignment of government-sponsored leading enterprises and factories contrasts sharply with the usual industrial structure. Ministries and municipalities have historically encouraged local self reliance, a strategy that surrendered manufacturing efficiency to community rule. Crossbar switch factories can still be found that also build assembly line machine tools and test equipment, as well as turning out postal delivery bags, sewing machine motors, household lamps, and whatever else was needed or independently profitable.

Leading research and production centers are meant to encourage R&D and factory floor specialization, coordinate talented personnel, and dissolve ancillary activities. If pursued with conviction, the "leading site" strategy may, by investing in organizations most likely to meet technological and commercial success, establish better linkage between research, competent factory management, and production.

Responsibility for actually implementing ELG–SPC policy rests largely with MMBEI. However, it has no clear bureaucratic mechanism to coordinate planners, R&D facilities, and factories in large-scale projects. Sector planning is rare and inefficient where it exists. The ministry can encourage limited association among affiliated factories that are managed largely as independent enterprises, but lack the managerial infrastructure to marshall major development efforts. Planning decisions are often made according to strict financial considerations—that is, who has foreign exchange—and not, in line with SPC "leading site" recommendations, according to carefully considered research and manufacturing efficiencies.

A case in point is a joint R&D and planned manufacturing venture for a small central office exchange between Italtel, the Italian national supplier, and MMBEI Research Institute 54 in Shijianzhuang. Following ELG's advice, the State Council limited the number of central office joint ventures to three. These are with Alcatel at Factory 520 under MPT, Siemens at Factory 738 in Beijing, and NEC at MMBEI factories in and near Tianjin. Feeling left out of the market, Italtel appealed to the government by proposing a joint project to design and eventually manufacture a small rural central office exchange. The Italian government made a soft loan to underwrite the project available, thus effectively maneuvering through restrictions established by the central government bureaucracy. In this way, a fourth switching venture was consummated without violating established rules.

MPT's manufacturing is managed by the Posts and Telecommunications Corporation (PTIC), presently a wholly owned subsidiary. Factories are being granted greater control over operations; in the future, the twenty-seven factories directly under PTIC and its 100 factory affiliates will be managed as increasingly independent enterprises. MMBEI's more than 1,000 factories have already been fully divested, and only R&D and overall strategic and production planning are guided by Beijing. The goal of decentralizing management, reflecting changes elsewhere in the economy, is to bestow responsibility for profit and loss to locations where work is done and to boost production incentives and efficiency.

5.3 Research and Development

China's emerging industrial policy for information technology calls for the commercialization of basic research combined with limited strategic alliances with foreigners. The state realizes the shortcomings of its domestic industry, as well as the importance of both moving its own R&D talents into the marketplace and acquiring technology and research methods from abroad. China, however, is keen to avoid the branch plant syndrome—assembly rather than true R&D—that it feels characterizes other newly industrialized parts of Asia. Furthermore, China wants to acquire foreign technology but keep its own R&D independent and developing. (This section is excerpted in part from Zita 1987, Chap. 6).

In the early 1990s the scientific and technological community is still reeling from the impact of the cultural revolution. Advanced research with the exception of certain military projects was brought to a halt between 1966 and 1976, a critical period in the global development of digital electronics. As a result, Chinese information technology is two or three generations behind the West. The engineers and technical workers who suffered the most have lost ten years of research and experience; they are now holding senior and middle management positions.

Recovery is complicated by deep-seated currents in Chinese science and technology in general. Science and basic research have traditionally enjoyed high status, business and applied science have not. One reason for the dichotomy is a distinction between a quest for knowledge and a search for practical application. Technology is the business of packaging the fruits of scientific endeavor, not of seeking truth. Like traditional Chinese military science, technological modernization with Chinese characteristics attempts to absorb Western technology without absorbing too much of its culture.

Deng Xiaoping has said, "We study advanced technology, science, and management to serve socialism, but these things do not by themselves have a class character." This is not quite true. Technology is a highly political activity, and the managerial systems that gave rise to it, and the social context in which it is employed, reflect highly particular organizational philosophies. To make the step into the information age, a technology management infrastructure has to be created, and it will have to be borrowed from abroad.

The general disdain for technology has several consequences. Advances in basic research do not effectively lead to product innovation. There is no adequate product development cycle from conception to approval, prototype, testing, and production for important developments. With no technology management, good ideas often never leave the labs. There are no formal mechanisms for the diffusion of innovations. Technical advances and transferred foreign processes and products often stay within the group first adapting them, leading to duplication of research efforts, wasted capital, and limited market penetration.

In addition, China suffers from poor allocation of trained personnel. Due to vertical segmentation of industries and long-standing policies regulating worker assignments, the R&D environment is not conducive to the cross-fertilization of expertise. The problem is compounded by increased competition and protectionism among newly privatized independent research institutes.

Both MPT and MMBEI, or their subsidiaries or affiliates, are attempting to develop optical electronics, digital microwave, and packet switching in addition to mainstay technologies; each is striving for technological breakthroughs to earn "leading site" status and the privileges that go with it.

5.4 Investment

With no appreciable rate base and no internal generation of funds, capital formation is among the industry's greatest challenges. Less than 1 percent of the state's fixed investment is committed to telecommunications; a severe shortage of foreign exchange curtails imports.

In 1987 MPT's net investment in telecommunications was $323.4 million, half of which came from bank loans, a sixth from MPT profits, and the balance from direct investment by the central government. State spending on the public network during the seventh five-year plan (1986–1990) will total $1.6 billion. By the year 2000, cumulative spending may total $21.7 billion (growing fastest in the late 1990s). Additionally, some $3–7 billion will be spent on private networks, coming exclusively from the relevant ministries.

5.4.1 Financial Considerations Affecting Industry Growth

Though telecommunications is typically a hugely profitable business, making money has consistently eluded the Chinese. Long-distance services generate the bulk of MPT's revenues and more than 40 percent of its profits. Local service revenues, which the ministry earns through yearend taxes, are nearly as large. Additional telephony services contribute little to the bottom line: telex is just barely profitable; data transmission is a cost center; fax has not yet matured; and feature group dialing (Centrex), teleconferencing, and other value-added features remain ambitious projections, and are only emerging in the biggest urban centers. The only sector to lose money in 1986 was the Postal Service, which, like European PTTs, is heavily subsidized by profits from telecommunications; the ministry maintains a consolidated balance sheet between the two.

MPT provides most of the financing for the national network, including long-haul cable and microwave routes, satellite earth stations, and tandem switching centers. Additionally, MPT meets up to 70 percent of the cost of combined tandem-local switching centers or cross-provincial transmission lines, with the local telecom authority putting up the remainder. Service revenues are distributed in the same proportions. By investing in combined local-toll equipment, MPT insures consistent trunking characteristics and maintains a presence as an equity shareholder in local network development.

More significantly, MPT makes available quotas of hard currency so local PTTs can convert local currency (yuan) to foreign exchange at the official rate. When MPT underwrites capital allotments through quotas, it automatically assumes an active role in contract negotiations and system selection. MPT foreign exchange quotas are a major force in purchase decisions though information regarding the size and conditions for the allotments is closely held.

International and domestic long-distance revenues are retained by the operating bureau and taxed at yearend as part of total earnings. Because international calls are paid to the local operator in foreign exchange, but taxed by the ministry in yuan, the ministry takes a loss with each international call. If an international call originates in a joint venture hotel, the local PTT is also paid in yuan.

Taxation is very complex. The overall burden is far less than what is typically collected from other industrial enterprises in China. The lower rate was established in light of telecommunications' high infrastructure costs and politically capped service rates. In the future, taxes will likely be managed by local agents; profits from local operators will be taxed by local governments, profits from intraprovincial services will be taxed by provincial governments, and so on. Some rationalization of international call revenues can also be expected.

A monumental shift is afoot in the financing of local networks. Before the early 1980s Beijing covered up to 60 percent of capital costs. This proportion will be drastically reduced in some areas to as little as 10 percent in the eighth five-year plan (1991–1995). (This point, and several that follow, emerged in discussions with authorities from the Guangdong Provincial Posts and Telecommunications Bureau in December 1987.) The balance in local budgets is slated to be made up by loans and foreign funding of one sort or another. The ministry, for its part, will seek to reduce its dependence on debt by making local operators responsible for their own borrowing. Table 5.1 compares Guangdong financing sources with World Bank programs in Africa.

The transition from appropriations to loans will slow local expansion. Just when plant costs are projected to soar, subsidies from the state will be cut, and a new cost—capital—will be added to the local load. The result is clear: Telecom operators collectively and individually are going to go sliding into debt. As inflation surged—the official figure was 18.5 percent for 1988, but economists estimate it was closer to 30 percent (Nakajima 1989, p. 2)—the central

Table 5.1. Distribution of Financing Sources*

World	Guangdong 5-year plans[b]		
Bank[a]	Sixth	Eighth	Source
45	20	10	Internal generation[c]
23	0	20	Bilaterla and commercial
15	0	0	Multilateral
13	60	10	Government
1	10	10	Subscribers
1	10	50	Other (incuding debt)

Source: World Bank and Tetra International estimates.

*Data are given in percentages

[a]Data are for ten projects in Africa, each about $500 million.

[b]The sixth plan covered 1976–1980; the eighth, 1986–1990.

[c]For China, revenue bonds as well as operating profits.

government restricted money supply, thereby tightening credit. Traditional sources of financing—fees, loans, and local investment—present special difficulties, and each must be addressed in turn.

5.4.2 Installation and Service Fees

In many regions—such as Beijing, Shanghai, and Shenzhen—the prevailing logic guiding installation fees is that new users must pay exorbitant charges to finance plant expansion; high fees effectively pay for the local loop. In Beijing the cost to initiate new service is about 5,000 yuan ($1340)—the equivalent of nearly five years of an average worker's salary.

Costly access charges address PTT's short-term capital needs, but they obscure long-term strategic interests. Common sense shows an operator's strongest source of revenues and highest potential for gains lie in services. Instead of squeezing new customers dry, Chinese PTTs would do well to encourage wider penetration of telephones and fund infrastructure expansion through a wider base of value added fares, particularly Centrex and private leased lines.

Not all areas are the same. Planners in Guangdong are striving to make initial connection charges affordable. Mindful of the traditional U.S. policy of universal service, Guangdong has initiated an ambitious, if modest by Western standards, effort to subsidize installation fees and pick up the budget slack elsewhere. The vision is admirable. Guangdong, of course, is one of China's richest provinces and so can afford to tariff installations on a sliding scale according to economic ability or contribution to the community. Installation costs in the late 1980s are shown in Table 5.2.

That some users pay more in the local network development phase is justified because everyone can be connected more quickly (Wellenius 1987, p. 43). A proposed alternative to offset initial user charges being considered in Guangdong is to reimburse some percentage of installation fees after a set number of years. In this way wealthy enterprises essentially make a mandatory investment in the local operator in order to receive local service.

Local service tariffs use the value of service concept and thus bear little relationship to cost. Basic rates for basic phone service (POTS) are consistent

Table 5.2. Line Installation Charges, Guangdong

Charge	Location
none	Remote
135	Urban[a]
539	Colleges, research & public welfare institutes
1,078	Enterprises

Source: Tetra International

Data, given in U.S. $, are for an unspecified year in the late 1980s, converted at the then-current exchange rate.

[a] For a household. This was equal to about 45 percent of an average worker's annual wage.

everywhere, since they were set by the State Pricing Commission using an odd calculus based on the size of the calling area. While administratively simple, the method obscures qualitative distinction between different types of technology: 10,000 lines of imported digital equipment costs a PTT three or four times as much as 10,000 lines of domestic analog equipment.

To build local revenues, PTTs need to pay closer attention to the variable rate scales inherent in different classes of service. If the price of a local telephone call were increased by a factor of three, to 15 fen (about 2.5 cents), the resulting revenues would be negligible and the public would be furious. A more promising solution is to promote penetration of PABX lines. In Shanghai in 1988 a standard local trunk earned the Bureau 12.5 yuan a month; a private branch exchange (PBX) trunk earned 70 yuan (about $19 at the time). Similarly, business services, such as leased and fax lines, IDD, cellular radio, and Centrex are expected to be introduced as soon as local economies can bear the cost.

Bill notification and collection are huge problems. Manual bill processing, minimal consumer credit and checking facilities, overcrowded payment centers, slow bill delivery, and account monitoring all contribute to tardy bill collection. The possible expansion of MPT's postal savings system, operating out of MPT's 14,000 local postal offices, could conceivably be employed to resolve aspects of these bottlenecks.

5.4.3 Domestic Loans and Foreign Borrowing

Local planners are nervous about the impending reliance on borrowed money, but they are positively shaken by the foibles inherent in transactions with the Bank of China (BOC). Few standards prevail.

As foreign corporations are well aware, BOC has its own, idiosyncratic way of doing things. It is even more fickle with Chinese clients. MPT bureau officers say they cannot rely on BOC to renew existing loans, let alone to insure a fivefold increase in borrowing in the near future. MPT bureaus must continually haggle with the bank over commitments, extensions, and terms of payment; they do not know with certainty if needed funds will be granted.

Moreover, terms of the money are anything but attractive. No low-interest loans exist in China. There are two options. Preferential loans, at 6 percent over ten years, made available by the Construction Bank of China, a part of BOC's system, is one. The state loans the money to the bank, which loans it to MPT. Such loans are awarded infrequently and apparently only to priority development projects. BOC loans, at 8 percent over three years are made through the Industrial and Commercial Bank.

Given the impending reliance on debt, planners feel that they are in a somewhat desperate predicament, and rightly so. Local PTTs are likely to get shortshrift; individually they will not have the same bargaining power the ministry does. Talks are underway with BOC to lengthen payback periods for infrastructure-related projects, including all telecom ones, but no fast action is anticipated. China's entire financial system is slated for thorough overall, but its specific direction is contingent on the strength and vitality of the central government's economic reforms.

In the meantime, PTTs make ample use of bilateral and development loans secured by foreign governments. Generally speaking, a government-finance, commercial or foreign affairs agency makes a state-to-state or state-to-province development grant for a minimum of 35 percent of total contract value, a level set by OECD-donor consensus; this money is then mixed with commercial or state bank loans. Such concessional financing (soft loans) insure the cost of capital to the Chinese customer is minimal. Concessional financing has evolved from a tactical advantage for foreign suppliers during 1985–1986 to become a sine qua non.

The amount of government-sponsored bilateral or trade assistance loans varies by government and industrial sector. Sweden underwrote telecom contracts during 1980–1989 worth $183 million. Soft loan telecom contracts with all countries combined during the period amounts to some $728 million. This represents the fourth highest area of infrastructure soft loan spending, after electric power, chemical plants, and railroads. The United States is the only major industrialized country that does not offer mixed credits to China, and US market shares reflect this.

Multilateral development loans have not been employed in telecommunications, but both the World Bank and Asian Development Bank are investigating projects. China, which once sought to keep its international debt service to a minimum, seems inclined to use nonaligned bank funds. The most likely course for borrowing will be co-financing among the banks and Chinese financial institutions—perhaps BOC, China International Trust and Investment Corporation (CITIC), or an all-new partner. Co-financing shares risks and eliminates third-party spending regulations. By contrast, a full-scale World Bank telecommunications investment project implies the bank determines precisely how funds are spent, something the Chinese are unlikely to accept.

Commercial credits from private foreign banks are likely to remain too costly in the foreseeable future. Likewise, supplier's credits, offered by big corporation's in-house financial services companies, are another, though even less attractive, option. Payback of supplier credit is customarily short at steep rates.

5.4.4 Local Investment

Depending on the political wind, MPT bureaus in larger cities may one day be allowed to obtain direct local investment from the sale of bonds and nontransferable public shares. The idea is to diversify fund sourcing throughout the local economy and to cushion dependency on financial institutions.

A first phase of raising local capital might be selling long-term bonds, as done effectively in Brazil, Thailand, and postwar Germany and Japan. With expanding budgetary power, local governments may find incentives and permission to boost public telecom spending by offering municipal utility bonds. PTTs might also require big customers with special network needs buy PTT bond as a condition for service.

To prepare for the financial challenges that lie ahead, local PTTs must become organized as formal corporations. *Corporation* here does not necessarily imply private enterprise. Rather, PTTs will need to establish stable organizational structures capable of diversified borrowing, debt management, and near- and long-range business planning. MPT is hesitant to relinquish control over local operations, and the ministry's centrist influence often muddles rational reorganization. Inculcating corporate structure in local operations long accustomed to haphazard bureaucracies bespeaks a revolution. Nevertheless, a management revolution may be required to safeguard PTT cash flow from collapse.

Taking on formal shareholders might also help curb corruption; however, payoffs for network favors are common. A local bureau may receive a truckload of sugar cane, several hundred pounds of cabbage, or simply cash as companies or city districts seek to buy insurance for their needs. A whole community may take part; *guanxi* (connections) are renewed, friends are bought, and, in theory, services delivered. MPT telephone installers are also notorious for extorting bribes.

5.5 The Role of Foreigners

In its dealings with foreigners, China, simply speaking, is seeking sophisticated hardware to upgrade existing facilities quickly and technological know-how to broaden the scope of the emerging industry. To obtain whole systems immediately, China has no choice but to buy the equipment outright. To garner manufacturing, managerial, and applied research expertise, China must over time develop long-term associations with foreign firms. The former relations are strictly commercial; the latter entail multiyear licensing contracts, elaborate training programs, and partnerships in joint business ventures.

The Chinese and foreigners typically view business alternatives through different lenses, and their priorities are set accordingly. The Chinese want to acquire technology, preferably without paying for it, while foreigners want to sell products, preferably without releasing the secrets of the technology.

5.5.1 The Market

The China telecom equipment market is tiny by world standards—an estimated $2 billion in 1991 (*Asian Business,* Nov 1991, p. 62). In 1985 total sales of foreign equipment amounted to $300 million, about what an RHC in the United States laid out in six weeks. By the year 2000, cumulative investment will be valued at less than the U.S. market in 1988 alone. Nevertheless, foreign competition in China is intense. Suppliers want to believe sales will one day take off, or at least see China a critical test for future sales to the entire developing world.

In the first nine years after digital switching systems were sold to China (PBX [Private Branch Exchanges] in 1979, and Central office, CO, in 1980) the market had hardly unfolded according to Chinese come-ons or Western hopes. Sales continued to be slow, and operating costs were high.

Competition is regulated. No foreign firm is allowed to garner "disproportionate" market share, based on commercial, diplomatic, and many other unpredictable considerations, without encountering stiff nontariff trade barriers. Fujitsu's impressive record in CO sales is a case in point. By mid-1986 the firm had confirmed orders of nearly 320,000 lines, more than twice that of Ericsson, the nearest competitor. Aside from installations in progress that year, no new shipments and no new CO sales had been made by Fujitsu through mid-1988. One reason for the standstill is the appreciation of the yen against the dollar, which has caused Chinese customers to balk at their reduced spending power for Japanese goods (the Chinese characteristically negotiate foreign purchases in U.S. dollars).

Equally significant has been deliberate Chinese market management. MPT has encouraged regional offices to demand renegotiated contracts. Currency fluctuations present an ideal opportunity for maneuvering, but the impetus for the Chinese position comes from a realization that their CO market was overwhelmingly controlled by a single foreign corporation, and worse, by the Japanese.

Opportunities for PABX imports are abundant as enterprises modernize office operations. The obstacle to the market fully emerging is the lack of formal distribution channels; PTT bureaus do not act as equipment resellers. The PABX landscape is changing dramatically, as eleven joint ventures or technology transfers with foreign firms have been approved. Analog PABX equipment, both new domestic and used from abroad, is commonly employed in public network applications. PBX in the local loop helps maximize line usage and, significantly, makes China's call-per-hour (line traffic rating) the highest in the world according to Ericsson executives.

Transmission equipment opportunities are more limited than those for switches. Stand-alone microwave sales grow increasingly rare in the public network, though many contracts are traditionally strong and have been augmented by a series of technology transfers from NEC. Imported radio links are periodically used in turnkey networks. Cellular radio potential is encouraging, although early configurations are small and expensive to operate. Party cadres and wealthy enterprises use cellular phones in at least four cities.

Wire and cable is a closed market. China's own capabilities meet domestic needs, in large part due to a huge joint venture with United Technologies' Essex. Fiberoptics spell better opportunities, but the market is spun with extraordinary complexities. Philips and Furukawa both have joint cable ventures, and Philips may be in line for another. The great majority of the cable produced in these plants is committed to high-priority long-distance routes. Many opportunities exist for provincial and municipal networks as aging paper-sheathed cables corrode and the price of fiber falls. The Chinese often buy fiberoptic cable through separate tenders; that is, a city's digital switch may be wired with another supplier's glass. This both escalates competitive stakes and creates after-sale opportunities for vendors who may have lost the primary contract.

5.5.2 Foreign Suppliers

Soft loans have meant sales. As a result, before 1990, Fujitsu and NEC were far and away the market leaders in CO equipment sales. Each had provided about 30 percent of the 1.6 digital public CO lines installed at yearend 1988; their shares of total telecom equipment were less. Alcatel's model E1OB exchange dominates the network in Beijing and its Shanghai joint venture is producing the System 1240. Ericsson, which has been selling in China since the 1890s, has a secure market position, particularly strong in the south. The US does not give concessionary financing and thus AT&T has a small market share.

Other established major players are Northern Telecom and Siemens. By earlier 1989 two newcomers, GTP (GEC Plessey) and Nokia (of Finland), had penetrated the market as well. Prices were stabilizing, and customer's needs and purchase preferences had crystallized: In 1986, most users were encountering digital technology for the first time; three years later they were more apt to know what they wanted and why. China imported just over $1 billion of telecom equipment in 1989 (*Far Eastern Economic Review*, Mar 7, 1991, p. 43).

In urban areas, some 60 percent of plant expenditures are made on central office technology; the remaining 40 percent goes toward transmission and the local loop. This proportion is roughly reverse in the countryside.

The value of CO imports had been steadily increasing prior to the June 1989 crackdown. Some 2.7 to 3.5 million lines were to have been shipped during 1988–1990, for a cumulative value of $600–750 million. Imported and domestic CO lines combined thus should have topped 6 million new lines, double what was publicly announced.

Instead, new soft loans were frozen after Tiananmen, which meant foreign suppliers generally could not sign new contracts. This was a windfall for Shanghai Bell (and thus Alcatel). Production went from 190,000 lines in 1988 to 405,000 (twice the original target) in 1989 and 600,000 in 1990. Another 600,000 lines were expected in 1991—double the number originally put under contract. This has helped Shanghai Bell take a 40 percent share of the market for public network telephone exchanges, with 5.5 million lines installed or ordered in twenty-four provinces. (*Asian Business,* Dec 1991, p. 14.) By November 1990 loans had been renewed and suspended contracts were revived. NEC had approval for domestic digital switch production before Tienanmen, and has indicated it will proceed. Siemens reached a joint venture agreement in November 1990.

In 1985 the MPT reportedly began seeking to reduce problems with system incompatibility and servicing by cutting the number of foreign suppliers to eight, and in 1990 the State Council was said to have issued a directive limiting future switching contracts to Alcatel, Siemens, and NEC. In each case, market access was tied to transfer of integrated circuit technology (*Business China,* Dec 24 1990, p. 185). However, neither restriction actually seems to have been put into effect.

Optical electronics and data communications sales are sharply limited by CoCom, the Coordinating Committee for Multilateral Export Controls, NA-

TO's watchdog for sales of high technology to socialist bloc countries. The U.S. Department of Defense is adamant about restricting sales of high-speed optical equipment, particularly systems capable of transmitting at rates higher than 140 Mbps, much to the chagrin of suppliers of optical test equipment, components, and systems. The rationale is that the Chinese should not, from a strategic point of view, be allowed to establish a network with intelligent distributed processing hypothetically capable of sustained operation even after repeated nuclear blasts. A similar argument stands for distributed data networking.

A further CoCom concern is communications security. With modern interception and deciphering techniques, analog transmission signals can be easily intercepted and read. With high-speed fiberoptic connections that do not radiate signal information, and digital switching, which manages traffic flows, electronic eavesdropping becomes increasingly difficult. A liberalization of CoCom guidelines may be forthcoming. Much satellite and earth station equipment, particularly that with spread-spectrum frequency management, is similarly affected.

5.6 Conclusion

Telecommunications in China is tracking toward profound realignment. While the MPT in Beijing has the last word in questions of policy, political power is being devolved to local authorities, state funding is being cut, and technological modernization accentuates the disparities between the information haves and the have nots.

The MPT faces a difficult contradiction. While it acts as China's supreme telecommunications authority, it is not a nationwide monopoly; the influence it holds over local networks is subject to swings in political pressure. Provincial PTTs are still functionally a part of MPT, although rifts between Beijing and provincial capitals over development strategies and administrative policy are common.

With increased decentralization, local authorities can take responsibility for profit, loss, and develop into sustainable business enterprises. Corporatization and diversified debt management can wean PTTs from the parental care of the state and heighten incentives for improved productivity. After an initial period of confusion, and perhaps even panic as funding commitments from Beijing are withdrawn, local PTTs could become a driving force in the establishment of municipal capital markets. Telecommunications infrastructure development may thus become directly aligned with community modernization and not simply be an extension of central government policy. These scenarios, however, are largely dependent on the central government's macroeconomic policy and plans.

Bibliography

Chen, Wei-Hua. 1988. "Growth Requires Better Communications." *China Daily Business Weekly* (June 27).

Nakajima, Seiichi. 1989. "The Chinese Economy in 1988." *China Newsletter* 79 (Mar.–Apr.).

Reardon, Lawrence C. 1991. "The SEZs Come of Age." *China Business Review* 18(6): 14 Nov./Dec.

U.S. Department of Commerce, National Telecommunications and Information Agency. 1986. *U.S. Telecommunications Team Report on November 1986 Visit To China.*

Wellenius, Bjorn. 1987. "Finding New Finance." In Bjorn Wellenius, ed., *Developing World Communications*. London: Grosvenor Press International.

Wen, Yanghui. 1988. "Adopting 'Second-Hand' Equipment to Get Optimum Economic Benefit." *China Telecommunications Construction,* p. 48 (Oct.).

Zita, Ken. 1987. *Modernising China's Telecommunications*. London: The Economist Publications.

———. 1988a. "Implications of China's Changing Rules for the Telecommunications Industry." Arthur D. Little, *Spectrum* (Mar.).

———. 1988b. "Telecommunications in China." Cambridge, MA: International Telecommunications Society (June).

6
Indonesia

JONATHAN L. PARAPAK

Indonesia is a nation of 6,000 inhabited islands (and even more uninhabited ones) sprawling over 5,000 km along the equator. The population of over 190 million makes it the fifth most populous country in the world. Tying the country together has benefited from modern means, especially satellites—Indonesia was the fourth country to launch a satellite-based communications system. That continues a long tradition—by the fifth century correspondence via letters was common: couriers walked, rode horses, or traveled by sailing vessel. This chapter looks at development of telecommunications in Indonesia; the next chapter expands on the structural changes that began in the 1980s.

6.1 History

Sitting astride the sea route between China and India, with a range of exotic products to contribute to trade between the two, the islands have been a trading crossroads for two millennia—and thus in a very real sense have long had good communications with the outside world. For most of its history what is now Indonesia was comprised of many small states, the more powerful ones based either on maritime trade or (on Java) on intensive rice cultivation. Islam was known in the islands by the thirteenth century, but it came to predominate on Java only in the sixteenth century.

Dutch occupation of parts of the archipelago beginning in the seventeenth century resulted in postal service, initially limited to official letters, between Indonesia and the Netherlands. Prior to the 1800s it depended on Dutch East Indies Company trading vessels, and thus cannot be described as regular. Due to uncertain delivery, original letters were normally followed by duplicates. In the 1630s, incoming letters were displayed at city hall after being recorded by the official confiscator. It should be noted that it was the twentieth century before the Dutch had consolidated most of what is now Indonesia, although Java was tightly controlled after 1830. Expanding influence, both geographically and in terms of degree of control, frequently came as a result of Dutch intervention in struggles between local sultans or over succession.

In 1746, the first post office was established in Jakarta (called Batavia by the Dutch) by the governor general of the Dutch East Indies Company. Its purpose was to guarantee the security of the postal service, particularly for letters to people involved in trade. In 1846, postal service was placed under the director of agriculture and civil warehouses. However, the government provided service only between Anyer and Banyuwangi. On all routes, people could send letters by any means they considered satisfactory. The post was made a government monopoly in 1862.

In 1876 the Department of Domestic Affairs took over postal management. The head of the postal service also held the position of Postmaster of Batavia. The Semarang and Surabaya post offices in Java and 17 others outside Java were each headed by a postal supervisor. Auxiliary post offices were headed by auxiliary postal supervisors.

In response to a suggestion by the East Indies Co, the Dutch government introduced telegraph service in 1856, linking Jakarta and Bogor. Initially, facilities were solely for government use. However, after completion of the Jakarta–Surabaya link and its Semarang and Ambarawa branches in 1857, use of facilities was made available to the public. Regulations on use of telegraphic apparatus were derived from those in the Netherlands. No one could offer telegraph services in Indonesia without government permission. The network expanded during the 1860s and early 1870s, and a telegraphist was placed in each post office. Postal and telegraph services were merged in 1875.

The government did not guarantee delivery of telegrams or provide compensation for damage, delay, or loss. Legal sanctions were applied to those who tampered with or destroyed telegraph facilities after the Indonesian Telegraph Administration signed an international agreement concerning the protection of submarine cables in Paris in 1884.

In 1909, a regulation allowing confiscation of telegrams whose contents were considered dangerous to colonial security, or contrary to law and order or to public decency was added. The power to confiscate was also applied to press, maritime, and semaphore communications, and to radio telegrams. Rules concerning secret codes, groups of figures, and abbreviations for official use were also incorporated in these regulations.

Telephone services had initially been organized by a private company in accord with a government decree of July 31, 1881. Inauguration of service in Jakarta took place on October 16, 1882, followed by the building of a line to Semarang in 1884. On September 20, 1906, the telephone network concession expired, and the networks for Jakarta, Gambir, Jatinegara, Semarang, and Surabaya (all on Java) were taken over by the government. Telephone services were merged with the post and telegraphy in November 1925.

6.1.1 Structure after 1945

At independence, Indonesia's telecommunications facilities were very limited. The country had inadequate equipment and lacked the expertise to operate it properly. The situation deteriorated further during the struggle against Dutch forces trying to restore colonial presence. Service gradually improved during the 1950s after the Netherlands recognized Indonesia's independence.

In 1961 the Department of Post, Telegraph, and Telephone changed its status and name, becoming PN Postel (PN indicates a state-owned company). This wasintended to provide more autonomy for the company's financial management.

The Dutch regulatory legacy was no longer considered appropriate because of technological developments, and in 1964 the government promulgated a new law revamping the system. As before, telecommunications was operated and controlled by the government. The law classified telecommunications into three categories: public, special (for government bodies), and those organized by nongovernment institutions.

Although government bodies can be granted special permits to organize telecommunications facilities for their own purposes, the MPTT regulates the control, installation, management, and use of the facilities. The civil aviation, radio and television services, and Pertamina Oil Company have facilities reserved for their exclusive use under these permits.

In 1965 PN Postel was divided into two corporations to manage its rapid development and expanded activities. PN Telekomunikasi took over telecommunications, and PN Pos Dan Giro took over the postal service. The MPTT was given responsibility for supervising their operations.

Further restructuring came as a result of several 1974 laws. PN Telekomunikasi was renamed Perumtel (Perusahaan Umum Telekomunikasi) and was designated the sole provider of public telecom services in Indonesia. Then, in 1991, Perumtel was again renamed, this time to PT Telekomunikasi Indonesia (PT Telkom) and its status changed from a public company to a limited company.

In 1980 the law was amended to make PT Indosat (Indonesian Satellite Corporation) the sole provider of international services. PT Indosat was founded in November 1967 as a subsidiary of International Telephone and Telegraph (ITT); its main responsibility was to operate and maintain the Intelsat standard-A earth station at Jatiluhur. The government purchased all the shares in December 1980.

Responsibility for the operations and technical requirements of each company is determined by government regulations; PT Indosat and PT Telkom plan, develop, and extend facilities and public telecom services in coordination with each other and with government guidance.

6.1.2 Regional and International Cooperation

The aim of national development is to realize a just and prosperous society in keeping with the material and spiritual focus of the state philosophy—"Pancasila" (Five Principles)—which espouses unity for the Republic of Indonesia as a free, sovereign, and democratic nation enjoying a peaceful, secure, well-organized national life within the world community.

The concept of cooperation among Association of Southeast Asian nations

(ASEAN) (Indonesia, Malaysia, the Philippines, Singapore, and Thailand) in telecommunications was first proposed in 1974 and moved rapidly towards realization. In this regard Indonesia has taken part in development of the ASEAN Submarine Cable Communications System. The ASEAN Philippines–Singapore (ASEAN P-S) cable was completed in 1979, and the ASEAN Indonesia–Singapore (ASEAN I-S) system commenced operations in April 1980. An ASEAN fiberoptic cable is being constructed and segments will be operational in 1995; it had originally been scheduled for completion in 1990. The cable will connect the capitals of each of the five countries to each other and to cables to Japan, North America, and Europe.

The submarine cable project aims to support the Intelsat system, particularly for high-volume routes like the one between Indonesia and Singapore. In keeping with the spirit of ASEAN, Indonesia offered to lease its Palapa satellite transponders to other ASEAN countries for their domestic telecom requirements and television broadcasting. Palapa transponders also have been leased by Hong Kong, Australia, New Zealand, Papua New Guinea, and several other countries.

The government's basic policy on international telecommunications is to support national interests. Links with new international carriers are established if they advance the national interest and benefit Indonesia. Implementation is on a case-by-case basis.

6.2 Service Levels

In April 1989 Indonesia had about 1 million phone lines. Clearly, density is very low—at 0.56 phones for every 100 persons, it is even less than the number in China. In Jakarta density is over 4 percent, and levels in four other urbanized provinces are much above the national average. Density in fifteen provinces is below average, while in the other seven it is near the average. This means the country faces a significant challenge in achieving density at the level recommended by the International Independent Commission for Telecommunication Development (IICWTD)—a telephone "within easy reach" of everyone or within less than an hour's walk.

Domestic direct dialing services is provided between 136 cities, and ISD service was available from eight Indonesian cities to more than 130 countries. The telex network has two international gateway exchanges and more than twenty-seven domestic telex exchanges, a 15,297-telex line capacity, and 14,087 subscribers. Gentex (telegraph circuit) service is available from more than 210 offices, and there are currently more than 660 telegraph offices.

Long-term development of infrastructure is oriented toward digitalization of the network. At the end of 1989, 160,000 lines, almost 18 percent of all lines, were digital. Packet switched public data communication is now provided from Jakarta, Bandung, Surabaya, Medan, Bogor, and Ujung Pandang to twenty-one countries.

Modernization has been occurring at a relatively slow pace. Geography and the huge population require enormous resources and monumental effort to deal with the great gap between supply and demand for services. Development continually seeks to balance requirements for expanding basic services for the general population and providing advanced services for the government and business—services necessary for involvement in the international economy.

6.2.1 Satellites

Indonesia's first satellite earth station for international communications was set up in 1969. Two domestic telecommunications satellites, called the Palapa A series, were launched by the government in 1976. This was the first domestic satellite system launched in a developing country and the fourth by anyone. In June 1983, a second-generation satellite, Palapa B-1, was launched, followed in 1984 by an unsuccessful attempt to orbit Palapa B-2. To replace the unsuccessful B-2, another second generation satellite, Palapa B-2P, was orbited in March 1987. Palapa B-2 was rescued and was relaunched in early 1990 as Palapa B-2R.

Palapa B-4 was the next satellite scheduled for launch, provided by Hughes Aircraft and launched by McDonnell Douglas. Manufacturing costs are $66.1 million, the launching fee is $44 million, and insurance is $21.5 million—a total of $132 million. Additional earth stations are being built, and the increased capacity will stimulate expansion of telephone, telex, and television broadcast services.

Plans are being made to launch a new generation of satellites, the Palapa-C series, in the later 1990s. Palapa has proven telecommunications infrastructure and can assist the development of the country's trade, industry, tourism, education, and the economy in general. It is also used for regional television broadcasting in Southeast Asia and for communications with neighboring countries including Malaysia, Singapore, and the Philippines.

6.2.2 Types of Services

Public telecom services managed by PT Telkom and PT Indosat consist of telephone, telegraph, telex, leased circuits, television, and facsimile, which are known as the traditional services.

New services are becoming available to the general public through the use of the existing public switched telephone network (PSTN). An international packet service gateway located in Jakarta was completed in 1984 to provide various new data communication services, including database access, electronic mail, computer time sharing, and applications for specialized fields such as tourism, banking and manufacturing.

Indonesia's increasing number of personal computers—estimated at 60,000 in 1988—occasions increasing use of data communications between personal computers and between personal and mainframe computers. An experimental videotex system providing tourist information was installed by PT Indosat at World Expo 1988 in Brisbane, Australia. However, the economic viability of domestic teletex and videotex services is questioned, especially when comparable services are already provided at lower costs through databases, bulletin boards, and electronic mail. These alternatives are accessible from personal computers through public switched telephone and public switched packet data networks.

6.2.3 Rates and Finances

PT Telkom tariffs consist of basic and special tariffs. Basic tariffs, determined by the MPTT, include the installation fee, monthly charges, and communication charges. Special tariffs are additional amounts that take into account the special circumstances of each region and district. These are determined by PT Telkom in accordance with guidelines issued by the government. PT Indosat rates are approved by the minister.

In principle, tariffs are to cover incurred expenses, while being in line with government policy, sensitive to the market, and providing equitable treatment to all users. The relative weight of each consideration cannot be measured definitively because each is subject to change.

Both PT Telekom and PT Indosat enjoy strong financial health. Budgets of state-owned companies must be in accordance with the government's approved plan and comply with government regulations. The three main components in budget allocation decisions are operation and maintenance costs, the level of dividends provided the government as the sole shareholder, and capital investment requirements. However both Telkom and Indosat have the authority to establish their own financial plans; their capital plans and budget decisions are excluded from the state budget. To expand operations and services or increase paid-in capital requires government approval, however.

Government control is exercised through an annual shareholder meeting, monthly meetings of their boards of commissioners, and government responses to regular reports. Both corporations are also audited annually by government auditors and state supreme auditors.

Day-to-day control of technical matters is exercised through the MPTT while the Department of Finance, as the actual shareholder, oversees financial matters.

6.3 Equipment Production

Telecommunications is one of nine ''strategic industries.'' Domestic telecom equipment manufacturing was started in early 1974 by Industri Telekomunikasi Indonesia (PT Inti) a state-owned company under the supervision of the Department of Tourism, Post, and Telecommunications. PT Inti was established in order to reduce dependence on imported equipment, guarantee timely availability, and ensure continuity of telecommunications development.

To better meet demand for equipment, PT Inti has had a cooperative agreement with Germany's Siemens since 1976 to assemble and manufacture EMD automatic telephone systems and desk telephones. In 1976, a technical cooperation agreement between PT Inti and Telephone Manufacturing of Belgium (now part of Alcatel) was also signed, covering production of crossbar automatic telephone exchange equipment, telephone sets, PABXs, and pair savers.

The company produces telecom equipment for both domestic markets and export. Its export products include small earth stations sold in Malaysia, avionic systems components sold in the United States, and fire control equipment sold in Italy.

Telecom equipment used in Indonesia includes various types from various countries. No restrictive policies apply; however, technical specifications standards recommended by the Consultative Committee on radio (CCIR) or the Consultative Committee on telephone and telegraph (CCITT) must be followed. In principle, terminal equipment is provided to users by the operating company. However, customers may purchase equipment from distributors as long as it meets standard specifications and has received type approval certification from PT Telkom insuring that it can be integrated into the network.

6.3.1 *Procurement Policies*

Procurement is regulated by a 1984 presidential decree that was later amended in 1988. The government buys equipment through open bids, limited bids, direct determination, or direct purchase. Indonesian contractors and foreign contractors in joint ventures with Indonesian contractors may participate in bid tenders as long as they submit a certificate of capability. If equipment must be imported, local manufacturers must declare that they cannot produce it. Installation, assembly, and other services must be carried out as far as possible by Indonesians or the Indonesian partners of the equipment suppliers. The obvious effect of this is to require involvement by an Indonesian national in virtually all transactions, if only as passive partner in a joint venture. PT Inti, who provided all of Telkom's and Indosat's switching equipment purchases in 1985–1990, is the largest domestic supplier.

6.4 Development

Telecommunications development is an integral part of the national development plan. The first five-year development plan started in 1969; the fifth, Repelita V, began in April 1989. The main objective of Repelita IV was to facilitate economic restructuring and self-sustained economic growth by reducing dependence on oil production, including creating job opportunities, and developing the manufacturing sector. Repelita V will continue development consistent with Repelita IV.

The telecom sector has been developed under these plans. The major sectoral objectives have been to provide telephone and telex services in the major cities, establish long distance direct dialing and international direct dialing in major cities, expand the networks to larger cities and remote areas through the satellite system, and develop new telecom services. As a result, 523,000 lines were installed during Repelita I through III. Repelita IV called for 950,000 new lines, including 200,000 carried over from Repelita III; actual installation was 350,000. Network coverage in 1989 was 2,069 subdistricts, 58 percent of the total number nationwide.

In its efforts to achieve its telecommunications objectives, Indonesia faces major challenges. At a macrolevel, slow economic growth is the first problem. In 1985 GDP grew only 1.9 percent due to the decline in oil prices and global demand, although the growth rate has subsequently risen and was around 6 percent in the late 1980s and 1990. A second problem is high population growth—2 percent nationwide, faster in the cities because of migration. This contributes to high unemployment.

Telecommunications users are classified into three categories: general public, business, and residential. It will be difficult to satisfy demand in all user categories by the year 2004 so a supply priority has been determined. At present, service is extended only to municipalities and regencies (counties), but not to all villages.

Several tentative long-term (year 2000) objectives have been considered. Telephone service would be provided to villages, but telex, with ISDN would be provided only in urban areas. A target density would be five to ten in urban areas and an overall national average of five (about nine times the level in 1989). Basic phone service would include long-distance direct dialing. Telegrams would be delivered within three days to even the more remote parts of the country—with one-day service in the areas where most people live. In early 1991 Perumtel's president publicly stated it would take 7 million lines to meet demand at the time.

To achieve these targets, several strategies have been adopted. Satellites are an important part. The system has been used for international services since 1969 and since 1976 for domestic services. Digital time division multiple access for international telecommunications was introduced in 1985 and in 1988 domestically. More digital communications satellite technology will come into service in the 1990s, including very small aperture terminals (VSAT), Intelsat's intermediate data rate (IDR), and Intelsat Business Service (IBS).

Much of Indonesia's previous development has been funded from exports of oil and natural gas products, but the early 1980s drop in global oil prices has significantly reduced this revenue flow. To achieve projected telecommunications development levels, the government has had to make a special effort to overcome a shortage of funds. One strategy has been to invite private participation in the financing and installation of telecommunications infrastructure and to allow recovery of private investment through revenue sharing schemes tied to specific projects. Another alternative is to obtain soft loans from foreign countries. These alternatives were both used as ways to achieve the target of adding 2.1 million lines during the 1989–1994 five-year development plan, many of which were installed by thirteen private companies rather than by Perumtel.

The plan originally called for 1.4 million lines, 600,000 from Perumtel and

800,000 from the other companies, but in part because of the November 1990 decision to double the size of a hardware order, the total was expanded. Under the contracts, NEC and a majority-owned subsidiary of AT&T will each provide 350,000 lines of digital switching equipment, including 150,000 to be produced domestically, at a cost of about $100 million. AT&T's partners are STET Spa (an Italian state company) at 20 percent, and Telefonica de España at 6 percent.

PT Telkom plans to add 15,200 telex lines (for a total of 32,850) and 75,000 pay phones. Other major projects during the 1989–1994 period include extension of telex terminal exchanges throughout the country, a rural and remote areas telecommunications network, a terrestrial transmission network, installation of time division multiple access, a tail link expansion connecting earth stations and telephone exchanges, submarine fiberoptic cable projects, a national data network, microdigital systems, and a national radio frequency monitoring system.

The list of proposals for fiscal 1988–1989 released by the National Development Planning Board (Bappenas) in June 1988 invited donor countries to provide financial assistance for several of these projects. These include improvement of the existing satellite channel, phase III of the remote area telecommunications system, the cross-Kalimantan digital microwave transmission system, a telephone outside plant maintenance center, computerization of the local network records system, extension of the data communication network, facilities for a regional training center, the trans-Sulawesi microwave transmission system, and development of a telecommunications research and development center.

6.4.1 Newer Services

Facsimile service was available through the public switched telephone network by the mid-1980s. By decade end, high-speed service based on the G-4 standard was provided in limited areas via leased lines and in the early 1990s it will be available through the digital PSTN.

Data communications services will be provided mainly through the existing packet switched network called Sambungan Komunikasi Data Packet (SKDP).

In August 1990 Citra Sari Makmur (CSM) began offering VSAT service for data transmission. In the first six months it signed only a dozen customers. By May 1991 thirty-five terminals had been installed and 100 were on order; seven of nine paying customers were banks. The company plans to offer integrated voice service as an alternative to Perumtel's network, but has experienced financial difficulties. In addition, it faces competition. In April 1991 Lintasarta—owned by Perumtel, Indosat, and several banks—received permission to market Vsats. Moreover, Salim, one of the country's largest conglomerates, has permission for an internal system using a transponder on Palapa (*Asian Business,* Apr 1991, p. 16; *Far Eastern Economic Review,* Jun 6, 1991, p. 42).

Cellular land mobile phone service has been expanded to cover cities within a radius of more than 40 km from Jakarta; it has been operational with a 10,000-

line capacity since 1986. In 1990 Perumtel announced it would operate two Motorola-supplied cellular networks with private investors. Radiopaging was introduced in 1987 and is available in several major cities. In 1989 competitors were allowed paging licenses in exchange for 15 percent of revenues.

All these service enhancements are designed to support ISDN development. In the initial stages, narrow band ISDN service will be provided by the digital telephone network. Full-scale introduction of wideband ISDN will not occur until after the year 2000. However, the service will be available to special subscribers on leased lines prior to then.

One measure of productivity is lines per employee. In 1974, Perumtel had eighty-six workers per 1,000 lines. By 1978 this had fallen to forty-eight, to forty in 1988, and thirty-one in 1990. The actual number of employees did not fall—in fact, it increased to 39,000 in 1988, but the number of lines rose much more quickly. The target is to have fewer than twenty employees per 1,000 lines by the year 2000. This target can be achieved by adopting automatic maintenance and operations systems and through the installation of digital equipment, which needs less maintenance and operation work.

6.4.2 Technical Development Strategy

Recent technological innovations must be carefully examined when making long-term telecommunications commitments. The nation's public networks will have to be integrated into ISDN in the future.

Indonesia's public network at the start of the 1990s included telephone, telex, packet switched public data, and nonswitched leased line networks. Before 1984, only 26 percent of exchanges were automatic. Until 1985 telephone networks primarily involved analog switching. Since 1980 all new switches have been digital. A tentative plan was for 54 percent of switches to be digital in 1989, 72 percent in 1994, 93 percent in 1999, and 100 percent in 2004. In 1990 about 60 percent of switches were analog.

Great distances and the fact the country is an archipelago make the most desirable backbone transmission system a combination of terrestrial and satellite transmission systems. Analysis shows the satellite system is more economical when links exceed 500 km. Roughly 36 percent of circuits were satellite in 1989. This is expected to fall to 25 percent by 2004 as the number of terrestrial circuits increases faster than new satellite circuits.

6.5 Development of the Electronics and Computer Industries

Rapid growth of electronics industries abroad has gradually affected Indonesia's domestic industry. Consumer electronics, manufacturing such products as radio, television sets and tape recorders, has been developing for some time, and there are now hundreds of assembly factories. Of particular importance is the development of telecommunications and office automation equipment manufacturing. PT Inti has moved beyond simple assembly, so that many items now have 70 percent local content.

In an effort to make optimum use of available satellite technology and at the same time promote the domestic industry, the government relaxed its policy on the use of satellite television receive-only (TVRO) terminals. There is currently a growing TVRO terminal-related business for residential and business use.

Locally made computers, particularly personal computers, are a booming business. Computer schools and training programs are emerging rapidly, offering opportunities for high school graduates to take short practical courses that enable them to obtain jobs with reasonable salaries. A substantial amount of foreign software is available, and local software houses are gradually developing to provide customers with tailor-made applications. Enactment of the Copyright Law in November 1987, will encourage the local software industry to develop more rapidly by discouraging pirating—which has been a major problem and complaint of foreign vendors.

A by-product of the computer hardware and software industries is the development of computer database providers. This has also been accelerated by creation of the SKDP Network, the public data communications network, in the mid-1980s. In the late 1980s tourist information was the principal type of database.

The government supports development of local manufacturing with various incentives to attract domestic as well as foreign investment. These incentives take the form of reduced import duties for raw components and concessions relating to initial capital investments. The government gives priority to locally made products in its purchase contracts.

6.6 Changing Telecommunications Policy

Historically, all public telecom services have been provided by state-owned enterprises. However, because increasing demand for new services and facilities necessitates large capital investments, gradual policy changes are taking place. The trend is toward allowing private enterprises to invest in certain types of facilities—as well as participate in their management and operation—and to recover their investment through a share of the revenues. This policy was formulated in the Telecommunication Law of April 1989. The new law basically retains the government monopoly over provision and operation of basic services. However, controlled private sector competition is allowed in provision of non–basic value added services.

Cellular and paging systems, including a citizen band communication network, have been licensed to private operators, as discussed earlier. Another example is the construction and financing of central exchanges in association with new residential areas being developed by real estate companies. There are five such licensed projects, involving some 100,000 lines. Typically, PT Telkom receives a percentage of revenue and outright ownership of the hardware after ten years. While this approach is occasioned by government budget constraints, it probably indicates a trend toward acceptance of limited privatization of telecommunications.

Private radio broadcasting stations exist in fourteen provinces. A kind of cable television service commenced operation in November 1985. It is a private subscription service on a UHF frequency band and has scrambling capabilities. Programming is mostly advertisements and entertainment. Because of limited coverage, it does not compete directly with existing state-run television broadcasting.

Indonesia has been a pioneer in using satellites, and they are now the mainstay of the system. This means the geographic area covered by the network will expand steadily. The original target of 1.4 million new lines in the 1989–1994 period was lifted to 2.1 million, in part a reflection of the fact 430,000 lines were installed in 1991—a sevenfold increase over 1990. The hope is to have world class (albeit low-end world class) telecom services for commercial users (and tourists) in Jakarta, Bali, and a few other major cities by the end of the 1990s.

Bibliography

Directorate General of Posts and Telecommunications. 1982. *History of Posts and Telecommunications in Indonesia,* 5 vols. Jakarta: Department of Transport, Communications, and Tourism.

———. 1985. *The Important Figures in the History of Posts and Telecommunications Development in Indonesia.* Jakarta: Department of Tourism, Posts, and Telecommunications.

JICA. 1987. *Telecommunications Systems,* vols 1 & 2. Jakarta.

Perumtel. *The Public Telecommunications Laws and Regulations.* Bandung.

———. (not dated.) *Telecommunications by the Year 2000.* Bandung.

7

Malaysia and Indonesia: Telecommunications Restructuring

VINCENT LOWE

The inability of their governments to provide sufficient capital to further the development of the telecommunications industry and meet public demand has provided most of the impetus for the restructuring that has taken place in Malaysia and Indonesia. This chapter looks at the development and implementation of structural changes, including an account of interest group reactions in Malaysia—public reaction in Indonesia has been limited.

Privatization, first announced in 1983, is the center of changes in Malaysia. Under two laws adopted in 1985 the property, rights, and liabilities of the Telecommunications Department were transferred to a private corporation wholly owned by the government; part of the new corporation was subsequently sold to the public. Things have been more complex in Indonesia: During the past thirty years wholly government-owned enterprises have been established and evolved so that by the 1980s there were four state corporations, each providing a different service—domestic, international, equipment manufacturing, and postal.

The countries are looked at separately. There are cultural similarities and historical relationships between the peoples of Malaysia and Indonesia. Forms of Malay are official languages in both, and Islam is the dominant, but not the only, religion. These commonalities, however, have not produced much in the way of similarities in development of telecommunications policies except at the most general level. The most obvious similarity is the close relationship between telecommunication firms and ruling elites. However, in Malaysia ethnicity is the principal element in this, while in Indonesia, it is more family connections.

7.1 Malaysia

After independence in 1957 Malaysia extensively reorganized the Telecommunications Department by replacing British expatriates with Malaysians as technical and managerial staff and decentralizing it into regionally autonomous units. The department is known as Jabatan Telekom Malaysia (JTM), its name in Malay. The next chapter provides more on the department's early history.

JTM's operations were profitable on a current basis, but earnings were not enough to cover the costs of developing an enhanced telecommunications infrastructure. The government covered the shortfall until 1986 when, as revenue growth slowed because of a recession, it too suffered a deficit. Substantial reductions were made in development sector expenditures, including a 6.4 percent cut in the communications development budget in the Fifth Malaysia Plan (1986–1990). Telecommunications expenditures were to have increased from M$2.9 billion during the period 1981–1985 to M$9.6 billion in 1986–1990, but it reached only M$3.9 billion instead.

It was against this backdrop of poor economic performance that the government launched a policy to privatize some of its departments. The first indication of this was in 1983 when JTM parceled out turnkey contracts for the installation of 1.78 million lines. Even more projects were subsequently contracted out. In 1987 JTM telecom operations were transferred to a newly created company, the first public sector organization commercialized with a view toward being privatized.

7.1.1 Legal Framework

During the postwar colonial and early independence periods the Telecommunications Act of 1950 provided the industry's legal framework. There were revisions in 1970, 1972, and 1977. The 1970 revision set out the exclusive privilege of the Malaysian government to establish, maintain, and operate telecommunications in the country. It also stipulated the rights of the telecommunications minister to grant licenses to any person or contractor deemed qualified to undertake contracts for installation, erection, and maintenance of telecommunications works in Malaysia.

The 1972 amendment authorized the establishment of a telecommunications fund, permitting financial semiautonomy to JTM. Such provisions (e.g., imposing commercial accounting) made privatizing JTM an evolutionary, rather than revolutionary, approach.

7.1.2 The Denationalization Process

The first direct step toward actual denationalization came in 1984 when the Arab Malaysia Merchant Bank was appointed to conduct a study of the financial implications of a transfer, in cooperation with London-based Kleinwort Benson (the same company that helped privatize British Telecom) and Hanafiah Raslan Mohamad Associates, a Malaysian accounting firm. The study, which has never been made public, set forth procedures for evaluation of assets as well as the accounting procedures involved in the transfer of assets and liabilities. Following the guidelines in the report, which was completed in 1985, was the recommendation that JTM be converted into a corporation called Syrarikat

Telekom Malaysia Bhd. (STM), to be wholly owned by the Ministry of Finance, effective January 1, 1987.

Two sets of laws were passed in 1985 to provide the legal framework for privatization. Amendments to the Telecommunications Act further listed the regulatory functions of JTM's director general, and the Telecommunications Services (Successor Company) Act legitimized transfer of JTM property, rights, and liabilities to STM. An international consulting firm, Arthur D. Little, was engaged in 1988 for the purpose of determining the organizational structuring of STM.

The next step was the granting of a license by the Ministry of Energy, Posts, and Telecommunications to STM on December 1, 1986. The license listed thirty-seven conditions. The initial duration of the license is twenty years, for which STM is required to pay M$500,000 on issuance. The annual renewal fee is not more than 0.5 percent of STM's gross turnover in its previous financial year. Other provisions include:

1. A committee representing the government would be set up to specify the financial provision for the development and maintenance of telecom services in rural areas (condition 2).
2. JTM's director general was to be notified by STM regarding charges, terms, and conditions of service not less than twenty-eight days before any proposal becomes effective. If the director general made any suggestions to change the proposal, it had to put into effect within twenty-eight days from the time of notification (condition 11).
3. The director general may direct steps to be taken to remedy any cross-subsidizing situation affecting the apparatus supply business or the provision within the country of value-added services (condition 13).
4. Rate increases were limited to being not more than the arithmetic mean of the annual increases in the consumer price indices for Peninsular Malaysia, Sabah, and Sarawak, using 1980 as a base. Notwithstanding this, rate changes are subject to approval by the Ministry (condition 18).

JTM became a regulatory body. Its responsibilities were to coordinate and control all telecommunications activities in areas such as frequency management, licensing, international affairs, rates, and tariffs. Part of STM was sold to employees and the public in 1990.

7.1.3 Views of Constituent Groups

The process has had its critics and met opposition from a number of sources. Three major groups involved—Parliament, unions, and academics—are taken up in this section.

7.1.3.1 Parliament

The government coalition held more than a two-thirds majority, so there was no question that any bill would pass. It was thus a situation where opposition views could be totally ignored except for eventual appeal to the electorate.

These were two ways that formal parliamentary procedures were strictly observed. Members were given only two days notice of the vote, and copies of the Amendment Bill and the Successor Company Bill were not actually distributed until the day before they were to be voted on (and passed) in the July 1985 session. No caucus was held with any members, or with the Malaysian Technical Services Union or National Union of Telecom Employees (NUTE). A formal debate, such as it was, took place. A motion to postpone consideration was defeated. Opposition views were predictable. Comments concerned whether a private monopoly would be efficient, and whether a profit-oriented company would act in the public interest.

Following denationalization of JTM, several members pressed the ministers with questions on whether privatization would benefit the rural populace, consumers, and workers, how the policy would be implemented, and new services. Policies must generally have a rural bias (rural, predominantly Malay votes are weighted five times more heavily than urban, predominately Chinese votes) so many questions centered on this.

Would rental rates for rural phones be higher? How quickly would damaged rural phone booths be repaired? What about STM's desire to have deposits reflect usage? There were also queries regarding cross subsidization of domestic and international calling.

Questions reflecting worker interests focused on lay-offs and the number of JTM workers who had refused to join the privatized company—499 out of 28,000. Of these, 102 decided to stay with JTM while another 397 opted for early retirement.

Consumer issues included billing complaints about overcharging, nonresponsiveness of STM toward complaints, new housing developments not served with telephones, illegal recording of telephone conversations, and commencement of detailed billing to various parts of the country.

Members also asked the minister of energy, telecommunications, and posts to come in on November 17, 1987, to explain the banning of satellite broadcasts. He gave three reasons. First, such transmission requires prior agreement and payment. Second, the government must differentiate between information that is useful, necessary, and important toward the building of a united nation and information that would poison the thinking of the people and affect the harmony and security of the nation. Third, the government feels it is necessary to restrict the possibility of the recording and retransmission of such broadcasts.

7.1.3.2 Unions

The unions were not pleased with the prospect of privatization. NUTE, with a claimed 22,000 paid-up membership, issued strike ballots and picketing was organized. The unions complained that regional directors harassed workers to agree to join STM. Union leaders took every opportunity to criticize privatization, and called for its postponement. Rumors, such as one claiming that 20 percent of staff would be let go, were used.

Prior to denationalization—in the Successor Act creating STM— the government gave JTM employees quite favorable terms. Workers were assured of employment on terms not less than those at JTM. STM had to employ every member of the JTM staff who chose to join the privatized company, and there was an assurance of no lay-offs for the first five years.

In 1985 the Pensions Act of 1980 was amended to give JTM civil servants the right to continue being eligible for government pension and other benefits should they opt for the privatization plan. Employees contributing to the Employees Provident Fund would have their contributions matched so their net salaries would remain the same. This meant the new company was required to contribute to the government Consolidated Fund at the rate 17.5 percent of their monthly salary.

7.1.3.3 Academics

Malaysian academics have been critical of the degree and extent of government involvement in the privatized STM. The Successor Act stipulates a long list of specific conditions, including one that is the catch-all statement that the energy, telecommunications, and posts minister has the power to ''give directions to the privatized company.'' Ongoing government control through appointment of the directors and chief executives is feared. Thus, the chief executive of STM resigned in protest over introduction of nontelecommunications executives (Vong 1987). (It is also believed that several people vied for the STM chairmanship before appointment of the present chairman.)

Another criticism is that a privatized but heavily regulated STM was not that different in practice from JTM: all that has been done is the conversion of a public monopoly to a private one. Indications are that the government intends to retain at least 30 percent of STM (Rita 1986, p. 82). Other observers claim privatization eliminates the social obligations and aims of the civil service providers. However, social obligations are imposed on STM under both the acts creating it and its operating license.

7.1.4 Rule Making

Besides transforming most of JTM into STM, Malaysia made three significant decisions in the mid-1980s with far-reaching effects on its marketplace. These relate to changes in rates, definitions, and boundaries between basic and value-added services—particularly regarding mobile telephone service, and the selling off of public pay telephones to a private company.

STM's charges for installing and maintaining telephones are subject to approval by the government. This has generally been on an ad hoc basis. That is, there are no formal procedures involving hearings or specified periods for comments by the public and other interested parties. Reconnection fees were increased to M$50 in to discourage late payment, which had become a major problem, and metered charges for local calls experienced a 30 percent increase in August 1985. These were seen as attempts to alleviate the M$4.5 billion in loan liabilities STM inherited from JTM (Noor 1988).

In 1985, the first cellular system—known as ATUR, for automatic telephone using radio—was introduced by JTM. Three years later STM replaced JTM. In partnership with Fleet Communications, a local company, STM also incorporated Celcom to provide a second mobile telephone network.

Although no public announcements have been made, it is surmised that a high-level decision was made that cellular will not be regarded as a value-added service, and thus will not be opened to competition. There is speculation that ATUR has spin-offs for extending telephone service to rural areas without laying cable. The system uses one mobile phone connected to a private branch exchange (PBX) capable of providing up to forty-eight extensions. The cost of renting such a system is comparable to that of an ordinary telephone system, but the call fees are high. The high rates for mobile service can provide a considerable element of cross subsidy for telephone systems in rural areas. Sapura Holdings has been licensed to develop the rural ATUR systems, with implementation costs borne by STM.

A not unrelated decision was granting a license to Uniphone to operate and maintain urban public pay telephones throughout Malaysia for fifteen years beginning in January 1989. The decision has been criticized by NUTE, which fears such cherry-picking will shrink STM's profit margin and thereby adversely effect union member salaries, bonuses, and other benefits. No details have been released on the revenue-sharing ratio between the two companies. It was charged that Uniphone claimed excessive metered charges from public pay phones, and that they did agree to pay STM M$1.5 million to avoid litigation (Rema 1989).

7.1.5 The 1983 Turnkey Contracts

The decision to award private companies M$2.4 billion in contracts to install 1.76 million lines throughout Malaysia during 1983–1988 had been seen as an early indication of the government's intention to privatize JTM. Upgrading of the network through these turnkey contracts was part of a five-year plan that also envisioned new value-added services being provided by joint ventures between local *bumiputera* (ethnic Malay) companies and foreign partners. All these firms were established by former JTM staff, and the four contracts were awarded without an open tender. However it was claimed, probably on the basis of the foreign partners' experience, that the recipients had proven track records.

The contracts were written to provide maximum help to the contractors. Thus, advances were made to them. In addition, no planning fees were charged for work already done, and, although materials had to be ordered from JTM's available inventory, transport costs were absorbed by the government.

Experienced or not, none of the contractors met the interim target for the end of 1985, let alone the overall completion date of the end of 1986. Moreover, the companies did not even come close to the contract price of M$1,433 per effective cable pairs (see Rita 1986, pp. 68–73). The government and others involved have been reluctant to release data on subsequent performance, but it seems not all the lines were installed, as shown in Table 7.1.

The contractors attributed nonperformance to several reasons. First, the

Table 7.1. Completed Telephone Lines by Contractor

Percentage of target for the year		Cumulative lines completed by yearend		
1985	1985	1986	1991	Contractor
72.5	49.7	220,918	333,330	Binaphone
55.1	37.8	167,728	336,176	Electroscon
70.1	48.1	213,475	292,063	Sri Com
83.9	57.6	255,562	485,277	Uniphone
70.4	48.3	857,773	1,446,846	Total

Source: Data for 1985 and 1986 are from *JTM Annual Report* 1986, p. 14; those for 1991 are unpublished STM data.

The target at yearend 1985 (the second full year of the project) was 304,669 lines for each company. The project was to be completed by the end of 1986, with each firm installing 440,000 lines for a total of 1,760,000.

agreement was signed in October 1983, but it was back-dated to January 1983, shortening the actual available work time. Second, they claimed three government departments—the National Electricity Board, the Municipal Council, and the JKR (the public works department)—were slow in approving permits for construction (*New Straits Times,* Sep 10, 1985). The contractors also claimed that JTM's staff opposed use of turnkey contracts because it saw them as an obstruction to the departments' expansion. This last reason would explain the contractors' claim that JTM held up the final detailed designs for a year (Krishnamoorthy 1985). For its part, the JTM claimed that their officers had to supervise the planning and installation work.

There were so many problems that the deputy minister of Energy, Telecommunications, and Post (ETP) announced JTM would revert back to using small cabling contractors instead of turnkey contracts. At the beginning of the contracts a hierarchy of committees had been formed for monitoring projects. It started at the regional level and culminated with an interministerial Steering Committee chaired by the ETP minister.

7.1.6 Upgrading the System

Not all the projects have been as contentious as the turnkey-line contracts and public telepay phone agreement. A number of others have been undertaken by local and foreign firms (often in joint ventures with domestics partners).

Work to upgrade the radio relay microwave system was begun in 1984 under a M$250 contract with Standard Electrik Lorenz (SEL) and ITT's German subsidiary (Knor 1988). A digital fiberoptic submarine cable project worth US$100 million to link the peninsula with Sabah and Sarawak was awarded in 1989 to a consortium of NEC and Fujitsu.

Paging systems were privatized in 1986. In 1987 telephone directory services were taken over by GTE in a joint-venture with the local Melewar Corporation.

In 1987, a M$5 million high-speed data network using a satellite was established to serve some thirty-eight banks and their over 800 branches. Packet and circuit switched data transport systems and telemail have also been introduced through joint-ventures of local and foreign companies (see, e.g., Raj 1988).

In March 1991, as part of a plan to expand and digitize the system, the government asked for bids on 4 million new lines. The process was very contentious and controversial. The winners were announced in March 1992—five of them, sharing equally in the M$2 billion (US$780) project.

One concern is that the systems will be incompatible—although one condition is that each supplier provide mutually interoperable equipment. However STM will have the expense of adapting the new equipment to the existing network, as well as making the various vendors' equipment work together. The low bid (made by Ericsson) must be matched by the other four.

Beyond technical considerations have been political ones. One of the successful bidders (Alcatel) has as a local partner a foundation affiliated with the Penang chapter of UMNO, the ruling political coalition, and the Finance Minister in particular. Because the Finance Ministry owns 76 percent of STM it had final say on the winners. It is known that the second-low bidder was Siemens, but it was not one of the firms selected, despite a recommendation from STM that it be included.

Sapura Holdings is the local partner of Nokia, the fifth-highest bidder but among the winners. One of Fujitsu's partners has ties to UMNO. Public controversy involving alleged favoritism surrounded another major tender in Malaysia in 1991 (gas turbines for the electric board).

7.1.7 Information Technology Policy

STM's research and development efforts in establishing ISDN are in line with the aims of the Malaysian Administration Modernization & Manpower Planning Unit (MAMPU) incorporated in the Prime Minister's Department. The unit started as the nucleus of the government's drive to automate its own data processing functions. In 1985 MAMPU was elevated to a national-level committee responsible for formulating, promoting, coordinating, and controlling computer technology policies for modernization, management, and national development.

STM is facing pressure to speed up implementation of an ISDN model. In a 1988 seminar on computerization for development,it was suggested that STM adopt a three-pronged approach (Mazlan 1988):

1. Provision of digital transmission and switching.
2. Introduction of basic ISDN services leading toward provision of specialized packet-switched and circuit-switched networks.
3. Services integration of packet-mode, broadband services and multimedia services.

Participants felt STM was much too slow in offering customers an integrated multipurpose network system. STM planned to introduce a pilot ISDN setup by the end of 1989 with commercial service to be available in 1992 (see Westlake 1989). The goal was partially met.

7.1.8 Domestic Companies

The dominant local telecommunications firms are Binaphone, Sri Com, Electroscon, and Uniphone—the four holders of the turnkey-line contracts discussed earlier. They have been quite unfazed by the brouhaha over those contracts, and their businesses have continued to expand. There is a good deal of interlocking shareholding among them and by UMNO, the dominant party in the governing coalition. Table 7.2 shows some of the relationships.

Manufacturing telecom equipment for domestic use in Malaysia is largely in the hands of Sapura Holdings, a private company. Sapura conducts research and development without funding from the JTM. In the late 1980s it produced a wholly Malaysian-made telephone set known as the S2000 series. There are two models, one aiming at the third-world market and the other—more sophis-

Table 7.2. Nongovernment Malaysian Telecommunications Companies

Holding Companies

AZH Holdings plc. parent of Binaphone.

Fleet Group. Wholly owned by UMNO, the dominant party in the governing coalition (see Seaward 1987). Holds 25 percent of Sistem Televisyen. Involved in Celcom—a joint venture with STM to provide mobile phone service (see Sabri 1988).

Sapura Holdings. Twenty-four subsidiaries, including Electronics & Telematique and Uniphone. Combined 1988–1989 revenues of M$200 million. 100 percent bumiputera. See Lee (1989).

The Four Majors

These were all established by former JTM staff members and shared equally in the 1983 contract for installing 1.76 million phone lines (see text).

Binaphone Sdn Bhd. Subsidiary of AZH Holdings plc. Owns 25 percent of Britarafon. Joint-venturer with Philips Electronics NV.

Electroscon Sdn Bhd. Joint-venturer (after first year of operation) with LM Ericsson.

Sri Communications Sdn Bhd. Joint-ventures on an each-job basis.

Uniphone Telecommunications Bhd (formerly Malayan Cables). Majority-owned by Sapura Holdings through the latter's control of Electronics & Telematique. Owns 20 percent of System Television. Has public pay telephone contact (see text). Joint-venturer with Sumitomo Denki Kogyo.

Others

Britarafon Sdn Bhd. Reportedly formed to purchase shares in STM when they were floated (Loh 1986a). Owned equally by Binaphone, Arab Malaysian Development, Electronics & Telematique, and British Telecom of London (see Lee 1986).

Electronics & Telematique (M) Sdn Bhd. Established in 1981 as an associate company of Sapura Holdings (see Lohn 1986b). Majority shareholder of Uniphone.

Sistem Televisyen Malaysia Bhd. The commercial television station in Malaysia: 20 percent owned by Uniphone; 25 percent owned by Fleet Group.

ticated—directed at first-world markets. The company anticipates capturing 1 percent of the annual market of 200 million sets (Lee 1989).

7.1.9 Overview

If profit making is the central criterion, the corporatization of Malaysian telecommunications could be considered a success, although STM did report a M$96 million loss its first year. This has helped make STM (also often called Telekom Malaysia in the press) became something of a stock market darling, with many foreign investors acquiring shares as a way of "playing" overall Malaysian economic development. The government has attributed STM's success to privatization models seen in Japan (described in Chapter 23) and the United Kingdom. One similarity between the United Kingdom and Malaysia is retention by government of a "golden share"—the power to veto policies. Overall, however, the Malaysian government has actually increased regulation and changed the industry into a private monopoly rather than a public one.

7.2 Indonesia

Indonesia's principal telecommunications entities, Perumtel (renamed PT Telkom after early 1991) and PT Indosat, are state-owned corporations, with no indication of an intention to infuse private capital into them. They are under the jurisdiction of the Department of Tourism, Post, and Telecommunications. Both are operated as private companies and subjected to income taxes like other corporations. However, the government exercises its role as the only shareholder through direct involvement in management and decision making. The companies are also required to consult with other government departments on technical and financial matters.

In 1980 PT Indosat, which had been an IT&T subsidiary operating a satellite system under a twenty-year license that had nine years to run, was nationalized and became the monopoly international service provider. At the same time Perumtel was made sole provider of domestic public telecommunications. (The preceding chapter provides more details on structure and history.)

Perumtel recently has been having trouble. By 1986 it had managed to complete only half of the projects carried forward from its third five-year plan (known as Repelita 3); these projects were to have been finished by March 1983. Moreover, only one of the nine projects stipulated in Repelita 4 was done. Work on three projects had not even commenced. According to Willy Moenandir, then Perumtel director general, funds allocated for telecommunications development were inadequate in both of the plans (Nasution et al. 1988b).

In 1984 there was an average of only 0.48 trunk lines per hundred people (compared with 5.7 in Malaysia). At yearend 1987 Perumtel faced a waiting list of 400,000, compared with 668,000 existing subscribers, at a time when there was a population of some 170 million (Nasution et al. 1988a).

These were undoubtedly factors in the government decision to implement build–operate–transfer schemes, contracting out, and obtaining ''soft'' loans from donor countries for telecom projects. All of these unfortunately appear to have done as much to create profit opportunities for suppliers as they have to improve the telecommunications system.

7.2.1 Equipment Procurement

PT Inti is the wholly government-owned principal supplier of such equipment as digital telephone exchanges, mobile phone units, and satellite stations. Throughout the 1970s and 1980s it had technical cooperation ties with several foreign companies including Siemens and Bell Telephone Manufacturing of Belgium (Hukill and Jussawalla 1989). Advanced equipment such as digital telephone exchanges, mobile phone units, and satellite ground stations were produced in Indonesia for local and foreign markets.

Presidential Decree 6/1988, which amended decree 24/1984, stipulated a more liberalized approach to procurement of telecom equipment. Beginning in 1988 Perumtel and PT Indosat were allowed to procure equipment through several alternatives, such as open or limited bids, direct determination, or direct purchase. Two requirements stipulated in the decree were the mandatory involvement of Indonesians in contracts awarded to foreign suppliers and a certificate of contractor capability. Direct procurement of equipment from foreign countries is only allowed if the technical specifications laid down by the appropriate regulatory bodies in the country of origin are fulfilled. However, a type approval from PT Telkom is required for any terminal equipment used.

7.2.2 Local Companies

Three major telecommunications firms are controlled through the Bimantara Group by Bambang Trihatmodjo, a son of Indonesian President Suharto. PT Elektrindo Nusantara (40 percent owned) produces equipment for PT Telkom under license from Hughes Aircraft, the company that provided the technical expertise for the Palapa A satellite in 1976. Cakra Nusa is a trading company, while Sattel Technology, based in the United States, conducts research and development to support Elektrindo's operations.

In 1987 Sattel purchased Indonesia's Palapa B2 satellite from Merritt Holdings, the underwriter for Lloyd's of London that became the owner after the satellite went into the wrong orbit when launched in January 1984. It was recovered by the U.S. space shuttle that October. Critics claim Sattel was incorporated to facilitate the sale and prearranged purchase of the Palapa B2 by Perumtel (see Nasir 1987).

In the late 1980s the government embarked on a ''user-credit system'' to install about 30,000 telephone lines in the cities of Jakarta, Solo, Surabaya, and Pontianak. These projects incurred sixfold cost overruns compared to those built by Perumtel.

A contract for installing 10,000 Ericsson mobile telephones was offered to PT Rajasa Hazanah Perkasa by Perumtel in 1985 without going through an open tender. The company subsequently failed because of the relatively low demand for mobile telephones in Indonesia (Ahmed 1988). This is somewhat ironic, as substantial later demand for cellular service has been driven by the general poorness of Perumtel's wireline system.

7.2.3 Parceling Out Business to Foreigners

In its 1988–1989 report Indonesia announced a list of ten projects to be offered to donor countries for financial assistance (see Chapter 6). Under this scheme, the donor countries would offer financial assistance in erecting and installing telecommunications networks, entitling them to annual profit sharing once the networks commence operation. This resulted in intense lobbying of government officials by foreign governments in support of their suppliers (*Tempo* Apr 1988, p. 26).

The 1990 awards to NEC and AT&T to provide switching and other equipment for 350,000 lines each require collaboration with local joint ventures. For both bidders, the soft loan credits are reported to exceed the value of the bids (for AT&T, $193 million in credits on a $103 million bid; for NEC, credits total $174 million, versus a $77 million bid). There is considerable speculation as to who will benefit. Considerable maneuvering took place to be the domestic partners—there was no open tender and AT&T and NEC did not have a free hand in the matter, which delayed the contracts and, therefore, the work. Companies associated with President Suharto's family subsequently emerged the choices—AT&T partners with PT Citra Telekomunikasi Indonesia (CTI), NEC with PT Elektrindo Nusantara. CTI is 25 percent owned by the younger brother of the minister of research and technology and was not formed until a few months before the contracts were awarded (see, e.g., *Far Eastern Economic Review* Jan 24, 1991, p. 41).

7.2.4 Overview

Telecommunications in Indonesia appears quite politicized, and the build–operate–transfer scheme and "soft" loans are anticipated to further aggravate the situation. Thus, the use of private investors to spearhead telecommunications expansion has resulted in preference being given to entrenched interests. Moreover, soft loans appear to be routinely used as facilitators to ensure equipment is bought from donor-country suppliers. With a liberalized approval mechanism for equipment and pressure to develop networks, a wide variety of gear has been put on the system. Despite equipment having to be approved by PT Telkom, varied standards and specifications have caused difficulties in network integration. This is a factor in the call-completion rate being below 30 percent.

7.3 Conclusion

One significant effect of restructuring in Malaysia is that it has reinforced existing bureaucratic capitalism. Many ownership interests ultimately lead to

UMNO, the major party in the governing coalition. Another observation is that these companies are owned by bumiputera. Ethnic discrimination is less obvious in Indonesia; however, politicalization is present.

A second observation is that the governments of both Indonesia and Malaysia retain tight control of their telecommunications provider. In Malaysia it is possible (but unlikely) that the regulator could be overwhelmed by STM's greater manpower and financial resources or lose its high level of political support. What appears fairly clear is that the system will continue to protect bumiputera interests. Indonesian insistence that foreign interests have local proxies is not too dissimilar. Any policy reforms must conform with local political and administrative styles.

In Malaysia, one possible future is to give STM the role of carrier and introduce competition among "big players." Another is to have controlled, restricted, or paced competition, as in the public pay telephone case. Whatever the eventual shape of the marketplace, it is fairly certain that commercial dynamism will be somewhat dampened by the political reality of having to give bumiputera stakes in new businesses and the requirement of providing services in (largely Malay) rural areas despite the low level of demand.

In Indonesia there was a belief that the build–operate–transfer scheme and soft loans from foreign countries would spearhead network expansion and relieve government financial constraints. Further "reforms" depend on whether local elites are in a position to benefit from them.

Bibliography

Ahmad, K. S., Harymurti, L. Djalil, and K. S. Djunsisi. 1988. "Bisnis Di Antara Kabel-kabel Telefon." *Tempo,* p. 30 (Apr. 30).

Fong, Chan Onn. June 1988. "The Development of the Malaysian Telecommunications System." Kuala Lumpur.

Hukill, Mdark A., and Meheroo Jussawalla. 1989. "Telecommunication Policies and Markets in the Asean Countries." Asian Mass Communication and Information Centre (AMIC), "Singapore Workshop on Telecommunications Infrastructure and Investment in the Asean Countries," pp. 16–26 (May 15–16).

Indonesia Development News. 1988. "Perumtel to Supervise Telecommunications Expansion" 12(2) (Nov./Dec.).

Khor, Eng Lee. 1988. "Backbone Network." *Investors Digest,* pp. 10–12 (Apr.).

Krishnamoorthy, M. 1985. "Delayed: Work to Install 1.2m Telephone Lines." *The New Straits Times* (Sep. 6).

Lee, Eddie. 1989. "Sapura Holdings: The Right Signal." *Malaysian Business,* p. 29 (Sep.).

Loh, Yee Liam. 1986. "Competing for Privatisation Cake." *Business Times* (Mar. 17).

Loganathan, S. 1989. "All System Go For STM." *Investors Digest,* pp. 12–13 (mid-Aug.).

Mazlan, Abbas. 1988. "Integrated Services Digital Network: Prospects and Challenges For Malaysia." International Seminar on Computerization for Development, Kuala Lumpur, August 18.

Nasir, Anwar. 1987. "Clear Connections: Indonesia Buys Back A Satellite from a Lucky Middleman." *Far Eastern Economic Review,* p. 69 (Apr. 30).

Nasution, A., A. Firmasyah, W. Muryadi, and I. Faridah. 1988a. "Menteri ini Menghadapi Soal Kecil yang juga Besar." *Tempo,* pp. 23–24 (Apr. 30).

———. 1988b. "Ini Bukan Perusahaan Jual Bir." *Tempo,* pp. 24–25 (Apr. 30).

The New Straits Times. 1985. "Third Parties Holding Up Telephone Project" (Sep. 10).

Noor, Adzman Baharuddin. 1988. "STM Squad Recovers $380,000 in Arrears." *New Straits Times* (Jul. 7).

Raj, Charles. 1988. "STM to Start Telemail." *News Straits Times* (Oct. 5).

Rema, Devi. 1989. "Uniphone, STM End Payphone Dispute." *Business Times* (Oct. 7).

Rita, Raj Hashim. 1986. *Privatisation of Telecommunications in Malaysia.* MPA thesis, University of Malaya, Kuala Lumpur.

Seaward, Nik. 1987. "Fleet Group Sails Into Heavy Weather." *Far Eastern Economic Review,* p. 59 (Sep. 3).

Teoh, Chew Chee. 1988. "STM Goes Digital." *Investors Digest,* pp. 9, 13 (Apr.).

Vong, Yamin. 1987. "Staff Unhappy Over STM Revamp." *Business Times* (Dec. 8).

Westlake, Michael. 1989. "Ringing the Changes: Malaysian Telecom Group Reports Increase in Earnings." *Far Eastern Economic Review,* p. 66 (Jul. 20).

Zabri, Adlin M. 1988. "Ericsson Submits 3 Bids to Celcom." *Business Times* (Jun. 11).

(*Investors Digest* is a monthly published by the Kualu Lumpur Stock Exchange. The *Business Times* and *New Straits Times* are daily newspapers published in Malaysia. They are in English. *Tempo* is an Indonesian weekly.)

8

Malaysia

FONG CHAN ONN

Rapid industrialization since 1970 has led to creation of a relatively advanced telecommunications system to address the complex needs of businesses for instantaneous information transmittal to and retrieval from points around the globe.

The development of telecommunications in Malaysia is traced and analyzed in this chapter, including its contributions to accelerated economic development. Some background is offered on Malaysia followed by discussion of telecommunications development during three time periods: preindependence, 1957–1970, and post 1970. Because privatization has become an important objective of the Malaysian government, this issue is analyzed at length, including the government's rationale and strategy for privatization.

8.1 Background

Malaysia has two major geographical parts: the Malay Peninsula and the northern part of the island of Borneo. Peninsular Malaysia, washed by the Strait of Malacca on the west and the South China Sea on the east, extends about 800 km from its northern border with Thailand to its southern end across the Johore Strait from Singapore. The states of Sabah and Sarawak make up Eastern Malaysia, some 650 km from the peninsula at their closest points and stretching another 1,100 km to the east. There are eleven mainland states plus the federal territory of Kuala Lumpur and Labuan. Total land area is approximately 330,000 km^2, 131,500 of which is on the peninsula.

Geography has made the peninsula a center of trade for millennia, and as such it has been repeatedly fought over and proselytized. Malacca's founding in the fifteenth century, succeeding Sumatra-based Srivijaya as regional hegemon, can be considered the beginning of an independent Malay history. The city served as both a cultural unifier of the Malay people as well as an outpost for the introduction of Islam, which its court blended with older Hindu–Buddhist influences. However, politically the region remained a number of small states—tributaries of Malacca, sometimes with intermarried rulers, but no real central control. The states—sultanates—that emerged in the fourteenth and fifteenth centuries form the basis of the states in modern peninsula Malaysia. The Portuguese sacked Malacca in 1511 and held it until driven out by the Dutch (with Malay help) in 1641.

Malaysia's modern history began with colonization of the peninsula by Britain, a process marked by establishment of Penang in 1786 and Singapore in 1819. These, along with Malacca and a few other British-controlled areas, became the Straits Settlements in 1826. This was followed by the spread of British influence, which became rapid after the 1874 Treaty of Pangkor. By 1896 most of the peninsula, Sarawak, and Sabah were under direct or indirect British rule, and a 1909 treaty with Thailand brought the northern Malay states into the British sphere. It should be noted, however, that the British governed in large part through Residents who "advised" Malay rulers.

Organized economic development took place under colonial rule. Tin was being exported to India by the fifth century, but systematic and extensive exploitation of the lands around Kuala Lumpur, Ipoh, and Taiping for this and other minerals did not begin until the mid-nineteenth century. Rubber was introduced from Brazil in 1877 and was planted on a large scale in the early twentieth century. These activities led to a tremendous demand for labor. To meet it, extensive immigration from China was encouraged, beginning with the 1848 Larut, Perak, tin discoveries. The rubber booms brought Tamil and others from India.

The Federation of Malaya—the eleven peninsula states—achieved independence in 1957. In 1963 Malaysia was formed by the federation of Malaya, Singapore, Sabah, and Sarawak. Singapore left the Federation as an independent country in 1965.

Malaysia's mid-1990 population was estimated at 17.5 million (growing at 2.4 percent—15 million in peninsular Malaysia, 1 million in Sabah, and 1.5 million in Sarawak. Malays and other indigenous peoples constituted about 59 percent of the population (a share that is increasing because of higher birth rates), Chinese 32 percent, and Indians 9 percent. Gross Domestic Product (GDP) was U.S.$43.1 billion (U.S.$2460 per capita) in 1990.

8.2 Preindependence Telecommunications

During the mid-nineteenth century, postal, telegraph, and telephone services were organized in one department under the control of the state auditor. In 1884, to meet the needs of the mining industry in the states of Perak and Selangor, the Telecommunications Department was placed under the direction of the superintendent engineer. In 1902, the department's functions were extended to cover the whole of the Federation of Malaya and the responsibility for operations was transferred to a department head. (The Federation included the four states across the middle of the peninsula, which had agreed to a federal government in 1896 with a capital at Kuala Lumpur.)

After World War II, as a step toward independence, Britain for the first time politically unified the entire peninsula (except Singapore) as the Malayan Union. As part of this, on April 1, 1946, the department was renamed the Malayan Telecommunications Department and given the specific mission of regulating and providing telecom services nationwide. Sabah and Sarawak became crown colonies, and their telecom services were separate. Legislation spelling out the department's powers was formally enacted as the Telecommunications Act of 1950.

8.2.1 Telephone Services

The first telephone was brought to Malaya in 1874 by the newly appointed British resident assigned to Perak. Telephone services were needed to enable him to better monitor the political and social unrest that was then prevalent due to fights among various mining clans in the region. The first exchange was completed in Kuala Lumpur in 1891. By 1907, six more had been built in Perak to cater to the expanding needs of the tin miners. As the number of subscribers increased, the need for an automatic exchange became apparent. In 1930, the first was completed, together with a new carrier system linking Kuala Lumpur, the federal capital, and Ipoh, capital of Perak. This was later extended to Singapore and other states. Demand for intercity services grew so rapidly that twenty automatic exchanges were completed in Selangor (the area around Kuala Lumpur) alone by 1933. However, during World War II hundreds of telephone and telegraph poles were uprooted, along with hundreds of miles of copper wire, which severely impaired the system.

In Sabah and Sarawak, the development of telephone services began in 1900 when the service was launched in three towns. Since then, service has expanded to meet the needs of loggers.

In 1947 there were an estimated 14,000 telephone subscribers nationwide. By 1957 the number had reached 61,000, with a backlog of 2,000.

In addition to expanding domestic service, there was dramatic growth in international calling. The first overseas telephone service began in 1936 with a single circuit linking Kuala Lumpur to Indonesia. The boundary station would then relay calls to Amsterdam and on to London. This arrangement was terminated for security reasons in 1939. After the war, international calls were routed via Singapore to London. To meet demand, more direct links with other countries, particularly to Asia and the Commonwealth, were established. In 1957, 24,100 international radiotelephone calls were made from Malaya and Singapore. Notwithstanding fairly extensive linkage, the relatively large number of interconnections required before a call could reach its destination amplified noise levels and reduced the quality of transmission.

8.2.2 Telegraph, Telex, and Other Services

Telegraph service was introduced to Malaysia in 1876 with the installation of a telegraph line between the British resident's office in Kuala Lumpur, the British assistant resident's office in Taiping, and the attorney's office in Matang, Perak. Private subscribers were allowed to utilize the link, and there was soon sufficient demand to make it a profitable government operation. This led to installation of a line from Kuala Lumpur to Malacca in 1886, which was later extended to Singapore and overseas.

Telegraphic equipment with high-speed typewriters was used to receive messages by the late 1920s. Next came a wireless connection between Kuala Lumpur and Kuantan on the east coast. By 1957, 1,125,053 messages were handled annually in the Federation and 654,648 in Singapore. Service was further improved by introduction of the multiple-type teleprinter exchange.

Two wireless stations were in operation in 1926—in Penang and Singapore—where they provided radio communication services with ships through the 500 kHz international marine frequency. Following unusually large floods in 1926, a radio warning system was set up to provide alerts along the Pahang River, the country's longest. This station eventually became the center for extension of the radio network. Stations were staffed by the Malayan Radio Amateur Society until it closed in the early 1930s. The Telecommunications and Post Department took over responsibility until 1942, when Radio Malaya, which the Japanese occupation forces had assigned the mission of maintaining radio stations for the police, took over.

The end of the war saw an eruption of communist insurgency, and the system of radio stations became a vital instrument in the government's fight for the hearts and minds of the people. By the end of 1957 the number of stations in the Federation was 667 fixed VHF stations, 287 mobile VHF, 62 fixed HF, and 168 mobile HF. The extensive system enhanced communication capabilities between security forces and the police, enabling them to better monitor the political and social situation. This contributed to the government's ability to isolate the guerrillas and, ultimately, terminate the communist insurgency.

In 1957 there was a Pan-Malayan teleprinter exchange network with 125 exchanges and 32 telex trunk circuits. Telex service became increasingly popular as its costs dropped to become one-third cost of an equivalent day rate telephone trunk call. The needs of service industries for written confirmation of messages further contributed to its popularity.

8.3 Telecommunications, 1957–1970

With independence, the organization of telecom services was reviewed. In addition, constitutional changes made both in the Federation and Singapore affected organization of Pan-Malayan telecommunications. Immediately after independence the government accordingly decided to:

1. Seek membership on the Commonwealth Telecommunications Board.
2. Become a Partner Government submitting to the Commonwealth Telegraphs Agreement of 1948.
3. Acquire the assets and take over operation of the Cable & Wireless installations in Penang.

4. Set up a Pan-Malayan Telecommunications Advisory Committee, in conjunction with the State of Singapore.

On September 1, 1962, the cable station in Penang was taken over. The local staff, absorbed into the Telecommunications Department, continued to staff the station. Also in 1962, Malaya became a participating member of the Commonwealth Telecommunication Board.

Because telecom services had originated and been developed under British rule, all the senior technical and managerial staff of the department were British expatriates; Malaysians were confined to roles as technicians, assistants, and clerks. At independence, Malaysianization of the department became a priority because telecom services were considered a foundation of national security and defense. To assist Malaysianization, three supplementary superscale posts were created in 1959 to expose Malayan officers to higher responsibilities and to enable them to gain valuable managerial experience. In 1961, the organization was expanded by separating responsibility for planning and operations at the director level.

With the formation of Malaysia in 1963, all telecommunications matters, including the Singapore Telephone Board, became a federal responsibility under the portfolio of the minister of works, posts, and telecommunications. V. T. Sambantan, the first minister, was a relatively senior minister and saw to it that the department was accorded the necessary budgetary allocations. In the case of Singapore, the director-general of telecommunications was accorded complete responsibility for the department, as opposed to the somewhat limited responsibility he exercised while Singapore was still a Crown Colony. The Posts and Telegraphs Departments in Sabah and Sarawak were merged into one regional entity under the direction of a regional director of telecommunications who was in turn responsible to the secretary-general of the Ministry of Works, Post, and Telecommunications.

Paralleling these organizational changes, human resource development and training were given particular emphasis. Many Malaysian technicians were encouraged to take the London City and Guilds Examinations locally to enable them to qualify as technical assistants. Classes were conducted by the department with British engineers often serving as course instructors. A number of technical assistants were sent to the United Kingdom to acquire professional qualifications as engineers, particularly diplomas in electrical engineering awarded by British universities and polytechnics. They were placed in assistant controller positions on their return and rapidly promoted to controller and director positions.

Over the 1958–1965 period, the government also sent about fifty of its best high school graduates to acquire degrees in telecommunications engineering under the Columbo Plan Aid Programme in countries such as Australia and New Zealand. The British government also offered scholarships to those chosen by the department for professional engineering courses in the United Kingdom.

As a consequence of this technical training assistance, Malaysianization proceeded rapidly. By 1965, the program was completed, and, for the first time, a Malaysian, Chew Kam Pok, a former technical assistant who had obtained his tertiary engineering qualifications in the United Kingdom following independence, was appointed director-general of telecommunications.

In January 1965, the department took over Singapore's External Services. After Singapore separated from Malaysia in September 1965, however, management was handed over to the Singapore government.

In 1966, a telecommunications training center was established in Kuala Lumpur with the support of the International Telecommunications Union (ITU) and the United Nations development program. The center provides in-service training to its technical staff to help upgrade their knowledge of telecommunications. Special emphasis has been placed on upgrading maintenance and installation capabilities. The center in its initial stages relied heavily on the Australian Post Office, which provided most of the instruction under a bilateral technical assistance program and offered training facilities in Australia. During 1966–1970, about forty Australian instructors held visiting appointments and another twenty Malaysian engineers received training in Australia. By 1970 the bilateral program was completed; the center has been manned since completely by Malaysians. The department is now able to train technical staff—including new hires and skill upgrading.

The organizational changes needed to cater to rapidly expanding demands following Malaysianization and consolidation were completed by the late 1960s. At the top is the director-general of telecommunications. Assisted by a deputy director-general and several directors, he implements telecommunications policy, which is formulated with his advice by the Minister of Works, Post, and Telecommunications. This group constitutes the directorate, generally known within the department as the "wise men." Executing directorate decisions are a group of controllers. Each is assigned a specific area of responsibility (i.e., exchange). Assistant controllers or above must be qualified engineers. (After privatization of Telekom Malaysia, many of these functions were transferred to the new company).

Technical assistants, who are graduates of polytechnics institutes, and technicians constitute the core of the maintenance and installation staff. Each installation or maintenance team, consisting of linemen and cable layers or installers, is headed by a technician under the supervision of a technical assistant.

Given the relatively advanced state of the infrastructure built by the British, an important objective of the new government was to maintain and enhance it so that the country would continue to have one of the best telecommunications systems in Asia. To achieve this objective, relatively generous budget allocations were made, allowing purchase of the latest equipment. As Table 8.1 shows, during 1966–1970 some 1.9 percent of the total national plan budget was spent on telecommunications.

In particular, the government adopted a very liberal attitude toward equipment imports. In all tender exercises the major criteria for awards were price and quality. Under this policy, foreign non-British suppliers began to make their presence felt. Ericsson became a major supplier of exchange equipment, while Fujitsu and NEC emerged as major suppliers for transmission equipment,

Table 8.1. Public Development Expenditure Allocated for Communications under Each Five-Year Plan, 1966–1995

1966–1970	1971–1975	1976–1980	1981–1985	1986–1990	1991–1995	Allocation
196	717	1,200	2,900	848	201	in M$ million
1.9	5.0	2.7	2.5	2.3	0.37	Percentage of Plan Total

Source: Malaysia Plan, various issues.
Includes broadcasting and information.

cables, and lines. U.S. manufacturers were distinctly absent, mainly because of the then relatively high value of the U.S. dollar.

8.3.1 Growth of Domestic Telephone Services

In 1959, increasing demand for telephone services in major towns brought introduction of a microwave system between Kuala Lumpur and Singapore. The system had 600 telephone channels. In 1963, it was extended from Kuala Lumpur to Penang, and later expanded to cover the east coast states. In 1962, subscriber trunk dialing—initially limited to Kuala Lumpur, Klang, and Port Swettenham—was expanded through a major trunk route connecting Kuala Lumpur to Singapore, Seremban, and Malacca. By the end of the 1960s, cables and microwave links continued to broaden coverage and enabled direct subscriber dialing among the major population centers in the peninsula, as well as operator-assisted trunk calls among all towns and villages.

The increasing number of subscribers and shortages of equipment, including telephone sets, were compounded by implementation delays for cable and exchange schemes. The waiting list for service rose from 2,075 in 1957 to 13,704 in 1970. To overcome this growing unmet demand, the department contracted Ericsson to establish a crossbar telephone exchange equipment assembly factory in Malaysia.

The department also negotiated with Mitsui to establish a cable manufacturing plant. Mitsui, however, felt local demand was too small to justify a plant. Further, Japanese imports of cables were facilitated by a generous yen credit from the Japanese government. Under the credit, all purchases of telecom equipment would be financed by a Japanese loan with an initial grace period of five years and repayment over fifteen years at 4 percent.

While domestic service developed satisfactorily in the peninsula, service in Sabah and Sarawak did not improve significantly in the early 1960s. Sabah and Sarawak continued to remain highly dependent on timber extraction with its comparatively minimal need for telecom services. The advent of political uncertainties caused by the Indonesian Confrontation during 1964–1967 (Indonesia's opposition to formation of Malaysia) accelerated development of telecommunications for security and strategic reasons. In 1965, the first stage of the South East Asia Commonwealth (SEACOM) Cable linking Kuantan in the peninsula and Kota Kinabalu in Sabah was completed. Internal services within Sabah were also given a boost as a result of this link.

Although the SEACOM cable improved linkage between the peninsula and Sabah, the connection to Sarawak was constrained by dependence on transmission via the Kota Kinabalu–Kuching radio link. In 1970, the Gunung Pulai–Gunung Serapi Troposcatter Link between the peninsula and Sarawak became operational, providing forty-eight telephone channels. With the opening of this link, subscribers in Sarawak gained access to national and international facilities in peninsular Malaysia. By 1970, the peninsula, Sabah and Sarawak were relatively well connected, which has contributed in no small measure toward the forging of a united Malaysian nation.

The number of subscribers grew to 176,000 by 1970 compared to 61,000 at independence. Annual growth in subscribers over the period 1957–1970 was only 8.5 percent. This can be attributed to the long delays experienced in the supply of cables and telephone equipment, the extensive reorganization of the department in the period immediately after independence, and the replacement of British expatriates by relatively inexperienced Malaysians as controllers and directors—which led to temporary declines in implementation capabilities of the department. The experience acquired by the new local staff, however, placed the department in a good position to expand in the post-1970 era.

8.3.2 International Telephone Service

A significant milestone for Malaysia in overseas communications was achieved in 1965 when high-quality cable telephone circuits were established to replace high frequency radio service. This was made possible by completion of the first stage of SEACOM. Telephone circuits were also established between Kuala Lumpur and Hong Kong and between Kota Kinabalu and Hong Kong. In 1966, SEACOM was extended to Guam to link with the Japanese-American Trans-Pacific Cable, making it easier to call Japan, the United States and the United Kingdom. Through SEACOM Malaysian overseas telephone service covered countries by 1970. International calling has increased from 23,000 calls in 1965, to 69,000 in 1969, to over 69 million in 1991.

8.3.3 Telegraph, Telex, and Other Services

Although international telex service was inaugurated in 1959, it was not until completion of the Kuala Lumpur–Kota Kinabalu and Kota Kinabalu–Hong Kong SEACOM cable in 1965 that they were widely used. Operations on radio telegraph and telex circuits were transferred to cable with radio circuits in Singapore serving as a standby. In 1968 the department began a program to renew teleprinters, many of which were obsolete. Many printers had been replaced by 1970, greatly improving system efficiency. A nationwide teleprinter broadcast network was installed for the Malaysian National News Agency. In 1970 the number of telex messages was 75,718 while telegrams numbered 1,172,182.

Pilot television service began in 1963, initially covering only the Kuala Lumpur area. However, this was quickly extended to practically all the west coast by 1964, and later to the east coast. By 1970 a satellite communications ground station was completed in Kuantan, making live telecasts from overseas possible. In addition, the department rented a transponder over the Pacific to provide transmission of television programs to Sabah and Sarawak from peninsular Malaysia, enabling live coverage of important events. Additional microwave radio links were also installed on the east coast, enabling improved tv transmission. By 1970, an estimated 70 percent of the population had good reception.

8.4 Telecommunications Since 1970

The 1969 racial disturbances in Malaysia led to the 1970 implementation of the new economic policy (NEP) with its two-prong objectives of eradication of poverty and restructuring of society to eliminate identification of economic functions by race; in particular, the policy sought to raise the economic status of Malays and other natives of the region, collectively known as bumiputras, to that of nonbumiputras, in particular, the Chinese. This ambitious plan was to be implemented in the context of a growing economy so no particular group would feel any sense of loss or deprivation. In an attempt to insure growth was actually forthcoming, the government undertook a strategy of accelerating industrialization by encouraging the inflow of foreign investments, as well as deliberate public sector interventions.

The ambitious economic growth and restructuring program required upgrading infrastructure, in particular facilities for telecommunications, as the hoped-for massive inflow of foreign investment implied demand would grow rapidly. However, the organizational structure completed in the late 1960s had not envisioned a dramatic shift toward export-oriented industrialization and services. Although the department attempted to expand its implementation capacity, it was still plagued by staff shortages, in particular of qualified technical personnel. To overcome this, many new recruits, particularly Malays, were hired and given crash courses at the Training Center.

Despite the increase in staffing, however, shortfalls in installation targets were still being encountered, and complaints from the private sector and the general public regarding the "inefficiency" of the department became commonplace. Numerous letters were published in the newspapers about the slow response to requests for lines, as well as delays in repairs and failure to deal with overcharging.

These problems were exacerbated by continued shortages of equipment and cables, and the complex process of land acquisition for telephone exchanges. Land is under the jurisdiction of the state governments. The department, being a federal department, must obtain approvals from state authorities for the sites where exchanges are to be built. Approvals take up to two years because the state authorities are generally reluctant to allocate land, as they do not derive any revenues for doing so.

Better coordination with the states was essential to overcoming such problems. In 1978, the department was placed under a revamped Ministry of Energy, Telecommunications, and Post and the responsibilities for "Works," including roads and other infrastructure, was taken over by a new ministry. An extensive reorganization of the telecommunications department was carried out with a view toward greater decentralization. The new organization was organized through the national headquarters, regional headquarters, and area offices.

The thirteen states were grouped into seven regions. More power regarding planning, implementation, and maintenance and operational management was given regional directors. The idea was to enable them to deal more smoothly and effectively with state authorities and to enhance the department's implementation process.

With greater regional autonomy, many new positions were created, especially at the regional levels. However, due to the shortage of qualified applicants, a large number of these could not be filled. At the end of 1981, of 35,423 approved posts only 26,965 had filled, a 24 percent vacancy rate compared to only 10 percent in 1980. Most vacancies were in technical areas, which severely constrained the expansion capability.

The inability of the department to meet the rising expectations and needs of a more industrialized economic structure, despite huge government allocations for provision of telecommunications, induced a complete reappraisal of the government's strategy.

It was generally felt the telecom sector was over regulated, so the department was not able to respond quickly. The 1950 Act empowered the department to be the sole entity in the sector, doing everything from purchasing and installing microwave transmission systems, telephone exchanges, underground cables, overhead cables, internal household wiring, and other equipment, to providing domestic and international telecom services. As a result of this strong monopolistic position, significant distortion occurred in pricing. While domestic calling rates were kept low, due to political pressures applied by the government, international calling rates were very high by international norms. For example, a three-minute station call to London cost M$34 in the early 1970s, about double the inbound rate. International telex calls were also high. Given the proximity of Singapore, where charges more closely reflected market prices, many businesses resorted to using Singapore as a rerouting center for international telex messages, causing a substantial loss of revenue.

In 1972, an amendment to the 1950 Telecommunications Act made the department operate on "commercial principals." All monetary transactions including revenue from services are now processed through the Telecommunications Fund, out of which the department's operating costs are paid. Any operating surplus is channeled to financing development projects. The amendment's objectives were to make the department self sufficient in terms of revenue for development, as well as to enable it to earn a targeted 6 percent return on its assets.

Instead of becoming self sufficient, the department had annual deficits of over M$1,000 billion by 1984. During 1976–1984 revenues increased fourfold, but operating expenditures went up five times and development expenditures increased eightfold. The result was a cumulative nine-year deficit of M$3.7 billion, covered by the government. The deficit would have been M$394 less had operating costs remained the same percentage of revenues as in 1976. Total development outlays over the period were almost M$6 billion.

The onset of global recession in the early 1980s, with its severe impact on government resources, led to drastic cutbacks in public sector spending. Cutbacks, however, had to be made in a way that did not affect the ability of the department to provide more and better service—which was considered crucial to the development of the country. Subcontracting was deemed appropriate.

In 1980, responsibility for internal household wiring was allocated to a large number of designated cable subcontractors. A major shift in policy toward suppliers of customer equipment was also announced. While the department had been the sole source and installer of CPE (including private telephone connecting equipment—PBX—telex systems, and cordless telephones), from 1980 several international manufacturers—including Ericsson and Fujitsu—have been authorized to supply and install end user equipment for the public. Their systems have to be approved by the department. Generally, however, the department approves anything compatible with its existing system, and has not been restrictive of the variety of offerings.

To improve procurement efficiency, the department decided in 1982 to select a number of manufacturers as long-term suppliers of switching equipment. It awarded a contract for the supply of digital switching equipment over the period 1982–1992—valued at M$1.23 billion—to two local–foreign joint venture companies, Pernas–NEC and Perwira Ericsson.

A major constraint on increasing capacity was the delay in getting cable routes completed. A decision was thus made in 1982 to subcontract cable installation, which is discussed later. In 1984 the government announced its intention to privatize the whole department, and this took place in 1987. The process is taken up later.

8.4.1 Service Growth

The major objective of the department was to reduce its long waiting list. During 1971–1975 numerous underground duct and cable routes were installed in major towns. This, however, did not solve the problem because there were still considerable delays in completion of the central office exchanges. In 1974 it was decided that temporary cabin exchanges would be imported from Japan as an expediency measure in certain areas, particularly in Kuala Lumpur, Johore, Ipoh, and Penang.

Crossbar main exchanges, introduced by Ericsson in the late 1960s, were found to be unreliable and limited in functions. In 1975 a strategic decision was made to phase them out in favor of newly developed electronic exchanges. The first was in 1976. This change facilitated introduction of international subscriber dialing (ISD) in 1979. Initially the service was available only to some exchanges in Penang, Klang, Kuala Lumpur, and Petaling Jaya. However, by 1989 ISD was available to over 80% of subscribers, enabling direct dialing to over 150 countries.

Progress in ISD was followed by similar progress in the expansion of subscriber trunk dialing (STD) within Malaysia. The 1981 completion of the Kuantan–Kuching cable allowed STD between the peninsula and Sabah and Sarawak. By 1985, the whole of peninsular Malaysia had STD. Similar progress was made in Sabah and Sarawak; it reached over 80 percent coverage by 1987. Introduction of STD not only improved service quality, it also drastically reduced the costs of making trunk calls, particularly between peninsular Malaysia and Sabah and Sarawak.

Malaysian Packet Switched Public Data Network (MAYPAC) and Malaysian Circuit Switched Public Data Network (MAYCIS) were introduced in 1984. These are separate switching networks designed for data communications using packet switching and circuit switching technologies, respectively. The following year, automatic telephone using radio (ATUR) was introduced, making Malaysia the first country in Asia to successfully launch the system nationwide. ATUR makes use of radio cellular technology and is particularly useful for residents in remote areas where installation of telephone lines is very expensive. Users are connected to the network via radio transmission through special stations and exchanges.

There has been a substantial expansion of the number of telephone subscribers, as shown in Table 8.2. Although businesses had been the main users,

Table 8.2. Telephone Subscribers and Backlog, 1970–1990

	Subscribers		Backlog	
Year	Number (in thousands)	Percentage Increase	Number (in thousands)	As Percentage of Subscribers
1970	176	—	14	7.8
1976	194	14.6	65	33.6
1977	228	17.1	76	33.6
1978	271	19.1	84	31.1
1979	325	20.0	106	32.5
1980	396	21.7	134	33.8
1981	489	23.5	150	30.7
1982	585	19.8	190	32.4
1983	700	19.6	200	28.5
1984	849	21.3	191	22.4
1985	959	13.0	332	34.6
1986	1,043	8.8	348	33.3
1987	1,132	8.5	297	26.2
1988	1,248	10.2	319	25.6
1989	1,336	11.2	347	25.0
1990	1,586	14.3	385	24.3

Source: Department of Telecommunications Annual Report, various years.

residential lines exceeded business lines by the late 1970s. By 1986 the ratio of residential to business lines exceeded 2:1. The telephone per capita ratio increased steadily, and by the early 1980s Malaysia was in the high telephone users group among developing countries.

Despite this expansion, a large demand for telephone lines remained. The backlog actually increased over 250 percent, going from 8 percent of existing subscribers in 1970 to a peak of 35 percent of a much increased base by 1985. Areas of unmet demand included new housing estates where cable routes and central offices were not completed on time, downtown areas where exchange line capacity had been fully exhausted, and rural areas where service remained economically unfeasible. In other words, almost everywhere.

The Telecommunications Department's objective was to have a network of 1.5 million telephones by 1985 and 2.4 million by 1988, with the ability to implement service within one week of application. The ultimate aim was to provide universal access by the year 2000. However, only about 1 million telephones were installed by 1985. Absence of private sector initiatives is considered a major factor in this. The department, known as STM after 1987, has a complete monopoly over the provision of telephone services. Rural subscribers, because of their sparsity, are financially unattractive customers. Rural per line installation costs are on average about two or three times those for an urban subscriber, and all users are charged essentially similar installation rates.

Data on revenues and expenditures of the department by urban and rural areas are not available. Discussions with the department indicated over 80 percent of its revenue is generated by urban subscribers; however, only about 60 percent of expenditures are incurred in urban areas. While rural subscribers contributed over 10 percent of revenue, they accounted for about 40 percent of total expenditures. Given the sociopolitical necessity of providing services to rural (primarily ethnic Malay) subscribers at "acceptable" rates, this urban-to-rural subsidy is perhaps unavoidable.

In 1991, in connection with the sixth (1991–1995) development plan, STM began a five-year M$5.5 billion (U.S.$2 billion) investment program. The intention is to almost double the number of subscribers, to 3.1 million, and have 25 percent density by 1995, with an emphasis on additions in rural areas. A domestic fiberoptic network and other equipment upgrades are also in the plan. Overall, the government expected M$11 billion to be spent on telecommunications during the plan period, compared to M$3.9 billion during 1986–1990.

8.4.2 Other Services and Competition with STM

Facsimile (called telefax) made its Malaysian debut in 1983. A 6,000 line SPC telex exchange was commissioned in Kuala Lumpur in 1982. It was installed as a local, national, and international switching center, with the last supplementing the existing system. In 1986, an international computerized telex exchange with 3,000 lines was completed. These developments led to a doubling of users from about 5,800 in 1982 to 11,200 in 1986.

Hutchinson Telecom of Hong Kong launched paging service in 1991, and thirty-two others have licenses. STM is not allowed to offer paging. Licenses have also been issued to twelve private mobile radio systems, and thirteen firms will be allowed to operate CT2 systems. Celcom began competing with ATUR (owned by STM) in 1989, gaining over 8,000 subscribers by mid-1990, compared to over 40,000 for ATUR. Celcom operates in the Kuala Lumpur and Johor Baru areas, while STM's system is nationwide. In pay telephones, Uniphone has been allowed to compete in urban areas. In February 1992 Skytel (M), a joint venture of Mobile Telecommunications Technologies, was licensed to develop a nationwide and international messaging system. Using Mtel's technology, alphanumeric messaging, voice mail, and transmission of international messages will be available.

8.4.3 Regional Collaboration

Regional collaboration in the supply of telecom services, particularly cooperation between Singapore and Malaysia, was born out of historical necessity. Singapore and Malaya have developed as a sociopolitical unit. During the colonial and pre-1965 independence periods priority was given to Singapore as a communications center for the peninsula. Although Singapore left the Federation in 1965, the basis for collaboration remained intact. Singapore is part of the domestic dialing network for Malaysia. Until the late 1970s, Singapore remained an important international dialing center for Malaysia.

Kuala Lumpur–Singapore is the most profitable route in the Malaysian network. The department consults its Singapore counterpart regularly on matters relating to changes as well as on the variety and type of services that can be offered over it.

The department also collaborates closely with Thailand on provision of telecom service between the two countries. In view of the close interaction between Penang and south Thailand, a special concessionary rate applies to calls between peninsular Malaysia and south Thailand. Special emphasis is given to insuring there is no congestion or delay in calls between peninsular Malaysia and Bangkok. Most of these calls move on a microwave trunk line up the Kra Isthmus.

Since early 1970 the department also interacted frequently with other telecom authorities in the Asia Pacific region, particularly those in Japan. Japanese manufacturers are major sources of technology, expertise, and financing assistance. This is clearly manifested by the many Japanese firms—ranging from cable and switching equipment manufacturers to makers of end user equipment—that have established subsidiaries or joint ventures in Malaysia. With the privatization of telecom services, their role in modernizing the system could become even more significant.

8.5 The Malaysian Electronics Industry

To provide perspective on the impact of international trends toward privatization on telecommunications in Malaysia, a discussion of the electronics industry, indicating the existence of a pool of domestic expertise in electronics and telecommunications, is appropriate.

Domestic electronics production began in Malaysia in 1966. Two years later, impetus for growth was given when the government designated it one of the priority, and premier, industries for investment incentives.

Part of the motivation for this was the realization that early starters such as Korea, Taiwan, and Hong Kong were likely to have their comparative advantage eroded by labor shortages, increasing wages, and higher costs. This optimism was also founded on the findings of a 1969 survey of the electronics industry in South East Asia by the Japan Electronics Industry Development Association. This found Malaysia possessed an abundant labor supply of generally good quality, subdued trade union activity, and low wages. A summary of findings on labor conditions is given in Table 8.3.

In 1970, when aggressive promotion of the electronics industry began, there were five joint-venture companies in operation. Four—Matsushita Electronics, Sanyo Industries, and Roxy Electronics Industries plus Toshiba—were with Japanese, one—Malaysian Lamps—was Philips. At that time, three more joint ventures had been granted approval. These companies were engaged in the production or assembly of television sets, receivers, radios, and other household electrical goods, as well as lamps. Employment in the industry since 1968 is summarized in Table 8.4.

Malaysia has emerged as a major exporter of integrated circuits and electrical appliances, beginning in the 1970s. In 1985, electronics contributed 15 percent of value added in manufacturing, up from 7 percent in 1978. An important feature of the electronics sector is its narrow scope. About 90% of the sectoral

Table 8.3. Labor Conditions in Selected Asian Countries, 1969

Condition	Japan	Hong Kong	Taiwan	Korea	Singapore	Malaysia	Thailand
Labor Availability	Shortage	Still available	Abundant, but	Abundant	Abundant	Abundant	Abundant
Quality of Workers	Good	Good	Good	Good	Good	Chinese are Good	Depends on training
Overall Average Monthly Wage (in 000 yen)							
	18–23	10–15	7–9	6.5–8	9–12	11–12	5–8
Electronics Industry Monthly Wage (in U.S.$)							
Minimum	50	28	19	18	25	17–30	14
Average	57	35	22	20	29	19–31	18
Maximum	64	42	25	22	33	22–33	22
Bonus[a]	3	1–2	1	1–2	1	1–2	1–2
Unions							

Source: Cheong et al. 1981.

For Malaysia, the higher electronic wages are for Kuala Lumpur area, the lower are for other parts of the country.

[a] Annual bonus in month's wages.

Table 8.4. Electronics Industry Employment (in thousands), Peninsular Malaysia, 1968–1990

1968	1973	1978	1983	1986	1990	
0.2	21.1	53.6	75.7	70.9	148.2	Total in Electronics
120.8	268.2	368.3	448.0	289.3	465.0	Manufacturing Total
0.1	7.9	14.6	16.9	24.5	31.9	Electronics as Percentage of Manufacturing

Source: Malaysia, Department of Statistics, *Industrial Surveys,* various years.
Includes radio and television sets, sound reproducing and recording equipment, semiconductors, other electronic components, and communication equipment and apparatus.

value added, and 86 percent of fixed assets, were in the components subsector in 1985. Other subsectors, such as manufacture of radio and television sets have been relatively unimportant. Electronics and electrical machinery and appliances rose from 48 percent of total exports of manufactured goods in 1980 to 57 percent in 1991.

In 1989 the country imported M$2.11 billion (U.S.$784) of telecom equipment and exported M$2.09, chiefly from foreign joint ventures. There were forty to fifty Malaysian-owned network and terminal equipment suppliers, with a combined annual output of M$50–100 million (*Far Eastern Economic Review,* Mar. 7, 1991, p. 43).

The government has given specific incentives to existing assemblers to upgrade their activities. In the late 1980s National Semiconductor established a wafer fabrication plant in Penang, and Intel seriously considered building another such plant. The Japanese have been less willing traditionally to transfer technology or otherwise do much more than provide ''women's work'' at their assembly plants, but this is slowly changing. Still, over the coming decade, electronics and computer-related components should be the country's major growth industry.

8.6 Turnkey Projects

The relative inefficiency of the Telecommunications Department as a government organization is demonstrated by its inability to clear backlogged requests for service. The heavy capital requirements building networks involves is another drain on the constrained revenue generating capacity of the government. As a result, since the early 1980s the government has been compelled to seriously examine privatization. (Although not explicit, it is assumed privatizing telecommunications would lead to an expansion of bumiputras participation in the corporate sector.)

In many respects the private sector has been involved in development of telecommunications in Malaysia for over two decades. Private corporations have subcontracted installation work and manufacturing operations since the early

1970s. In 1973 the National Trading Corporation, known as Pernas, formed a joint venture with NEC—Pernas NEC—to manufacture and install computerized digital switching equipment. Since then, Pernas NEC has extended operations to semiconductor components and multiplexing equipment. Also in 1973 Ericsson formed a joint venture known as Perwira Ericsson with the Armed Forces Fund Board to manufacture switching equipment. Since then Perwira Ericsson has diversified into manufacture of PABXs and mobile telephones.

In the early 1980s Marconi, a GEC-controlled company was awarded a contract to locally assemble pulse code modulation (PCM) transmission equipment. In 1986, Komtel, a subsidiary of Sapura Holdings, a major local telephone manufacturing concern, was awarded the contract to develop a paging system for the private sector using a numeric/alpha display system rather than a voice message system.

Seeking to overcome delays in getting the cable routes completed, it was decided in 1982 that the department's planning and installation of cable should be subcontracted. This was also intended to reduce the financial burden on the department and enable it to reassign its cable laying work force to other functions. Because the subcontractors are not required to follow government procedures, they can purchase required materials more rapidly. It was thought that this would accelerate construction.

The major contracts for the planning and implementation of outside plant network projects during 1983–1988 were awarded to four bumiputra contractors. All four were established by former employees of the department who had either retired or resigned to establish corporations specifically geared toward securing contracts. Each has a foreign partner.

The work was valued at M$2.5 billion (about U.S.$1.1 billion at the time) and sought to provide an additional 1.76 million telephone lines by 1988, which would have increased total telephone lines to 2.4 million. The plan was ambitious, and Datuk Leo Moggie, minister of energy, telecommunications, and posts, declared that, "By 1985 we hope to have 1.5 million subscribers. We have to install many more lines. We will move fast. We are convinced that there is no other way to achieve these targets but to get the private sector involved on a turnkey basis" (*New Straits Times,* Oct 15, 1983).

The contract was distributed to a spatial basis as follows (foreign partner in parentheses):

Binaphone (Philips): Kelantan, Terengganu, Negeri Sembilan
Electroscon (Ericsson): Pahang, Malacca, Johore, Kuala Lumpur (part)
Sri Communications (various): Penang, Kedah, Perlis, and Sabah
Uniphone (Sumitomo): Selangor, Sarawak, Kuala Lumpur (part)

Specifically, the aims of the contracts were to:

1. Provide cable pairs to enable the department to clear all backlogged requests by 1985.
2. Install a main cable network capable of meeting demand for the next five years.

3. Install a distribution cable network capable of meeting demand for the next twenty years.
4. Install a duct and manhole system capable of meeting demand for the next thirty years.

Effective coordination by the department was crucial to monitor the contractors and to insure that subscribers were connected with telephones provided by the department once cable work was completed. To achieve this, a project management group was to meet with the contractor in each region on a monthly basis. Overall monitoring was by a steering committee chaired by the minister.

8.6.1 Evaluation of Turnkey Projects

There has been much criticism of the turnkey projects. The contractors were selected without a public tender, leading to speculation inflated costs were shouldered by the government. More important, in 1991 over 310,000 of the lines (18 percent) had still not been completed and not one of the contractors had even come close to meeting the targets.

Each side blames the other. It is generally agreed that the major factor has been delays in approving plans submitted by the contractors, but there is contention as to who is responsible for this. All plans have to be approved by the department, and implementation bottleneck of the preturnkey era have not been overcome. The public works department and local authorities also have to sign off on plans. There have also been shortages of materials and manpower. In addition, contractor staff was relatively inexperienced, but this was somewhat overcome by 1985 when more foreign staff were recruited for design and implementation.

There has also been a basically negative attitude among department staff toward the turnkey projects. Those remaining in the department regarded their former colleagues at the contractors as a privileged lot who were essentially doing their old jobs at much higher wages. Employees were also conceptually opposed to the system, as it took away the glamorous part of their operations—planning and engineering. Most importantly, however, this attitude prevailed because the turnkey system had, in their eyes, obstructed promotional prospects: The department now no longer needs to expand. These attitudes are alleged to have resulted in deliberate delays in approving plans submitted by the contractors.

8.7 Telecommunications Department

Although privatization was debated widely from 1982, it was only formalized as a government policy in 1985 (see Malaysia 1985). However, it was clear as early as 1983 that the department was targeted as the first public sector organization for full privatization. This was consistent with emerging policy objectives to:

- Reduce the financial and administrative burden of the Government
- Promote competition, increase efficiency and productivity
- Stimulate private entrepreneurship and revitalizing the economy
- Reduce the involvement of the public sector in the economy
- Increase bumiputra participation in the modern corporate sector

Many objections to privatization were encountered among workers of the department. A study was conducted by the government on feasibility and implications. It recommended that the operational functions of the department be transferred to a corporation initially controlled by the government, but that this later be floated on the stock exchange so private investors would ultimately be the major shareholders. While the new corporation would be responsible for its own financial requirements, regulatory functions would be in government hands under a restructured telecommunications department. This followed the model adopted by the British in 1984 very closely.

Following this, the National Action Council recommended setting up a company wholly owned by the government. The proposal was formally approved by the cabinet on March 6, 1984. Subsequently, the Arab Malaysian Merchant Bank was appointed as lead financial adviser to the government. Its main tasks were to set up the terms of reference for the privatization process and to make recommendations on the transfer of the department's operations and assets to the new corporation.

Like many such exercises the world over, this was plagued by problems. There were issues related to asset valuation, organizational structure, and uncertainties over how to transfer workers without infringing on their legal rights as government employees. There was a rift between the workers and the government over the latter's proposal to transfer employees to the Malaysian Administrative Modernization Planning Unit and then reassign them, on a seconded basis, to the privatized company as a way to protect their privileges as government employees. The workers' union claimed that this contradicted the government's previous position, which had been to let employees decide whether they would like to stay in government service or join the new corporation.

Because of all these issues, it was only in July 1985 that bills on privatization of the telecommunications department were tabled in Parliament. By August 1985 the Telecommunications (Amendment) Bill 1985, the Telecommunications Services (Successor Company) Bill 1985, and the Pensions (Amendment) Bill 1985 had been passed.

The Amendment Bill effected the necessary changes to the legislation governing telecommunications and enabled the restructured department to become a regulator while assigning operational functions to the new corporation. The Successor Company Act contained provisions enabling the government to set up a company as the initial step toward privatization. The company, named Syrarikat Telekom Malaysia (STM), is held by the government through the Minister of Finance (Incorporated) Act 1957. The act also provided for compulsory employment by STM of staff who opted to join the private company and assigned the assets, rights, and other liabilities of the department to STM.

The Pension (Amendment) Act provides that government servants will continue to enjoy pension benefits even after privatization of their departments. This served to reassure workers of the other departments on the privatization list.

On January 1, 1987, the Telecommunications Department was formally privatized, and its operational functions taken over by STM, with Tan Sri Mohammed Rashdan Baba—formerly the chairman of Guthries Corporation, a large plantation company with wide connections to the manufacturing and service sectors—as it first chairman and Daud Ishak, the former director-general of the department, as its first executive managing director.

In its first year of operations, STM restructured tariffs on both domestic and international calls to more closely reflect relative costs. Domestic calling rates were increased by 30 percent while international call charges were reduced 15–50 percent. This did not result in the expected increase in the number of international calls. Installation and disconnection charges were raised, to the distinct displeasure of users. Subscriber-dialed international calls, per six seconds at the peak rate, cost from M$0.30 (to Thailand) to M$0.54 (to Japan). Because there is a high probability calls to Indonesia will not be completed, the person-to-person rate is a 71 percent premium over the usual person-to-person surcharge.

While services have generally improved, the length of the waiting list has not become shorter. Unimpressed by this performance, the government abolished the post of the executive managing director in September 1987 and expanded the post of the chairman into executive chairman with Baba becoming chief executive.

The corporation adopted a more market-oriented approach toward clearing the waiting list. Mobile offices were set up in all new housing estates to register potential subscribers with the aim of providing them service within twenty-four hours where there are existing lines, and within six months where existing lines were completely utilized.

Informal conversations with government officials involved in the exercise indicate several areas of dissatisfaction with the model of privatization adopted.

First, the 1987 transition merely involved transfer of a monopoly from one government entity to another. There was no private infusion of capital. For all intents and purposes, STM is just another public enterprise. Under such circumstances, the extent to which private sector culture and mentality can be instilled is doubtful.

Second, because STM retains a monopoly on the domestic and international network, it is unclear to what extent STM will be compelled to improve the efficiency of its delivery system. It is only in the peripheral areas of network services—such as public telephones, paging, and data base services—that third-party vendors are allowed. They must operate under license from STM, which severely controls the extent to which they can respond to market signals.

Third, no one has resolved the issues related to the trade-offs between profit motive and social responsibility. While focusing on increasing telephone density in urban areas, rural networks have been essentially neglected. There have been increasingly frequent complaints from rural users about STM neglect, and it is imperative that the corporation formulate and implement a coherent and just policy of cross subsidy between urban and rural subscribers. The ATUR program is a clear example. The initial rationale was that it would bring telephone services to isolated rural settlements that have not been linked to the microwave network. However, expensive installation charges for ATUR mean it is currently utilized only for business on car phones. (Installation charge for an ATUR telephone is M$6000, about 1.5 times 1987 per capita income.) Some form of cross subsidy program should be formulated to enable the truly needy isolated rural settlements—such as those in the east coast of peninsular Malaysia, Sabah, and Sarawak—to benefit from ATUR.

Telekom Malaysia was listed on the Kuala Lumpur stock exchange in November 1990, after 13 percent of the shares were sold to the general public. Stock also was sold to STM employees and to bumiputra institutions. As a result, the government (with the institutions) holds 81 percent of the shares. The company earned M$563 million (U.S.$203) in 1990.

8.8 Prospects

Malaysia will remain an open economy with increasing emphasis on private sector participation, particularly from foreign investors, in implementing economic growth. To be attractive to investors, Malaysia's infrastructure facilities, especially its telecommunication network, will have to remain among the best among developing countries. In the Fifth Malaysia Plan, 1986–1990, the sector was allocated a massive M$9.6 billion for capital expenditures. This is over and above what STM can provide.

With this extra allocation, telecommunications in Malaysia will continue to expand and improve. In the late 1980s services such as ATUR, facsimile, Datel, MAYPAC, and MAYCIS were introduced. These have met with good public acceptance and utilization. Further changes in services will be effected. The earth satellite station in Kuantan will be replaced by a digital fiberoptic submarine cable, upgrading the quality of international services. A Kuantan–Kota Kinabalu fiberoptic link was commissioned in 1989, and construction has begun. When completed it will give the country, in conjunction with a link to Manila or Hong Kong, broad access to the U.S. west coast and Japan. A new digital microwave network will be installed by the early 1990s, resulting in establishment of ISDN and Intelsat Business Service (IBS). The IBS medium can transmit digital voice, data, fax, and video teleconferencing at speeds of 5,600 to 2 million bps.

An ambitious program for provision of digital circuits on land was underway in 1992. Involved are 4 million lines, which will cost some $M2 billion (U.S.$780 million). When finished in 1997, 80 percent of the country's transmission network will be digital. These circuits are capable of transmitting up to 2 Mbps with lower error rates than 9,600 bps on existing analog circuits. To complement the digital distribution network, the microwave network will be upgraded at a cost of M$460 million. This five-year turnkey project has been described as the final phase in the department's master plan to extend, upgrade, and modernize domestic and overseas telephone services. The projects involve the cooperation of experienced international contractors. To the surprise of many five companies were each awarded 800,000 lines in March 1992. As with the earlier turnkey project, political and other controversy surrounds the plan. (See the discussion in Chapter 7.)

Malaysia's ability to attract foreign businesses and remain an efficient place to work and live depends on easy availability of telecom services. It thus is essential that clearing the waiting list and providing telephone, telex, and facsimile services within several days to those who want it be achieved in the foreseeable future. Steps to do this include better coordination between STM, its contractors, the restructured telecommunications department, and other authorities.

In addition to emphasizing improvement in the quality of services, the government should also extend its scope of coverage. In fact, expanding telephone coverage by the year 2000 should be made a primary goal. To do this the government could consider liberalizing the sector and allowing more third-party vendors into value added services. For example, specific regional companies could be given permission to provide services, particularly to rural areas, with the government providing fiscal incentives to enable companies to recover their costs and make a profit.

Whatever happens in Malaysia, the government is going to be involved. Real privatization in which companies can compete on price and service within broad regulatory bounds is simply not going to happen anytime soon, if ever. The perceived social necessity of protecting ethnic interests is simply too strong. Government still thinks it must make policy in very specific areas—if not indicative planning by a Malay elite with the omnipotence attributed to Japan's MITI in its heyday, then at least continued attempts intended to make Malays the economic equals of Chinese and Indians through a tilting of the playing field. Thus, as telecommunications is merged into the broader concept of information technology (IT), some see it as imperative that the Malaysian government formulate a comprehensive national policy specifying the strategy for the adoption and assimilation of IT into its manufacturing, plantation, and service activities.

For several centuries, Malaysia's rulers—whether British or Malays, sultans or civil servants—have assumed ethnic Malays cannot or will not help themselves economically: things have to be made easy for them. There is an emerging entrepreneurial class that recognizes this for the insult it is. Liberalization of value added services and other areas of telecommunications that have been shown elsewhere as capable of supporting many relatively small competitors just might give them the opportunity they really need.

Bibliography

Cheong, Kee Cheok et al. 1981. *Comparative Advantage of Electronics and Wood-Processing Industries.* Tokyo: Institute of Developing Economies.

Kualu Lumpor Stock Exchange. 1988 *Investor's Digest,* Kuala Lumpor (Apr.).
Malaysia. 1985. *Guidelines on Privatization.* Kuala Lumpur: Jabatan Telekom.
Malaysia, Jabatan Telekom. 1986. *Sejarah Telekommunikasi Di Malaysia* (History of Telecommunications in Malaysia). Kuala Lumpur: Jabatan Telekom.
Rita, Raj Hashim. 1986. ''Privatization of Telecommunications in Malaysia.'' MPA Thesis, University of Malaya, Kuala Lumpur.

9

Pacific Island Nations

MEHEROO JUSSAWALLA

The concept of a Pacific Telecommunity is rapidly becoming a reality with the emergence of a vibrant region of enormous opportunity in business and equally significant potential in the conduct of everyday life. Communication and transportation technologies have globalized commerce and are integrating the countries and remote islands of the Pacific into a vital, synergetic region.

Pacific island nations (PINs) occupy less than 550,000 km^2 of land (about the same as France) scattered over fifty times as much ocean. The combined 1991 population of the Polynesian, Melanesian, and Micronesian islands was around 6.3 million (somewhat larger than Hong Kong), 62 percent of them in Papua New Guinea, which means only some 2.4 million elsewhere (slightly fewer than in Singapore). Many of the islands are north of the equator, although all are commonly called South Pacific; tropical Pacific is more accurate. They range in size from Papua New Guinea (PNG) which represents about 83 percent of the region's total land area, to Nauru and Niue, which are single raised coral islands. They all have growing populations, limited arable land, balanced budget problems, export earnings typically dependent on a single commodity, and burgeoning foreign aid dependency. These development problems have exacerbated the growing restiveness among the young, whose rising expectations are unlikely to be met.

The 1980s saw the rise of more assertive and nationalistic Pacific Island polities. National government is a relatively new concept to these islands. In 1962 Western Samoa became the first to achieve independence; Vanuatu was the last, in 1980. The forms of government vary, but they are largely modeled after those of the former colonial ruler. The governments and residents of the PINs are struggling to chart their own destiny. However, as they are well aware that as at best minor players in a global economy the task is, indeed, daunting.

Pacific Island countries, through such regional and international bodies as South Pacific Forum (SPF), South Pacific Telecommunications Development Program (SPTDP), and International Telecommunications Union (ITU), and through attendance at annual conferences such as the Pacific Telecommunications Council, have made it quite clear that telecommunications improvement is a high priority. It is equally clear that because of limited resources, financial costs as well as opportunity costs present formidable obstacles requiring negotiation throughout the decision making process.

This chapter first analyzes the relationship between development and telecommunications, then takes up some of the common problems facing PINs, particularly financing development. The second section examines the existing structure and services of specific PINs, with special emphasis on the larger ones, Fiji, PNG, Guam and the Federated States of Micronesia. Restructuring the regulatory regimes of their PTTs is touched on in the third section.

The cost-effectiveness of satellites makes them a very important part of telecommunications for nations widely dispersed across numerous islands. The fourth section details the satellite options available and clarifies some of the supply-push evident in the region as the result of vendors from industrially advanced countries. The final section rounds up the discussion with a look at market expansion in the Pacific based on innovative technology and the changes in regulatory policy that such expansion entails.

9.1 Telecommunications and Development

The relationship between telecommunications and development has to be explored within the social, economic, geographic, and environmental context of the country for which policies are being devised. Numerous correlation analyses made in the 1970s suggest that telecommunications is a necessary part of development infrastructure and as important a component of investment planning as roads, power supply, and irrigation dams. The relationship between per capita GDP and teledensity was first demonstrated by Jipp (1963, pp. 199–201) and has since been known as Jipp's Law. While this provides a general insight into the relationship between economic well-being and telecommunications, correlation does not prove causality. However, most impact analyses have been based on Jipp's Law.

There have been studies in the 1980s to establish an interaction between telecommunications and other sectors that emphasized a specific production-function attribute of telecommunications. Among these, Gille's model (1986) provided a theoretical framework for examining the dynamics of telecommunications supply and demand. Originating from the work of Machlup (1962) and applied to the U.S. economy by Porat (1977), information sector analysis attempts to identify the contribution of the primary and secondary information sectors to gross national product (GNP). (The primary information sector comprises those industries producing or distributing information goods and services. The secondary sector refers to such activity *within* industries and government. Their contribution to GNP is, by definition of GNP, measured by their value added.) The impact of the information sector on the growth of 10 Pacific region countries has been examined by Jussawalla, Lamberton and Karunaratne (1988, pp. 15–63).

Although there is a correlation between the number of phones per capita and

GDP, the direction of causality (if any) is not clear. There are many other factors—such as management expertise, government and business organization, and adequacy of transport—that contribute to progress and prosperity. It is not easy to quantify the direct or indirect benefits of any telecommunications system. This is particularly true of externalities that are indirect and not related to the cost of investment. Studies of value-added by information-related activities to GDP in advanced countries indicate the contribution can be anywhere between 30 and 50 percent.

That there is a link between telecommunications and development is evident; the question is what it is and how it works. The benefits for education, emergency services, and social interaction that telecommunications confer are no longer questioned. The intangible benefits lie in shrinking distances, reducing the disparities between rural and urban areas, and enhancing the quality of life. Dramatic advances in information technology are opening up new opportunities for cultural and socioeconomic development unknown in the past.

Still, telecommunications is often considered a luxury compared to agriculture, water supply, and roads. This attitude is a blind spot in the planning process. The use of telecommunications in generating higher incomes and improving standards of living give it the characteristics of a public good capable of conferring direct and indirect benefits for society as a whole. For example, in PNG the Maitland Commission estimated 5 percent of telephone calls from rural and remote areas are for emergencies and medical reasons. In the South Pacific, the Peacesat network has been used to summon medical teams to deal with cholera outbreaks and to coordinate emergency assistance after typhoons. Losses incurred through lack of communication in some less-developed countries (LDCs) have also been estimated as being 110 times higher than the cost of providing adequate telephone service.

There is a clear-cut need for the development of economical thin-route (low traffic density) systems that will meet the urban as well as rural and remote communication needs of the islands. Developments such as demand assigned multiple access (DAMA) technology, very small aperture terminals (VSATS), and high-powered satellites designed for the Pacific may assist in making a thin-route system that is more efficient and complete and less costly in the long run. This is so because satellite communication per unit of message sent is cost insensitive over distance, and it is less costly than terrestrial channels for thin-route networks.

9.2 Generic Issues and Problems

Public telecom services in most PINs are provided by institutions that are part of government administrations or semi-autonomous enterprises. They generally have a monopoly for all domestic communication services and are responsible for government and regulatory functions such as licensing of private networks and radio spectrum management. Fiji has privatized.

The chief reason that these countries exercise tight control is the fact that while the fixed investment is high, marginal costs decline with additional subscribers, yielding increasing returns to scale. The multiproduct nature of the monopoly enables it to derive economies of scope and scale. This, in turn, means the rate of return to the economy as a whole is high, sometimes 15 percent or more. Unfortunately, the returns from these monopolies are used to subsidize losses in the postal system or other government agencies; the profits are not ploughed back into modernizing or refurbishing equipment, with consequent deterioration in services.

Local, long distance, and international telephone services account for 80 percent of the sector's investments and revenues. Telex and facsimile services are being added even as telegraph and HF radio are being phased out. Data transmission is catching on with the introduction of Intelsat SuperVista services. Most of the islands invest less than 1 percent of their GDP on telecommunications. In 1986 investment by PTTs in telecommunications as a percentage of GDP was Fiji, 1.12 percent; PNG, 0.95 percent; and Federated States of Micronesia (FSM), 1.4 percent. This is inadequate to meet demand, so new applicants having to wait several years to obtain a connection.

Facilities in the PINs are inadequate even for the low levels of economic activity that currently characterize them. Internal networks between islands and between coastal and inland areas are poor, and high-frequency (HF) radio systems that are still in use are constrained by infrequent contact schedules and unfavorable atmospheric conditions. Teledensity is low. In general, services are clustered in their urban areas. Table 9.1 gives an overview of the region.

Urban–rural dichotomy is quite apparent in the archipelagic island countries. The bulk of the population is in rural areas where there is a chronic shortage of telephones. The PINs, with a total regional density of little more than three telephones per 100 persons, have an average urban–rural dichotomy of approximately 5:1, 7:1 including PNG. PNG rural callers often must travel 7 km or more to reach a telephone. The PTTs are anxious to keep up with digital technology and the emerging ISDN that business communities in the major urban areas desire, but this is only a remote possibility. At the same time, the PTTs are pressed by their respective governments to look for low-cost solutions to meet the POTS needs of the bulk of their largely remote and scattered population.

The institutional setup prefers urban over rural networks and gives preference to those parts of the business sector that actively contribute to the economy, such as trade, utilities, banking, and government administration. In the mid-1980s nearly 85 percent of telephone calls and revenues were generated by production and distribution systems in the economy.

Following the 1973 meeting of the SPF and the establishment of the South Pacific Bureau for Economic Cooperation (SPEC), the interconnection of the metropolitan centers of the island countries to the international telecommunications network had been of paramount importance in order to maintain trade links with major basin partners.

In 1921 Fiji became the first PIN to be connected with the global submarine cable network. With development of microwave radio, other islands in the Pa-

Table 9.1. Pacific Islands Telephones, 1990

Population (in thousands)	Telephones	Density[a]	Wait List[b]	Government Ownership[c]	Country
18.5	2,540	13.7	245	60	Cook Islands
112.0	2,400	2.1	700	100[d]	Federated States of Micronesia
726.0	68,532	9.4	11,500	100[d]	Fiji
68.8	1,130	1.6	133	100[d]	Kiribati
42.1	1,193	2.8	1,500	25[e]	Marshall Islands
9.1	1,600	17.7	160	100	Nauru
2.3	390	17.2	—	100	Niue
3,600.0	73,068	2.0	1,491	100[f]	Papua New Guinea
329.0	5,976	1.8	130	60	Solomon Islands
95.8	3,984	4.2	680	100[d]	Tonga
8.6	150	1.7	60	100	Tuvalu
159.8	6,480	4.1	88	51	Vanuata
182.0	4,335	2.4	2,600	100	Western Samoa

Source: South Pacific Forum *Regional Telecommunications Report* 1991.

[a] Telephones per 100 people.

[b] Length of waiting list for telephone service.

[c] Percentage of telephone operations owned by government.

[d] The telecom authority has been "corporatized" to operate as a state-owned enterprise, generally as part of a process that will lead to privatization.

[e] In December 1991 75 percent of the Marshall Islands National Telecommunications Authority was sold to the public; the government retains a 25 percent share.

[f] The Post and Telecommunications Corp has been an "independent" body since 1982, but the PNG government owns all the shares and appoints all eight directors.

cific developed microwave links, but, for the most part, the availability of communications is proportionate to the distance from urban centers. By 1980 it had become apparent that development without a rural focus was accentuating an already exacerbated urban–rural dichotomy. Therefore, in 1983 the SPF established the SPTDP to upgrade national telecommunications using suitable satellite options and to coordinate the provision of reliable service to rural areas. SPTDP emphasized that undersea cables reach only metropolitan port centers while thin-route communication were consequently being neglected.

As telecommunications becomes more technologically complex and pervasive, it is becoming increasingly difficult to manage. This manifests itself in the lack of skilled local technical and middle and higher-management personnel in most PINs. These countries see a need to change this situation.

9.2.1 Financing

Financing is an important issue. There is a wide misconception among PIN governments that telecommunications is a lucrative business able to subsidize other government functions. Use of the telco as a taxing agency for capital rationing should be out of the question for these island nations. Their mission should instead be to provide basic universal service at low cost to remote areas and to pass this burden on to corporate users through higher tariffs on urban usage. In some of the PINs, even this causes income inequity since urban incomes are too low to bear the cost of subsidy. While privatization may help reduce the cash-cow role, it is unlikely to bring rural areas within the ambit of the networks without government intervention and subsidy support. The major need is to simplify the relationship between the PTT and the government and, as in the case of PNG, contribute to government revenues. The tax base is too fragile even for cross subsidies between classes of services. Taxing company or high-income classes of users is precluded because of their political support of the government.

A major hurdle to infrastructure development is the capital-intensive nature of the technology and the foreign exchange allocations required. One option may lie in sharing hardware costs such as the leasing of transponders on Palapa by ASEAN countries or leasing Intelsat transponders for domestic use.

A further problem is that a large part—generally at least 60 percent—of telecommunications infrastructure investment must be in scarce foreign currency. Export earnings are low and returns on telecommunications investment are received in local currency. Several factors help mitigate this challenge. Telecommunications generate wealth by contributing to economic activity. A dynamic rate of innovation in the industry rapidly drives down both fixed and operating unit costs. Equipment suppliers are aware that LDCs represent a potentially large market, which motivates them to offer favorable credit terms. Additionally, because markets in technologically advanced countries are characterized by surplus supplies, LDC buyers are in a strong position to obtain equipment at lower prices. Nevertheless, LDCs need concessionary financing.

For a commercially oriented operation, users are expected to pay their own way and contribute to general overhead and profit. This is possible in metropolitan centers and for international services, but not for satisfying the communication needs of rural dwellers.

A high percentage of the population in the Pacific Islands do not have access to basic telecom services, which should increase the relative value of new systems reaching them. Even so, the gap between the information rich and information poor may continue to widen simply because the problem is larger and runs deeper than economics. The gap is a function of the utilization of information flows, which in turn is determined by variables like history and culture.

One financial consideration is the fact that cost allocation to different services is a problem that defies solution, especially when a monopoly has to be sustained.

Intraregional trade should not be based on comparative advantage because the islands produce and export similar commodities. Their regional markets need to be integrated with global ones, a development necessitating a reliable communications system.

The major sources have been bilateral government aid, supplier credits, commercial banks, and multinational financial institutions. All of these are limited or come with strings attached. The Japanese government and hardware suppliers have shown interest in funding since the mid-1980s. Both the World Bank and the Asian Development Bank have been reluctant to lend for such technologies as satellite networks, earth stations, and broadcast transmitters. In any case, the World Bank allocates only 2 percent of its loans to telecommunications. Australia is a major contributor—it funds over half of PNG's investment budget. The 1987 coup in Fiji occasioned withdrawal of lending agencies based on fears of political instability and expropriation.

In the World Bank scheme of things about 45 percent of project financing has to be generated by the government or the borrowing entity and it expects the tariff structure in the borrowing country to generate funds for networks expansion. This is next to impossible in the low-income economies of the PINs, where incomes are not high enough to provide revenues for both expansion and cross subsidy even in urban centers.

Under the SPTDP plan, automatic exchange equipment is now being purchased for local networks along with small digital telex exchanges. The UNDP/ITU Manpower and Training Needs Program is providing each PIN with its training requirements for ten years (through the mid-1990s). It is arguably cost effective for PINs to invest in the latest digitized exchanges as this saves on manpower and maintenance costs. Current equipment is generally outmoded and obsolete, and was supplied at the expense of the former colonial government, so scrapping it is not a "loss."

Financial constraints on telecommunications development have also been mitigated by satellite technology. With appropriate organization, PINs can have economical, financially affordable networks. The only missing factor seems to be organization at the government and private sector levels capable of providing the financial and technical inputs needed to use the technology successfully.

Table 9.2 provides data in projects and programs planned in PINs during the 1990s.

9.3 Service and Structure in Specific PINs

This section looks at four PINs—Papua New Guinea, Fiji, Guam, and the Federated States of Micronesia—in some detail, and provides brief background on a number of others. Most PINs have PTTs as monopoly national carriers. They all use Intelsat through earth stations in each country for international communications as shown in Table 9.3.

9.3.1 Papua New Guinea

Located north of Australia and just below the equator, most of PNG is on the island of New Guinea, which it shares with Indonesia. This main island is divided from north to south by a high chain of mountains, giving its two halves vastly different cultures. PNG was granted independence from Australia in 1975. With a population of 3.9 million (July 1991) and a total area of 462,000 km^2

Table 9.2. 1990 Estimates of Expenditures on Telecommunications Projects by Pacific Island Nations to 2000*

Central Office	Outside Plant[a]	Outer Islands	Satellite[b]	Total	Country
740	1,000	2,567	5,600	24,907[c]	Cook Islands
—	—	—	—	41,000[d]	Federated States of Micronesia
3,507	3,987	4,543	4,857	16,894	Kiribati
4,790	6,045	3,796	4,600	19,231[e]	Marshall Islands
86	247	—	—	333[f]	Niue
1,570	864	341	1,570	4,345	Tonga
448	424	240	2,524	3,636[g]	Tuvalu
—	—	6,200[h]	3,200[i]	9,400	Western Samoa

Source: South Pacific Forum *Regional Telecommunications Report* 1991.

*Expenditures given in U.S.$ thousands.

[a]On main islands or in major urban centers; outer islands are in column 3.

[b]Facilities for international and domestic service. Some satellite expenditures are included as outside plant or outer island networks.

[c]Includes $15 million from the Asian Development Bank that will be used mostly for outer island facilities; a detailed breakdown is not available. Excludes compensation to Cable & Wireless for the takeover of its facilities by Cook Islands Telecom in late 1991; the amount is in arbitration.

[d]Loan from U.S. Rural Electrification Authority to FSMTC; breakdown not available.

[e]Includes $18 million loan from US Rural Electrification Authority.

[f]Excludes $510,000 being sought for various purposes, including expanding satellite station capacity from six to twelve international circuits.

[g]To 1998.

[h]Includes a 13-m Standard A earth station under Lome III (the EEC regional aid program), which in 1991 replaced a Std B station.

[i]Proposed project using a Japanese grant to provide 261 phone lines in fifty villages.

(slightly larger than California), PNG encompasses 1,300 islands. PNG has concentrated its telecommunications networks in Port Moresby, the capital.

Domestic operations are a state-owned monopoly of the Posts and Telecommunication Corporation (PTC). The PTC operates as a commercial entity with a ministry in charge of overall regulatory functions. As such, it makes money: For 1985, the return on investment was 9.1 percent with operating profits of 7.1 million kina (U.S.$1 = 1.0 kina in 1985, 1.05 in 1990; unless otherwise indicated, data are in U.S. dollars).

The prospects for liberalization or privatization are mixed. Demand in the thin-route areas is not backed by purchasing power so the beneficiaries of competitive prices are likely to be the urban elite and owners of mines and plantations. Expatriates are more likely to gain from privatization than are local inhabitants. This argues against privatizing PTC.

Services reflect early concentration in establishing a basic network that services mainly government (24 percent of telephones) and business users (47 percent). During the 1990s a surge in demand is expected from the rural, loosely organized cash crop sector. It is primarily involved in the cultivation of coffee

Table 9.3. Intelsat Earth Stations in the Pacific Islands

Country	Location	Type	Installed	Owner
Cook Islands	Rarotonga	Std B	1980	CITC[a]
Federated States of Micronesia	Yap, Chuuk, Pohnpei, Kosrae	Std B	1983	FSMTC[b]
Fiji	Wailoku	Std A	1975 (Decommissioned 1987)	
	Vatugaqa	Std A[c]	1987	FINTEL
Kiribati	Tarawa	Std B	1983	Kiribati Telecom
Marshall Islands	Majuro, Ebeye	Std B	1983	NTA[d]
Nauru	Yaren District	Std B	1975	Directorate of Telecoms
Niue	Alofi	Std D1	1989	Post & Telcoms
Papua New Guinea	Port Moresby	Std B	1985	Post & Telecommunications Corp (PTC)
Solomon Islands	Honiara	Std B	1975	Solomon Telekom Co. Ltd. (STCL)
Tonga	Nuku'alofa	Std B	1975	Cable & Wireless
Tuvalu	Funafuti	Std D1	1990	Telecoms Dept
Vanuata	Port Vila	Std B	1979	Vanitel
Western Samoa	Afiamalu	Std B	1980 (Decommissioned 1991)	
	Maluafou	Std A[c]	1991	Post & Telecoms

Source: South Pacific Forum *Regional Telecommunications Report* 1991.

[a] Cook Islands Telecommunications Corp (CITC) took over the operation of the international gateway from Cable & Wireless in 1991.

[b] FSMTC purchased the four earth stations, which were originally installed and operated by Comsat, in 1988.

[c] New 13-m Intelsat Standard A.

[d] The National Telecommunications Authority (NTA) purchased the two Comsat earth stations in 1987.

in the Highlands area of the main island, where the bulk of PNG's population resides.

The backbone of the network is a series of microwave links providing communications to the main urban centers. Consisting of over 100 towers with solar-powered repeaters on mountaintops throughout the rugged Owen Stanley Range, they are in many cases accessible only by helicopter. The towers are popular targets for attack by villagers with grievances or land compensation claims, as this is one of the few concrete ways they can express their anger with the government. The entire network is thus quite vulnerable. The towers vary in age from obsolete Italian equipment installed prior to 1971 to French equipment installed in 1985.

Telephone service is provided through fifty-two exchanges (1988), the majority of electromechanical designs are no longer manufactured and are thus difficult and expensive to maintain. They provide telephone service to nearly 30,000 subscribers, the majority in urban areas.

National and international STD are available to 97 percent of subscribers. Rural subscribers too remote to connect to the network directly are served by

HF radio telephone links through number of control centers that are manually connected to the network.

International telecommunications are provided by approximately 125 submarine circuits through a connection at Lae and seventy-four satellite circuits through an Intelsat B standard earth station located at Gerehu (near Port Moresby) that became operational in 1985 and was financed by an EEC loan under the Lome II Convention. PNG also utilizes the Intelsat station for direct telephone circuits to twelve domestic destinations.

Perhaps the most obvious change during the 1980s relates to the relationship between the government and the PTC. In June 1982 the then Department of Public Utilities was established as a commercial statutory authority whose functions and objectives were specified in an Act. The immediate impact of this was to free the corporation from the normally bureaucratic budgetary process. Assets and liabilities were transferred to PTC. Private sector expertise at management and board levels was secured, and the government limited its board representation to no more than two of eight members. In addition to its responsibility of providing service, the corporation was also required to operate more commercially.

While there were reservations about some of the elements of the relationship with the government, the impact on PTC has been quite dramatic. The "public sector" mentality has given way to more commercially oriented decision-making processes. This is reflected in profits before interest and taxes, which increased from 6.0 million kina in 1984 to a record 17.2 million in 1988. Increased profitability was achieved with a minimum of tariff increases, and these were below the inflation rate over the same period. Operating costs increased more slowly than revenues, and PTC was able to secure greater utilization of its assets. PTC's financial rate of return increased from 7.8 to 14.8 percent. PTC's capacity to undertake investments, secure commercial loans and service its debts likewise increased.

From nil receipts in 1983, the government received K32.5 million from corporate taxes, dividends and repayment of government advances over the next five years. Dialog between the government and PTC improved dramatically, especially with the submission of and agreements on a five-year development plan for PTC. A host of programs are contained in the plan, including village pay phones, digitalization, trunk and junction upgrades, mobile radio, and data communication.

The number of telephone subscribers grew from 25,179 in 1983 to 30,993 in 1988 while international subscribers increased proportionately from 10.9 to 12.6 percent. Nontraditional services were introduced—facsimile, satellite, mobile, data, and postal money orders.

PNG utilizes the Intelsat system to provide direct telephone circuits to twelve destinations via the Standard B earth station in Gerehu. Inmarsat facilities are available for five remote site applications using land-based, ship-type earth stations. These have been successfully used to support major infrastructure projects in remote centers such as the Misima and Porgera Gold Mine and the British Petroleum oil exploration.

During the early 1990s PTC plans to expand the use of satellite facilities to include rural applications through Pacstar. This project was developed in response to expanded user demand for regional and domestic communications services in the Pacific. A major advantage of the system is its use of highly concentrated spot beams and of smaller, inexpensive, on-premises earth stations at cheaper overall transponder cost for a given bandwidth. The system also uses C band that is largely unaffected by tropical rain and does not require ground stations to track the satellite. The Pacstar project was expected to be operational by 1991–1992, but remains only on paper.

9.3.2 Fiji

Fiji is the second largest of the PINS with 744,000 people (1991), not quite one-third of the non-PNG regional total. Among the PINs, it has a relatively diversified industrial base. It is able to supply textiles, paper, chemicals, and metals to the domestic markets. Sugar accounts for almost half of its exports; another 10 percent of export earnings come from seafood, and 9 percent from copra and coconut oil. Tourism is another major determinant of Fiji's prosperity. Indians who originally came to work sugar plantations slightly outnumber ethnic Fijians. Just before the 1987 coups, Fiji's eighth development plan (1981–1985) had provided a blueprint for coordinating national resource allocation with the objective of diversifying the economic base.

The Department of Posts and Telecommunications (DPT), under the Ministry of Works and Communications, handles domestic service. Fiji International Telecommunications Ltd. (FINTEL)—51 percent owned by the government and 49 percent by Cable & Wireless—has provided international service since its inception in 1976.

Fiji has the most advanced system in the region; automatic direct dialing exchanges are available on the main island of Viti Levu, and the outer islands are reached either by microwave trunk telephone or HF radio. Since independence in 1970, the major objectives of DPT have been the expansion and upgrading of telecom services to meet public demand and the improvement and expansion of services to rural areas. Nonetheless, the telephone system is antiquated and stretched to its limits. Moreover, it has only slowly increased telephone connections, resulting in a large unfulfilled, registered demand.

This is partly due to the high level of rehabilitation work necessary after typhoons—five in 1986 alone—and partly to the difficulty experienced in obtaining a suitable make of telephone to replace the model that was widely used previously but is now no longer manufactured. New systems have to be able to withstand sudden climatic changes and power failures.

Major developments envisioned during the 1990s are an annual investment growth of approximately 5 percent, which means an increase of 26,000 lines in exchange capacity, a 14,500 increase in subscribers to reduce the waiting list for telephone service, a 150 increase in telex subscribers, and the acceleration of development and improvement of services in rural areas. The increase in exchange capacity includes opening new automatic exchanges in urban and semiurban areas and the upgrading of the existing trunk network with the provision of alternative routes to avoid disruption in times of natural disasters.

DPT estimates 73 percent of the capital budget requirements will be generated from internal revenues while the remaining 27 percent will come from loan finance and development grants. However, this ambitious development plan has been put on hold due to the adverse economic conditions resulting from the military coups of May and September 1987. With a prolonged drought that adversely affected sugar planting, a 30 percent devaluation of the Fiji dollar, a decline in investor confidence, a substantial drop in tourism (Fiji's major foreign exchange earner), nonrenewal of expatriate contracts for skilled and managerial personnel, and the withdrawal of a majority of Fiji's overseas direct aid as an official expression of donor disapproval of the coups, developments will likely lag far behind projected growth rates and remain sluggish into the 1990s, although both tourism and sugar rebounded in 1989. (For more on this, see Jussawalla and Ogden 1989.)

9.3.2.1 International Connections

From 1963 to 1964, for technical and strategic reasons, the Pacific section of the Commonwealth round-the-world cable system, Compac, was required to have a shore-based repeating station in Fiji, adding to the island's long-established international links. Compac has given Fiji high-quality links to the rest of the world.

An Intelsat Standard A earth station was built at Wailoku in 1976. In 1987, this then-outmoded station was replaced by a smaller Standard A one in Vatuwaga, with an expected life of fifteen years. Both stations have been built, maintained, and operated by FINTEL.

FINTEL also operates the submarine coaxial cable terminal at Vatuwaga. FINTEL's largest single investment, $7.59 million, has been in the Anzcan trans-Pacific cable, which is owned by twenty-two organizations from fourteen countries with an overall installed cost of more than $300 million. The cable links Fiji to Australia, New Zealand, Hawaii, and Canada. Worldwide telecommunications are provided through direct circuits and utilization of several overseas switching centers. In addition to telephone, telex, and telegraph services, FINTEL also provides leased circuits, data, and facsimile (bureaufax) services.

9.3.3 Guam

Historically, Guam's economy exhibits a boom or bust pattern dependent on military spending levels and on Japanese government policies on trade, how much money its citizens can take abroad, and frequency of flights; hence, it is largely dependent on external revenue and political decisions outside its control.

The public sector employed approximately half the labor force from World War II until 1989. In view of cutbacks in federal spending and a desire to obtain Commonwealth status with the United States, which would eliminate many federal jobs, employment opportunities need to be created in the private sector, which, at present, appears unable to absorb any major labor shift. However, the basic underpinning of U.S. military expenditures remains strong, and Guam was in a growth phase in the late 1980s.

The Guam Telephone Authority (GTA) became an autonomous statutory agency owned by the government of Guam in 1974. GTA inherited the telephone system created by the U.S. Navy (which administered the island until 1950 when Guam became an Unincorporated Territory of the United States). It was inefficient, undercapitalized and in need of upgrading. Work commenced through the use of a $27.5 million loan form the U.S. Rural Electrification Administration (REA). The result was an 8 percent increase in the number of primary lines in use and a subscriber base of 28,973 (1987).

Due in part to poor planning and underanticipated growth in the Aganan and Tamuning areas, the exchange capacity of these government and business districts was quickly overloaded, resulting in a thirty-minute wait for a clear line into or out of these areas at times during peak business hours. In 1985 total revenues increased by 6.5 percent, but total installations decreased by 0.4 percent, with the reductions coming primarily from business and government lines.

The situation has not improved much since 1985, although GTA received an additional $24 million REA loan to fund projects to increase network capacity and versatility. Some of the planned developments in the 1990s include digital central offices in Agana, Tumon and Talofofo. Automatic number identification digital systems in all central switching offices and extensive line extensions into semiurban and urban subdivisions are also being planned. In anticipation of increased demand for high-quality transmission, a fiberoptic cable loop is planned to connect central switching offices.

Guam, due to its strategic position, has excellent worldwide telecom service through undersea cables (TPC-2&3, and the 40,000-circuit TPC-4 fiberoptic cable that became operational in 1989) and satellite. Overseas connections are provided by RCA Global Communications (purchased by MCI in 1987), which owns and operates the Pulantat earth station facilities, and by Island Telecommunications and Engineering Corp (IT&E), a discount resell service. RCA installed a $3 million international digital telephones switching system in 1985 to increase call capacity some 350 percent. When connected to the local digital system being installed by GTA, subscribers and RCA customers will be able to direct dial without entering authorization codes.

A proposal by McCaw Space Technologies to construct, launch, and operate the Celstar International Satellite System will help keep Guam a hub of international communications, as discussed later.

9.3.4 Federated States of Micronesia (FSM)

A report by the U.S. National Telecommunications and Information Administration (NTIA 1986) focused on the cooperative provision of telecom services to the islands of Micronesia by a nonprofit organization called Sky Channel (Pacific) that was based in Guam. In August 1986, Sky Channel was awarded a $2 million grant from the Public Telecommunications Facilities Program of the U.S. Department of Commerce and the NTIA to conduct a feasibility study for the improvement of broadcast services throughout Micronesia.

The approach taken by Sky Channel has been to provide broadcasting capacity as well as specialized point-to-multipoint interactive services for Micronesia. The needs determination conducted in the 1980s placed heavy emphasis on instructional uses and provision of health, social services and public safety networks via radio or television.

When it was part of the Trust Territory of the Pacific Islands (TTPI), the telecommunications system in FSM consisted of little more than a group of HF radio links between Saipan (the administrative center) and the outlying districts, supplemented by very small telephone exchanges in the administrative centers and unreliable radio links to the outer island. Dial telephones were introduced around 1970 and international and interstate calls were nearly impossible to make because of overcrowded lines and extremely poor transmission quality. FSM has about 108,000 people (1991) and 100 islands totaling 702 km^2.

The FSM Telecommunications Corporation (FSMTC) was established in 1981 as a public, statutory corporation and became fully operational in 1983 with consolidation of the telephone and telegraph subsystems in each state under its operating organization. FSMTC operates as the sole provider of telecom services within the FSM with the exception of the state-operated radio broadcasting stations. Telephone subscribers total only 1,350 (1986) and are restricted to the main administrative centers of each state.

The exchanges are in varying stages of degradation and in need of modernization, particularly in Pohnpei and Truk, where most of the population resides. Moreover, these exchanges are saturated, making it impossible to serve more subscribers even though there is a demand for additional lines. The exchanges are for the most part mechanical relay switches that are no longer manufactured. Parts required to maintain them are typically cannibalized from other units. In some cases, as in Kosrae, newer, electronic exchanges are in use. However, these units were designed primarily as hotel switchboards and are therefore not equipped with the operational functions essential for proper management, control, and service.

Services to almost all the inhabited outer islands are provided primarily by means of HF single side-band (SSB) radios. This system is divided into three state subsystems with control centers located at the Pohnpei, Truk, and Yap stations.

In 1982 an agreement was entered with Comsat for provision of Standard B satellite earth stations in Yao, Truk, Pohnpei, and Kosrae. The stations became operational in 1983 and linked the FSM to the Intelsat system, providing international connections through stations in Pohnpei and Truk. This has resulted in very high-quality interstate and international telephone service, although international STD is still not available.

Because the terrestrial systems connected to the satellite stations are of poor quality, FSMTC has developed and begun to implement a comprehensive system expansion aimed primarily at providing state-of-the-art service nationwide by the early 1990s. Major funding is being provided by an REA loan of about $41 million at 5 percent interest for thirty-two years.

The program is expected to expand the subscriber network to accommodate around 7,000 users served by thirty-one digital exchanges, including replacement of the four existing exchanges. Complete replacement of the existing terrestrial system is planned, including the installation of approximately 1,300 km of cable, 650 km of microwave carrier links, twenty-one satellite earth stations in the outer islands, and a large amount of digital radio transmission and subscriber premises equipment.

9.3.5 The Smaller PINs

In the Solomon Islands, Cable & Wireless provides international carrier service through 51 percent-owned SOLTEL. The government owns the other 49 percent. SOLTEL has been operating since 1978. Domestic service is provided by the PTT both through automatic telephone exchanges and HF radio telephone links to all the islands. There are domestic and international (via Sydney) HF radio telephone links.

Since 1984 Tonga has provided domestic services through the Tongan Telecommunications Commission, which is responsible for all domestic services and is part of the Telegraph and Telephone Department. International services are provided by Cable & Wireless under a ten-year franchise to operate the Intelsat earth station. The franchise expired in 1992. The government obtained satellite orbital slot approval from IFRB and in 1991 Tongasat was in business.

Vanuatu (the New Hebrides) Postal Department planned to develop a telephone trunk network in 1970 at a time when only two manually switched HF channels were available between Port Vila and Luganville. Its telecommunications requirements were met by a conglomeration of private, public and government low grade radio networks, mostly HF, developed to meet the needs of commerce, shipping and aviation and public requirements. The government hired W. D. Scott and Co. to provide a study of demand for the islands, and their report was submitted at the end of 1978. The spine link between Vila and Luganville has been upgraded and served by automatic telephone exchanges while HF radio transceivers are used for inland communications. Overseas radio telephone service is provided with links to Australia, Fiji, New Caledonia and, Hong Kong. VANITEL (Vanuatu International Communications Ltd.), established in 1979, is jointly owned by Cable & Wireless and Cables et Radio (France).

In Western Samoa, both domestic and international services are provided by the General Post Office (GPO). Automatic telephone services are supplied in Apia, the capital. HF radio is used to link New Zealand and American Samoa. The Afiamalu earth station is currently overloaded and is scheduled to be expanded.

There are significant disparities in domestic services available in the various current and former U.S.-administered islands. Capacity is adequate in American Samoa. In the former Trust Territory, except for the Northern Marianas, service is poor and falls short of current demand. This is especially so in the Marshall Islands, although it has included provisions for upgrading telecommunications in its long-term plans (OTA 1987). Telephone penetration per capita is low, and there exists a wide disparity between the services available in urban areas and those available in rural areas. Equipment is outdated and poorly maintained. In April 1985 a team from the U.S. government and telecommunications industry visited three of the four former trust territories—the FSM, the Republic of the Marshall Islands and the Republic of Palau—to evaluate existing facilities and to assess the requirements for telecommunications development of these islands (NTIA 1985).

9.4 Institutional Restructuring

The problems faced by these PTTs stem from lack of financial and administrative autonomy, while government ownership prevents tariffs from reflecting costs. On average, an additional telephone line requires a $2,000 investment, over half going to imported equipment. PTTs are denied permission to raise funds in capital markets despite the profitability of their enterprise. Government ownership often results in poor management, high operating costs and poor maintenance of equipment. On the whole, the gap between supply and demand for basic services continues to widen.

The South Pacific Commission (SPC) is well aware that shifts in the institutional setup are called for. A distancing of the PTT from the government and the introduction of private enterprise and mixed or joint ventures with the state will provide greater flexibility and efficiency. The real need has arisen because these islands are opening up to data communications with services like Intelsat Vista and SuperVista, Intelnet, and VSAT technology. Consequently, administrative procedures designed for government are not useful for managing technology intensive, dynamic services. If PTTs are still required to operate like petty fiefdoms, benefits of value-added networks and fiber optics to link rural areas will not be available to them.

In its long-range development plan the SPTDP wishes to avail itself of digitization and sophisticated systems. It has a sizable investment program that calls for greater autonomy and commercial practices for the organizations supplying these services. Such autonomy will enable them to operate more independently within a broad framework of government policy and regulatory guidelines. The main argument favoring autonomy is that the commercially operated supplier of basic and enhanced services will be able to reinvest in telecommunications rather than subsidizing other government agencies. Fiji set a lead in this direction by permitting 49 percent of FINTEL's shares to be owned by Cable & Wireless.

Such an institutional restructuring allows the monopoly supplier to concentrate on the larger task of improving and installing a countrywide infrastructure of basic services and mobilizing additional resources for expansion. Such a policy will generate higher incomes and employment and also help attract investment to remote areas and stem the tide of migrants to the urban centers.

In the PINs, there is an acute shortage of thin-route telephony. This reflects the fact a subsidy has to be worked into the tariff to provide service in remote regions. Charges should take into consideration the fact telephone tariffs must offer a viable alternative to HF radio, which has so far been used, even if not very satisfactorily.

9.5 The Role of Satellites

Satellite communications in the Pacific Region started with Intelsat's launch of Lani Bird in 1966. In 1989, there were twenty-five commercial satellites, excluding military systems, and more than eighty transponders serving the Pacific Basin.

Most operate in the C band frequency, but it is likely that newer generations will use the higher-frequency KU and KA bands. In general, the higher the frequency, the greater the capacity to carry information. The location of the satellite in geosynchronous orbit and its on-board power dictates the specifications of receiving equipment. Signal loss is greater for satellites using higher-frequency bands, while greater on-board power enables use of smaller, cheaper earth stations for transmitting, and receiving signals.

Pacific nations operate sixteen domestic satellites in addition to maritime satellites in the Inmarsat system. The Pacific Basin has become an ideal showcase for satellite communications. Flexible interconnections in satellite systems render the technology more cost effective than terrestrial systems, while providing the capability to reach geographically difficult terrain and remote islands at the same time. Intelsat uses circuit multiplication equipment so that low income countries can lease circuits at low cost.

Satellite communication is cost-insensitive to distance; earth stations can be located in remote regions because they are not dependent on extensive availability of power and transportation facilities and maintenance costs are lower than for terrestrial systems.

Within a satellite network, savings accrue when circuit capacity is shared among earth stations on a demand assigned multiple access (DAMA) basis. This means that when a transponder is leased, many earth stations can share the operation on a demand basis instead of having to pay dedicated or long-term charges. Such a system is useful for linking remote thin-route areas where traffic is low. Compared with terrestrially carried long-haul systems, satellites can accommodate communication nodes or usage points anywhere within a region to meet changing traffic patterns.

9.5.1 Intelsat

Intelsat has fifteen satellites in its global system, providing two-thirds of total overseas telecom services and linking 170 countries and territories throughout the world. Internal long distance in twenty-seven countries use leased or purchased capacity from Intelsat. Each year the system handles about 1 billion calls.

There are certain characteristics of the Intelsat network that make it attractive to PINs. It provides global interconnectivity by providing 1,700 earth station-to-earth station pathways, even though over half of these collectively generate only 10 percent of Intelsat's revenues. In other words, low-income users benefit to a greater extent from this interconnectivity.

Thin-route services have gained international attention during the last decade as being particularly necessary for the island nations of the Pacific. Intelsat currently permits the use of small earth stations that are very different from its Standard A and Standard B stations, which call for investments the PINs cannot afford. Stations range in diameter from 75 cm to 4.5 m and cost from $2,500 to $250,000. To assist development of infrastructure in the Pacific Basin, Intelsat has devised several services for small antennas that are not as financially burdensome for these island nations.

The single channel per carrier system (SCPC) first evolved in the international system of Intelsat and was then applied by Intelsat to provide low-cost communications for domestic use. It has been widely used by the Palapa system for difficult-to-reach areas that face problems of climatic severity. Intelsat's domestic leases provided 5,000 SCPC systems in 1986, most of them for thin-route communications.

It can be claimed that Intelsat is the most effective conservator of the Clarke (geosynchronous) Orbit, thereby helping LDCs in two ways. First, they may find it redundant to stake a slot in space and to compete for a suitable one when they may not be able to afford its use in the near future. Second, they can get the same benefits of satellite communications for domestic and international uses at a much lower investment than if they had to commission a satellite for using their allotment of the orbital arc, as Brazil, India, Indonesia, and Mexico have done. The economies of scale and scope available to Intelsat enables the system to offer services at declining unit charge.

The most useful service provided by Intelsat is Project SHARE (Satellites for Health and Rural Education). It was started in 1984 as a sixteen-month experiment during which Intelsat donated free use for health care and education. It can be of great value to the social and economic development of small, growing economies that want to integrate their rural and urban areas and reduce the constant flow of migrants to metropolitan centers. It is used by the University of the South Pacific, founded in 1968 in Suva (Fiji), for audio links to campuses on other islands.

9.5.2 ATS-1 and Follow-Ons

There are many competing vendors of satellite systems to the Pacific Islands. The locals were already conversant with satellite technology and tele-education from more than two decades ATS-1 use. The ATS-1 Satellite linked Peacesat at the University of Hawaii with several of the PINs until August 1985, when the experimental satellite, loaned by NASA, went out of orbit.

Since 1985 SPEC has been examining several alternatives for a replacement. One alternative has been to shift NASA's Geosynchronous Orbiting Environmental Satellite (GOES) 2 and 3 to 162° East. This option is beset with various bureaucratic procedural hurdles, as GOES is administered by National Oceanic and Atmospheric Agency (NOAA) and the antennas would have to be realigned to service South Pacific Islands.

Another replacement option is use of the tracking data relay satellite system (TDRSS). The satellite's current use by NASA is for collecting weather and scientific data relayed to earth stations mostly located in the United States. The advantage of TDRSS is that it is equipped with moving antennas and three satellites that provide links with twenty-four other satellites. Each TDRSS satellite carries seven communications antennas, four of which can be steered by ground control. In the event of torrential rainfall, these can be moved from island to island to reduce the loss of signals. A 1986 study done by Westinghouse found that the satellites available for use in the Pacific—GOES 2 and 3—are not capable of performing meteorological functions and could be used for data exchange only.

A third possibility is the Pacific Marisat satellite. The islands would be treated as ships anchored in harbor if Marisat service were used and good quality communications would be available at reasonable cost for the ground segment. The disadvantage is that Marisat may not be a good channel for video transmission for television purposes.

9.5.3 Other Ventures

Japan is emerging as a leader in space technology in the Pacific. In 1977 the Japanese experimented with direct broadcast satellites (DBS) by launching the Sakura system. In 1983, CS-2A and CS-2B were launched carrying signals in the C Band. In 1985 the Japanese Satellite Company placed an order with Hughes Communications for construction of a private system called the Jcsat consisting of KU-band satellites for voice, television, and data using small antennas. Japan had captured half of the $1.5 billion market (cumulative) for ground terminals by 1985. It has also designed a launch vehicle called the H-1, a rocket using an indigenous guidance system with a liquid fuel engine. Its fully indigenous rocket, the H-11, is under development.

Since 1982 Japan has been contemplating a regional satellite for the Pacific Islands. The Research Institute for Telecommunications and Economics (RITE), together with Mitsubishi Electric, has conducted studies. Their focus has been on rural and remote thin-route needs for voice, video, and high-speed data. Coverage should not include the ASEAN countries as they are already linked to Palapa.

Aussat is a major player in the developing part of the Pacific. This Australian system has satellites at 156 to 160° East and is made up of three identical KU band satellites accessible by a wide range of earth stations. The South Pacific

Islands may be able to lease transponders for their intra- and interisland requirements, and may choose do so depending on how cost effective AUSSAT is, as well as the quality of the service offered. Whether existing earth stations can be used with Aussat or whether a new network is needed depends on the availability of services from competing vendors and the types of technology offered. OTC of Australia has a hub in Sydney that Pacific Island's VSATs can use for links with Intelsat.

Another emerging player on the Pacific scene is McCaw Space Technologies, based in the United States, where its parent is the major cellular phone player. It has plans to introduce a separate system consisting of two RCA satellites in the KU band: Celstar I, positioned over the Pacific Ocean at 170° East, and Celstar II, over the Indian Ocean at 70° East. The system intends to provide international voice, video, and data satellite services on a selective, noncommon carrier basis (not connected to the public switched network) for business users on either a sold or long-term lease basis. On-board matrix switching circuits will be able to switch any uplink channel in any beam to a downlink channel in any beam. A Guam facility will be used to control telemetry on both satellites as well as to double-hop traffic from one Celstar satellite to the other.

Pacstar is the satellite system that PNG is contractually locked into. ITU, under its present rules, requires that applications for orbital slots be made only by sovereign states. As a commercial organization TRT (now Pacific Telecom) arranged to route the Pacstar application through the PNG government. Pacstar is designed to offer domestic and regional communication using two satellites. One of these will cover Fiji, the Solomons, Japan, Southeast Asia, and PNG. The other will extend to Hawaii, California, and French Polynesia. PNG will receive capacity on one satellite free of charge, but it will not own the system. Pacstar was planned for launch in 1991–1992. However, it has not materialized and PNG leased a transponder on Palapa in January 1991 for its domestic services. If it becomes operational, Taiwan is interested in using Pacstar. Pacstar's failure is one example of how vendors can mislead developing countries.

9.6 Conclusion

The PINs face the dilemma of whether industrialization should precede telecommunications networks. Investment in telecommunications has to marry economic return to social returns, which is difficult to achieve under current conditions of foreign exchange scarcity. It is not acceptable for a single foreign supplier to take over the markets on a regional basis because of political and cultural conditions. Even so, for historical reasons, Cable & Wireless is the largest single supplier of services. There is no production of telecom equipment in the PINs; it is mostly purchased from funds received from overseas direct aid. The area's major suppliers and investors are from Australia and New Zealand. Mining is present in PNG, but it has not improved the living standards of people in the peripheral regions.

PTTs are finding the benefits of natural monopoly eroding as a consequence of new products and services, rising R&D costs, and shorter cycles of innovation. Fiji's domestic telephone monopoly and the PNG PTC are changing the rules, permitting greater play of market forces. Bureaucracies justifying themselves on grounds of customer protection, and reflecting and effecting static organizations and rigid structures, must address change. Still, despite the movement toward deregulation elsewhere, PIN PTTs are still powerful, and companies seeking to increase their share of Pacific markets will have to deal with them.

The only technology that allows room for private ownership is VSAT, if they are available at low cost and can be owned and operated by rural communities. Because they are digitized, maintenance and operation costs of these receive-only earth stations are much lower than Intelsat's A, B, and D stations. They could initially be used for data transmission and for information on health, agriculture, fishing, disaster warning, and relief to the rural areas. Voice and video channels can be introduced at a later stage.

Cable & Wireless' influence in the Pacific is being challenged by France Telecom. British Telecom, Pacific Telesis, and AT&T are all vying for developing-country markets. This gives decision makers in those countries wider choices of products and prices, and it spurs suppliers to offer credit. If this trend continues, governments will face increased pressures for open access, and PTTs will find their domestic markets attractive to foreign equipment and service providers. Consequently, PTTs may become more protective of their home markets and tighten regulatory pressures. This resistance will result in negative externalities for domestic business users and individuals. The internal dynamics of all this point to a reduction in the monolithic powers of PTTs.

The major crunch comes from financing. Bilateral government aid, supplier credits, commercial banks, multinational financial institutions and, in some cases, countertrade are relied on. It is difficult for the countries reviewed in this chapter to obtain concessional funding and loans for telecommunications on preferential terms. Intelsat's Vista and SuperVista, Intelnet, and Project Share are options being used. U.S. overseas development funds and Australian aid have provided relief, along with REA grants in Micronesia for network development.

The problem of financing is aggravated by rigid control over supplies of services by the public sector. For the Pacific Islands, the problem is being solved by pooling resources into the SPTDP. Training requirements are being met by ITU and the United Nations Development Program (UNDP). The fact remains that all purchases of equipment have to be made in foreign currency, whereas the rate of return on investment accrues in local currency. This makes it difficult for LDCs to meet loan requirements and simultaneously set aside funds for expansion.

The one advantage is that suppliers competing for larger shares of the global market are offering favorable credit terms to the island nations. Pacific Satellite offered a free transponder to PNG on Pacstar, Australia is offering reduced tariffs on Aussat III, New Zealand offered a free television channel to the Cook Islands, and Japan has offered free use of its weather satellite, Hemawari, to

Fiji for hooking up the University of the South Pacific to sites off its main campus.

The future promises a number of innovative options and heralds a degree of competition in services hitherto unknown. The opportunities are exciting and the risks are considerable. Because the geography of the region is island-based, there is a wide divergence in economic development and natural resources, and there will be risks involved.

Bibliography

Gille, Laurent. 1986. "Information, Telecommunications, and Development." In International Telecommunications Union, *Growth and Telecommunications*. Geneva: ITU Publications, pp. 27–61.

Jipp, A. 1963. "Wealth of Nations and Telephone Density." *Telecommunications Journal,* pp. 199–201 (Jul.).

Jussawalla, Meheroo. 1984. "Socioeconomic Policy Issues of New Telecommunications Technology." In S Rahim and D Wedemeyer, eds., *Pacific Telecoms*. Honolulu: Pacific Telecommunications Council, pp. 111–19.

Jussawalla, Meheroo. 1987. "Papua New Guinea: Taking Time to Decide" *Intermedia* 15 (4/5): 63–65 (Jul.–Sep.).

Jussawalla, Meheroo, Lamberton, and Karunaratne. 1988. *The Cost of Thinking: A Study of the Information Economies of Ten Pacific Region Countries*. Norwood, NJ: Ablex.

Jussawalla, Meheroo, and Michael Ogden. 1989. "The Pacific Islands: Policy Options for the Telecommunications Investment." *Telecommunications Policy,* pp. 40–50 (Mar.).

Machlup, Fritz. 1962. *The Production and Distribution of Knowledge in the United States*. Princeton, NJ: Princeton University Press.

Maitland Commission, Report of the Independent Commission for World Wide Telecommunications Development. 1985. *The Missing Link*. Geneva: ITU. (The Commission was established by the ITU and is generally called after its chairman, Sir Donald Maitland.)

National Telecommunications and Information Administration. 1985 "Reports on an Assessment of Telecommunications Requirements in the Islands of Palau, FSM, and the Marshall Islands." July. Processed.

Office of Technology Assessment. 1987. "Study on Telecommunications in the U.S. Pacific Islands." Processed.

Ogden, Michael. 1988. "Issues and Interest Groups in Pacific Island Nations." An unpublished paper prepared for USIA.

Pelton, Joseph, and Patrick McDougal. 1986. "New Services and Regulatory Environment: The INTELSAT Global Satellite System in Transition." Atwater Conference on the Global Information Economy: Risks and Opportunities." Montreal, Canada, Nov. 3–7.

Porat, Marc U. 1977. *The Information Economy in the U.S.*, vols. 1 and 2. Washington D.C.: Department of Commerce.

10
The Philippines

THOMAS G. AQUINO

The Philippines is one of the world's few major archipelagic nations. As such, the ability of a government to wield influence throughout the islands has depended to a large extent on its communications network.

There were thriving commercial centers long before Ferdinand Magellan arrived in 1521, but precolonial political units were small. There are still some eighty-five indigenous languages and dialects. The Spanish brought macropolitical unification of the islands and Roman Catholicism—and stopped the spread of Islam, pushing it back to the southern islands. In 1898 the Philippines declared independence, but the United States assumed control from 1900, following the Spanish–American War, until 1942. A second republic was founded in 1946.

A legacy of Spanish rule is the view that public utilities are probably better off under government direction and ownership, still a prevailing view in most of continental Europe. Hence, it is not surprising to find government in the business of providing telecom services—or at least willing to provide them. In contrast, U.S. rule introduced the idea of private entities providing services. The interplay of these philosophies continues to influence the development and structure of the Philippines' telecom sector in important ways.

10.1 Economic and Political Background

About 95 percent of its 66 million people (mid-1991) and land area of 300,000 km^2 are on eleven islands; Luzon and Mindanao together have almost two-thirds the land area. However, all told, there are over 7,100 islands stretching some 1,800 km along the coast of Southeast Asia. The country is divided into thirteen regions, and has seventy-three provinces and sixty cities. Filipinos are of Malay–Polynesian descent, although they have mixed with Chinese, Spanish, and American peoples. Pilipino (a form of Tagalog spoken natively in the Manila region) was made the national language in 1939, while English is the language of commerce and industry; little Spanish remains.

An economic recovery began as an offshoot of the urban, popularly supported, and bloodless uprising that brought Corazon C. Aquino and a new set of national leaders to power in elections held in February 1986. The uprising brought a complete change in the political order. Constitutional authoritarianism was replaced with democracy, a government similar to that provided for by the 1935 Constitution (which had been drafted as part of preparation for independence during the period of U.S. rule.)

Even while a new economic plan was being prepared, several policy initiatives designed to restore dynamic market forces were undertaken. Monopolies in the sugar and coconut industries were abolished, import liberalization was pursued, and privatization of publicly owned corporations was adopted.

Numerous challenges confront the Philippines, many of which affect the nature and extent of telecom services. Because of the capital-intensity of modern telecom equipment, one of these challenges is the nation's huge external debt, U.S. $28.4 billion in 1990. It continues to be a heavy burden and a drag on growth. Principal and interest payments claimed almost one third of export earnings in 1988. Restructuring of production sectors to make them more export oriented has been routinely delayed by intractable problems related to inefficiency, ineptitude and corruption in government.

Income distribution also continues to be a serious problem. About 55 percent of households, some 32 million people, live below the poverty line of approximately U.S. $140 in monthly income. In 1985, 10 percent of the population had income more than fifteen times that of the poorest 10 percent. On average rural incomes are less than half that of urban ones. Insurgency is a serious concern in certain areas of the country. Unemployment, a major cause of the discontent that spawns insurgency, is being gradually alleviated by improvement in domestic military capability and positive developments on the economic front.

GNP of U.S.S.$45.2 billion (1990), about $700 per capita, grew at 5–6 percent in real terms in the late 1980s, but it had slowed to 2.5 percent in 1990. Consumer spending remains strong, sparked by wage adjustments and renewed government spending since 1986. Inspired by investor confidence, the economy continues to register favorable developments so direly needed to sustain growth and change, although growth was not as rapid, and inflation was higher, at the end of the decade than it was earlier (see, e.g., Cruz 1989).

Economic recovery has placed the telecom sector under close examination. As a facilitator in the exchange of data and information needed in the movement of goods, people, and ideas, the sector and its potential have become a subject of interest among users and policymakers aware of its role in fostering economic development.

10.2 Telecommunications History

There were well-established lines of foreign as well as domestic communications under the Spanish. In the latter part of the nineteenth century, regular mail service between Manila and Barcelona was provided by Compania Trans-Atlantique, a steamship line subsidized by the Spanish government. There were ships each way at least once every four weeks. Other lines provided links with North America, other parts of Europe, and points in between, as well as Australia. Interisland mail to the main government centers was carried by La Compania Maritima. Four mail lines were established by the Spanish government to cover North Luzon, South Luzon, and the southeastern and southern islands.

In 1872 the first telegraph line was installed by the colonial government to link Manila and Cavite, a province southwest of Manila. Eventually, three main lines emanated from Manila to the ends of Luzon. The northwest line followed the highway toward Laoag in Ilocos Norte; the second line ran northeast along the highway to Aparri in Cagayan; and the southern line ended at Sorsogon. There were forty-nine telegraph stations on Luzon, nine on Panay, four on Negros, and three on Cebu, connected by 2,818 km of line.

In 1880, a submarine telegraph cable was installed from Hong Kong to Bolinao, Pangasinan. A land line connected it to Manila, some 250 km away. (For more on this see Santos 1986, pp. 25–27.) This link was operated by Eastern Extension Australia and China Telegraph Company Ltd. The same company was granted the concession to lay three submarine cables connecting Manila to the Visayan trading centers of Panay (Capiz and Iloilo), Negros, (Bacolod) and Cebu.

10.2.1 The U.S. Era

Six months after the Visayas cables went into operation, war broke out between the United States and Spain. The Hong Kong cable was cut during the siege of Manila Bay and remained inoperable until the end of the hostilities. When the United States assumed control of the archipelago, Eastern Extension restored operation to both its interisland and overseas cables.

The government telecom service centered its operations on the telegraph network established by the American occupation forces. In 1906, a government reorganization transferred administration of telegraph services from the Philippine Commission (the highest governing body in the colony, composed entirely of Americans) to the Bureau of Posts. Meanwhile, mail service was expanded throughout the archipelago.

Due to the increased American interest in the archipelago and Asia, facilities for communication expanded considerably. The San Francisco–Philippines submarine telegraph cable project began in 1901 and was placed in service in 1903. The Manila-based Philippine Islands Telephone and Telegraph Company was started in 1905, serving 500 subscribers; Cebu Tel & Tel followed in 1914, among others (see Arce 1986).

With development of the country's internal communications system, the interisland submarine cables of Eastern Extension were closed in 1917. On November 28, 1928, the Philippine Legislature granted a fifty-year franchise to a new corporation called the Philippine Long-Distance Telephone Company (PLDT), which was run largely by American management and whose majority stockholder was British Columbia [Canada] Telephone. By 1930, PLDT had acquired the assets and franchises of the various local telephone systems. It initiated long-distance service to various parts of the country, starting with service between Manila and Iloilo in 1931. Two years later, overseas radio-telephone service was established between the Philippines and the United States and other parts of the world. There were approximately 28,579 telephones in place after the PLDT's first decade of operations.

10.2.2 Postwar Era

Immediately after the war, U.S. military authorities took over a system in which only 10 percent of the original facilities were still operational. In 1947, operations and maintenance of the telephone exchanges were turned over to the newly created Government Telephone System of the Bureau of Telecommunications (Butel). An Executive Order implemented as part of the 1947 government reorganization abolished the Bureau of Post's Electrical Commission Service, which was among other things the implementing agency for telecommunications, and the telephone system was returned to PLDT. By 1949, the company had restored service to 12,000 telephones and reopened its exchanges in the south. During the 1950s, independent telephone companies sprang up outside Manila and were interconnected to the toll network of PLDT.

It was not until 1953 that PLDT recovered its prewar level of 33,712 stations. After 1953, PLDT started converting provincial manual exchanges to automatic or dialing systems. In 1956 PLDT's majority owner, British Columbia Telephone, sold itself to General Telephone and Electronics (GTE). GTE began divesting shares to Filipino investors in 1967 because of the slated 1974 expiration of the Laurel–Langley Agreement, which had granted the US trade concessions for an eighteen-year period beginning in 1956 (see Taylor 1964, pp. 208–19).

In 1964, the Philippine–Guam Submarine Cable System (TPC-1), spanning 1,468 nautical miles, was landed at Baler in Quezon province. The initial capacity of the coaxial cable was 128 message channels, including telephone, telex, telegraph, and data traffic. Installed and operated by PLDT, it was the first system to bring high quality voice circuits to the Philippines.

Satellite capabilities were added to the country's telecommunications system in 1967 when Philippine Communications Satellite Corporation (Philcomsat), a joint venture between the government and private investors represented by the Philippine Overseas Telecommunications Corporation (POTC), installed and operated a portable satellite station. This was later replaced by a Standard A Station connecting to a Pacific Ocean satellite. In 1970 it was complemented by another A Station that connected it to an Indian Ocean satellite. Philcomsat vested POTC to manage and operate its earth station communications facilities (646 circuits). The charter of Philcomsat as a carriers' carrier, not regulated by any government agency, covers its operation and maintenance of earth stations and leasing of satellite circuits only to international common carriers.

In 1968, PLDT initiated a service improvement and development program budgeted at 700 million pesos (U.S. $180 million at the time) over a period of four years. All remaining manual provincial exchanges were subsequently converted to dial operation. Also, PLTD's microwave toll network was extended to the southern island of Mindanao. At that time, 184,782 lines were in service.

In 1969, the first domestic tropospheric scatter system was installed and operated by Oceanic Wireless Network (OWNI) under the supervision of Eastern Telecommunications Philippines (ETPI), the successor of Eastern Extension. Its capacity of ninety-six circuits operating to Taiwan was replaced a decade later by a submarine cable.

10.2.3 The 1970s and 1980s

To broaden the ownership base and secure additional capital funding sources, PLDT implemented a subscriber investment plan (SIP) in 1973. It requires applicants seeking to acquire new telephone lines, transfer existing lines, or upgrade lines to invest 10 percent cumulative convertible preferred stock in PLDT. Because PLDT stock is freely traded, a subscriber can immediately sell the shares. Table 10.1 shows the required amounts.

PLDT awarded Siemens a contract in 1977 to provide and install 60,000 electronically switched lines. At the same time, it launched a ten-year 110-million peso rural telecommunications development program (RTD) to assist government rural development initiatives. By the end of 1978, PLDT's installed telephones in service had reached 496,266.

Additional submarine cable systems were installed. The Okinawa–Luzon–

Table 10.1. PLDT's Required Investment Under Subscriber Investment Plan (SIP)

Metro Manila	Elsewhere	
New Installation		
147	95	Business, private line
95	71	Business, party line
86	62	Residential, private line
43	88	Residential, party line
Transfer Service		
38	29	Business, private line
29	24	Business, party line
29	24	Residential, private line
24	14	Residential, party line

Source: Philippine Long-Distance Telephone Co.

These levels have been in effect since 1988. At yearend 1991 US$1 = 26.3 pesos. Amounts given are in U.S.$.

Hong Kong (OLUHO) cable became operational in 1977 with 1,600 circuits to Okinawa and 1,840 circuits to Hong Kong. This U.S. $55 million venture was undertaken by ETPI, Cable & Wireless of Hong Kong, and KDD of Japan. In 1978 the ASEAN Philippines–Singapore (ASEAN P-S) cable, installed by ETPI at a cost of U.S. $60 million, became operational with 1,380 circuits. Additionally, in 1980 the 480-circuit Taiwan–Luzon (TAILU) cable became operational; it is jointly owned by ETPI and International Telecommunications Development Corporation (ITDC) of Taiwan and operated by ETPI.

Domestic Satellite Philippines (Domsat) started operations with a capacity of 176 circuits nationwide in 1978. Domsat is a privately owned company that has contracted eleven earth stations and has an agreement with PT Telkom of Indonesia, the operator of the Palapa satellite system, for its space segment needs.

Up until the late 1970s, efforts to upgrade and enlarge specific capacities led to a fragmented network for telecom services. Duplication, the inadequacy of backbone routes, and concentrations of equipment and facilities, indicated the need for a long-range plan to integrate and develop the network. As the sector's implementing agency, Butel is responsible for maintaining backbone telecommunications networks to assist the private telecom sector. However, it is also a public utility that operates telegraph and telephone systems in government offices and in municipalities and cities throughout the country, although it is not the exclusive provider of telecom services to the government even in Metro Manila. Butel (renamed Telof in 1989) has its own network, with a central exchange in its Metro Manila headquarters.

Government planners also realized that appropriate infrastructure has to be developed if telecommunications is to help stimulate social and economic development at both national and regional levels. As envisioned, the emerging system will consist of a single homogeneous national network that will pave the way for orderly and progressive development within the limits of existing resources. Alongside this is a goal to improve the nationwide regulatory system, which oversees compliance by both public and private operators with government standards for efficient service and monitors related national security interests (see NEDA 1977, pp. 285–305).

To ensure that both strategic direction and regulatory needs are addressed, the Ministry of Transportation and Communication (MOTC) was created within the executive branch in 1979 by Executive Order. A five-year (1978–1982) national development plan containing a new focus on telecommunications and a ten-year (1978–1987) telecommunications sectoral plan underwent major revisions starting in 1980 in light of the second oil shock. Higher inflation rates and pressure on international reserves altered the government's public investment portfolio, which included telecommunications, other infrastructure, and various industries. Nonetheless, in 1983 MOTC was able to obtain official approval for a long-term (1984–2000) telephone development program. However, the protracted domestic political crisis that ensued delayed completion of World Bank studies, a precondition for funding, until 1985. (see *Bulletin Today,* Nov 17; 1985).

10.2.4 Legal Foundations

Historically, there has been minimal government concern for interventionist telecommunications policies, as evidenced by the absence of major legislation to guide its evolution and development. Legislation was generally limited to the granting of franchises. During the 1970s, administrative policies and guidelines were, if not simply reactive, then superficial, and favored companies with strong ties to the Marcos administration. It is generally assumed PLDT was protected from competition, as were other companies, owned or controlled by cronies of the president (see, e.g., Friedland 1988).

Only when the Aquino administration took power did a need for direction in telecommunications development appear to have been given consideration. An executive memorandum order created the National Telecommunications Development Committee (NTDC) to coordinate with the legislative branch and the private sector regarding resolution of issues regarding telecommunications development. DOTC issued a department circular in 1987 as an official guiding policy for the development of telecommunications.

10.2.5 Postal Service

The postal sector has remained a government monopoly. Postal services covered approximately 95 percent of all major settlements through 1,654 post offices by 1977. Nonetheless, service remained insufficient due to a host of problems. Over 90 percent of post offices—and there was just 1 per 26,000 persons—were either in rented private homes or rooms of municipal halls that were inadequate for mail processing as of 1990. Postal manpower was at a level of 1 employee per 3,100 persons in 1987, the volume of mail handled was 800 million pieces. Management of postal logistics was poor, and mail dispatching and delivery were unsynchronized. Improvements in mail service depend to a large degree on an honest, well-managed, disciplined work force, as well as government financial outlays.

10.3 Industry Structure

The telecom sector comprises private and public entities; Table 10.2 summarizes the major companies. PLDT, the dominant company, has mixed Filipino and foreign ownership—its shares are listed on the American Stock Exchange in the United States. It is controlled by a group of Filipino businessmen (see Friedland 1988, p. 70). In early 1990 two companies were licensed to compete with PLDT in carrying toll traffic.

Two separate entities provide satellite communications. Philcomsat operates direct voice–record–data circuits to eleven Pacific Basin countries and twenty countries in the Indian Ocean area through Intelsat. Domsat carries mainly "live" or "real time" broadcasts from and to any part of the country. Due to the marginal use of its facilities, it has suffered financial losses (see Table 10.3).

Table 10.2. Major Telecommunication Firms

Abbreviation	Full Name, Founding or Franchise Date, 1990 Revenues (in Million Pesos), Services, Territory
Voice Carriers	
PLDT	Philippine Telephone & Long Distance Co (1928); P12,839; Principal domestic local and long distance carrier; international calls from United States and Europe.
PT&T	Philippine Telephone & Telegraph (1962); P346; Domestic voice and record carrier.
RCPI	Radio Communications Philippines (1950) Biggest telegraph operator; also telex.
Telof	Telecommunications Office (was Butel until 1989) Government-owned tel & tel provider.
Satellite	
Domsat	[Domestic Satellite]; relays broadcasting signals.
Philcomsat	Philippine Communications Satellite Corp; P551; IRC; Uses Intelsat.
International Record Carriers (IRCs)	
CAPWIRE	Capitol Wireless, Inc (1962); P76; International record carrier.
ETPI	Eastern Telecommunications Philippines, Inc (1967); P1,000; International calls from Hong Kong, Taiwan; connects to PLDT, opened own gateway in 1991; successor to Eastern Extension; also common carrier.
GMCR	Globe-Mackay Cable & Radio Corp (1928); P631; International record carrier.
Philcom	Philippine Global Communications (1977); P936; International calls from Japan, Korea; connects to PLDT, given permission for own gateway in 1990.
Others	
OWNI	Oceanic Wireless Network, Inc (1959); P76; Common carrier.

Source for revenue: Mahal Kong Pilipinas Foundation, *Philippine Company Profiles,* 1991 edition.

10.3.1 Telephone

The telephone sector consists of more than sixty mostly private telephone companies. Many of these operate within a single town. PLDT has well over 90 percent of total access lines. It operated forty automatic telephone exchanges in Metro Manila and seventy-one local exchanges throughout the rest of the country. In thirteen important urban centers there is more than one telephone company. (These are all 1988 data.) The national telephone toll network had 271 central exchanges at the end of 1989.

Availability is low. There are thirteen provinces, including three subprovinces and eight cities, that still have no telephone service. Of the nation's approximately 1.1 million main stations (1990, excluding military service), close

Table 10.3. Major Telecommunication Firms: Financial Data, 1983–1990*

1983	1985	1987	1989	1990	Listed by Gross Revenue
					PLDT
12,569	17,048	21,264	26,583	36,224	assets
3.7	3.8	2.4	1.2	1.2	debt-equity ratio
2,504	4,718	6,591	9,459	12,839	gross revenue
398	779	1,362	2,080	3,187	net income
					PHILCOM
379	667	896	896	1,115	assets
1.0	1.1	1.1	1.6	3.0	debt-equity ratio
282	493	688	847	936	gross revenue
108	231	270	374	458	net income
					ETPI
672	923	1,334	1,451	1,770	assets
0.9	0.6	1.5	0.4	0.3	debt-equity ratio
280	437	528	668	1,000	gross revenue
105	180	159	229	531	net income
					GMCR
272	388	512	611	800	assets
0.2	0.1	0.9	0.7	0.6	debt-equity ratio
186	276	342	372	361	gross revenue
85	132	108	79	52	net income
					PT&T
457	578	752	951	1,139	assets
1.3	1.7	1.4	0.9	0.6	debt-equity ratio
97	187	235	312	346	gross revenue
10	7	5	29	29	net income
					OWNI
—	50	87	132	173	assets
—	0.3	0.2	—	—	debt-equity ratio
—	35	48	57	76	gross revenue
—	17	14	18	21	net income
					CAPWIRE
36	48	67	96	121	assets
1.8	0.7	0.6	—	—	debt-equity ratio
14	34	47	66	76	gross revenue
−4	−8	10	11	9	net income
					PHILCOMSAT
—	984	942	1,948	2,006	assets
—	0.4	0.2	—	—	debt-equity ratio
—	578	546	910	551	gross revenue
—	298	234	258	148	net income
					RCPI
122	117	—	—	—	assets
1.5	2.5	—	—	—	debt-equity ratio
95	115	—	—	—	gross revenue
2	−12	—	—	—	net income
11.1	18.6	20.6	21.7	24.3	Pesos per U.S.$

Source: Mahal Kong Pilipinas Foundation, *Philippine Company Profiles*, 1988 and 1991 editions.

*Data in million pesos, except debt–equity ratios.

— = not available.

to 80 percent are in the National Capital Region (Metro Manila) comprising four cities and thirteen municipalities. Baguio, Cebu, Bacolod, Iloilo, Davao, and Cagayan de Oro account for another 8.5 percent. Metro Manila and these other urban areas are only 13 percent and 4 percent respectively, of the total population. Penetration is about 1.6 percent nationwide, but over 10 percent in Metro manila (1990). In 1988 there were 147 exchanges on Luzon (which included Manila), thirty-two in the Visayas islands, and thirty-seven on the southern island of Mindanao. Table 10.4 shows the country's telephone density.

Legislation proposed in the late 1980s reflects increased awareness of an acute need for telecom services. One bill proposed public telephone and telegraph systems in every town or municipality. Another was to provide 200 million pesos for installation, operation, and maintenance of public telephones in every municipality. The Municipal Telephone Act was passed in February 1990. It required each of the 1,050 towns then without service to have at least one public station by mid-1993. The stations are to be capable of both voice and data transmission. All told, some 143,000 lines were planned for 1990–1993 (*Telephony*, 1990 Oct 20, p. 40).

Extelcom has received government permission to provide mobile telephone service in Metro Manila using the PLDT network. PLDT has gone to court to

Table 10.4. Telephone Main Lines and Density, by Area, 1987

Main Lines	Density	
733	1.31	National
527	7.37	Metro Manila
62	2.96	Other urban areas
144	0.31	Rest of country
		Region (major city)
178.9	0.45	Ilocos (Bagio)
3.5	0.13	Cagayan
34.5	0.61	Central Luzon
30.7	0.42	Southern Tagalog
9.6	0.24	Bicol (Legaspi)
27.9	0.54	Western Visayas
32.2	0.75	Central Visayas (Cebu)
6.6	0.21	Eastern Visayas
7.9	0.27	Western Mindanao (Zamboanga)
6.1	0.19	Northern Mindanao
24.8	0.63	Southern Mindanao (Davao)
3.8	0.14	Central Mindanao

Source: NEDA, "Medium-Term Investment Program, 1988–1992," June 1988.

More recent detailed data are not available. At the end of 1990 there were nearly 1.1 million lines nationwide according to PLDT, which made density just over 1.6. The 1989 figure is about 988,000 lines. Data are in thousands and lines per 100 people.

prevent this, arguing it is not fair for another company to use facilities PLDT has paid for.

10.3.2 Telex, Telegraph, and International Record Carriers

Of the four private international records carriers (IRCs, i.e., companies that lease circuits or channels, or telegram exchange and facsimile services), two have telephone correspondent status. ETPI operates four submarine cable systems out of Currimao, Ilocos Norte. ETPI and Philcom have enjoyed telephone or voice correspondent status since the 1930s and 1977, respectively, by virtue of their indefeasible rights of user (IRU) access to the cables.

Competition among IRCs is keen and aggressive. All of them operate their own telex switching exchanges and are linked to international telex switching centers around the world. They also provide a range of services based on modern circuit, message, and packet-switching technologies including store-and-forward and other enhanced features. In addition to telex, the IRCs offer telegram and mailgram, leased telegraph and data circuits, facsimile and Bureaufax, data access and other packet-switched data communications services, press bulletin service, Datel, satellite television broadcast, and electronic mail.

PLDT has a digital telephone switching exchange (Siemens EWSD) for both national and international traffic with a capacity of a 1,230 international circuits and ninety-six digital switchboards. An ARM Crossbar and an ESK Relay Metric with 600 and 345 circuits, respectively, are complementary exchanges. All three on the CCITT standard 5, signalling with the digital exchange having system 7 capability.

In 1986, Philcom held 42 percent of the telegraph market while GMCR was the leader in telex with a 34 percent share. The tariffs charged by these companies are uniform and regulated by the NTC.

Among the nine domestic record carriers, Philippine Telegram and Telephone (PT&T) and Radio Communications of the Philippines (RCPI) both operate domestic telex switching exchanges. In 1986, they dominated the market of 2,254 subscribers. PT&T is a private company set up in 1962; during the height of intense competition in the early 1970s, its losses prompted a restructuring of operations. It discontinued telegraph services and shifted its focus to telex and leased channels systems. It makes use of an NEC-NEDIX 103B telex exchange that has a capacity of 3,000 lines expandable to 30,000 lines. In 1983, it introduced the computergram service featuring delivery in three hours through a Computer Oriented Message Switching Exchange (COMET) utilizing a VAX 11/750 computer with C&W Incotel software. PT&T's integrated digital network (IDN), completed in 1984, makes available high-speed digital data communication within the Metro Manila area and other major cities. The companies Digital Data Service (DDS) introduced in 1986, offers low-speed (300 baud), medium-speed (up to 9,600), and high-speed (beyond 2 Mbps) data communications service between Manila and major provincial cities. In 1989 PT&T launched its national Datanet service utilizing a Siemens EDX-P packet-switching exchange, the country's first public-switched data network (PSDN). PT&T's Datanet service connects nine major cities.

Established in 1950 through a congressional franchise under a legislative act, RCPI is the country's biggest telegraphic carrier, controlling roughly 65 percent of the market. It operates a stored-program controlled telex exchange (an ELTEX II) and provides line capacity for 1,500 telex subscribers with such capabilities as abbreviated dialing, store and forward systems, multiaddress calls, and automatic advice of call duration. The other domestic carriers are Universal Telecommunications Service, Clavecilla Radio Systems, BFC Communications, and Federal Wireless. These companies operated 1,913 telegraph stations and 142 domestic telex stations as of yearend 1989.

Despite the presence of these private companies, the government through Telof operates telegraph facilities at 1,522 stations located primarily in rural areas. Most privately owned stations are in commercially viable areas or urban centers while Telof generally operates in low volume areas. Nevertheless, there are still 198 isolated towns and municipalities without telegraph service.

10.3.3 Industry Finances

Providing telecom services is generally a capital-intensive business. Technological advances made much of the installed base obsolete in the 1980s. This means local telecommunications tend to involve foreign firms and institutions—for both capital and equipment. Foreign equity contributions often take the form of equipment and technology transfer.

Because the Philippine economy is still recuperating from economic crisis, it is considered almost impossible to set up an entirely locally financed and equipped telecommunications venture. The majority of companies, especially the international carriers, accordingly have foreign partners. ETPI is associated with Cable & Wireless, GMRC is partly owned by ITT Communications Services, and part of Philcom is owned by RCA Global Communications. Companies associated with multinationals show stronger debt–equity positions than such 100 percent Filipino firms as PT&T and RCPI.

PLDT deserves special mention. Although it has some foreign ownership, it is controlled by a group of Filipinos. Its debt–equity ratio is the highest of all major firms due to its ambitious expansion projects, which have been financed largely by loans. The government has declared a moratorium on debt repayments. Some of PLDT's foreign debt is government-guaranteed. Because of the moratorium, the company is not entitled to soft-loan financing offered through regional governments by industrialized countries. PLDT, however, has worked out refinancing agreements and new loans from various international banks and lending institutions. This demonstrates its comparative advantage in terms of access to funds relative to other purely Filipino-controlled telecommunications companies.

A government-mandated 12 percent ceiling annual return on investment applies to all telecommunications firms. The government, through the NTC, also sets rates. Rates in Manila are given in Table 10.5.

Table 10.5. Telephone Service Rates, Metro Manila

Rate	Service	Rate	Service
	Residential		Key System
148–185	Private	446–584	Main Equipment
103–138	Two-Party	126	Per trunk termination
29–59	Extension	146–168	Key telephone (each)
	Business		PABX–ESK
404	Private	3,838–6,398	Equipment package
295	Two-Party	336–554	Trunk line (each)
55–73	Extension		

Source: Philippine Long-Distance Telephone Co.

Rates vary with type of equipment (e.g., desk handset, wall handset, extension with or without bell, rotary or pushbutton, etc). These rates, given here in pesos, have been in effect since 1988.

10.4 Regulation

Telecommunications is considered a public utility. The Department of Transportation and Communication (DOTC) is the entity of the executive branch directly responsible for the promotion, development, and regulation of the entire telecom sector. Its areas of concern are basically: (1) formulation of telecommunications policies, guidelines, and systems at all levels of government; (2) regulation of communications activities, and determination and collection of fees; (3) assessment, review, and direction-setting in communications research by government and other institutions; and (4) establishment and administration of comprehensive and integrated programs for communications.

In mid-1987 DOTC Circular 87-188 laid down a set of policies that called for:

1. Establishment of nationwide communications services subject to competitive and regulated market entry.
2. Provision of at least one integrated and reliable nationwide transmission facility for voice, record, and data services interconnecting all major cities and towns.
3. Promotion of state-of-the-art and cost-effective technology (e.g., digital transmission).
4. Conformance of all installed equipment with accepted CCIR or CCITT recommendations.
5. Rationalized management of radio frequency spectrum.
6. Interconnection of at least one reliable nationwide marine coastal communications system (i.e., ship-to-shore) with public networks.
7. Operation of mobile radio communications by franchised carriers interconnected to the public network.
8. Rationalization of all government transmission networks.

9. Promotion of entry of private enterprise to areas presently served by DOTC's telecom services.
10. Connection of customer provided equipment to public networks subject to guidelines.
11. Encouragement of domestic manufacturing of telecom equipment.
12. Comparatively efficient interconnection for all public carriers.
13. Any and all value added services for public use subject to grant of certificates of public convenience and necessity from NTC.
14. A policy change concerning the "international gateway" to better address the public interest and national security.

The National Telecommunications Commission (NTC) is the regulatory agency within DOTC and has quasi-judicial powers. It implements all the policies and plans of the department regarding communications. With the exception of military telecommunications installations, the commission exercises jurisdiction over, supervises, regulates, and controls all telecom services in the country. Some of its powers include: (1) issuance, revision, suspension, or cancellation of permits to operate facilities; (2) determination of rates; (3) allocation of frequencies (4) regulation of equipment importation; and (5) adjudication of legal issues.

It may be logical to expect an independent regulatory mechanism to be in place due to the sector's public service features. Up to the early 1980s, however, government's role was limited to the imposition of rules and regulations relative to rates, frequencies, certificates of public convenience and necessity (CPCNs), and franchises. (CPCNs and franchises are prerequisites for a private company to install and operate a telecom system; they are granted by the NTC.)

The NTC's role in standardization of equipment is a development of the late (post-Marcos) 1980s, and no major concrete steps have been taken. Regulation primarily revolved around infringement and violation of a set of bureaucratic rules and regulations that in practice had little to do with how well service was provided. Supervision was primarily reactive and did not attempt to influence the industry to develop in line with any coordinated national telecommunications program. It should be noted, however, that in the early 1990s the NTC has been trying to assert a stronger presence in the sector both as regulator and initiator of telecommunications development.

In addition, there were no mechanisms in place to reconcile national interests with the interests of private business. Even Philcomsat was not governed by any regulatory agency until 1987. It is partly government-owned, but it has not always been the case that government regulatory agencies upheld the public interest. (Thus, for example, there have been instances where regulatory bodies have appeared to show undue preference for the interests of regulated government-owned entities to the detriment of regulated private entities.)

Likewise, no regulatory mechanism could ensure competitive discipline and prevent abuse of monopoly power associated with the economies of scale common to telecommunications activities. It should be noted that PLDT's telephone service has been subject to frequent public criticism, mainly because of its "take it or leave it" attitude, which reflects PLDT's indifference to the quality of its service.

For quite a long time, standards necessary for interconnection among carriers were seldom spelled out and often did not exist. This reflected a weak understanding and poor recognition at higher government levels of the dynamic changes in technology. In such an environment, and where private corporate behavior has been chiefly influenced by large reactive regulatory supervision, any initiative to deliver efficient products and services, including promotion of standards, will be limited.

Heavy government intervention during the latter part of the Marcos regime rendered many of the usual advantages of private enterprise largely ineffective. Regulation is meaningful and effective only as long as government has a clear vision of what it wants and crafts rules to guide business toward that state of affairs. Policy visions only emerged in the 1980s. These were strengthened by public recognition of the important role of telecommunications in their lives.

By the early 1980s strengthening the NTC was recognized as necessary, and thus expected. Rationalization and standardization of rates was clearly identified as a task to be undertaken jointly with the finance Ministry and Congress, although leaving NTC the authority to grant franchises and establish the specific tariffs was considered desirable.

10.4.1 Ownership

Ever since U.S. rule, private enterprise has been the guiding philosophy of development. However, in areas where private telecommunications companies do not operate, government through Telof is present. At least 60 percent of invested capital must be owned by Filipinos in order to obtain fifty-year operating authority. Under the 1987 Constitution, participation of foreign investors in the management of any public utility is limited to their proportionate share of its capital. Moreover, all executive and managing officers must be Philippine citizens.

Although the NTC grants the necessary franchise for any telecommunications entity to operate, Congress has the right to amend, alter, or repeal such franchises when the common good—universal access to telecommunications facilities and information—dictates.

Because telecommunications is very capital intensive, recovery of investments often takes a time. Financial resources are very limited, and operators frequently have difficulty in raising capital to expand or upgrade facilities. The Philippine capital market is virtually nonexistent, and the government itself does not have the means to assist private operators directly—the government's deficit was about 2.3 percent of GNP in 1989. The few telecom operators possessing strategic advantages (e.g., links with big business groups—including banking, insurance, real estate, and commercial center development—or association with a multinational telecommunications firm such as Cable & Wireless, RCA, ITT, etc.) find themselves in a powerful position vis-à-vis regulatory bodies.

10.5 Interests the System Seeks to Protect

The historic events of 1986 dramatically altered the country's political course and produced new perspectives on how the telecom sector relates to national interests. A widely held view prior to 1986 was that the state-owned domestic and international communications systems were under control of Marcos and his close associates (see Sussman 1981).

Authentic adherence to private enterprise is now stressed more than was the case during the twenty years prior to 1986, and the public expects more firms to enter the telecom sector. Areas where conditions call for monopoly power through exclusive permits will have to be better justified to an interested public.

A strong sense of nationalism permeates the Congress and will influence legislation affecting foreign direct investment. Thus, while important as sources of needed technologies, multinationals in their roles as minority owners of telecommunications entities—particularly satellites and as record and voice carriers—face greater scrutiny than before.

In terms of telephone usage, metered local service will probably not be introduced. As far back as 1983 NTC was ordered by MOTC to meter local telephone service immediately where possible—for example, when subscribers were served by electronic switching equipment. However, the current unmetered practice appears strongly rooted within the system. Any move to push metering would require well-defined benefits and improved levels of service.

10.5.1 Political Parties and Major Interest Groups

Political parties have proliferated since the restoration of democratic processes in the country following the EDSA revolution. Foremost among the groups that brought the present leadership to power are the Pilipino Democratic Party and Lakas ng Bayan (PDP-Laban), the United Nationalist Democratic Organization (UNIDO), and two factions of the Liberal Party. Political groups formed by leaders from the previous regime are the Nationalist Party of the Philippines, New Society Movement (KBL), and the Nacionalista Party. Key leaders of PDP-Laban and Lakas ng Bansa formed the Laban ng Demokratikong Pilipino (LDP) party, the administration's party, in 1988.

Given the fluid delineations between political groups, it is not easy to identify their main positions or leanings. A few things are clear, however. There is at least a rhetorical commitment to providing better telecom services, a goal agreed on by all concerned social sectors through the process of democratic consultation. For example, the LDP included a specific agenda in its economic platform calling for improved rural communications systems (LDP 1988, p. 4).

Certain areas in telecommunications are of particular concern to interest groups. A key issue for consumer groups is SIP, the telephone subscriber investment plan. The Philippine Consumers Foundation (PCFI) petitioned NTC to suspend it. The NTC dismissed the group's plea in 1988 after hearing arguments from PLDT and the Subscriber Investors Association. The plan had been modified previously to allow installment payments for SIP. Prospective telephone subscribers make a downpayment of 30 percent, with the balance in six equal monthly installments. SIP opponents have made no serious suggestions as to how system expansion could be financed absent the program.

Interests of telecommunication carriers, suppliers, dealers, manufacturers, and professionals are represented under the umbrella organization of the Philippine Electronics and Telecommunications Federation (PETEF). This private industry group has worked closely with the government on policy directions—including study groups that drafted eight implementing guidelines that are a part of DOTC Circular 87-188. So far, however, only the circular's guidelines on customer-provided equipment have been approved (see *Business World,* Jul 21, 1988).

Attaining increasing levels of overall efficiency is another concern. At NTC's suggestion, the Telecommunications Users Groups of the Philippines (TUGP) was formed in 1987 to provide government with a clearer picture of the requirements of large users, such as major banks (see *Philippine Daily Globe,* May 2, 1988, and *Business Star,* May 13, 1988). TUGP has publicly stated its opposition to monopoly providers beyond those necessary for an orderly national telecommunications system. TUGP views competition among international record carriers positively (*Business Star,* May 31, 1988). The group specifically noted it was more concerned about PLDT's monopolistic—''take it or leave it''—attitude than its monopoly per se. It wanted PLDT to be measured against a set of stringent standards and penalized if it fails to deliver acceptable levels of performance.

There is also a Manila-based network of institutions known as People in Communication (PIC) that seeks to use media for the service of human development and social transformation. While acknowledging telecommunications as an instrument for national development, it believes universal service and telecommunications development should be compatible with culture and tradition. It advocates industry-balancing commercial objectives against the telecom needs of a developing country. PIC's components are the Asian Social Institute Communication Center, Communication Foundation for Asia, National Office of Mass Media, New Day Publisher, Philippine Educational Theater Association, Radio Veritas, and Sonolux Asia.

10.6 Reforms and Disagreements Within the Government

Under the 1987 constitution the state recognizes the vital role of communications and information in nation building as a matter of national policy. How this role is discharged depends to a large extent on what methods the government considers best to maximize the potentials of telecommunications. Congress plays a role because it has the power, when it deems the common good requires it, to intervene and amend, alter, or repeal franchises granted by NTC. Any disagreement between Congress and the NTC may require resolution by the Supreme Court.

In spelling out its policies in Department Circular 87-188, DOTC has defined the parameters that will govern the growth and development of the telecom sector in the 1990s. However, Congress might try to micromanage. A number of bills have been introduced calling for various studies of issues that have already been addressed. Worse are numerous bills to accelerate provision of services at local levels—which could totally disrupt attempts to systematically develop needed networks within very severe funding constraints. Unless DOTC is quick to point out its program of action, disagreements over the proper time frame for delivery of public services could arise.

The NTDC was established to facilitate resolution of immediate development, regulatory, and other issues affecting telecommunications. It serves as an organized forum for interaction among the principal telecom authorities and other major interest groups to assist DOTC. The chairman of the Senate Committee on Public Services as well as the chairman of the Committee on Transportation and Communications of the House serve as consultants. NTDC's principal tasks are to:

1. Facilitate consultation and interaction between the relevant authorities and private sector parties affected by issues such as the degree of competition, the role of the various carriers, and timebound objectives for increasing telephone density.
2. Act as a central forum where operators, users, and government can meet to discuss vital problems and arrive at solutions despite disparate interests.
3. Facilitate formulation of mechanisms for accelerating development of the sector.
4. Provide the venue for consultation on development of a comprehensive regulatory framework and implementation of necessary regulatory reforms.
5. Expedite formulation of short-term measures to improve service availability and quality quickly.

10.6.1 Impact of Reforms Abroad

Deregulation and greater competition in some industrialized nations, notably the United States, appear to have strengthened public views that monopolies should be disciplined by exposure to market forces. The current assumption of some within government is that producing better telecom service must occur at the expense of PLDT (see, e.g., *Philippine Star,* Jun 8, 1988, and *Manila Bulletin,* Aug 15, 1988). At the same time, it is argued—primarily by those opposed to PLDT's virtual monopoly—that PLDT's nationwide franchise should be a strong enough incentive for it to engage in continuous development even of unprofitable areas. The general view is that its inability to meet this expectation seems to be reason enough to revoke its exclusive license to operate nationally. Of course, not much is said about PLDT's performance under previous governments, which were partly to blame for not having exerted enough pressure on PLDT to operate in new areas. To a large extent corporate behavior is the result of a given business environment. Policy directions to resolve this issue might be to confine PLDT to its current operations and open other areas to competitors or confine PLDT to trunk business.

10.6.2 Exceptions to the Telecommunications Monopoly

Under Circular 87-188, it would seem to be only a matter of time before most areas of the industry will be open to competition. This may contradict plans for network development, but this possibility is addressed using several policy guidelines. One guideline specifically states that there should be "at least" one nationwide backbone facility. In effect, the avenue for additional integrated facilities has been defined. PLDT operates four international gateways; EPTI started its own at a cost of about U.S. $5 million in early 1991. PLDT fought granting EPTI the right to do this. Like EPTI, Philcom received approval for a voice gateway in early 1990, but had not started service by the end of 1991. Other companies have circuit leasing arrangements. In February 1990 PLDT cut overseas rates 20 percent to discourage new entrants (international traffic has provided 60–65 percent of PLDT revenues).

Groundwork has been laid for competition in CPE and the area of value added products through a set of March 1988 guidelines. Previously, only PABXs, key systems, and wireless telephone sets were allowed interconnection to the public network. More items will be allowed once implementing rules are issued. Cellular telephone subscribers generally own their telephones, which need not be bought from the telephone company.

In its application to the NTC to install and operate an international gateway, ETPI, through Cable & Wireless and seven local investors holding a 40–60 percent ownership interest, formed Digital Telecommunications Philippines (Digitel) to construct a new nationwide telephone network. Initially about 2.7 billion pesos (around U.S. $130 million) will be invested over a five-year period to install 57,000 single-line digital telephones in Bulacan and Pampanga, two provinces just north of Metro Manila. To begin operations, the company must secure congressional approval for its franchise as well as NTC approval. From its application in September 1987 through 1990 Digitel appeared at twenty-six NTC hearings with only provisional authority to operate in one city and province (*Asian Business,* Jan 1991, p. 42). Digitel intends to offer rates 15–20 percent lower than PLDT's. Philcom has also proposed an international gateway (*Philippine Star,* Jun 9, 1988).

10.7 Emergence of the Electronics Industry

Unlike some of its Asian neighbors' flourishing electronics industries, Philippine electronics is still embryonic. Demand for electronics goods, now mostly satisfied by imports, is on the rise. The semiconductor firms that exist cater exclusively to the export market. Meantime, trade and industry officials are grappling with the strategic problem of how to induce substantial growth in the industry.

The crisis years of 1983 to 1985 led to a substantial drop in foreign investment inflows. Studies toward the end of the Marcos regime highlighted several reasons for the failure of initiatives to make the Philippines an attractive place for multinationals engaged in technology intensive activities. These included: (1) lack of access to the domestic market for consumer electronics, (2) poor availability of materials and components that offset lower labor costs, and (3) red tape and bureaucratic delays that create significant obstacles.

Almost all computer hardware is imported. Demand for computers surged in 1986 after removal of a 1983 restriction on their importation. Fierce competition, especially in the banking and government sectors, was observed between IBM and NCR, the established leaders, and rivals Philips, Nixdorf, and Unisys, aggressive latecomers. As of 1986 (the most recent user survey) 226 mainframes, 472 minicomputers, and 14,050 microcomputers were in operation in the country.

The software industry bucked the economic doldrums of 1980–1985. The number of firms grew from twenty-three in 1981 to fifty-one by 1985. Software development—including design and coding—for local and export markets boomed in the late 1980s. In 1990 exports by companies registered with the Board of Investments, the country's investment promotion office, reached U.S. $7.8 million.

No substantial link exists between major semiconductor firms and electronic firms with the potential for local manufacturing or assembly of electronic products including telecom equipment. In fact, all integrated circuits manufactured by multinational subsidiaries must be exported under legislation formulated during the late 1970s. They have to be reimported for domestic use.

As markets and production capacities for electronic products increase, a closer link could be initiated because rising labor costs in the newly industrialized Asian countries make the Philippines attractive. Thus, in 1988 several locally owned companies set up to assemble circuit boards for communication equipment for a Manila-based Japanese telecommunications firm (see *Business Star*, June 2, 1988).

In an attempt to promote the electronics industry, a comprehensive long-term plan for the development of information technology was drafted in early 1986 by a task force composed of computer experts from the private and government sectors. Called the Strategic Program for Information Technology (SPRINT), it was presented in 1987. The plan focused on expanding computer system use prior to manufacturing promotion. Thus, SPRINT's main features included a call for massive expansion of the installed base of computer systems through fiscal incentives to users and suppliers (i.e., lower tariffs).

Although the local electronics industry at its early 1990s level of development can offer only minimal help to the development of telecommunications, the Electronics Sector Plan (ESP 2000) envisions that the telecommunications industry will employ substantially more support from the electronics industry by the year 2000 (see *Electronics Towards the Year 2000,* a joint sectoral study undertaken in 1988 by BOI and industry).

Only Electronic Telephone Systems (ETSI) is taking visible steps to aid telecommunications development by manufacturing telephone sets and exchanges for the domestic market. Other local companies have focused on exports. Their usual arrangement is for equipment to be assembled locally using imported components. However, the local market potential for industrial electronics, including telecommunications components, is apparent. DOTC included a provision in its guidelines concerning the telecommunications industry (Circular 87-188) encouraging "development of a domestic telecommunications manufacturing sector . . . particularly in the manufacture of electronic and communications equipment and components to complement and support the expansion, development, operation, and maintenance of an efficient nationwide network."

10.7.1 Equipment

Support industries for local manufacture of telecom equipment are either absent or underdeveloped. As a result, most equipment is imported. A major constraint on continuous upgrading is high tariffs. For example, wireless equipment faces a 50 percent duty plus 25 percent advance tax. In addition, the Central Bank requires payment of 50 percent of duties before import letters of credit are approved. These levels were implemented to protect the few domestic vendors, generate revenue, and limit demands on foreign currency reserves.

Local equipment manufacturing started in 1977 when PLDT began a major expansion of its system under a U.S.$56 million contract to Siemens, and another U.S. $544 million contract with Siemens for electronic switching equipment to supply 220,000 lines by 1982. ETSI, a joint venture with Siemens, was established in 1982 to manufacture telephone instruments with 45 percent local content. At that time, annual demand was estimated at 25,000 units. In 1987, demand was estimated at 80,000 units and annual growth was projected at 25 percent. ETSI officials claim 100 percent local content for telephone instruments would be economically viable if demand reached 300,000–500,000 units a year. Siemens switching equipment had reached 20 percent local content in 1987 (*Business World,* May 13, 1988). With deregulation of CPE in 1988, demand growth does not necessarily translate into ETSI sales as there has been an influx of telephone instruments from foreign sources.

Three things are likely to be major influences on development of local manufacturing. One is a policy to encourage domestic manufacturing. In the 1988 Investment Priorities Plan of the Board of Investments, telecommunications and information handling equipment have been accorded pioneer or nonpioneer status. (Pioneer status gives a firm several incentives under the investment code. Status is achieved by using a type of technology or offering a service that is new and untried in the Philippines, or manufacturing goods not being locally produced on a commercial scale.)

Investment proposals to manufacture telecom equipment locally may qualify for incentives after proper registration with the One-Stop Action Center for Investments of the Department of Trade and Industry (see BOI 1988). Start-up investments required by firms offering telecom services are normally estimated to be four to six times expected annual income (Maglalang 1988, p. 63).

A second has been the PLDT X-5 expansion project (1989–1991), estimated to have cost U.S. $350 million through 1991, including U.S. $173 million for local components. It has involved the installation of additional facilities for toll traffic and business subscriber local service and was intended to partially cover residential subscriber requirements up to 1991. Roughly 130,000 SPC digital lines were involved. Siemen's was the prime contractor. Part two of the X-5 program (1991–1993) involves over $510 million for 355,000 additional digital lines.

Third is the NTP for 1990–1994 involving tranches for groups of regions. In NTP Tranche 1-1 (1990–1992) a digital transmission route using radio–cable spurlinks will be constructed in central and southern Luzon (Regions III-V) affecting forty-two municipalities and involving 61,209 lines. Tranche 1-2 (1992–1994) involves a digital transmission network covering eighteen municipalities and involving 30,500 lines in the Visayas (the islands between Luzon and Mindanao, Regions VI-VIII). Mindanao (Regions IX-XII) will receive a transmission network connected to Luzon, covering twenty-five municipalities and involving 45,800 lines under Tranche 1-3 (1992–1994). These investments are estimated at U.S.$369 million. Average cost per line before taxes and other conventional capitalized expenses (such as interest during construction) was estimated at U.S.$2,215 (E&T 1988, pp. 33–42).

While both PLDT and Telof have major expansion plans that will affect equipment and supplies procurement, the source of financing might not fully translate into support for local manufacturers: the Luzon project will receive Japanese funding; the Visayas project, French development assistance; and Mindanao, Italian. NTP originally assumed funding by a single financial institution. However, due to constraints imposed by government budgetary ceilings and loan ceilings at financial institutions over the prescribed implementation period, the government had to resort to multisourcing. If requirements for compatibility and integration into one telecommunications backbone are not effectively enforced, local manufacturing may not expand from present levels.

10.7.2 Communications and the Service Sector

Starting in 1975, the hosting of international conventions and increased tourism created a tremendous demand for telecommunications. Access to government credits enabled convention centers, hotels, and resorts to be built all over the country. Pressure to install the latest telecommunication facilities mounted. At the same time, the country was being promoted as a site for regional headquarters (*Business World,* Sep 23, 1988). The immediate implication was the need for reliable infrastructure—particularly telecommunications. However, the worldwide situation following the second oil shock and crises in the Philippines discouraged both tourism and headquarters siting.

10.8 Long-Term Prospects

The Philippine telecommunications industry is undergoing many fundamental changes. Conditions that prevailed over a long period of time are gradually being challenged by new thinking, policies, and views. Filipinos have an acute awareness that the nation's telecom services have become quite backward in a fast-changing, communications-intensive world. Popular enthusiasm to embark on projects to develop an industry considered by many to be inefficient has spread, and the government, for its part, wants to accelerate the delivery of basic services. Expansions of the system will undoubtedly be digital. It is hoped that the cost will not be so great that it will delay the installation of required interconnective equipment and services. It is unclear what sort of physical interconnection will be needed at exchanges where regional, provincial, or municipal systems meet the backbone system. As PLDT and Telof expansion plans get underway, this delineation should emerge.

The franchise issue is also of critical importance. The often-used yet powerful reason for entertaining entry is inadequate service in areas where entrants will operate. Will this encourage local telcos to become regional ones or encouraging new entities to absorb smaller ones?

Pricing is a major issue, although it is not mentioned in the policy guidelines. Rate modifications to reduce cross subsidies from long-distance and international service to backbone and local operations have occurred and such efforts to adjust rates should continue in order to reduce biases for or against various types of services. As development of the domestic side (toll switching and local distribution) is as important as the international side, alternatives may have to be explored. Having different operators for the different segments, allowing for more focused use of capital and attention of management, may be one such alternative.

Finally, policies to develop competition to replace existing monopoly elements may provoke misinterpretations of the government's dealings with a private sector monopoly. This issue has to be addressed. The government has given explicit signals to PLDT to make local distribution in Metro Manila and international gateway operations as its core businesses. Such concentration may enable the company to manage expansion better. PLDT's problem is funding: To develop its franchise areas requires tremendous amounts of relatively cheap capital (soft loans) that only government can obtain and provide from concessionary lenders. Because the government, of necessity, has become less inclined to partake of such arrangements, PLDT might be required to focus more on core business.

The thrust of DOTC since the late 1980s encourages participation of private enterprise in a regulated, fair, and competitive environment in order to accelerate the development and expansion of an efficient and adequate telecommunications infrastructure. There is reason for guarded optimism that service will expand and improve.

Bibliography

Arce, Ernesto C. 1986. "The Telephone Sector: Present and Future Challenges." *Electronics and Telecommunication.* Metro Manila: PETEF (Philippine Electronic and Telecommunications Federation).

Board of Investments (BOI). 1988. "1988 Investment Priorities Plan." Based on Memorandum Order 166, April 5.

BOI. 1988. *Electronics Towards the Year 2000.*

"British, Filipino groups Form New Telephone Firm." 1988. *Philippine Star,* p. 6 (Jun. 9).

"ETSI to Hike Local Content of Phones." 1988. *Business World* (May 13).

Cruz, Bayani. 1989. "Banks See Positive Year Despite Economic Slowdown," *Business Star,* vol. 2, No. 136, pp. 10–11 (Jan. 23).

Department of Transportation and Communication. 1987. "Circular 87-188" (May 22).

Electronics and Telecommunications 1987–1988. 1988. "The New National Telephone Development Plan" (May).

Friedland, Jonathan. 1988. "PLDT's Number Change." *Far Eastern Economic Review,* pp. 68–71 (Oct. 6).

"IMI Subsidiary Boosts Capital." 1988. *Business Star* (Jun. 2).

Lakas ng Demokratikong Pilipino. 1988. "Economic Program."

Maglalang, Laura. 1988. "Strategy for the Development of Regional Telecommunications." Unpublished masters thesis, Center for Research and Communication (CRC), Manila.

Manila Bulletin, 1988. "DOTC Welcomes 2nd International Gateway," p. 36 (Aug. 15).

"Move for Telecom Master Plan in the Offing." 1988. *Business World* (Jul. 21).

National Economic and Development Authority. 1977. *Five Year Philippine Development Plan 1978–1982.* Manila.

Philippine Daily Globe, 1988. "Telecom Users Bat for Better Service," p. 11 (May 2).

"PLDT Monopoly Tops Telecom Group Agenda," 1988. *Philippine Star,* p. 12 (Jun. 8).

"Regional Facilities Cheapest in Manila." 1988. *Business World* (Sept. 13).

Santiago Jr., Raul M. 1987. "In the Crucible of Philippine Politics." Paper presented at a seminar-workshop sponsored by the Center for Research and Asian Institute of Journalism on *A Working Agenda for the New Congress,* Aug 15, 22, and 29.

Santos, Ernesto. 1986. "The Philippine Submarine Cable System." In *Electronics and Telecommunication.* Metro Manila: PETEF.

Sussman, Gerald. 1981 Jun. "Telecommunication Transfers: Transnational Corporations, the Philippines and Structure of Domination." Third World Satellite Center, University of the Philippines, Dependency Series 35.

Taylor, George E. 1964. *The Philippines and the United States: Problems of Partnership.* New York: Praeger.

"TUGP Forms Pressure Group." 1988. *Business Star* (May 13).

"Users Groups Opposes PLDT Monopoly." 1988. *Business Star* (May 31).

11
Thailand

SRISAKDI CHARMONMAN

Thailand, in the late twentieth century, is one of the third wave of Asian countries (after Japan and the ''four dragons'') to engage in widespread industrialization. Even before this time, Bangkok had become an important world city, home to the regional offices of many international organizations. This has created tremendous demands on the country's infrastructure, including telecommunications. Unfortunately, Bangkok has an inadequate telephone system for a city of such size and importance. The situation is such that poor service nationwide has been considered a major impediment to investment—particularly by foreigners—and a major stumbling block in trying to encourage development outside the Bangkok area.

Solving the problem is simply a question of will and money. The expectation is that foreign capital—including soft loans from the governments of countries whose companies want to sell equipment to Thailand, as well as some direct funding of networks by the firms that will be using them—will reduce the problems. This, however, will be aimed primarily at business users. It is impossible to judge when residential users throughout the country will be able to obtain service without a long wait.

The government introduced the telephone and telegraph into Thailand (called Siam until 1939 and in 1946–1948). By law and regulations, government agencies have had a monopoly on them since the nineteenth century, but this began to change with the 1991 announcement of new franchises—one for Bangkok and one for up-country.

11.1 History

Thais date the history of their country from 1238 when an ethnic Tai state was established in part of what is now Thailand. They took the name Thai, meaning free, to distinguish themselves from Tai groups still under foreign domination. The boundaries waxed and waned—at various times parts of what are now Laos, Cambodia, Malaya, and Burma were under Thai suzerainty. At other times Thai empires were toppled by invaders, most recently by the Burmese in 1767. The country had been reunited by 1776, and in 1782 the current dynasty was founded, along with a new capital at Bangkok. Although parts of the country were ceded to the neighboring British and French colonies in the nineteenth and twentieth centuries, the two foreign powers recognized the usefulness of a buffer between them, so the Thai core area remained independent. The country was an absolute monarchy until 1932, although central government influence was limited except in the capital region until the late nineteenth century. The monarchs, if not their ministers, were generally open to ''modernization''; Mongkut (Rama IV, 1851–1868) and Chulalongkorn (Rama V, 1868–1910) are particularly noted in this regard.

Until the late nineteenth century, three message delivery systems were in use, each for a different category of user: royal, town councils, and merchants and monasteries. Important messages, such as the king's letters to provincial governors, were delivered by royal messengers, ''traveling commissioners,'' who went directly from the town of origin to the town of destination carrying a royal seal or sword that empowered them to order any and all town councils along the way to provide them with transportation and anything else required to execute their duties.

Town councils had a special group of messengers called ''fast runners,'' who carried official messages in bamboo tubes slung over their shoulders. Though called runners, they traveled on horses, elephants, boats, rafts, and buffalo carts, as well as on foot. Merchants or monasteries employed their own messengers or used traveling traders to carry messages, which were written or verbal.

11.1.1 Postal System

During the reign of King Mongkut (1851–1868) the royal monopoly of commodity trade was abolished and in 1855 the country opened its doors to foreign trade by signing the Treaty of Friendship and Commerce with Britain (Bowering Treaty). A large number of foreign merchants, therefore, came to Bangkok and several consulates were established.

In 1867 the British Consulate was permitted to operate a postal service for foreign countries. However, it was only one way: from Bangkok to Singapore. Those wishing to send letters beyond Singapore had to buy British stamps (imprinted with a ''B'' for Bangkok) from the British Consulate. Letters went to Singapore by British merchant ships. The Singapore connection was used until 1882.

In 1875 the first Thai postal stamps were ordered from Thomas de la Rue Company in London. In 1881 it was suggested to King Chulalongkorn that an official postal system be established. The king directed Prince Bhanurangsi to collaborate with Henry Albastor, an advisor from England, to draw up a plan. The first set of postal rules and regulations was promulgated two years later. They originated in a letter from Alabstor to Prince Bhanurangsi, who attached them to his letter to King Chulalongkorn, who approved them. Among other things, the document outlined the responsibilities of the post office, penalties for depositing dangerous material or garbage into any postbox, and the privileges of postmen.

The headquarters building for the Postal Department was officially opened on August 4, 1883. The department was under the command of Prince Bhanurangsi, now known as the father of the Thai postal system. The first post office was established nearby. The total area receiving postal service at that time was less than 10 km^2. The Thailand Postal Administration was admitted to the Postal Union on July 1, 1885.

11.1.2 Telegraph and Telephone

In 1869 two Englishmen received permission to set up a British company to install and maintain a telegraph system for various provinces of Thailand, with a connection to Penang in Malaysia. Unfortunately, they could not complete the work.

In 1875 the Thai government assigned the Ministry of Defense the project of installing a telegraph line from Bangkok to Samutprakarn, a fortress town on the Gulf of Thailand, a distance of 45 km. The line was later extended using submarine cable to the lighthouse in the sea a few kilometers from the mouth of the Chao Phraya River. The line's primary purpose was to report on vessels crossing the river's bar to and from Bangkok. The first telegraph office in Thailand was established at Saranromya Palace in Bangkok.

The country's second telegraph line was installed in 1878 running from Bangkok to Bang Pa-in Palace, about 46 km to the north. This line was later extended to the old capital of Ayudhya, another 32 km. Like the Samutprakarn line, it was used only for official purposes.

The first telephone system was built in 1881 for the same use as the first telegraph system six years earlier: Facilitating Samutprakarn reports on ship movements in and out of Chao Phraya River. The person in charge of surveying and installation was an Englishman.

At the same time the Postal Department was formally created in 1883, a Telegraph Department was established by royal decree. All the telegraph facilities maintained by the Defense Department were handed over to the new department. A telegraph line was installed from Bangkok to the province of Karnchanaburi for connection to Burma in 1884, but it was later discontinued because of difficulties in repair and maintenance. Between 1895 and 1897 three domestic trunk lines were installed, connecting all major towns to Bangkok.

In 1886 telephone service was moved from the Ministry of Defense to the Post and Telegraph Office (PTO). PTO expanded the system and offered services to the public. At that time, a Magneto system with local battery served sixty-one numbers. In 1887 private citizens were allowed to use private telephone handsets without having to pay rental fees. By 1899, subscribers were allowed to use telephone service twenty-four hours a day.

The first foreign telegraph connection, from Bangkok to Saigon via Phnom Penh, was completed in 1883. A line to Rangoon was completed in 1897, and one to Penang came into operation in 1899. Thailand joined the International Telegraph Union in July 1885.

The PTO was split into departments for posts and for telegraph (including telephone) in 1891, and was recombined in 1897 as the Post and Telegraph Department (PTD). In 1906 a foreigner named Collman was appointed director general. He served until 1909 and had considerable influence over telecommunications in Thailand.

The military played an important role until the mid-1920s. In 1911 the navy started using a radio telegraph system for communication with foreign countries. In 1912 telegraph codes for the Thai language were announced as the result of work done by the Defence Ministry together with the Railway Organization and PTD. In 1914 the navy was given royal permission to establish the first radio telegraph station. The navy transferred all the equipment and personnel at its radio telegraph stations to PTD in 1926. Since then, military-related telecommunications have been more or less separate from commercial telecommunications.

Many early telephone and telegraph lines were funded by Thai individuals on behalf of the government. In 1897 the Governor of Lumpang (modern Lamphun, in the north near Chieng Mai) donated personal funds to install a telegraph line from Lumpang to the adjacent province of Phrae; Prince Ratanaburi of Lumpang later donated funds to install a line from that town to the Tern district. (Districts—there are over 400—are the next local administrative level down; the seventy-three provinces each have the same name as their capital). In 1903 the princes of Chieng Mai made personal donations to install a 36-km telegraph line. In 1905 the governor of Udorn organized a group of people to work on installing a telephone line from Markkang district to the province of Nakhon Phanom, some 245 km away. In 1919 the population of Sukhothai province donated money for the construction of a telephone exchange.

Telephone service remained very localized. In 1911 a system was installed in the Mueng Li district of the Payap region, covering a radius of 38 km, to combat thieves and robbers in the area. In 1927 more exchanges were added, and telephone service was expanded from Bangkok and Samutprakarn to include a few more provinces.

By 1921 there were 243 PTOs, 6,787 km of telegraph lines, and 10,168 km of wire under the PTD, plus radio telegraph service in several locations operated by the navy. During the late 1920s radio telegraph stations were established in fifty districts and provinces where lines had not been installed. By 1930, therefore, the cities and major towns were linked with Bangkok and, in some cases, directly with each other.

Realizing the extent of demand for telephones, the government in 1935 ordered two exchanges (totaling 3,500 lines) from General Electric Company of England. They were installed in 1936 by Thai engineers, who had been sent to England for training, under the supervision of English engineers. In 1937 radio telephone service was established in the provinces for communication with Bangkok using AM, HF systems designed in part by PTD. The same year, two more exchanges were added to try to cope with the demand for service.

The Japanese occupied Thailand during 1942–1945, and pressured the government into declaring war on Britain and the United States—although the Thai minister in Washington refused to deliver the declaration and the United States never declared war on Thailand. Still, Bangkok was bombed by the Allies. The war disrupted international services and destroyed some facilities. In 1946 radio telegraph service was resumed or started from Bangkok to London, Singapore, Shanghai, Hong Kong, and Stockholm. Other connections followed.

The first new exchange since 1937 came in 1951, with others in 1952 and 1956. At that point the number of subscribers reached 12,920.

A telegraph machine capable of handling Thai-language messages was invented in 1953. English–Thai telex machines designed in Thailand and manufactured in Japan were installed in 1957. Telex service between Bangkok and Geneva was started in 1959 and leased telegraph circuit service in 1960.

An innovation in making the government accessible to the people was made in 1962 when a telegraph service was offered allowing complaints to be sent to the government without charge to the sender. Table 11.1 provides a summary overview of the chronology of Thai telecommunications before the 1980s.

11.2 Structure

Three organizations are responsible for telecommunications: the PTD, the Telephone Organization of Thailand (TOT), and the Communications Authority of Thailand (CAT). PTD is supposed to be responsible for rules and regulations, CAT for the operation of post, telegraph, and related services, and TOT for telephone services. In practice, their authorities overlap.

All three answer to the Ministry of Communications, which is also responsible for transportation. A deputy permanent secretary and a deputy minister, who also sits in the cabinet, are responsible for the three operations. PTD is a government department headed by a director general. CAT and TOT are state enterprises, and as such are supposed to operate like private companies.

Members of the ministry and PTD are government officials. All government officials serve in one of eleven classifications under the same salary scale no matter which department or ministry they belong. PTD employed about 400 in 1989. CAT and TOT staff are not government officials, but are instead employees of state enterprises. Their salary scale is about 30 percent higher than that of government officials.

TOT is responsible for domestic telephone service, international service to Laos and Malaysia (with which Thailand has land borders), and leased circuits for domestic point-to-point transmission of voice, telegraph, radio, and television. It is governed by an eight-member board and its managing director is appointed by the cabinet. Board members are usually from MOC and other government agencies. In 1989 TOT employed about 17,000.

CAT is responsible for the postal service as well as telegraph and telex, telephoto and facsimile services, domestic radio–telephone links to some isolated areas, international telephone service for countries not served by TOT,

Table 11.1. Chronology of Thai Telecommunications before 1980

1875	First telegraph service
1881	First telephone service
1883	Postal Department and Telegraph Department established
1883	International telegraph available to Saigon via Phnom Penh
1885	Post Office Law, covers telecommunications as well
1897	Telegraph Department combined with Post Department to become the Post and Telegraph Department (PTD); laws governing telecommunications revised.
1914	Telegraph Act
1920	Radio Telegraph Law
1929	First permanent radio station
1929	Private citizens allowed to own radios. (This had been prohibited under the 1914 Telegraph Act.)
1931	First over-the-air radio station
1935	Radio Communications Act
1936	Long-distance telephone service to Tokyo
1936	Four-wire system replaced by a two-wire system
1937	Automatic dial telephone
1939	Radio telegraph service to Shanghai
1954	Telephone Organization of Thailand spun off from PTD
1958	Telephone service to Taipei via Hong Kong
1959	IBM computers installed at TOT for billing purposes
1963	Telex service to Japan
1967	Ship-to-shore services
1971	PTD installed NEC cross bar switch with fifty international circuits
1971	Radio paging service started with 200 units
1972	Car phones introduced with fifty units
1974	Bangkok phone numbers changed from five to six digits
1976	Bangkok phone numbers changed to seven digits
1976	Communications Authority of Thailand spun off from PTD
1977	Push button telephones introduced by TOT
1979	Facsimile service

and international leased circuits. The eight members of its board, as well as the governor (chief executive officer), are appointed by the cabinet. The board usually includes the deputy permanent secretary of MOC and the director of TOT. Other members come from the Department of Defense, the office of the prime minister, and other organizations approved by the cabinet. In 1989 CAT employed about 20,000 in thirty-six divisions.

Overlapping functions among various organizations is characteristic of the entire history of telecommunications in Thailand. However, the trend has been toward consolidation of like activities in a single entity. Thus, up-country telephone service in eleven provinces, together with ten exchanges, 1,600 numbers and 1,213 employees were transferred by PTD to TOT in 1960. All the remaining up-country telephone services were transferred from PTD to TOT in 1961. In 1964 TOT was given responsibility for the Telecommunication Center at Krung Kasem donated by the U.S. government. The center was used for long-distance connections in the fourteen provinces in the central, northeastern, and eastern regions.

11.2.1 Legal Foundations

Five major laws form the legal foundations for post and telecommunications in Thailand. They are the Telegraph and Telephone Act of 1934, the Post Act of 1934, the Telephone Organization of Thailand Act of 1954, the Radio Communication Act of 1955, and the Telecommunications Authority Act of 1976.

The two 1934 acts were primarily a result of the shift from an absolute to constitutional monarchy. However, the Telegraph and Telephone Act also sought to respond to growing business uses for telecom services, which had made the previous law inappropriate. The act was quite detailed in many ways, particularly regarding definitions of terms. As before, PTD was given a monopoly over telegraph and telephone business inside Thailand. Under the act, the Minister sets rates and some other regulations. An amendment in 1974 exempted the Ministry of Defense, thereby allowing it to set up its own telecommunications system. The Post Act superseded the 1897 law. The PTD remained the monopoly provider of postal services, and set rates under the act. The act was amended in 1940 to change the definition of a few terms.

The Radio Communication Act of 1955 superseded Radio Communication Acts of 1935, 1938, 1940, 1947, 1948, and 1954. Anyone importing radio communication equipment had to have permission from the government. The PTD, Public Relations Department in the Office of the prime minister, Ministry of Defense, and other government units subsequently announced were all exempted. A 1961 amendment exempted all government units automatically without having to name each unit in ministerial regulations.

The Telephone Organization of Thailand Act of 1954 separated the telephone section from PTD and made it a government enterprise called TOT under MOC. The Communication Authority Act of 1976 established CAT as a government enterprise. Prior that time the activities taken over by CAT had been handled by PTD directly. Most telegraph stations were in post offices, and most international telephone calls had to be booked and made from a post office, so it was logical to keep these services in the same organization.

11.2.2 Financial Health

As a government agency, PDT's budget is approved by Parliament, and the Department is not allowed to spend more than that. Allocations are used to prepare policies and plans and enforce rules and regulations, which are PTD's functions. PDT does not have any income of its own, but money from radio communication licenses, rental of telecommunications equipment, and fines for late renewals of licenses—all of which must be remitted to the government treasury—can be considered PTD income.

As a state enterprise, the underlying philosophy of CAT is to be profitable, and it is. CAT is allowed to pay bonuses to employees. In what can be considered payments in lieu of taxes, state enterprises remit a part of profits to the government treasury. CAT has been providing an increasing amount—although the amount relative to gross (and net) profits is lower than in the early 1980s.

TOT has obvious needs for capital beyond even its gross profits, let alone the share it retains. However, there is a severe overall shortage of funds relative to what the government wants to spend, and telecommunication needs must be balanced against these other demands. Thus, because of the government's aggregate ceilings on foreign debt levels, TOT could add only about 200,000 lines per year in the late 1980s, despite a reported half-million line backlog in 1990 and (from a different source) 1.2 million in 1991.

11.3 International Cooperation

Bangkok emerged in the 1960s as a major international airport as travel between Europe and East Asia increased. As such, the country has developed extensive aeronautical communication services and surveillance radar that is connected by microwave and satellite to other regional airports, including Bombay and Kuala Lumpur. The services are provided by Aeronautical Radio of Thailand (Aerothai), a government-owned company created in 1963 to take over a company formed in 1954 by the foreign airlines using Bangkok. In 1967 Thailand was designated the Southeast Asian telecommunications center for weather information.

In 1978 membership in the Asia Electronic Union (AEU) was transferred from the National Research Council of Thailand, a member since 1969, to PTD. PTD set up the Electronic Association of Thailand in 1980. In 1979 PTD became a member of Asia–Pacific Telecommunity (APT). The headquarters of the organization was opened in Bangkok in 1982. Thailand hosted the thirteenth meeting of signatories of Intelsat in 1983. The country has been part of the Asian–Pacific Postal Union (APPU) since 1972.

Thai students have gone to study telecommunications in the United States, Japan, Canada, England, and other countries. In Thailand, there are also a few telecommunications training centers. A training institute is organized every year for members of Asian countries, with financial support provided by the Japanese government.

11.3.1 Submarine Cable and Satellites

Thailand belongs to several consortiums that have cables both in the region and to Western Europe. The most important is the Philippines–Singapore Maintenance Agreement (which actually covers countries from Japan to Australia). Thailand's holding is 5 percent. In 1986 the cabinet gave permission for Thailand to invest in ASEAN Cableship Private Ltd. (ACPL), a company responsible for the maintenance of submarine cables. Thailand initially put about U.S.$280 thousand in this project.

In 1966 Thailand, through PTD, became the forty-ninth member of Intelsat with a 0.1 percent share. This was increased to 0.48563 percent in June 1984. GTE International was hired to construct the earth station at Sriracha in Chon Buri Province, southeast of Bangkok. It became operational on April 1, 1968. In 1971 Sriracha Earth Station I was tuned to Intelsat IV-F8 and Station II connecting to Intelsat IV-F1 became operational.

In 1979 the cabinet approved a MOC request to rent transponders on Indonesia's Palapa satellite. In 1980 the Thai Supreme Command through PTD also contracted to use Palapa. A joint earth station was approved by the cabinet in 1981 and completed in 1984. In addition, CAT constructed a main station at Sriracha and a substation along the Gulf of Siam. Fourteen substations had been constructed by 1982.

In mid-1991 the government granted Shinawatra Computer an eight-year monopoly and a thirty-year license to launch Thaisat. The first of two satellites was to be in orbit by 1993, with the government shifting usage to Thaisat as its leases expire on Palapa and other systems. The new system will address television as well as domestic communication needs.

11.4 Development, Services, and Rates

By 1978 Thailand had 146 telephone exchanges with 362,150 numbers, including 32 exchanges with 277,918 numbers in Bangkok (population over 4 million at the time) and 114 exchanges with 89,232 numbers up-country. TOT's staff numbered 10,221. TOT introduced 100 unstaffed public telephone booths for the first time.

Thailand's teledensity in the 1980s was well behind the "four dragons" and Malaysia, but it was ahead of Indonesia. Most phones in the region are in government or business offices—not homes—so density figures based on population have limited meaning. In the mid-1980s 38 percent of Thailand's telephone lines were residential, and TOT estimated that only 25 percent of the population had "access" to phone service.

TOT and CAT began providing mobile telephone and paging services in competition with each other in 1986 and each has licensed a second provider. TOT uses a VHF (around 440 MHz) system while CAT's is UHF (800 MHz). Because it is as difficult to get through traffic in Bangkok as it is to complete a telephone call, mobile service has been immensely popular.

Telex service was made available in 1962, but only six customers applied that year. In 1963 users were allowed to use their own telex machines rather than renting them from PTD. Telex subscribers can also use teleprinters to send and receive inland and international telegraph messages, using CAT circuits. Domestic calls can be made in Thai or Roman characters. In 1988 only thirty-six of the seventy-three provinces of Thailand had telex services; further expansion is still unlikely. The cost of telex is relatively low compared with telegrams but it is relatively high compared with fax.

Anyone wishing to have a telex must submit an application to CAT and release an existing telephone line or submit an application to TOT to obtain a new telephone line for the telex. In some areas, it turns out that not enough lines are available. CAT, therefore, has installed its own lines to serve its telex users. Because these lines are dedicated to telex, they do not directly compete with TOT. However, if TOT is supposed to handle the telephone network, it can be argued CAT should not be allowed to install its own network.

11.4.1 Data Communications

Apart from using voice-grade leased circuits, there are three types of data transmission over telex lines for both domestic and overseas communication available. These are International DATEL Service, International Data base Access and Remote Computing Service (IDAR), and Super Telex Transmission.

Data transmission was initially not explicitly included in the law governing telecommunications. When banks started using on line terminals, they usually leased telephone lines from TOT to connect with their mainframes. In the early 1980s CAT interpreted the law to mean it was allowed to charge fees for those using the lines for data transmission. A few banks received invoices but refused to pay, saying that CAT did not provide any service beyond a regular telephone line, which was already being paid for. CAT countered that telephone lines were supposed to be used for voice transmission, not for data. The banks appealed to PTD, which eventually ruled that the banks did need not pay CAT.

11.4.2 Rates

It is expensive to get a telephone. It is particularly expensive if one does not want to wait a long time. The waiting period was up to seven years in 1990 in Bangkok, but a nonrefundable, noninterest-bearing "deposit" of 60,000 baht ($2,250) can reduce that to about a half year. In addition, there is an active black market. It is not uncommon for people to apply for telephones in the full expectation of selling their place, and brokers will buy connection rights from failing businesses and others for resale—at prices up to $5,000. The queue-jumping "deposit" requirement began in 1968 as a way to fund system expansion. The amount was initially 15,000 baht (then equal to U.S.$750).

On the other hand, basic monthly rates have been relatively low, starting from $2 for maintenance. Subscribers may own their own (approved) telephone instruments. There is a flat charge for each local call. Long distance is very expensive: $0.24–0.72 per domestic minute and $1.20–5.80 international.

11.5 Networks and Exceptions to Telecommunications Monopoly

Thailand has experienced over a dozen coups and coup attempts in the postwar period, as well as the bloodless 1932 coup against government ministers that led to the establishment of a constitutional monarchy. Many of these involved rebel troops from up-country, moving on orders initiated in the capital. Therefore, the government does not allow any private parties, even commercial banks, to own private telecommunications networks. However, there are alternatives that give essentially the same result as a private network, and these may be considered exceptions to the monopoly.

Government agencies and ministries, such as Defense and Interior, have set up their own systems in competition with CAT and TOT for ministry-related activities. Universities are another exception. There are both government and private universities. Government universities may undertake projects—such as setting up a fiberoptic system for data communication with other universities. International associations also can be given exemptions. As of 1988 SWIFT was the only example in this category. Of course foreign embassies, notably the United States, may and have set up their own networks. Reuters Monitor Service has entered a contract with CAT to offer a service which, by law, can be offered only by CAT, TOT, or PTD.

The final exception is networks established through donation. A private party can install a system for its own use, donate it to CAT, TOT, or PTD, and then rent it back for a nominal fee. As an example, a television company signed a contract with the government to run a television station for thirty years by donating more than U.S.$10 million to establish a network, including satellite earth stations and microwave stations. All such plant became government property, the donor receives the concession to operate the stations by paying additional fees to the government. Bangkok Bank is another example, discussed below.

11.5.1 International Database Access (IDAR)

CAT began offering this service in April 1983. The number of subscribers has grown steadily from ten in 1983 to more than 100. IDAR's database and remote computing services are available from Telenet, Tymnet and a few university networks.

The equipment required to use IDAR include a sixteen-bit microcomputer and a modem. Sharp successfully bid to provide the microcomputers. CAT was renting them at U.S.$88–136 per month, depending on transmission speed, in 1987. At U.S.$88, the rental for two years easily equals the purchase price of a comparable microcomputer from Taiwan. Most customers, therefore, have purchased their own equipment and rent only the modem from CAT—subscribers may not use their own modem.

11.5.2 Society for Worldwide Interbank Financial Telecommunications (SWIFT)

SWIFT is a cooperative company created under Belgian law with an international headquarters in Brussels. It is wholly owned by its member banks, which in 1988 numbered more than 1,200 in fifty-seven countries. Members can be connected to the SWIFT network either by the use of a dedicated minicomputer handling only SWIFT functions or by the use of a direct SWIFT link (e.g., the bank's central computer is attached directly to the network).

The bank's computer is linked to the country's regional processor, which is owned and operated by SWIFT, usually via a leased telephone line, at a transmission rate of 2,400 bps. The regional processor is normally linked to the operating center via satellite channels. For Australia and Asian countries, the operating center is located in Culpepper, Virginia.

11.5.3 Banks

TOT cannot provide enough telephone lines to its customers, especially residential users. Nonetheless the banking industry occupies a large portion of the network. Several banks have their own systems based on TOT and CAT networks as well as other international networks accessible through CAT. The banks have been ready and willing to establish their own networks and to release their leased lines for public uses. For example, some banks wish to install their own microwave networks. However, under present regulations they have been prohibited from doing so. Bangkok Bank's network will be discussed as an example because the bank is the largest commercial bank in ASEAN.

In 1982 Bangkok Bank Ltd. used a fleet of vans to transport cash between branches. Dispatching was done through CAT without encryption, allowing the schedules to be monitored by anyone who cared to listen. A private radiotelephone broadcasting system with encryption was recommended, with an equipment cost of about U.S.$3 million. The thirty-three-story bank building could be used as a transmission site. However, Thai law prohibited a private concern from owning broadcast equipment. After negotiation, the bank donated the system to PTD and rented it back for a nominal fee.

Bangkok Bank's international telecommunications network is based mainly on Intelsat. Bangkok is connected through the Pacific Intelsat to Singapore, Hong Kong, Taipei, Tokyo, and SWIFT in Culpepper, Virginia. Singapore is connected through the Indian Ocean Intelsat to Brussels, which is, in turn, connected to London as well as through the Atlantic Intelsat to Culpepper. The bank has over 330 branches, 216 of which are on-line through the mainframe network. The links are provided by TOT as leased telephone lines, microwave or a combination. Transmissions to and from the head office can be made at 4800 bps, and at 2400 bps otherwise. Average daily volume in 1987 was 180,000 transactions and the peak was 320,000. Most of this was posting demand deposits, with some loan management and other accounting.

In 1987 Bangkok Bank had 123 ATMs, ninety-three in Bangkok and its clearing area and thirty in sixteen other provinces. Monthly volume (mostly withdrawals) was U$60 million for the system, with the average ATM handling 400 transactions each day. In 1988 the bank installed twenty-two more ATM machines in twenty-two other provinces.

11.5.4 Interuniversity Network

In January 1986 an interuniversity computer network called ATUNET (AIT-Thailand Interuniversity Network) was established by the Asian Institute of

Technology (AIT). The AIT-RCC (Regional Computing Center) consists of two mainframes, an IBM 3083 and an IBM 3031. AIT is located about 42 km north of Bangkok. A microwave link has been installed to connect it with the nearest TOT telephone exchange.

Packet switching is not available in Thailand, so the nodes within the country must be connected to ATUNET either through microwave or telephone line links. Dial-up lines may be used, but the reliability of this is not good and leased lines are therefore recommended—however these are only available from the Bangkok area. Satellite links, such as those using Indonesia's Palapa, can also be used, but the user must install an earth station.

ATUNET has been linked with university networks in North America and Europe, including BITNET and EARN (European Academic and Research Network). Linkage has been established via the European Nuclear Research Center (CERN) in Geneva. ATUNET is also linked to ARPANET, CSNET, MAILNET, ACSNET, IBM'S VNET, and UCCPNET.

11.5.5 Reuters

Reuters service was introduced in Thailand in 1983, but it was handled from Singapore. The network carries four principal services. These are news, quotation retrieval (which transmits prices from exchange floors), the Reuters Monitor family of services for the financial, commodity, shipping and energy markets; and the Reuters Monitor Dealing Service.

Most Reuters customers in Thailand are banking and financial institutions, but CAT has not allowed them to take full advantage of the last two services. Specifically, Thai users are not permitted to contribute or enter information into the system. This restriction partially defeats the system's purpose. For example, a bank in Thailand wishing to convey information about the rate at which it is willing to trade currency must telex or phone its branch in Singapore or Hong Kong. That branch then enters the information into Reuters system. Some claim the reason CAT does not allow entering information in Thailand is that it wants to derive revenue from the resulting telex and long distance calls.

11.6 Procurement

In the early 1990s there were more than 100 companies in Thailand providing telecommunication and related equipment and services. The larger ones are branches of multinational corporations such as Ericsson, National, NEC, and Philips.

Because all three telecommunication organizations in Thailand are government organizations, their procurement policies must follow rules and regulations on procurement issued by the Office of the Prime Minister. Some of the important points of basic policy are summarized here. There are five types of procurement: small, cost-survey, bidding, special, and special-case.

As the name *small procurement* implies, the price must be relatively low (in the late 1980s, up to around U.S.$800). The procurement officer simply obtains a quotation from a qualified supplier and gets approval from the head of the organization. The next level is *cost-survey procurement,* for items up to about U.S.$16,000. Two committees of three persons each must be established; one to ask for quotations from at least two suppliers, and the other to inspect the merchandise or service.

Bidding must be used when the cost is relatively high—more than U.S.$16,000. At least two, generally four, committees must be established. No overlap of members is allowed between committees, because the purpose is to ensure outside review of previous steps. The steps are to accept the bidding documents, open them and check for qualifications, evaluate the bids, and to inspect and accept the merchandise or service. If only one bid is received or none of the bids meet the requirements, then the bidding must be canceled and a new bid called or special procurement requested. The lowest bid is usually selected, but exceptions are allowed. If the bids all exceed budget, negotiations can be undertaken, starting with the lowest bidder.

Special procurement includes auctions, emergency situations, secret service contracts, government-to-government transactions, purchases directly from manufacturers, and when all the previous other methods have failed. Authority for special procurement is vested with the board of directors of government enterprises; for government departments, approval may have to be obtained from the minister or the cabinet, as the case may be.

Special case procurement means there is a law or a cabinet decision stipulating what to do. Many of these simply require government agencies to use each other as sources. For example, certain local transportation services must be procured from the government's Express Transportation Organization. Telephone equipment for government agencies must be obtained from TOT. Other situations relate to financing, which may require procurement from companies of the loan provider's nationality. The Japanese tie a large percentage of their "aid" in this way.

The government sometimes requires a foreign source to involve a Thai counterpart so that someone local is available to take care of operation. If a government agency wants to hire a government university, college, or another government agency, then the contract can be entered directly without having to call for bids.

11.7 Build–Transfer–Operate Strategy

In the late 1980s Thailand had insufficient capital to expand the network, but recognized the need do so in order to sustain economic development, including attracting foreign investment. Thus, it was decided in late 1989 to allow TOT to join with major international companies to build new capacity. Bids were invited to provide and maintain 3 million lines—over twice the number in service at the time. The new systems were to be in place by 1996. They would be operated jointly with TOT for a period of years—up to thirty—with revenue sharing allowing TOT's partners to recover their costs, which were estimated at $5–6 billion.

To comply with Thai law, which was not being changed to facilitate the arrangement, TOT would technically own the system. Moreover, TOT's director of corporate planning stressed that this was a "temporary" strategy to satisfy backlogged demand, necessitated by a shortage of funds rather than a desire to joint venture or introduce competition.

Initial implementation moved quickly. Five groups had submitted proposals by May 1990, and the probable winners were named in the press in August. In September the speculation was partially confirmed. The Charoen Pokphand (CP) Group, the country's largest agroindustry group, was named the winner. CP's consortium included British Telecom. Contrary to expectation, CP got all 3 million lines. It had been thought it would get the 2 million Bangkok portion while a Japanese group would get the 1 million up-country segment.

The decision was immediately protested by members of the government, who felt it should have been debated by the cabinet. However, after some haggling, a contract signing was set for February 26, 1991. As part of the review process, a World Bank team suggested the basis of the proposal was inappropriate: build–own–operate was deemed more appropriate. TOT did not like this at all because it meant CP would be competing with TOT rather than simply expanding its capacity. In any case, the signing did not take place; there was a coup on February 23.

The new government undertook a review. CP had no previous experience in telecommunications, and British Telecom was merely CP's advisor rather than a partner in the bid—a fact that had given many observers pause from the beginning. It was also felt the revenue percentage retained by CP was excessive: 84 percent in Bangkok; 78 percent up-country. Nonetheless, TOT sought to proceed. However, the review committee could not decide on a negotiating stance and disbanded in May. The entire matter became a major political issue within the government. (For additional details see the *Bangkok Post Weekly Review,* which is in English, for May 31, 1991, p. 1; Jun 21, p. 20; and Jun 28, p. 1, as well as its Jun 30 midyear Economic Review issue, pp. 26–27; also, *Far Eastern Economic Review* Jun 13, 1991, p. 22 and p. 73, and Jul 4, p. 42.)

Things may yet change again; however, in January 1992 it was announced that Nynex, a Baby Bell, and TelecomAsia would be strategic partners in the Bangkok part of the project. Nynex would acquire an equity interest in Telecomasia, which is part of the CP group, and appoint many of the key personnel. Nynex is also interested in participating in the up-country network. CP gave up the right to bid on these lines as part of keeping those in Bangkok, thereby returning the overall project to its original conception in this regard.

11.8 Conclusion

Telecommunications in Thailand has come a long way from the first telegraph service in 1875 to satellite and computer services in the late twentieth century. However, it has a long way to go in the face of demands for more quantity and better quality. There is hope that the agreement between TOT and Charoen

Pokphand will help alleviate some of the problems. The arrangement is also a possible prototype for other countries (Indonesia is doing something similar) that do not wish to fully privatize their telephone networks, but lack the capital or other elements necessary to expand and upgrade them.

Bibliography

The following are in Thai. All are published in Bangkok.

Communications Authority of Thailand. 1983. *One Hundred Years of Telecommunications in Thailand*. Prayoonwong Press.

———. 1983. *Postal Services in Thailand*. Prayoonwong Press.

———. 1983. *Telecommunication Services*. Porsamphan Press.

Post and Telegraph Department. 1970. *Satellite Telecommunications*. Co-op Press.

———. 1973. *90 Years of PTD, 1883–1973*. Army Press.

———. 1980. *97 Years of PTD*. D & S Press.

Telephone Organization of Thailand. 1979. *25 Years of TOT*. Aksornsampan Press.

III

ADVANCED NETWORKS IN TRANSITION

12
Australia

TREVOR BARR

In a vast but sparsely populated country, Australia's telecommunications system has been driven by the goal of overcoming distance and breaking down isolation. Australia is a continent of 7.7 million km^2 with a population of just over 17 million (mid-1991). However, the population is highly urbanized, with 43 percent in urban Sydney and Melbourne. The prime development focus for telecommunications has long been to provide a nationwide grid, as well as international communication gateways.

In many ways the system is remarkably sophisticated by international standards. The benchmark policy principle—to provide universal service at affordable prices—has resulted in Australia having one of the highest telephone densities in the world, with tariffs reasonably priced by any international comparison and a network well connected to the rest of the world. Australia has also developed and maintained advanced, viable local equipment production.

12.1 Bicentennial Communications: 1788–1988

Australia has always shown a propensity to take up new communications technology quickly—from the first telegraph service in the 1850s, through radio telegraph in 1912, radio broadcasting in the 1930s, and television in the 1950s, to facsimile and cellular radio services in the 1980s. The notable exception is television: Australia still had no cable or pay television services in 1992.

Telegraph was operating in Australia within a decade of its inception in the United States. The initial service in 1854 clicked its Morse code messages over the 19 km between Melbourne and Williamstown. Poles with single iron-wire lines linked Sydney, Melbourne, and Adelaide by 1858. A year later the remarkable achievement of linking Melbourne to Tasmania by submarine cable across Bass Strait was completed. Charles Todd undertook to construct a 2,900-km line between Adelaide and Darwin in the 1870s. Using information chronicled by the explorer John Stuart, who had made the first crossing of the continent just eight years earlier, and traversing some of the harshest environment in the world, Todd completed the project in 1872. Australia does offer a rich history of extraordinary telecommunications pioneers.

Within a year of Bell's 1876 patent, successful telephone trials were conducted over telegraph wires in Australia. Robison Bros in Melbourne installed Australia's first commercial telephone to link their Flinders Street office and their South Melbourne foundry, using the prestigious "1" telephone number they retained until the 1920s. In 1880, just two years after the world's first commercial exchange opened in Connecticut, Australia's opened in Melbourne. The first interstate trunk route was between Melbourne and Sydney in 1907. In their early days both telegraph and telephone services were expensive, distance dependent in pricing, and primarily available only to business organizations.

When the Australian states agreed to a federation in 1900, the Constitution vested responsibilities for postal, telegraphic, and telephonic services with the national government. With federation came formation of the Postmaster-General's Department (PMG), which was granted authority to establish, erect, maintain, and use stations and appliances for the purpose of transmission and receipt of wireless messages. Moyal's classic history of Australian telecommunications (1986) details the early problems of harnessing the disparate postal, telegraph, and telephone services into a central administration.

A Royal Commission in 1908 investigated the management, finance, and organization of PMG, and complaints about the services. Moyal reminds us that many of the vexing issues in Australian communications—dissatisfaction with postal services, network services, cross subsidies for remote areas, and interstate rivalries—are as old as the Commonwealth itself.

Radio stations 2CF and 2BL commenced broadcasting in New South Wales (Sydney) in 1923, and in 1932 the Commonwealth government established the Australian Broadcasting Commission (ABC) along lines similar to the British Broadcasting Corporation (BBC). A two-tier system emerged: There was the publicly funded national ABC, plus private stations and networks for both radio and television. Commercial Channel Nine in Melbourne initiated the first Australian telecast with the 1956 Olympic Games, but national commercial networks did not emerge until a decade later.

In 1960 significant technical changes fashioned modern telephony, including introduction of crossbar exchanges, automatic trunk dialing, and digit numbering. Telex, which became automatic in 1966, was introduced in 1954.

The government approved broadly stated objectives for telephony development that were to set the foundation for long-term national objectives. Known as the Community Telephone Plan for Australia (1960), it directed the postmaster general to progressively improve new connections service, increase the number of automatic telephones, improve distant transmission standards, and enable telephone users to dial any other subscriber within Australia. Governments after World War II were usually a coalition of two conservative parties, the Liberal Party and the Country (now National) Party. The Country Party naturally showed special interest in communications for its rural constituents, and it often held the PMG portfolio.

In 1972 the first Labour government in twenty-three years swept to power on a theme of "time for change." The election heralded institutional reforms, including significant ones for communications. In keeping with international trends at that time, responsibilities for postal and telecom services were divided between an Australian Postal Commission (Australia Post) and Telecom Australia, which was established as a statutory authority under the Telecommunications Act 1975.

These changes were the outcome of the Vernon Royal Commission in 1974. Vernon recommended separate, government-owned, corporations, each administered by a board of seven commissioners empowered to determine the conditions of service and pay for its own staff. The government maintained the prime policy role, with a minister and a Department of Postal and Telecommunications (DPT).

Responsibility for overseas telecommunications became a politically contentious issue in the mid-1970s. The Overseas Telecommunications Commission (OTC) had been established in 1946 as a commonwealth business enterprise through the merger of the international telecommunications division of Cable & Wireless Ltd. and Amalgamated Wireless Australia Ltd. Though the Vernon Commission showed unanimity on the recommended division between postal and telecommunications institutions, it split on merging OTC into Telecom. Chairman Vernon dissented because he did not believe planning would be optimized by a single authority. The Whitlam Labour government (1972–1975), with strong Telecom trade union support, continued to push for amalgamation, but a hostile senate (the upper house) voted against it. OTC maintained its institutional independence from Telecom until the 1990s.

In 1981 the coalition (Conservative) Fraser government confronted the monopoly issue by establishing the Davidson inquiry with a prime term of reference being "the extent to which the private sector could be more widely involved in the provision of existing or proposed telecommunications services."

The Davidson Report (1982) concluded that Telecom ought to develop a stronger commercial orientation, which could best be achieved by conversion to an incorporated company—Telecom Australia Ltd.—that was still, however, wholly owned by the government. The report, however, did challenge Telecom's extensive monopoly powers. It advocated allowing private operators to maintain terminal equipment and wiring in customers premises, as well as interconnection of private networks to the public switched network, with resell of excess capacity. Davidson, however, was well ahead of his time.

Members of the National (Country) Party heeded the speculative publicity about increased rural phone charges, and the recommendations languished in the political maneuvering leading up to the 1983 general election. The election brought to power a Labour government led by former trade union leader Bob Hawke, which initially assured Telecom's status quo. However, arguments about competition and liberalization were destined to resurface.

Telecom also had to face what it perceived to be a major threat from satellite technology in the 1980s. The catalyst for introduction of a domestic satellite system came from Channel Nine's commercial television network proprietor Kerry Packer. Packer knew such a system offered him both alternative distribution and the prospect of increased audiences, and the potential to break the terrestrial monopoly. His request, albeit progressively modified by government inquiries and objections, essentially became national policy in 1979. At the time of the announcement, the institutional questions concerning terrestrial and satellite control were unresolved. Instead, a "tyranny of distance" speech was delivered to Parliament by the minister of postal and telecommunications, who justified the decision on the grounds that "it is all too easy to overlook, or remain blissfully oblivious to, the plight of those of our fellow countrymen who are seriously disadvantaged by a lack of communications services and communications dependent services."

Governments of different political shades subsequently grappled with conflicting claims from many vested interests associated with the system. They also agonized over finding ways of making the national satellite authority, Aussat, both politically acceptable to Telecom supporters and economically feasible. The Fraser administration (1975–1983) proposed Aussat as a commercial company, with 51 percent public ownership. However, under successive Hawke governments (1983–1991), with their strong union power base, 75 percent of Aussat's shares were held directly by the minister for communications on behalf of the Commonwealth; Telecom took 25 percent. Telecom–Aussat's arranged marriage was widely perceived as being only temporary, because the forces of liberalization will inevitably bring major changes. Indeed, in 1991, competition in network service was introduced (see Sec, 12.8 for an extensive discussion).

12.2 Contemporary Issues

Australian telecommunications needs to be set in the context of the national political economy. Australia was labeled "the lucky country" by polemicist Donald Horne in 1964, who warned that Australia (in the mid-1960s) urgently needed to take stock of its institutions and policies and examine its attitudes to politics, business, the arts, the cities, and the country. For Horne, Australia "lived on other peoples ideas" in "a nation more concerned with styles of life than with achievement." According to the rules, he warned, "Australia has not deserved its good fortune."

For an economy heavily dependent on agricultural produce, energy, and raw materials, the downturn in world prices for iron ore and coal in the 1980s meant that Australia's economic luck had run out. In addition, protectionist policies adopted by major agricultural trading countries, especially the European Community and the United States, resulted in loss of some major overseas markets.

Australia's information industries typify the economic, trading, and industrial problems of the nation as a whole. At a time of growing consumption of information goods and services, the import dependence for information components, devices, and software packages is staggering. Senator Button, minister for industry, technology, and commerce crystallized the problem in his "1987

Information Industries Strategy'' document. ''Presently, Australia has an information industries trade gap in excess of $4,000 million which is forecast to approach $10,000 million by the early 1990s. In the absence of strong action to redress fundamental industry weaknesses, trade in information products and services will be the largest contributor to Australia's trade deficit in the next decade.'' (Throughout the chapter, unless noted otherwise, figures are in Australian dollars.)

A series of policies in the 1980s attempted to address these structural economic problems. Early on, the Hawke administration introduced measures aimed at the general revitalization of industry, including a 150 percent tax concession for R&D, accelerated depreciation of plant and equipment, and venture capital opportunities through the creation of management investment companies. These policy measures were seen as catalysts for change rather than as ongoing industry subsidies, and the government eventually phased them out.

The sense of national economic urgency has occasioned government attempts to foster a more entrepreneurial, outward looking, export-oriented and productive culture. The Department of Industry, Technology, and Commerce has negotiated a series of Partnership for Development agreements with major transnational corporations having Australian subsidiaries. The companies have promised certain levels of R&D and exports. Ericsson, GPT and Northern Telecom are among the participants. The Telecommunications Act was amended to enable Telecom to engage in joint industry ventures; a subsidiary company, Telecom Australia (International), was established in October 1986 to engage in international consultancy and project management. By 1990 it had a staff of 200 and contracts in thirty-two countries—including a widely contested one to manage Saudi Arabia's network. These are all responses to the international trading crisis and indicate an awareness of the urgent need for Australia to find its place in economic globalization.

Debates about privatization and deregulation across the whole spectrum of economic activity gained added impetus in the 1980s. The Fraser government embraced the rhetoric, but not the reality, of Friedman's 1970s monetarism and small government. Thus, at the change of office in 1983, Hawke inherited a record federal government deficit. His administration acted somewhat out of character for a Labour Party, whose members saw their prime role as redistribution of political and economic power to remedy perceived injustices of capitalism. The Australian dollar was floated, the banking system deregulated, with the introduction of foreign banks, and government austerity cut into social welfare programs. Trade deficits have continued and aggregate foreign debt (public and private) stood at around U.S.$140 billion in 1991. GDP in 1990 was U.S.$255 billion, growing at just 1.6% adjusted for inflation.

The issue of privatization of government-owned enterprises has been agonizingly divisive for Labour. Public ownership of major enterprises is deeply embedded in the Australian social fabric. Soon after Labour's re-election in July 1987, Hawke floated the prospect of possible selloffs, an action initially widely regarded within the Labour movement as ideological treachery. Gradually, debate centered on whether to privatize particular enterprises and on the crucial question of how public enterprises could raise the substantial capital required to remain technologically and commercially competitive.

Reelected to a fourth term in March 1990, the Hawke government's major policy thrust became microeconomic reform. Telecommunications became the centerpiece of a review of the efficiency of major industries. The issue of Aussat's debt could no longer be avoided. The range of vested interests and strongly held convictions about change led to a special Labour Party conference in September 1990. There, Labour took the plunge and agreed on a modest privatization program: Commonwealth Bank, Australian Airlines, and Quantas (the domestic and international air carriers, respectively). Kim Beazley, minister for transport and communications, argued for merging Telecom and OTC, and selling Aussat to someone who would then be allowed to compete with the combined publicly owned entity. Paul Keating, then Finance and Deputy Prime Minister, later Prime Minister, favored using a privatized OTC as the basis for a carrier to compete with Telecom. In the end, a duopolistic competition model was introduced; Telecom and OTC were merged into one company called AOTC (see Section 12.8) and Aussat was taken over by Optus, an international consortium.

12.3 Institutions and Services

The Australian information industry is a significant sector of the economy, with $20 billion in annual revenue, which accounts for about 5 percent of GDP. The telecommunications component—about $12 billion—had been dominated by the three public carriers—Telecom, OTC, and Aussat. Information technology markets are typically dominated by the major computer corporations. There is also significant private sector involvement in customer premises equipment and value added services.

Responsibility for policy rests with the Australian government, specifically with the minister for transport and communications. With the election of the third Hawke government in 1987, the former Department of Transport was amalgamated with Communications. The Department advises the Minister on all matters relating to provision of telecom services and is the primary architect of national policy. It also manages the radio frequency spectrum pursuant to the Radio Communications Act.

For broadcasting, a quasi-judicial statutory agency, the Australian Broadcasting Tribunal, has authority under the Broadcasting and Television Act 1942 over license renewals and investigations, as well as issues relating to standards. Though this body performs many of the functions of its U.S. equivalent, the Federal Communications Commission (FCC), it does not play a significant policy advisory role. Following a spate of takeovers in broadcasting during 1987, the chair of the Tribunal, frustrated by the its inability to review rapid ownership changes adequately, described the authority as a "toothless tiger." Its powers and functions are under review by a standing committee of the House of Representatives.

The Department of Transport and Communications (DTC) plays a far more significant role in policy formulation than does the Tribunal. In 1988, for example, DTC had no fewer than nine inquiries or reviews underway concerning major policy issues relating to telecommunications and broadcasting. These include an overall review of the regulatory environment, an inquiry into introduction of cable and pay television services, a review of national public broadcasting policy, an analysis of the prospects of community television, and an overhaul of the Broadcast Act. These are internal, departmental inquiries; there are no established mechanisms to enable broad-based public participation, and the likelihood of fundamental policy changes from this plethora of inquiries is problematic. In many ways the statutory authorities concerned with the provision of services maintain de facto policy authority for the Australian telecommunications environment.

Domestic telecom services had been provided by Telecom, which was responsible for the terrestrial public switched network, and by the domestic satellite carrier Aussat. International services were provided by OTC, which was also a member of Intelsat. Before their merger both Telecom and OTC were public corporations established under acts of Parliament. Aussat was set up as a commercial company, governed by the company's Memorandum of Articles of Association, with responsibilities according to the Satellite Communications Act 1984.

The next three sections describe the major carriers prior to their mergers and the restructuring of the industry announced in 1990 and undertaken in 1991. The now-merged Telecom and OTC continue to offer the same services described later. (Green 1990 is a summary of the situation through early 1990.)

12.3.1 Telecom

Telecom was Australia's most profitable public corporation, the telephone company Australians love to hate. It was by far the largest of the three authorities with \$8.9 billion in 1990–1991 revenues—more than five times OTC's and almost fifty times Aussat's. OTC, however, was relatively more profitable.

The Telecommunications Act 1975 vested authority in Telecom "to plan, establish, maintain, and operate telecommunications services within Australia." No person may construct, maintain, or operate telecommunications installations within Australia unless authorized by Telecom or otherwise permitted to do so by the Telecommunications Act 1975, the Wireless Telegraphy Act 1905, or the Broadcasting and Television Act 1942. No one could attach a line or item of equipment to the public switched network unless authorized by Telecom.

It must be stressed, given the country's size and population distribution, that providing universal services at affordable prices for all Australians has long been the benchmark policy objective. The proportion of households with access to a telephone has increased from 62 percent in 1975 to 90 percent in 1986. Australia ranks eighth in the world in terms of telephones per capita. Cross subsidy arguments for rural Australia were consistently used to defend the monopolistic positions of Telecom and Australia Post.

Plain old telephone service had been at the heart of Telecom's business. However it moved toward being a more broadly based network service provider, recognizing VAS is the biggest growth area. Indeed, Telecom's management since the early 1980s acted with a sense of urgency over the need to diversify its product lines and to organize management structure around selected target customer groups.

Business demand for data transmission has continually outstripped capacity. Requests for two new data transmission services introduced in the 1980s, Digital Data Service (DDS), a leased circuit service, and Austpac, a packet switched service, quickly exceeded expectations. AOTC also offers a national videotex service, called Viatel, which is similar to the UK's Prestel. Since Mobilnet, Telecom's cellular operator, began in March 1987 there has been sustained demand for mobile services.

12.3.2 OTC Ltd.

OTC was a corporate body, wholly owned by the Commonwealth of Australia, to which it paid a dividend ($46 million for 1987–1988). Treasury long regarded OTC as a cash cow, with a resultant reduction of investment funds available to the commission, much to the dismay of its senior management. Though OTC's annual turnover was only a fraction of Telecom's, its profit growth as a percentage of revenue in the 1980s was usually the best of Australia's nine major statutory corporations. Revenue reached $1.2 billion during fiscal 1988 with a record pretax profit of $190 million; these reached $1.7 billion and $276 million, respectively, in 1991.

International telephone calls were OTC's principal service and revenue earner. Access to over 200 destinations is available; about 93 percent of all calls were dialed direct in 1991. IDD rates have fallen in real terms every year since direct dial was introduced in 1976. Newer business offerings include an ISDN-based service, private network services, and an advanced fax product. OTC was the largest investor in the PacRim fiber network.

The company was also active outside Australia. It developed a surprisingly successful strategy: OTC built satellite facilities to be paid for from operating profits. The first, linked to Intelsat, was in Ho Chi Minh City, Vietnam, in 1987. The U.S.$700,000 cost was recovered in a year. This led to a larger dish and one in Hanoi and ultimately, in late 1990, to a ten-year contract worth $250 million to develop Vietnam's international telecommunications infrastructure. Cambodia and Laos have also become customers. (A summary of OTC activities in Southeast Asia is *Communications International,* Oct 1990, p. 10.) Other poor and formerly communist countries approached OTC. In addition, OTC went so far as to form a consortium with Southwestern Bell to bid for New Zealand Telecom.

12.3.3 Aussat

Aussat Pty Ltd. was incorporated in the Australian Capital Territory in 1981 for the purpose of establishing, owning, and operating the national satellite system. It was conceived as a commercial company, unlike statutory corporations Telecom and OTC, and operates under the Satellite Communications Act 1984. The company had a high debt to equity ratio (22:1 in 1991) and consistently lost money—except in 1988–1989. Senior staff publicly argued that regulatory limitations on the types of service it could offer impeded significant private network product development, and thus profitability.

The first two Aussat satellites were launched in 1985 by NASA. A similar, third satellite was launched by Ariane rocket in September 1987. In total, the Australian satellite system has forty-five transponders: each satellite has 15 transponders of 4 $\times$30 watt and 11 $\times$12 watt capacity. The first two satellites exhausted their positioning fuel during 1992–1993, making room for a second generation (L-band). Primary services have been the relay, distribution, and assembly of commercial and national television programs, as well as direct to home satellite broadcasting, and a national aviation network. There are about 100 earth stations in remote areas.

12.3.4 Equipment Manufacture and Supply

Despite the comparatively small size of the market, most of the world's major communications companies operate in Australia. Local subsidiaries of Ericsson and Alcatel–STC dominated the market in the 1980s. The industry is heavily dependent on Telecom, which has a network with leading-edge technologies in digital switching, mobile telephones, packet switching, and ISDN. As the de facto manufacturing policy architect, Telecom's purchasing practice has sought world tenders for infrastructure development, with associated Australian manufacture and support. Thus, because Telecom decided in 1975 to standardize on Ericsson's AXE switch, Alcatel made it under license in Australia. In 1990 Telecom decided to buy Alcatel's System 12 switch as a second switch type.

The communications equipment industry in Australia involves about 200 companies and has annual sales of over $1.5 billion. It exports about $250m per year and employs well over 10,000 people. There is about 70 percent real local content in major products. It makes a considerable variety of advanced products, including digital switching technology, submarine fiberoptic systems, PBXs, mobile radio telephones, and transmission equipment.

With this product range and the advantage of a weak Australian dollar, modest export performance, especially at a time of phenomenal international telecommunications growth, has been seriously questioned. Policy analysts have attempted to encourage structural adjustments in CPE manufacture in the hope export spin-offs will occur. Australia, however, faces substantial cost disadvantages as a world manufacturer, and economic ''drys'' point to the undesirability of dependence on Telecom's local preferential purchasing policy. Some argue that without Telecom support there would be almost no equipment manufacturing in Australia. Export performance has emerged as one of the most serious areas for wholesale policy review.

12.4 Subsidies

Telecom always interpreted its primary role as being delivery of nationwide telecommunications services at uniform, affordable prices. Section 6 of the Telecommunications Act 1975 describes the general complexity of demands on the common carrier, in requiring Telecom to "perform its functions in such a manner as will best meet the social, industrial, and commercial needs of the Australian people and make its telecommunications services available throughout Australia for all people who reasonably require those services." Hence, Telecom had long argued it could subsidize mandated but uneconomical services only as a monopoly.

Within certain service categories, uniform pricing is charged irrespective of location. Charges for local calls do not vary according to duration or time of day. In 1988 Telecom floated the prospect of introducing timed local calls, only to receive such a hostile public reception during a disastrous-for-Labour by-election in Adelaide that Prime Minister Hawke personally intervened to instruct Telecom management to immediately drop such plans.

Also, connection charges bear little relationship to substantial differences in the costs to Telecom, particularly for rural versus metropolitan installations. Evidence tendered to a House of Representatives standing committee on expenditures in 1986 claimed installing a typical metropolitan residential phone cost $1,500, as opposed to an equivalent rural service cost of $6,000.

The major source for subsidies is the STD network for intermetropolitan trunk calls. Telecom argued that these funds do not come from the government, and asserted "prices are still competitive in long distance traffic compared with other major developed countries." Little data are available about the extent of the cross subsidy. Evidence tendered to the Prices Surveillance Authority in 1984 by Telecom forecast annual long-distance revenues for 1984–1985 of about $1.5 billion with direct costs of about $0.45 billion.

The crux of Telecom's charging strategy was to levy on long-distance calls and business users in order to hold down prices for local calls, residential users and rural subscribers. Telecom has long maintained that such redistribution of income is socially desirable and would be threatened by any change that allowed private entrants who only wished to provide services on the highly profitable intercapital city routes. The "cream skimming" argument long carried great weight politically. Also, of course, the central place of the Country (now National) Party in coalition governments from 1949 to 1972, then again from 1975 to 1983, meant support for rural subsidy.

The Davidson Committee, which reported on telecom services to the national government in 1982, highlighted many anomalies in cross subsidy arguments. Davidson argued that uniform pricing had the effect of placing the financial burden for support of socially desirable objectives on selective groups of customers, rather than spreading them over the whole community. His committee challenged Telecom's pricing premise that low telephone rentals were provided for low income families. Davidson argued, that some wealthy families, for instance, enjoyed their membership in the nonbusiness residential group, while "struggling businesses who provide employment pay discriminatory high telephone rentals." Similarly, Davidson argued that low-income metropolitan families subsidized wealthy rural subscribers.

Other services internally cross subsidized by Telecom but regarded as desirable public interest functions have included losses associated with directory assistance, public pay phones, and the telegraph. Again, data regarding the extent of these losses are hazy. The only budgeted direct government subsidies apply to specified pension and welfare rental concessions, for which the national government reimburses Telecom annually.

Critics, Davidson included, have asked whether it was the proper task of Telecom management to make their arbitrary judgments about crosssubsidization priorities, or whether these were matters for the political judgment of elected parliamentary representatives. They further ask why cross subsidies are not clearly identified as such, with their amount readily available for public scrutiny.

There have been great variations in estimates of the extent of cross subsidy. Telecom provided estimates each year, and they have been around 6 percent of total revenue. In its presentation to the Hutchinson policy review included a subsidy figure for community service obligations "of the order of $1 billion" (Telecom 1987, p. 6). There are, of course, great difficulties associated with the methodology of estimating the exact extent of the cross subsidy, involving issues much deeper than mere disputes about accounting conventions. Ergas, in an OECD report, points out,

> It is nonetheless extraordinarily difficult, if not impossible, to rigorously determine the extent of cross subsidization inherent in a given structure of telecommunications prices. This is because joint costs, which account for some 40 percent of the total costs of a telecommunications system, are by definition difficult to allocate, and even specification of the type of cost which should be used to evaluate the cross subsidy is intensely disputed.

12.5 Private Networks and Interconnection

The essence of Australian policy on common carriage was that Telecom was the exclusive provider of domestic services, except for satellite communications provided by Aussat, and for specifically designated private networks approved by Telecom. Monopoly carriers tend to restrict access by competitive carriers where it is against their interests, and strict policies apply wherever private networks interconnect with the public common carrier. Telecom has felt obliged to move in some way towards a more flexible policy position since the late 1980s, both on private networks and on interconnection, as a result of new and increasing demands for diverse services, as well as organized political pressure.

Private networks are essentially the province of large organizations with geographically dispersed operations, such as banks, oil and mining concerns, retailers, and some state government utilities, such as the State Railway Authority of New South Wales. As well, two states, Queensland and Victoria, established their own telecom authorities, Q-Net and Vistel, to facilitate internal state government communications. Queensland chose to sell Q-Net to a major media group, Bond Corporation, in early 1988. Early 1992 found New South Wales (the most populated state) and Queensland seeking bids for private systems that included telephones, data terminals, mobile radio and video services. Contenders are joint ventures, generally including foreign communications firms.

12.5.1 Aussat and Networks

Aussat held a unique position in Australian private networks. Its prime business is to sell private leased satellite services to large users, notably broadcasting organizations in Australia. It could permit a customer or set of customers constituting a common interest group to operate or share a leased satellite circuit. Before 1991, however, Aussat could not permit customers to interconnect with the public terrestrial network unless the terms and charges are acceptable to Telecom.

What Telecom management feared was that private networks would make progressive inroads into common carrier traffic. Defence of the status quo was also made on grounds of threats to national interest. That is, ran the argument, private network expansionism has the potential both to degrade the technical standards of the network, and to evade justifiable contribution to the subsidies that make possible continued extended network construction and desirable though unprofitable services.

12.5.2 Policy

Telecom had argued that "private networks policy cannot be determined in isolation: It is heavily dependent on the approach taken to common carriage and the location of public network service boundaries" (Telecom 1987a, p. 11). Telecom safeguarded its position by defining specific categories of approved private networks as outlined earlier as well as by other regulatory measures, including those relating to facilities ownership, maintenance, and third party carriage.

Private ownership of private network facilities is normally limited to CPE (i.e., PBXs and terminal equipment). Transmission plant and equipment must be provided by and leased from Telecom unless exempted under Section 94(2) of the Telecommunications Act. Exemptions include transmission restricted to private property, public utilities, and licensed radio transmission. Not only must network attachment equipment meet approved technical standards, but maintenance by Telecom staff was mandatory for interconnected CPE. Only from the late 1980s have PBX manufacturers and vendors been permitted to offer maintenance directly to customers.

It is important to quote the grounds on which Telecom justifies this (Telecom 1987a, p. 2).

1. Telephone subscribers using the trunk network for calls make a contribution to the establishment, development, and operation of a national asset—the public telephone network, through the payment of a social contribution included in STD and other trunk call tariffs. Operators of private networks extending over a trunk distance should pay this share when they wish to interconnect to the public network to gain the benefits of using that network.
2. Local call areas in Australia, particularly the capital city calling zones, are priced on the basis of unit call fees and represent an undervalued resource, bearing in mind their size and scope and the high level of connectivity and other forms of utility involved. The value of connection from outside to local call areas is reflected in STD charges for PSTN users and, in the case of private network operators interconnecting with the network, in interconnect charges.

Telecom argues that "sharing must stop short of full-blooded third party carriage, otherwise the concept of common carriage has no separate meaning" (Telecom 1987a, p. 12). Telecom does not permit any carriage of third party traffic on behalf of any other body.

Many critics, including Newstead, formerly of Telecom's senior management, believes that the underlying premises of such positions are suspect. He questioned the validity of the interconnection charge of $2,000 per circuit, based on the concept of mandatory social contribution. The foundation for the charge, he asserted, was simply what the market would bear.

12.6 Formulating Policy

From the late 1970s, successive Australian governments have struggled to develop policy responses to the waves of technological change in telecommunications. This has been part of an essentially irrational political process, generally characterized by short-term pragmatic responses and the shelving of decisions in complex areas. A small group of policymakers in Canberra hold the balance of power in the process because the Commonwealth (federal) government holds the constitutional authority for broadcasting and telecommunications. A few politicians—members of the government of the day—together with their advisors and senior bureaucrats, with selective inputs from labor and commercial interests and a few other pressure groups, determine major public policy decisions in this Westminster system. Unlike the United States, the courts have not been a focal point for policy change.

Australian governments have generally lacked an integrated frame of reference grounded in social investigation, strategic planning, and a careful assessment of options. This is true despite the plethora of public inquiries and major reports that dominated the communications field in the 1980s. One group was responsible for the telecommunications Task Force Report, sometimes referred to as the Hutchinson Report, after the head of telecommunications policy for the Department of Transport and Communications.

The Hutchinson review, in accord with the ministerial directive, focused on four strategic policy issues: (1) the nature and extent of monopoly powers needed for telecommunications carriers; (2) the extent to which private sector involvement should be allowed or encouraged; (3) the extent to which carriers need to be restructured and relieved of government constraints to operate effectively and competitively; and (4) the way in which the industry should be regulated.

The principal submissions were from the major interest groups: the carriers, the Australian Telecommunications Users Group (ATUG), Australian Information Industry Association (AIIA), the Australian Telecom Employees Association (ATEA), Bond Media (a private conglomerate), and the Business Council of Australia.

The players were inevitably caught up in the tide of deregulation. In an industry experiencing exponential growth, pressure for greater market liberalization in the face of an entrenched public common carrier was bound to be politically significant and an ongoing force for change. The major respective positions and pressures for change can be summarized as follows.

Organized select business interests wanted a greater share of the action and growth, and they argued for substantial diminution of Telecom power and a more laissez-faire regime. Arrayed against them were three groups with strong vested interests in the old ways, but aware that the status quo was probably not sustainable. Foremost of the three was, of course, Telecom. Whatever the future regulatory rules were to be, Telecom management intended to maximize its incumbent advantage and thereby maintain its prime market power and position.

Trade union interests—which have considerable political power in Australia, especially in the highly unionized telecommunications industry—wanted to insure that an industrial status quo would be maintained. Also, there are traditional equipment suppliers with a guaranteed market in the PTT under local content regulations; they are keen to see no major structural change. The task of the Hutchinson inquiry was to weave its way through this thicket and produce a workable blueprint for Australian telecommunications.

12.6.1 Positioning

In the lead up to the Hutchinson inquiry, Telecom had signaled its awareness of the need for a new regulatory regime as a response to major structural, technological, and market changes. It was acutely aware that, at a time when the prime minister had canvassed the possibility of the privatization of certain government enterprises (although because of strong opposition he quarantined Telecom), Telecom had to be seen as an efficient and accountable publicly owned enterprise. Managing Director Mel Ward argued that when Telecom was established as a statutory corporation in the aftermath of the 1974 Vernon Inquiry, competition was not an issue. The central issue then was the relationship between Telecom and government, and the government of the day had constructed a series of management controls on a number of aspects of Telecom's operations.

Telecom called for the government to give it a much freer hand in terms of management accountability. Ward called for abolition of restrictions on levels of capital borrowing and on conditions for the employment of staff, as well as reductions on the terms of external approval needed for real estate, building activity, and purchasing policy:

> If Telecom is to be able to respond efficiently, expeditiously and viably in the new and increasingly competitive environment, then this plethora of external management controls must be removed and the enterprise allowed to manage its business and to be accountable to its owners and its shareholders on results. (Ward 1987)

Telecom's submission to the Hutchinson inquiry (Telecom 1987b) asked the government to continue to exercise its primary role in determining overall policy for the industry. For Telecom,

> Efficient industry outcomes cannot be achieved without due consideration of national policy objectives. Telecom's existing charter implies clear national policy objectives with respect to the provision of universal, nationwide access to telecommunications services and to equity between metropolitan and nonmetropolitan areas within Australia. (p. 2)

Telecom held that in the framework of restructuring there needed to be a clear separation between national policy, regulation of the industry, and business operations of enterprises in the market.

Telecom agreed with the view that policy administration should be handled by an agency independent of Telecom and other operators. It conceded to the business interests argument that Telecom could not expect to be both a player and an arbitrator. The Telecom submission suggested (pp. 4–5) that the regulator oversee national performance standards, service quality, pricing rules, and community service obligations.

12.6.2 Subsidies

One of the most contentious issues for the Hutchinson inquiry was the issue of cross subsidy. Telecom contended that a community service obligation (CSO) arose "where a government requires a business enterprise to carry out activities which they would not elect to provide on a commercial basis, or which could only be provided commercially at higher prices." CSOs, according to Telecom, involved universal access, pricing for nonmetropolitan users, and local manufacture. Its estimate of the net cost of CSOs, a term it preferred to cross subsidy, was on the order of $1 billion per annum.

Telecom derived financial advantage from its public sector exemptions from tax and other trading liabilities, and also from direct government subsidies, such as for pension and other welfare phone rental concessions. Telecom maintained CSOs were in the national interest and asserted that if a more competitive telecommunications model was to emerge, then "all competitors should share, directly or indirectly, the responsibility for underwriting the CSOs."

The incumbent giant called for any significant change on few other issues. The submission claimed it was inappropriate to consider full liberation of CPE supply and maintenance and that provision of value added services was "already quite openly competitive.'' Private network competition should not, according to Telecom, be used to ''introduce network competition by stealth,'' sharing must stop short of full third party carriage, and the interconnection policy ought to remain untouched. Hence, its public position made few policy concessions, though privately Telecom senior staff conceded major changes were inevitable. Of course, it did not miss an opportunity to seek gain. It cheekily floated a possible takeover of the two other major common carriers, OTC and Aussat. Its carefully constructed submission ended with a plea that Telecom ''must retain the ability to diversify its operations in line with industry developments and opportunities.''

12.6.3 The Advocates of Change

The forces of deregulation and liberalization were represented by a well-organized professional lobby, the ATUG. It essentially sought fundamental change involving a major diminution of Telecom's power and prime market position. Central to ATUG was that the national government had to recapture its policy authority rather than to allow Telecom so much leeway. For ATUG, new policy ought to be formulated around the notion of a two-tier regulatory structure, with Telecom confined to a monopoly of the basic voice network, while all enhanced services ought to be open to competition. The group also advocated removing Telecom's maintenance monopoly on PBX and interconnected private networks.

ATUG called for an independent regulator, although it failed to fully define the responsibilities of such a body. The principle asserted by ATUG was to the situation where Telecom ''calls most of the shots'' (Rothwell 1988, p. 4). ATUG's call was essentially for a procompetitive model not only in the area of CPE and VAS, but also for third party carriage at least by Aussat. How, asked ATUG, could Aussat compete with ''one hand tied behind its back, its only competitor sitting on its board and holding a 25-percent stake?'' They called for Telecom's holdings to be sold to the private sector. They added that third party traffic and capacity resale are reasonably possible and should occur. Later, during the 1990 debate, ATUG went further, advocating breaking Telecom into separate companies for basic, mobile, and value-added services.

The AIIA, representing computer corporations interests, also advocated major deregulation. It appealed for greater productivity and efficiency, which it considered would be achieved by dismantling the Telecom monopoly. The AIIA's two central planks were that (1) strategic policy making power should reside with the government and (2) the regulator should be independent (see Moumic 1988).

Taking a very different stance among other business interests were the larger members of the Australian Electrical and Electronic Manufacturers Association (AEEMA) and its partner organization, the Australian Electronics Industry Association (AEIA). Clearly, members and corporate suppliers such as Alcatel–STC, Ericsson, Fujitsu and NEC thrived on a market structure where their primary customer, Telecom, held a monopoly on first-phone installation and

PBX. Such interests could be threatened by a major shift in market power, and the local equipment industry would be threatened by introduction of an open purchase market. (AEEMA and AEIA positions were reported in *The Australian* of March 14, 1988, under "Manufacturers Speak Out on the Role of a Regulator.")

12.6.4 Unions

Labor unions, particularly the Australian Telecommunications Employees Association (ATEA), confronting the tide of business protest against Telecom practices, argued for retention of the existing regulatory order on the grounds it was essentially efficient and equitable. The union's secretary, Musumeci (1988), asserted the drive for a new regulatory regime was merely the outcome of certain players wishing to change the rules so they could make handsome profits. He even claimed the essential outcome of the U.S. and U.K. deregulatory experiments was that costs for the majority had risen while costs may have fallen for a select group of users. Retention of the status quo—particularly Telecom's monopoly on installation of first phones and all PBX maintenance—of course also protected ATEA-member jobs. ATEA advocated a universal tariff to maintain the concept of universal service.

The ATEA's other basic assertions were that change might result in complex and costly legal wrangles, especially since telecommunications markets cannot be neatly segmented into basic monopoly areas and competitive areas, and that deregulation could have adverse effects on trade balances (Musumeci 1988, p. 4). ATEA merged with the Australian Telephone and Phonogram Operators Association in early 1989, giving the combined union some 34,000 members.

Union views count, as ATEA/ATPOA has a strong political position—unlike its counterparts in the United States and United Kingdom—in part because it is a major contributor to the Labour party. ATEA dramatically affected policy in the late 1970s with major dislocations of the national network. It had also been critical in shaping the rules by which Aussat could operate.

12.7 Into the 1990s: Reform and Response

A set of complex forces involving new technologies and unprecedented demand for services forced fundamental adjustments to the institutional and regulatory framework of Australian telecommunications. By international standards, however, these changes have been longer in gestation and slower in evolution than in most advanced economies.

It is useful to crystallize the major questions that need to be addressed.

1. What is the most desirable model for ownership and control?
2. How will policy makers overcome the incongruity of seven pieces of legislation in this field, ranging from the Broadcasting and Telecommunications Act 1946, to the Telecommunications Act 1975, to the Australian Broadcasting Corporation Act 1983?

3. How ought the telecommunications environment be liberalized in its regulatory and operational framework?
4. How should responsibilities be allocated within the operational environment, and then properly overseen? Is there a compelling need for a new and independent body to administer the economic, social and technical functions?
5. What are the legitimate social obligations of the various players?

12.7.1 A New Framework

How have the Australian government and the players in the industry responded to such vexing issues? In May 1988 Senator Gareth Evans, then the minister for transport and communications, tabled *Australian Telecommunications Services: A New Framework*. Though the document announced considerable policy changes for Australian telecommunications that were easily the most far reaching since Telecom was established as a statutory corporation in 1975, it did not propose radical change. Australia was not about to move toward a privatized, highly deregulated policy model. a shift toward deregulation in gradual steps was essentially advocated. By Australian criteria, the document represented a significant shift of direction. By international standards it was incremental and politically cautious.

The *New Framework* declared a restructuring of the regulatory environment on its first page. This would provide for "continuing authority for the existing carriers to be the sole providers of basic network facilities and services," though there would be "increased scope for competition in the provision of network terminal equipment for connection within customer's premises" and "full scope for competition in the provision and operation of value added services."

Much to its credit, and in response to mounting pressure, the government outlined what it described as "newly articulated objectives" for Australian telecommunications policy. The objectives were to:

1. Ensure universal access to standard telephone services throughout Australia on an "equitable" basis and at "affordable" prices, in recognition of the social importance of these services.
2. Maximize the efficiency of the publicly owned telecommunications enterprises—Telecom, OTC, Aussat—in meeting their objectives, including fulfillment of specific community service obligations and the generation of appropriate returns on investment.
3. Ensure the highest possible levels of accountability and responsiveness to customer and community needs by telecommunications enterprises.
4. Provide the capacity to achieve optimal rates of expansion and modernization of the system, including introduction of new and diverse services.
5. Enable all elements of the domestic industry (manufacturing, services, and information provision) to participate effectively in the rapidly growing Australian and world telecommunications markets.

6. Promote the development of other sectors of the economy through the commercial provision of a full range of modern telecommunications services at the lowest possible prices.

Retention of public ownership of the common carriers had never really been in doubt, and the Hawke government duly declared (p. 3) that the government, "will retain all its rights and obligations as the ultimate owner of Telecom, OTC, and Aussat, and will insure that they operate consistently with the long-term national interest."

The government argued natural monopoly advantages for retaining the status quo in the basic network!

> The existing monopoly has permitted pricing structures to embody the internal cross subsidies that have been used to sustain the important general policy of providing universal access to standard telephone services at uniform affordable prices. (p. 4)

Existing monopolies were maintained for public switched data, text, video, and ISDN, as well as for leased circuits and even mobile telephones! So much for the rhetoric of deregulation that surrounded the inquiry.

The government declared that it had "recognized a need to insure that community service obligations are met to standards that are subject to government scrutiny, within cost parameters that are determined by government in a national resource allocation context" (p. 5). In particular,

> Telecom will be required to obtain the approval of the Minister for Transport and Communications for its plans to meet its community service obligations, including the associated levels of costs and cross subsidy that will be involved. The approved plans will then be set out explicitly in Telecom's corporate plans. (p. 7)

The spirit was to replace forms of government control with greater accountability by the public carriers. in keeping with new "level playing field" policies, it was stressed that carrier management would be freed to face competition, and that the government would focus on "bottom line performances" for public telecommunications agencies. Chapter 7 was given over entirely to "Freeing the Carriers From Government Restraints." Telecom and OTC lobbying also won on that issue!

The report identified delineation of the boundary between VAS, currently open to competition, and the basic switched voice and other reserved services, which had monopoly protection, as a key regulatory issue. The government decided, "Value added services will be open to full competition. There will be a licensing arrangement administered by an independent regulatory authority to insure that value added services do not intrude on the monopoly services reserved to Telecom and OTC" (p. 7). Telecom, OTC, and Aussat will be required to maintain separate accounting records for their value added services and "their VAS charges will be required to reflect the standard tariffs for associated use of monopoly facilities and services."

As far as the majority of Australian telephone users were concerned, the most urgently needed reform was to get vastly improved service for equipment installation and maintenance. The Australian media have long delighted in running a constant barrage of attacks on Telecom's alleged incompetence in the installation of basic telephone services and billing. The report argued for a progressive dismantling of established Telecom responsibilities for CPE and for greater competitive opportunities.

Despite strong trade union opposition, the government decided new service arrangements would apply from January 1, 1989, including opening PBX maintenance and allowing competitive supply of standard-feature phones for second and subsequent instruments.

The government went further in June 1989, passing legislation that opened even more areas to competition, including inside wiring, most CPE, and VAS. In addition, a price-cap plan was implemented for most services.

12.7.2 Austel

These provisions were to be administered by what was probably the most innovative policy announcement of the Evans document and the June 1989 legislation, establishment of a new independent regulatory authority to be called Austel (formally, the Australian Telecommunications Authority). This was proposed as a single specialized telecommunications agency, independent of the carriers, and answerable to government through the minister for transport and communications. In many ways, Austel was seen as the linchpin for implementing the government's blueprint for change. It was to have five major functions.

1. Maintenance of system standards. Statutory responsibility for insuring that quality and safety are protected and that interoperability is maintained throughout the public network.
2. Protection of the monopoly. The authority to administer the provisions defining the boundaries of the carrier's monopoly over specific facilities and services.
3. Protection of competitors. Where competition is permitted, Austel will promote fair and efficient market conduct, including identifying possible breaches of the Trade Practices Act.
4. Protection of consumers. Austel administers price control arrangements and specific universal service provision conditions.
5. Promotion of efficiency. Austel will monitor and report on the efficiency and adequacy of monopoly operations by Telecom, OTC, and Aussat, particularly with respect to Telecom's fulfillment of its community service obligations.

In assessing Austel's role, its first chair, Robin C. Davey, described it as facilitating, noting, "Regulators stop things; we're here to make things happen" (*Telephony,* Feb 26, 1990, p. 37). Austel set developing reciprocal agreements with other countries on equipment standards as a top priority. This will help domestic producers export, as well as promote competitive pricing in the local market.

12.8 The Restructuring

On November 8, 1990, the government announced a radical restructuring of telecommunications. After some debate and a number of amendments forced by opposition parties, relevant legislation—seven bills—was passed in May and June 1991. Telecom and OTC were to be merged, becoming Australian and Overseas Telecommunications (AOTC). Aussat, the debt-ridden satellite carrier, was to be sold "at a price determined by tender."

The new framework can be summarized as:

1. A fixed-network duopoly, licensed to supply a full range of domestic and international services. The two will be AOTC and the newcomer that takes over Aussat (which is to be the foundation of the new service).
2. The duopoly ends June 30, 1997, when full network competition will be permitted.
3. Each duopolist receives a mobile telephone system license; a third mobile operator will be selected by the end of 1992 to begin operations in 1993.
4. Domestic and international telecom services may be resold.
5. There will be full competition in public access to cordless telecommunications, subject only to Austel technical standards and class licensing.
6. A "universal service obligation" exists and is to be shared among the carriers.

Labour does not intend to privatize AOTC. The Liberal and National parties, on the other hand, appear set to move toward a Thatcher-type model involving the sale of substantial public enterprises, including AOTC, should they be elected in a general election.

Although there was some skepticism as to whether the conditions a newcomer would face would attract bids, two major international groups did contend. However, in the end the one led by Hutchinson Telecom withdrew. The government granted a license to Optus Communications to be the second carrier in November 1991. Optus will pay the government $800 million and invest some $3.1 billion (U.S.$2.5 billion) by 1997 to build its network. The payment will be used primarily to retire Aussat's $600 million (U.S.$469) debt and tax liabilities. Beyond continuing Aussat's services, the new company expected to be in business very quickly—initially reselling cellular service and providing long-distance between Sydney and Melbourne. Its own digital GMS cellular network was to be in place by 1993, and nationwide long-distance was expected by 1996. Optus is owned by an international consortium consisting of a majority shareholding by local Australian investors, as well as BellSouth and Cable & Wireless (see Table 12.1).

Only "line links" are reserved to the duopolists—defined as any means of carrying communications electronically, be it cable, satellite or microwave. Resale of capacity on leased lines and installing switches by resellers is permitted.

Table 12.1. Telecommunications Firms in Australia and Their Owners, 1992

The Duopolists

Australian & Overseas Telecommunications (AOTC)

Formed by merger of Austalia Telecom and the Overseas Telecommunications Commission. Government owned.

Optus Communications Group

Owned 24.5 percent each by Bell South (United States) and Cable & Wireless plc (United Kingdom), 51 percent by Optus Pty Ltd. Optus Pty Ltd. is 49 percent owned by Mayne Nickless Ltd. (a transportation, health services and securities firm); other owners are Australian Mutual Provident Socy (19.6 percent), AIDC Telecommunications Fund (19.6 percent) and National Mutual Life Assoc of Australasia Ltd (11.8 percent). (One of the conditions of Optus being licensed was that domestic ownership be over 50 percent within five years. Thanks to AIDC, which is controlled by the government, this was true immediately.)

AOTC will be subject to a price cap, and it and Optus will have to provide unlimited local calls to residential users.

In the now open environment, other international players are investing. For example, AAP Information Services, a local national news service, is joint venturing with MCI (of the United States) and Todd Corp (New Zealand) in a business-user oriented virtual private network service patterned after MCI's VNET. OTC and Telecom were both investing outside Australia before their merger, and continue to see themselves as major regional players. Indeed, in a May 1991 speech to the National Press Club in Canberra, then Telecom managing director Mel Ward said "Telecom/OTC could reasonably expect a customer base of 50 million lines in service, most of it overseas, with significant network operations in the region."

In response to many forces demanding change and during the country's deepest recession since the 1930s, the new framework has been offered to the country by the Labour Party, whose founders would have thought it inconceivable their successors would privatize and deregulate. The new paradigm for Australian telecommunications centers around choice and competition.

Bibliography

Albon, Robert. 1987. *Competition And Competitiveness In Telecommunications*. Centre of Policy Studies, Monash University, Melbourne.

Bureau of Industry Economics. 1988. *Regulation of Private Communications Networks In Australia,* Information Bulletin, AGPS 11.

Blampied, Chris. nd. *Telecom Australia: Cross Subsidies And Taxes*. Impact Project, University of Melbourne, Melbourne.

Button, John. 1987. In *Australian Technology Magazine,* p. 2 (Sep.).

Davidson Report, *Report of the Committee of Inquiry Into Telecommunications Services in Australia*. 1982. 3 vols. AGPS.

Davies, Anne. 1988. [Articles in] *The Australian Financial Review* of May 26, May 27, and May 31. The last, ''Evans Prickly Path To Reform'' is especially important.

Ergas, Henry. 1988a. ''What Is The Problem and Why Does It Matter?'' In *Competition And Monopoly In Telecommunications,* Seminar proceedings, Monash University, Melbourne, Feb. 9.

Ergas, Henry. 1988b. ''Telecommunications—Is The Wave A Dumper?'' Paper presented to ATUG annual conference. April.

Geller, Henry. 1988. ''Australia's Communications Futures.'' Address, Queensland Institute of Technology (May).

Green, Jeremy. 1990. ''Unwitting Vote Against Any Short Sharp Shocks.'' *Communications International,* p. 31 (May).

Horne, Donald. 1964. *The Lucky Country.* Penguin Books.

McKay, B. 1988 ''Telecommunications Manufacturing Outlook For Australia.'' Conference Paper for Institute of Engineers Australia, Bicentennial Electrical Engineering Congress. April.

Moyal, Ann. 1986. *Clear Across Australia.* Melbourne: Nelson.

Moumic, R. 1988. ''Telecommunications Industry—Rules for the Future.'' Paper presented to ATUG annual conference (Apr.).

Musumeci, Mick. 1988. ''Rules for the Future—The ATEA Perspective.'' Paper presented to ATUG annual conference (Apr.).

New Framework = Statement by the Minister For Transport and Communications. 1988. *Australian Telecommunication Services: A New Framework.* AGPS (May 25).

Newstead, Tony. 1988. ''Private Telecommunications Networks.'' Paper delivered to Australia's Communications Futures Conference, Queensland Institute of Technology, Brisbane (May).

Rothwell, Wally. 1988. ''Rules for the Future—A Users Perspective.'' Paper presented to ATUG annual conference (Apr.).

Telecom. 1986. *Telecom: The Facts* (Aug.).

Telecom. 1987a. ''Interconnection of Private Networks With Telecom Networks.'' *Policy Conditions* (Sep.).

Telecom. 1987b. *Government Telecommunications Policy Review,* p. 2 et seq. (Dec. 1).

Ward, Mel. 1987. ''The Telecommunications and Information Industry and Economic Development.'' A speech to Science And Industry Forum (Jul. 7).

13

Hong Kong

JEROME J. DAY, JR.

To understand Hong Kong's telecommunications policy and practice requires an understanding of Hong Kong's uniqueness in terms of the broad parameters and institutional frameworks within which these matters are worked out compared with those elsewhere.

First, Hong Kong is not a country. Rather, it is a British Crown Colony headed by a governor appointed by the Queen. In historical practice this means that the governor and other senior officials have been "vetted" for their posts by the British government in London. While the Hong Kong government has enjoyed the authority to legislate law for Hong Kong, to interpret this law through an independent judiciary, and to administer it, British constitutional and administrative practices naturally have predominated. (In the run-up to 1997 this has changed somewhat; see, e.g., Miners 1991).

It is not clear if the government pays much attention to the views expressed in public consultations regarding its policies, although it assiduously seeks such consultation. The direct election of legislative council members took place for the first time in 1991. Britain, however, has not developed a local representative government apparatus. This would surely have provoked China, which would have seen it as a preliminary process leading to eventual Hong Kong "independence." Although the government in Beijing may have been willing to tolerate a Hong Kong firmly administered by Great Britain, it would never have tolerated an independent Hong Kong—or "three Chinas"—considering its concern with Taiwan and the "two Chinas" problem.

The vast majority of people in Hong Kong have traditionally remained carefully apolitical. The result has been an implicit social contract whereby the government has had public acquiescence if not genuine support, and the public has had fairly unfettered pursuit of personal prosperity. In this political environment, Hong Kong's government was unusually free to set policy. Telecommunications was not an issue the public cared much about, and because it has been set by a colonial government, policy making has usually been carried out in private. This has made it difficult to discern the balancing of interests and to weigh the prospects and likelihood of changes.

The government has espoused a minimalist role—what has been described as one of "positive noninterventionism" by long-time government official Sir Philip Haddon-Cave (1989) when he was Hong Kong's Financial Secretary in 1971–1981. The effect of this is extreme reluctance to initiate government action unless the need is quite apparent and it is the only way to deal with an issue.

Another aspect is small geographical size. The practical consequence is that Hong Kong does not require domestic long-distance services. Thus, one of the major considerations in deregulatory debate and practice elsewhere does not exist. The Hong Kong Telephone Company system offers a unitary and fairly seamless territory-wide service. Until 1990 there were three telephone dialing and numbering areas, but they were only for convenience and organization of switching.

Household and business plain old telephone service (POTS) is charged on a fixed monthly basis per telephone (albeit charges are higher for business telephones than for residential ones). Installed equipment could not handle usage-based rates. Quite simply, however, there has never been sufficient dissatisfaction with the current system to warrant the added costs.

Hong Kong's role as an international trader and entrepot center, both for China and other countries of the region, has generated a very large transportation and financial services sector, as well as a highly developed tourism industry. This means there are substantial international telecom requirements. The business services and communications sectors overshadow manufacturing and require a large-capacity, high-efficiency system.

Until the mid-1980s, telecommunications policy, such as it existed, was largely based on U.K. practices as interpreted by Hong Kong Telephone, where management has been predominantly British, and the Cable & Wireless group, which was British owned. Because the business community and other private groups have considered service they receive superior to that generally available in Southeast Asia, there has been little initiative for them to seek a role in policy making.

Another unusual feature of Hong Kong's geography is the land form. The numerous islands are mountainous with rocky soil conditions. The urban area along certain seafront corridors is densely developed, but there are extensive sparsely populated hillsides—"green-lung" areas. As a consequence, telecommunications distributions systems are rather unusual. To achieve nearly complete coverage, radio and television systems make extensive use of multiple transmitters and repeaters.

Cable wire distribution systems must be installed almost entirely underground. Terrain and congestion make trenching difficult and disruptive. Though these would ordinarily be matters of small import, they took on special significance in debates over the introduction of cable television and the requirement of building an entirely new distribution network.

13.1 History

Telecommunications development largely paralleled that in Britain with a few years lag. The first submarine telegraph cables were laid to Hong Kong in 1871 by a company that later became part of C&W. The telephone was introduced in 1877.

By 1882 the first public telephone service, with fifteen subscribers, became operational. By 1905 this venture had become the China and Japan Telephone Company and had been granted a twenty-five-year license. In 1925 the company was taken over by the newly formed Hong Kong Telephone (Telco), to which the government granted the sole right to supply and operate telephone service for a period of fifty years. In 1975 the company's exclusive franchise to operate a domestic telephone network was extended for twenty years to 1995.

In the meantime, international telegraph services continued to be provided by the precursor companies of C&W. C&W began offering fixed wireless services by 1939, and the company entered arrangements with Telco to provide international telephone services. Before these were developed to any significant extent, they were interrupted by World War II.

In 1948, C&W was granted exclusive license to provide all of Hong Kong's external services and link them with the domestic network of Telco. By 1981 the parent Cable & Wireless plc had established C&W Hong Kong (C&WHK) as its local subsidiary, holding 80 percent, with the Hong Kong government having the other 20 percent. C&WHK exercises the exclusive license for international telecommunications facilities and services, which expires in 2006.

Telco and C&WHK, both private companies as far as the Hong Kong government is concerned, have been the exclusive telecom services organizations in Hong Kong. However, the British government was the sole shareholder of C&W from 1946 until it was reorganized as a public limited company and offered to the public beginning in 1981. Though C&W is privatized, the British government retains effective control through a "golden share" that confers a veto right over whatever the company may do. As a de facto instrumentality of the British government, the Hong Kong government has historically not concerned itself with the local activities of this organization since 1946. Similarly, before 1946, British colonial governments would not normally intrude into the local activities of "home country" companies. The local government's role was to aid and abet the interests of such companies.

Telco had historically been independent of C&W, although the two shared a close working relationship. Additionally, the Hong Kong government held shares in the company. During 1983–1984 C&W purchased a majority holding in Telco, ending its independence. This effectively placed Hong Kong's domestic telephone services and international telecom services under the same corporate management.

In early 1988 the corporate structure of these companies was rationalized when a new holding company, Hong Kong Telecommunications Ltd. (Hong Kong Telecom, HKT), was set up. It holds a 100 percent interest in Telco and a 100 percent interest in C&W Hong Kong. Additionally, it holds interests (usually 100 percent) in a group of nonregulated telecom services companies that had been deregulated subsidiaries of the regulated Telco.

HKT is owned in turn by the former owners of Telco and C&WHK. These owners, and their initial shares of the new holding company are C&W plc (United Kingdom) 79.5 percent, Hong Kong government 11 percent, and public shareholders 9.5 percent. In December 1988 the first two sold part of their holdings in a global stock offering (and shares began trading on the New York Stock Exchange).

In March 1989 an entity of the Chinese government, China International Trust & Investment Corp. (CITIC), bought a 20 percent stake from C&W (part of which had been purchased by C&W from the Hong Kong government). CITIC got the shares at a 15 percent discount to the market and put up less than 20 percent in cash. At that point, ownership of Hong Kong Telecom was about 58.6 percent C&W plc, 20 percent CITIC, 18 percent public, 3.4 percent Hong Kong government. The government later sold its remaining stake.

From the companies' point of view, once C&W owned Telco, a holding company with clear separation of regulated and unregulated activities made sense. Also, as a matter of localization, the new structure enabled C&W's Hong Kong operations to be placed more clearly under a Hong Kong–based company, Hong Kong Telecom, whose shares are traded on the local exchange. Such localization is quite commonly seen as desirable and expedient in view of the changes anticipated in 1997.

There are indications certain individuals within the government were not altogether happy with the amalgamation into a holding company because it tended to lessen competition, albeit in different sectors of the market. Clearly, Post Office authorities would have preferred to regulate independent companies.

13.2 Legal Foundations for Regulation of Telecommunications

The Telephone Ordinance includes provisions for the determination of rates chargeable by Telco. The Scheme of Control restricts the rate of return from basic telephone service that can be distributed to shareholders to no more than 16 percent of shareholder funds. Of the excess above 16 percent, 80 percent must be transferred to the development fund and 20 percent to capital reserves. Telco asked in 1991 that this system be replaced by a price cap system—deregulating rate of return in exchange for keeping rate increases below the inflation rate.

The Licence and Regulatory Agreement allows C&WHK to provide exclusive international telecom services in return for a 7 percent royalty on receipts payable to the Hong Kong government, which also has right of approval over the company's charges. The royalty percentage is subject to variation at five-year intervals during 1988–2006.

The legal foundations for telecommunications in Hong Kong are laid down in the Telephone Ordinance (Chapter 269 of the revised edition 1985) and the Telecommunications Ordinance (Chapter 106 of the revised editions 1983.) The

Telephone Ordinance concerns governance of relations between the Hong Kong government and Hong Kong Telephone. However, it does not embody a regulatory philosophy. Both ordinances are quite antiquated in terms of today's needs, despite minor revisions in 1983 and 1985. They are basically pre-World War II legislation.

Little of Hong Kong's past telecommunications development is of much relevance to the 1990s situation. For example, at the end of World War II, the population of Hong Kong was less than one-tenth of what it was forty-five years later, with most of the increase being refugees from China, mainly during the 1950s. Most of the telephone network has been built since the mid-1970s, during which period Hong Kong has enjoyed very high economic growth, and high growth of per capita income.

13.3 The 1980s and 1990s

The situation in telecommunications policy and practice is fluid and evolving rapidly. In contrast, the environment was very stable until about 1983: Telco had a monopoly on all domestic services except radio and television, while C&W monopolized international services. Broadcasting was and continues to be privately owned and operated under government license and is largely supervised and regulated outside the framework of the Post Office authority, except for technological issues relating to frequencies, transmission power, antenna location, and so on.

At one point there were over a half dozen radio-paging companies in operation—they are licensed on a competitive basis—but then there was a period of considerable consolidation. The field exploded again in the late 1980s, with twenty-nine operators in 1991.

Domestic radio and broadcast television issues are relatively unimportant. They are discussed here briefly because they provide insight into prevailing attitudes in favor of competition where such competition does not obviously infringe on Telco's C&WHK's franchised monopolies of "traditional" telecom services. The development of new services on a competitive basis, such as paging, also indicates a tendency to interpret previously granted "telephony" and "telegraphy" franchises in a restricted way, rather than as encompassing all types of new services that may fall under these rubrics.

13.3.1 Regulation

Prior to 1983 regulation consisted largely of oversight on telephone rates, dealing with questions of adequacy of services, and assuring adequate response to the rapidly rising demand and expansion of business and housing. The government's public housing estates and new town developments required a great deal of close interaction between the government and the telco. The government concern was largely that Telco could generate capital to finance required expansion.

Partial deregulation began in 1983. Influenced by such moves elsewhere, a decision was made that certain telephone services would be offered to the public in competition with other vendors. Prices and other aspects would largely be regulated by market forces. This was a major policy departure, the complications and ramifications of which were largely unforeseen.

It is unclear whether deregulation was instigated by the government and the telecom authorities or by Telco itself because both have claimed that they exercised the initiative. What is clear, from the lack of significant external debate and the speed with which the whole process occurred, is that both the authorities and the company were very favorably disposed to the concept from the outset, and that they quickly reached agreement on the nature and scope of deregulation. No one was opposed to it, although few people outside the telephone company and government were even aware of the matter. Most who were perceived it as a matter between the government and Telco.

It was agreed that CPE would cease being a telco monopoly. Subject to receiving permission-to-connect to the public network on the basis of adhering to technological standards, phones, key-lines, PABXs, and other CPE would be supplied, installed, and maintained competitively. In addition, various value added services, such as cellular, would be provided competitively.

Telco set up two unregulated subsidiaries, Communications Services Ltd. (CSL) and Integrated Business Systems Ltd. (IBS), and numerous other domestic vendors quickly appeared. The bulk of Telco's product lines and unregulated services are handled through CSL. These include data message switching, View Data, cellular mobile radio telephone, and telephone directory advertising.

The importance of similar deregulatory action in the United Kingdom should be emphasized. Deregulation was quickly accepted by the government and business community (to the extent it was aware) simply because it had become British government policy and because it was in accord with Hong Kong's laissez-faire proclivities. Thus the authorities' reasons for pursuing deregulation are fairly clear.

At the same time, Telco saw provision of CPE as a very lucrative activity whose full profit potential could not be realized for stockholder benefit because of the return on equity lid placed on company operations. To some degree, the subsidiaries have a market advantage because everyone knows they are "part of the telephone company." All companies in the group use the same logo.

Despite the higher profits HKT has garnered from deregulation, points of friction have been alleviated because greater choice is now afforded, and services have been improved. Friction had been relatively minor, and it was mostly among business telephone services users. They wanted greater freedom to meet their telecom requirements along lines developed in the United States and emerging in the United Kingdom.

There have been at least two unexpected consequences of this deregulation. The first is a growing awareness on the part of commercial organizations that there are alternatives, which calls for active management of telecommunications by a new generation of managers. This works to the disadvantage of monopoly regimes. Second, the very success of this initial deregulatory effort has led to the question that, if a little deregulation is good, would not more be better?

A growing awareness of service improvements and cost reductions available in a competitive environment is leading managers, some elements of the general public, and independent services and equipment vendors to call for more deregulation. At the same time, greater competition and its benefits are being deprecated by the suppliers of franchised services (Telco and C&WHK).

Further changes have occurred in the institutional structure. When CSL and IBS were set up, certain other company activities were also taken into subsidiary outside the regulatory scheme. Significantly, all the internal data processing and information systems operations moved into a new company, Computasia. Computasia sells data processing services to the telephone company group as well as computer software and processing services to organizations in Hong Kong and abroad.

Another company, Telco Properties, has been set up for developing and managing the real estate. It is not clear to what extent it will sell services to Telco. All the unregulated operations became Telco sibling companies in 1988 when all became subsidiaries of the Hong Kong Telecom holding company.

Computasia and Telco Properties are mentioned because, from the regulatory point of view, their operations raise issues regarding pricing of services sold to the telco, which become part of the cost basis of regulated services. The possibility of cross subsidization between regulated and unregulated portions of operations is obviously a vexing one. If it occurs, the quality of the competitive environment suffers, as does the public interest. The official Telecoms position is that such cross subsidization does not occur. The official regulatory authorities position is that they will not permit it. Despite these protestations, adequate mechanisms do not exist to know. Further, there have been disputes regarding cross subsidization between Telecoms and the authorities regarding rate proposals and transfer prices between Telecoms regulated and unregulated subsidiaries.

The government exercises regulatory oversight under the Telecommunications Ordinance, which is rather antiquated in terms of the different technological environment at the time it was written. For example, it envisages multiple competing uses and users of limited resources, such as the radio frequency spectrum, and gives the telecom authority power to grant licenses and thereby control the allocation of such resources. It does not envisage the possibility of granting exclusive licenses (franchises), and the consequent need to regulate a monopoly, even though monopoly is what occurred. Not renewing, or revoking, the license is the authorities only statutory power, plus whatever conditions are included when granting the license.

Policy, such as it is, is an accretion of ad hoc decisions made in response to specific situations and taken with insufficient consideration of broader consequences, need for broader consistency over time, and adherence to well-articulated and acknowledged goals and specifically stated objectives.

This means de jure regulation is difficult or impractical to implement. A greater degree of de facto regulation must be imposed, which in practical terms involves a greater use of "moral suasion" than may be desirable. While there appears to have been few specific cases of adverse consequences resulting from this in the past, there is concern that in anticipation of 1997, Hong Kong might be better served if it now implements a rule of law rather than a rule of men.

13.3.2 Interests the System Seeks to Protect

Current regulatory practice apparently seeks to protect the broader interests of the community while insuring that private providers accrue the financial resources needed for rapid expansion and development. These interests can be vaguely defined simply as universal telephone service at nondiscriminatory rates that do not yield an excessive return to the provider. More specifically, the primary community interests are the need for advanced services, a responsive system, and the infrastructure to insure Hong Kong's role as an international trading and financial center. This implies that low rates for households are not an interest that can be pursued to the serious detriment of services for business and commercial organizations.

Having said this, it should be emphasized that there is currently little substantial evidence of conflict or disharmony in the interests being served and protected. There is a perception that perhaps, if anything, business and commercial rates subsidize households. However, it is difficult to argue that there is substantial cross subsidization of domestic services. International services are a much more complex matter and are considered later. (Mueller 1991 discusses these issues.)

13.3.3 Development of the Network

The fact that an appropriate balancing of interests and rates has been obtained is evidenced by the rapid development of the network and services. In the mid- to late 1980s Hong Kong was developing new towns in formerly rural areas, which led to massive population shifts. The new developments primarily require household telecom services.

Demand for business and commercial lines has also grown substantially, predominantly in areas served by the existing network. As new, larger, commercial premises are developed by pulling down old buildings, the existing network also needs redevelopment. However, this demands fewer new resources than expansion into new areas.

The telephone numbering system was revamped at the end of 1989, eliminating area codes. The network is growing in technological sophistication. Most interexchange trunks are fiberoptic. The first digital Centrex service outside the United States, called *Citinet*, was introduced in 1989. All exchanges have digital switching capabilities, which means all subscribers have access to advanced service offerings based on this capability. At the end 1989, thirty-one of sixty-nine exchanges were 100 percent digital switch, and 50 percent of all exchange lines were digital. By the end of 1991 90 percent of lines were digital. The system is obviously generating financial resources to enable this rapid expansion and development of the network at affordable rates without overt cross subsidization (see Table 13.1).

13.3.4 The Telecommunications Authority

The postmaster general is the telecommunications authority (TA). He administers the telecommunications and telephone ordinances. The Post Office, an activity of the Hong Kong government (one of the few in the world to consistently run a budget surplus), appears to be adequately resourced to perform its mission. In fact, the government tends to fund its activities in a rather stringent fashion in order to maintain relatively low taxes and improve the attractiveness of Hong Kong to domestic as well as multinational business investors. From the early 1980s on the government has financed its various services and departments through fees and the like to the maximum extent practicable. While the TA is independent and financially healthy, it is not empowered to take an activist role in regulation. Indeed, such an activist role would contradict the government's view of the minimalist role it should play anyway.

Initiatives in development lie with the private sector, and the TA often appears to hold up implementation of new technology and services. The 1983 deregulation was reactive rather than proactive; government authorities hardly realized what they were getting into. This stems in part from the authority's inability to assess new technology and service proposals. In general, the staff of Telecoms and other vendors are better connected to sources of information on industry trends and technology than the very small technical staff at the Post Office. Local authorities are often out of touch with the latest developments worldwide, in part because there is no substantial telecommunications research activity or manufacturing activity in Hong Kong. This makes them excessively dependent on information from vendors and insufficiently equipped to assess it independently.

Table 13.1. Telecommunications in Hong Kong

1980	1990	Number of
1.3	3.2	Telephone subscribers (millions)
1.5	2.4	Exchange capacity (millions lines)
31	54	Teledensity (phones per 100 people)
		Cellular and mobile phones
—	95	Total (thousand)
—	1.67	per 100 people
		International
43	582	calls, outward (thousand minutes)
53	537	calls, inward (thousand minutes)

Source: Hong Kong 1981 for 1980. Cellular data are for May 1990, reported in *Far East Economic Review,* Aug 2, 1990, p. 40. Other 1990 data are from *Hong Kong Telephone Annual Report.* 1991.

13.3.5 Services and Rates

The major role in development coming from private initiative has produced a steady introduction of new services. This is especially true with paging, cellular telephones, CT2, and a full range of value added services—particularly data communications, information and packet switched data network services. While international telex services was growing at a high rate for many years, facsimile has grown even faster since the mid-1980s, and it is now commonplace for domestic as well as international communications.

Data transmission services, such as Telco's DataCom, are also being developed. DataCom includes Datapak, a local packet switched and circuit switched service including access to international public data networks; dialed data transmission over the switched telephone network; facsimile support services; 2-wire and 4-wire private circuits; and 56 Kbps and 1.5 Mbps digital services. T-1 and E-1 (2.048 Mbps) are also available.

Basic service includes unlimited territory-wide calling, but excludes the telephone, which may be rented from Telco or provided by the subscriber. In January 1991 the annual cost of a business line was U.S. $120, and U.S. $86 for a residential line. As with general inflation, these are up sharply from 1987, when the rates were U.S. $102 and U.S. $66, respectively. Prices for leased lines vary with the length and speed of the connection. Telco discourages their use and aggressively prices Datapak to encourage customers to prefer it. Rates in this area have fallen, so in 1991 an E-1 domestic circuit cost about the same as one at only 56 Kbps in 1987, and the slower speed was some 75 percent cheaper than it had been.

13.3.6 Procurement Policies

Telco and C&WHK enjoy complete freedom to purchase equipment from whomever they choose. In earlier years they purchased the great bulk of it from British firms. In the 1980s there was a substantial shift to Japanese suppliers—Fujitsu and to a lesser extent NEC—along with the acquisition of minor supplies and equipment from local manufacturers. There has been a great deal of collaboration within the Hong Kong Telecom group on equipment procurement, but no collaboration with organizations outside the group.

In the past the authorities did not collaborate on development. However, involvement is occurring regarding permission to connect previously unapproved equipment to the public network. No customer-supplied equipment can be connected to the network until type approval and "permission to connect" is granted by the telco. Once approval is obtained, vendors can freely market the items, and the telco will connect it. Obtaining approval, however, has sometimes involved a lengthy process, and outside vendors have complained that Telecom's subsidiaries got approval more readily than they. Without addressing the merits of the charge, it is agreed by all parties that the authorization process should reside with an independent body. The Post Office is being pressured by Telco to relieve it of the responsibility, but no new mechanism or organization had been set up as of early 1992.

13.3.7 Regional Collaboration and International Services

As a British colony, Hong Kong's diplomatic relations are handled by London. Working relationships with other telecom authorities result almost exclusively from C&WHK's commercial relationships with other international carriers and the national authorities with whom the international carriers interact. A special case is collaboration with China, which will be considered in the next section.

C&W working relationships are quite extensive, deriving from the company's long history in the region and its status as the former "official" carrier of the British Empire and later, to a lesser extent, the British Commonwealth. Hong Kong has long been a hub for the C&W network in Asia, which includes extensive undersea cable connections to other parts of East Asia and links to Pacific and Indian Ocean satellites.

Hong Kong generates such a high rate of incoming and outgoing international calling activity that its international telecom services require special consideration. Rates are largely set by the carriers and international conventions and are less regulated than domestic services. In conjunction with the important role that these services play in Hong Kong, this has generated some friction and is likely to generate pressures for competition in international services provision in the future.

The major issues arise from toll-splitting arrangements. Dissension is caused by the difference between what Hong Kong pays foreign carriers to connect its outbound calls to their ultimate destination, and what Hong Kong receives for connecting inbound calls. This is a set of bilateral, generally intergovernmental issues. It is also a relatively minor problem compared to the fee split between Telco & C&WHK. Because they are sibling companies, it is obviously in the group's interest for Telco to receive a relatively small fee, since the company is subject to regulation on profits and rate of return. On the other hand, it is to the public's advantage for Telco to receive relatively large fees. There have been disagreements between Telecoms and the authorities on this matter, and the issue has generated pressure to open the international market to other carriers besides.

In late 1990 Hong Kong and Singapore agreed to abolish accounting rates for calls between them. The originating carrier will keep all revenue from outbound calls. Traffic between the two is roughly equal.

13.4 Change

Hong Kong has developed a very substantial consumer electronics industry. It has been export oriented, often manufacturing to contract based on designs from by overseas buyers. The local electronics and computer industries have a very high level of foreign and multinational participation, and there has been a good deal of technology transfer. Even more important, there has been an awareness of the possibilities that these technologies afford. Additionally, local manufacturers and exporters of products such as feature telephones, portable telephones, and citizen band equipment have exerted pressure on local authorities to authorize the use of these and similar products in Hong Kong.

Historically, the government has played a very active role in promoting development of export markets, but it has not been particularly receptive to the introduction of many of the same products into the domestic market. This attitude has changed with the advent of deregulation of telecom services. Nevertheless, while the government has been very supportive of the electronics industry through its other agencies, it has not seen a role for itself in promoting development of telecom services that could provide opportunities for the local electronics industry.

It should be pointed out that the domestic market is small, and many of the manufacturers care very little about it. Indeed, they are often prohibited by their buyers from distributing locally. Many of the products are produced for the U.S. market and are not currently admitted to the U.K. market. Examples have been feature telephones, cordless telephones, citizen band radios, and radio-controlled models and toys.

13.4.1 The Role of Government

There appears to have been little debate or disagreement within the government over telecommunications policy until the late 1980s. Even if there had been, it would be difficult to discern. It has been considered inimical to the authority of the colonial government and inconsistent with the British parliamentary tradition of collective cabinet responsibility to repeat policy differences. Any indication of lack of harmony would discredit the government in the perceptions of more traditionally minded members of the community. This is changing, but only slowly.

Public satisfaction with telephone services is high. Not surprisingly, therefore, the public—individuals and businesses—has been indifferent to policy matters. It is well pleased with the broadening array of services and the declining cost of international calling. Penetration is substantial—88 percent of households have telephones. Consequently, other governmental bodies have also seen no need to take a great deal of interest. The ease and rapidity with which deregulation of CPE occurred illustrates the low profile of telecommunications issues.

13.4.2 Positions of Other Groups

Lack of public concern with deregulation is a consequence of a general lack of significant bodies for creating public awareness, such as political parties, strong labor unions, or pressure groups. This may seem to be contradicted by the intense interest in the cable television controversy. However, in that instance the public was aroused primarily because entertainment was at issue, not because of any genuine interest in policy.

A Telecommunications Users Group does exist within the structure of the Hong Kong Management Association. However, it functions more as a club of individuals with a common interest in telecommunications management and specifically renounces any role as a pressure group. Its members were only casually aware of the deregulatory moves in their early stages.

Domestic business telephone services are economical and convenient. Because charges are based on a fixed monthly fee, shopkeepers freely allow customers and even passersby to use their telephones. Directory assistance is also free. This service is so attractive, convenient, and responsive that many people use it in preference to looking up numbers.

Hong Kong also enjoys very good international services. International direct dialing (IDD) from household and business telephones (providing the subscriber has established an IDD account) and declining rates have greatly broadened access. As a consequence, there is neither pressure nor a strong stance on telecommunications policy development attributable to households or small domestic businesses.

The larger business community draws useful distinctions between domestic and international services. Business organizations are generally satisfied with both. However, larger multinational corporations, particularly those very dependent on international services, often express frustration with Hong Kong's lag in providing the range of services available in North America and Europe, for example, and with high costs and bureaucratic procedural matters compared to those in other countries. By and large, this stance and attitude has been effective in pressuring policy development along lines of greater flexibility and responsiveness. The government's outlook has changed from "This is the way it is because this is the established policy" to "This is the way we are going to change our policy because it will lead to improved services that will be attractive to international companies."

13.4.3 Growing Exceptions to Service Franchises

The 1983 deregulation did not explicitly open the market to VANs, but it did lead to a change in the market environment that enabled VANs to develop. To be specific, data VANs have never been prohibited in the domestic market, because Telco's exclusive franchise has been interpreted as extending only to domestic switched voice services since the early 1970s. International value added data networks, however, may run afoul of the C&W exclusive franchise for "telegraphy." With the entrance of new players offering CPE, other players, offering domestic value added data network services, appeared.

As a result, there has been a vigorous growth of value added services since the early 1980s, and the new players have demanded an even more open market. Among them are local companies such as Hutchinson Telecoms, Chinatel, and Tricom. International companies include several Baby Bells, notably Nynex and Bell South. Northern Telecom has entered the market, as has British

Telecom. All this new entry has publicized deficiencies in the range and quality of services and raised questions regarding the extent to which monopoly franchised services are a desirable policy regime as opposed to a more broadly competitive environment.

These debates are unfolding in an era of increased public interest and involvement in conflicts between franchised and competitive service providers. One example is a conflict between Telco and a cellular provider on fees for interconnect and call delivery services.

In general, the existence of independent vendors has caused the two monopoly providers to effectively tolerate many questionable practices, and even outright violations of their franchises. This occurs particularly over cases of attachment of unauthorized devices to the public network and installation of satellite television reception equipment. It also occurs in private network developments where very fine distinctions are drawn on whether or not third party services are being offered in violation of the franchises.

Private leased lines are readily available and are tariffed at speeds up to the T-1 level. Several such lines are in operation. There is also at least one higher-speed link provided by Telco on its fiberoptic network for video services between the two racecourse sites operated by the Jockey Club. In general, however, Telco discourages the use of private lines and the development of private networks. Because most leased lines and private network development are for data rather than voice, the company's preferred solution is that customers connect to its packet switched data network.

Nevertheless, there is widespread use of private lines. Because Hong Kong serves as a regional headquarters for many multinational companies, extensive leased-line facilities are used to connect local offices to C&WHK. Additionally, many multinationals use Hong Kong as a communications switching center for their own international private networks, which further increases the number of local private lines required.

Telco does not have a franchise for data services, so there is extensive reselling of data services using the company's leased lines. In particular, there are no restrictions on development of private data networks or on public data networks. However, there would likely be problems if a VAN wanted to offer public voice services such as a voice mail box.

VAN vendors are not required to go to the telephone company to obtain circuits. In principle, they can obtain a government license to operate their own network. Through 1991, however, there was only one noteworthy example of a group wanting to build networks for public use instead of obtaining lines from Telco—cable television.

Value added markets include financial information services, airline reservation services, public event reservation and ticketing services, electronic point-of-sale funds transfer services, credit card authorization services, and so on. Finally, there are extensive data networks serving the banking system's branch network with teller terminals and automatic teller machines (ATMs). By and large, these private networks use leased lines or the public packet-switched data service provided by Telco.

Because these networks often serve a group of disparate users whose only common characteristic is their need for the type of service offered, the existence of semi-private VANs constitutes a form of officially sanctioned third-party resale. This is exemplified by one of the ATM networks, which provides services for several independent banks, by an electronic point-of-sales network that serves a large number of retail establishments with customer-direct bank debit services, and by an educational network organized by a consortium company that sells network services to five tertiary educational institutions.

There are other cases of private networks using facilities separate from Telco. The electric power companies have their own facilities, and the statutory authority operating the mass transit railway system has its own communications lines, as does the railway statutory authority. There has been speculation that in a more deregulated environment both might offer public telecom services to clients along their respective right of ways. This is purely speculative—no action is pending to explore this possibility.

Numerous providers offer services that compete with those of Hong Kong Telecom subsidiaries. In the late 1980s several of these were voluntarily bought out and consolidated under Hutchinson Telecoms, a subsidiary of one of Hong Kong's largest business groups—Hutchinson Whampoa. The group is controlled by Li Ka-shing, arguably the wealthiest individual in Hong Kong. Hutchinson Telecom itself is not large compared to Hong Kong Telecom. However, the group—with its substantial interests in property, housing, trading, shipping (container terminal operations), one of Hong Kong's two electric power and distribution companies, and so on—is a very viable contender. As discussed in the next section, at one point it looked as though it might take on Hong Kong Telecom almost across the board.

During the 1980s most of Hutchinson's inroads were in paging and, in partnership with Motorola, cellular. CSL a subsidiary of Telco, started carphone service in 1984. A joint venture of Pacific Link (the third cellular provider), Singapore Telecom Intl. and U.S.-based Mobile Telecommunication Technologies (MTEL) received a license in November 1991 to run a voice message and paging system. MTEL is providing the technology for the system.

Hong Kong is a unique cellular market. It has the densest cell system and number of users—1,350 per square kilometer in 1990—in the world, as well as the highest proportion of users and probably the largest green-light time per user. Hutchinson has over half the market. (For more on cellular see, e.g., *Telephony*, Dec. 24, 1990, p. 17 and *Far Eastern Economic Review*, May 9, 1991, p. 55).

Telco estimates that half the international calls made from Hong Kong are faxes. Here, too, Hutchinson is providing competition—in a joint venture with AT&T Network Services—and more providers are anticipated. Under agreements among Hong Kong, the United States, Japan, and the United Kingdom made in early 1991 both fax and data services are to become more competitive. Using lines leased from Telco, the venture is seeking to in effect bypass the local loop, tying directly into AT&T-owned international lines. British Telecom is involved in a similar venture.

13.5 Cable Television Development

In 1985 the government announced its intention to call for proposals for cable television service in Hong Kong. Hong Kong has had two privately owned, commercial television broadcast companies; each offers a channel in English and one in Chinese. There is no dual language programming (that is, simultaneous broadcast of English and Chinese voice channels). In the 1960s, the early days of television in Hong Kong, there was a small cable network, but it converted to broadcast and developed into what is now the second broadcaster.

The limited choice of viewing material gradually led to public pressure for increased variety. In addition, development of a large tourism industry created pressure for increased variety and a range of programming that could be offered to hotel guests. The large, affluent, and influential foreign community has also sought more programming from their home countries, particularly given the vague perception that "it's all up there on the satellites anyway" and could somehow be made available.

While the government was well-intentioned in making its announcement, in retrospect it appears very little thought was given to what was being embarked on. In view of public demand for a wider choice of programming and the knowledge that cable provides this in the United States, and that moves were being made to introduce cable in the United Kingdom, it seems the local authorities simply decided Hong Kong should have cable too.

Several consortia were formed to offer all or part of the new services. It quickly emerged that a number of organizations only wanted to offer programming over any cable distribution facility, but two wanted to offer both the distribution network and programming.

At this point, what the government originally thought to be simply an entertainment and television matter, not greatly different from a decision to license a new broadcast tv station, developed into a full-blown policy matter. Should Hong Kong have a single regulated telecom services network or should it have two (or more) networks? What degree of regulation should be placed on each?

13.5.1 The Protagonists

Cable Television Hong Kong (CTHK) was a partially owned subsidiary of Hong Kong Telecom and, as such, a sibling of Telco. Other partners in the CTHK consortium include the Swire group (involved in property, trading, and Cathay Pacific Airways), Golden Harvest (a local Hong Kong film maker) and Viacom (a U.S. cable television company). The essence of CTHK's plan was to provide programming services using a distribution system to be put in place by Telco.

Telco's distribution system was to be based on its high-capacity fiberoptic network, billed by the company as one of the most highly developed urban fiberoptic networks in the world. At some point distribution from the exchanges to individual subscriber premises would shift to conventional coaxial cable television technology, presumably using the underground ducting systems in place for telephone cables.

The company maintained that its network could easily be upgraded to the bandwidth requirements necessary for cable for three reasons. First, main trunk capacity between exchanges is largely through installation of fiber optic cable. Second, additional cabling could be installed easily because underground duct space is available as a result of ongoing replacement of large-diameter, multi-core, copper wire with small-diameter fiberoptic cables. Third, already in place underground ducting systems, which reach most subscriber premises for telephone services, can also carry television coaxial cable, thereby minimizing the costs and disruption associated with trenching work to install cables.

Telco thus projected itself as the territory-wide provider of a switching and distribution network offering voice, data, and wide-band video services. All of this was to be developed consistently with Integrated Services Digital Network (ISDN) standards as they evolve. As the government franchised carrier for all types of voice, data, and video services, Telco would provide distribution services for any company that wanted to use the network to deliver television programming services.

The Hutchvision consortium was led by Hutchinson Whampoa and included British Telecoms, which is developing and operating cable service in the U.K. The company proposed building a new distribution network, it did not think it feasible to offer programming over a network installed and operated by the telephone company. Network operations and programming services activity were to be vested in the same organization. Hutchinson indicated that it would distribute programming only for organizations that wanted to be involved just in programming. Hutchinson was beginning to look like a vehicle for developing a second domestic common carrier.

13.5.2 Policy Issues

Several interesting perspectives began to emerge from observers of the cable television issue. Telecommunications issues began attracting widespread public interest and front page coverage in the newspapers for the first time. Many observers had doubts about the economic viability of cable in Hong Kong. Many points were raised:

- It is expensive to produce programming. Dissatisfaction with programming by the two broadcast stations is partially the result of neither seeing its way clear to finance substantial local production or pay up for better programming from abroad.
- Local demand is primarily for Chinese language material, specifically in Cantonese. China's national dialect is Mandarin; for materials not produced in Cantonese, dubbing or subtitling is necessary.
- The Hong Kong viewing market is small and exhibits large income disparities. The less wealthy segments, which provide the bulk of the potential audience, cannot afford to pay much in fees. Because of their low purchasing power, advertisers will not pay much to reach them. On the other hand, the relatively small class of wealthier households are less

inclined to television viewing, and may prefer to rent tapes to play on their VCRs.

What it came down to, many observers concluded, was that cable cannot be economically viable unless the delivery system is used for more than just cable. It is relatively straightforward for Telco to offer cable. This is because much of its wide-band network is already in place, or must be put in place as the company moves towards ISDN. Cable television would necessitate only marginal costs to expand the existing network and operate the video delivery service. At the same time, as a regulated public utility, the telephone company is virtually guaranteed a fixed rate of return on shareholder funds. If incremental investments for cable television cannot be made to yield adequate revenues and returns, then this will have to be made up by cross subsidization from other services.

The network would have to be used for other services for Hutchvision to be economically viable. In other words, Hutchinson would need to develop as a second domestic common carrier to achieve revenues commensurate with installation costs. This could be quite lucrative, depending on how prices are set and whether price competition develops between the Hutchinson and Telecom networks.

Related to this is the fact the Hutchinson group is interested in international telecommunications, satellite communications, and domestic telecommunications in China. The group clearly has a strategic goal of becoming a major player in the Asian telecom arena. The C&W group, Telco's parent, is also seeking a major role in China as well as a continuation of its major role elsewhere in Asia. Both organizations perceive that becoming a major player in China requires a strong position in Hong Kong. This requires involvement in domestic and international voice, data, and video services.

The overriding issue in the debate on who should build the Hong Kong cable television network was the question of what it would cost to build.

Telco has argued that it has most of the necessary infrastructure in place, including 80–90 percent of the underground duct space required to pull new cable into all existing buildings in Hong Kong. Hong Kong already has a great deal of urban congestion and roadway interruption caused by a high level of construction work. Anything that avoids more disruption gains a fair measure of public support.

Hutchinson, on the other hand, has argued that traditional methods of trenching are unnecessary. Specifically, the use a great deal of hand-dug deep trenching—as done by Telco and the electric companies because of the rocky soil conditions and problems with power machinery disrupting services already underground—was not necessary because its cable would not have to be laid so deeply.

13.5.3 Government Reaction

What government authorities initially viewed as an entertainment and television programming matter has developed into a full-blown policy matter. The Hutchinson proposal to build what would inevitably become a second telecommunications network threatened the licensed and regulated telephone company network. On the other hand, given Hong Kong's preference for competitive regimes, advantages were perceived in breaking the Telco monopoly.

When matters of important public policy arise in which there are substantial clashes of interests and potential for controversy and charges of government favoritism, the usual government practice is to call in consultants from outside Hong Kong. In this case, the government contracted Booz-Allen Hamilton.

The initial report was delivered in the spring of 1988. The contents have been given to the major parties (HK Telecoms and Hutchinson), but they were not publicly released. The consultant was required to develop policy alternatives and to elucidate their ramifications. In the broadest terms, these would include the economic aspects, the benefits, the disadvantages, the technological aspects, the impact on competition, the advantages and disadvantages of competition in telecommunications, practices elsewhere in the world, and so on.

13.5.4 The Government's Decision

In early July 1988 the government announced that a separate network would have to be built by the successful tenderer, and that HK Telecoms could only have a small equity participation (15 percent) in the new network. Separate licenses would be required for the network operator and programming organizations. Concern was expressed that if Telco built the network, telephone subscribers might end up subsidizing it. The government expressed the view that by requiring a second network and by limiting HK Telecom participation in it, overseas bidders would be more likely to submit proposals. (Subsequently, at least two large U.S. organizations and one Japanese one indicated interest in building or operating the network.)

The July 1988 announcement was a shock to HK Telecoms. Although it was widely speculated that the government could not shut out Hutchinson's bid to expand into the local telecommunications market, neither could it shut out HK Telecoms. According to these views, the worst HK Telecoms could expect was for the government to give both contenders a green light to go ahead if they wanted. Neither would get an exclusive license. In response to these speculations, Hutchinson had indicated it could not build a network unless it had an exclusive franchise; HK Telecoms had more or less indicated that video services were part of its ISDN strategy irrespective of whether there was another cable television network and operator in Hong Kong. In other words, the CTHK venture would go ahead, if granted a license, even if Hutchvision were also licensed.

Given the likelihood that both groups would be given a go-ahead to fight it out in the market, indications of an accommodation between the two began to emerge. At the higher levels of the parent organizations, cooperation was becoming evident. The principals had previously become involved in joint proposals for domestic telecommunications systems in China (including the AsiaSat satellite system, which Hutchinson hoped would be allowed to serve Hong

Kong), and the principals of the Hutchinson group purchased 5 percent of C&W. Cynical pundits foresaw a common Hong Kong phenomena: When competition between two parties devolves into threats to their individual profitability, they join forces. Apparently this was a factor in the government's decision to limit the equity participation of HK Telecoms in the network. Notably, however, the July 1988 statement did not say that network operators would be limited in their equity participation in HK Telecoms (or its parent C&W).

Then, in a move that again took many by surprise, in August 1989 Hong Kong Cable Communications (HKCC), which had belatedly entered the bidding process, beat out Hutchinson, which had pulled out amid government demands for an additional U.S. $130 million commitment, in addition to the U.S. $508 million the company had already agreed to lay out to build the system. Costs were estimated by the government at H.K. $5.5 billion (U.S. $705 million) to complete the system and cover possible cost overruns, despite the fact actual outlays were probably going to be nearer the H.K. $4 billion Hutchinson was committing.

Members of the HKCC consortium were Wharf Holdings, a major local commercial and residential development firm (28 percent ownership), Sun Hung Kai Properties, a large Hong Kong–based construction and development company (27 percent); US West, a Baby Bell (25 percent), Coditel SA, one of Europe's largest cable companies and a unit of Belgian trading company Tractebel (10 percent), and Shaw Brothers, the premier Asian film and television production and distribution company (10 percent). U.S.–based Paramount Pictures was an adviser and consultant.

Under the terms of the franchise, the network was to reach 75 percent of Hong Kong's 1.5 million homes within five years (by 1995) and 97 percent within ten years. The completed network envisioned 1,200 km of cable underground, and another 1,000 km within high-rise offices and flats. Enough fiber was to be installed to allow fifty-nine channels. The initial cost to the Hong Kong viewer was expected to be about U.S. $15 a month, plus an installation fee of U.S.$30. For that, viewers would have a basic package of twenty channels, including news, drama, variety, and movies—all heavily tilted towards the Cantonese speaking population. Two English channels were expected. Analysts did not expect the system to be profitable until its at least seventh year of operation.

The award put HKCC in a position to issue a well-prepared challenge to build a second telecommunications network when the telephone monopolies of C&W expire in 1995 (local telephone) and 2006 (external traffic). In addition to video services, the network could provide computer-to-computer communications and interactive information retrieval services to rival those of Telco. HKCC was licensed to provide all nonfranchised telecom services, which means essentially anything but voice service.

It is evident that the initial government announcement to license cable stumbled into an area with major policy implications. In this, the government had neither an adequate policy framework to guide development nor an adequate policy-making infrastructure and organization to cope with the situation as it developed. Further evidence of this comes from a debate in the government on other matters, not telecommunications issues, but ones that suggest a lack of policy coordination in a different aspect of cable television. These are licensing, programming standards, and related matters for the two existing broadcast television stations, which are under a government body separate from the Post Office. Debate is developing on how television programming will be licensed and monitored within government, and specifically whether the same regulatory body will handle both broadcast and "narrowcast" (cable) television. One would have hoped that an issue such as this would have been anticipated and policy coordination worked out earlier.

13.5.5 The Government's Cable Decision Is Made Irrelevant

If it was the government's intention to get a second local wireline network via the cable franchise, its own subsequent actions prevented that from happening. This is so even though the government says it does want to encourage a second network.

C&W and Hutchinson, along with the Chinese government's CITIC (which owns 20 percent of HKT) had formed a consortium to launch satellite—called AsiaSat. It went into orbit in April 1990, the first privately owned satellite in Asia. Hutchvision announced plans two months later to base a pan-Asia DBS project in Hong Kong using AsiaSat. HKCC immediately recognized this as a threat to its cable system. The government suggested that the two groups work something out, and stepped back to see what happened.

In fact, however, government decisions largely determined the outcome. In August the Executive Council said no license was necessary to connect several television sets to a satellite dish—unless it was a communal dish—and further rules would be forthcoming. This would effectively kill extensive DBS in Hong Kong and Hutchvision ran full-page ads saying if that were indeed the result of final policy, the company would move its headquarters elsewhere.

By November the government had in effect relented and Hutchvision said it could provide DBS service by the end of 1991. HKCC consortium members, already dickering among themselves, abandoned the cable project, and the government graciously said it would not impose any penalty on them for withdrawing. The result was that six years elapsed, and in the end, no cable television or second telephone network.

It is an interesting question whether or not Hutchvision would have looked to DBS if it had gotten the cable award. AsiaSat was underway while the cable controversy was dragging on, and adding DBS to it made sense—although this seems to have happened only after Hutchvision did not get the cable franchise. At some point C&W realized DBS would probably kill cable, and thus reduce if not eliminate the prospect of a local loop competitor. Thus, the cynics who during the cable controversy predicted Hutchinson and C&W would make a deal were somewhat right—although certainly not in a way anyone seems to have expected.

In mid-1992, the government issued a new Pay-TV Law intended to, once again, spark the development of cable television. This plan provided a three-year exclusive franchise after which new companies could provide competitive cable television service. However, Star TV's success would be a major concern for the new operator.

13.5.6 Star TV

Hutchvision's system, which is aimed at 2.7 billion people in thirty-eight Asian countries, is called Star TV. Test broadcasts began in May 1991. In August DBS twenty-four-hour sports programming (through a deal with U.S.-based Prime Network) began. From September on there has been a twenty-four-hour music channel (provided by MTV, customized for an Asian audience). In early 1992 there were five channels. A major remaining rub is Hong Kong's refusal to allow broadcasts in Cantonese; Mandarin is used. (Because of its Hong Kong headquarters, Star programming is subject to local regulatory control.)

Star—from Satellite TV Asia Region—can reach 2.7 billion people. It controls ten of AsiaSat's twenty-four transponders—and the others are also spoken for, so there will be no competition until there is another satellite, which will be in the mid-1990s. (For more on Star see, e.g., *Far Eastern Economic Review*, May 2, 1991, p. 42, and May 30, p. 60. For Hutchinson and its telecom ambitions see, eg, *Asian Business*, 1991 Nov, p 29.)

13.6 China's Attitude

Many changes are in store in Hong Kong leading up to and immediately after 1997. While China has not previously been much different from any other Asian country with which Hong Kong maintains telecommunications links, this is no longer the case. Working relationships have developed between authorities in the Hong Kong Post Office and their counterparts in Beijing and Guangdong Province (which adjoins Hong Kong). Likewise, HKT and C&W Hong Kong have developed good working relationships with Chinese authorities. Currently, for example, they are involved in cellular telephone operations in China; the Hutchinson Group is similar. There has been greatly increased communication between Hong Kong and China, particularly with the special economic zones in Guangdong. (In 1991 almost one-fourth of HKT's revenues were from China traffic.)

Hong Kong has a much more sophisticated telecommunications infrastructure than anywhere in China, as well as more experience with advanced technology and the international system. It is Hong Kong's hope, and particularly the hope of the management at HKT and Hutchinson, that these strengths can make Hong Kong a base for transfer of telecommunications technology and services into China. As long as it is clearly understood that local authorities and private organizations are not usurping the role of national authorities in Beijing, there is a fair chance that Beijing will be willing. This inevitably means telecommunications policymaking in Hong Kong will be carried out with increasing sensitivity to views in Guangdong and Beijing. In any event, the march toward 1997 means Hong Kong's telecommunications relations with China will not be characterized by the same degree of independence or autonomy that they have been.

China will probably be quite happy to engage in a wide variety of joint venture projects with Hong Kong–based companies, as it is already starting to do, to allow them to develop and operate China's own domestic telecommunications system, and to provide international services. Still, China will regulate its communication system in accordance with its own practices.

13.7 Conclusion

Indications are that China prefers dealing with a large number of competing firms. C&W, along with its Hong Kong Telecom subsidiary, and Hutchinson are simply the first two substantial players. There is room for more, but they will first have to establish positions in the Hong Kong market. To this end, Hong Kong has been liberalizing the rules for entry into its local telecom services market.

An issue here is how much, in the interests of providing a base for more viable players in China, Hong Kong will further break HKT's monopoly on domestic and international distribution services. This will inevitably mean that the Hong Kong regulatory system will come to rely more on competition and market forces. Indeed, the government began a telecommunications policy review in late 1991. There are indications that the international franchise may be thoroughly eroded long before its expiration in 2006. Talk includes allowing new international voice carriers liberalized use of private circuits, in effect allowing bypass of Telco and thus reducing the value of its local franchise.

What happens after 1997 is little more than speculation. The head of China's Hong Kong and Macao Office of the State Council told the press in December 1990 that Beijing has to be consulted before any post-1997 franchises are granted. Still, Hong Kong companies clearly are acting as though they will be allowed to continue to exist and even have a major role in Chinese telecommunications. This may be whistling in the dark—but it could also be self-fulfilling.

Bibliography

Hadden-Cave, Philip. 1989. "Introduction." In D. G. Lethbridge, ed., *The Business Environment of Hong Kong*, 2d ed. Hong Kong: Oxford University Press.

Hong Kong 1991, a Review of 1990. Published annually by Hong Kong, Government Information Services.

Miner, Norman. 1991. *The Government and Politics of Hong Kong*, 5th ed. New York: Oxford University Press.

Mueller, Milton. 1991. *The Hong Kong Telecommunications Industry: A Case for Liberalization*. Chinese University of Hong Kong Press.

Ure, John. 1989. "The Future of Telecommunications in Hong Kong." *Telecommunications Policy*, December, vol. 13, no. 4, pp. 371–378.

14
Singapore

EDDIE C. Y. KUO

Singapore is a compact city-state of not quite 3 million people and limited natural resources. The economy has long depended on the city's natural harbor and location at the intersection of major trade routes, which helped it develop as a center of trade, shipping, and finance. Due to its small size, the economy is very internationally oriented; its external markets are larger than its domestic markets. (Good brief studies of Singapore are in Krause et al, 1987.)

Singapore was a British colony until 1959. For over thirty years Prime Minister Lee Kuan Yew and the People's Action Party (PAP) maintained an efficient government and political stability. Yew retired as prime minister in November 1990; PAP continues in power. The government believes in long-term planning and implements its development programs as planned. Its strategy can be described as pragmatic and interventionist. As an important component of this, the government makes a conscious effort to attract foreign investment, with the result that up to 70 percent of investment in manufacturing is by foreign firms.

During the 1960s and 1970s Singapore succeeded in an ambitious industrialization, focusing on labor-intensive, export-oriented industries. By the late 1970s and early 1980s Singapore faced severe competition from other developing countries with more competitive labor supplies and wage levels. To sustain growth Singapore had to restructure and embark on a high-value-added, technology-based, knowledge-intensive strategy.

A "second industrial revolution" was thus proposed in 1979 and implemented in the 1980s. It focused on development of banking and financial services, tourism, and high technology. This led to increased internationalization of the economy and greater dependence on multinational companies. Development of information technology and the telecommunication infrastructure played a key role in facilitating the restructuring.

The Telecommunication Authority of Singapore was designated the key institution to pioneer Singapore's information age by developing an efficient telecommunications network incorporating the latest information technologies. Singapore Telecom's role is made all the more significant because Singapore's strategic physical location has been diminished in a relative sense by modern transportation and communications. Its role as an entrepot has also declined and its competitive edge is now based on its ability to become an international center for telecommunications and services in the region (see Fong 1991).

This chapter presents three major aspects of telecommunications in Singapore. After a brief history, including the evolution of Singapore Telecom, the status of conventional services is discussed. This is followed by a section on new technologies being planned and installed. Finally, analysis of relevant policies, including new policy directions for the 1990s, are presented. (Singapore's fiscal year is April 1 to March 31 and is written as, for example, 1991–1992.)

14.1 History

Telephone service in Singapore began in 1879, three years after Bell's invention. The first exchange, with a simple standard manual switchboard of fifty lines, was operated as a private exchange by Bennet Pell, the local manager of the Eastern Extension Telegraph Company. Three years later, Pell's exchange was bought by the Oriental Telephone and Electric Company (OTEC), which had been registered in 1881 in London and held licenses to develop the telephone patents of Bell and Edison in countries east of Malta. For more than seventy years OTEC's government-granted license enabled it to dominate the colony's domestic telephone service.

In 1955 the colonial government established the Singapore Telephone Board (STB), which took over OTEC's assets and operated the network as a public undertaking. In the meantime, Singapore's international service was provided by The Telecommunication Authority of Singapore (TAS). This structure remained unchanged through the brief period 1963–1965 that Singapore was a part of the Federation of Malaysia as well as after its independence in 1965.

In April 1974 STB and TAS were merged to integrate and streamline the domestic and international telephone networks. The resulting entity was called the Telecommunication Authority of Singapore. (The Authority was referred to as TAS before 1974, then as Telecoms. In April 1989 the name Singapore Telecom formally replaced TAS. Generally, the form of the period is used; when more than one period is discussed, TAS is used.)

Under the Telecommunication Authority of Singapore Act (1974), the new Telecoms held exclusive franchise to provide, operate, and maintain national and international telecom services in Singapore in accordance with international standards and practices. It is the sole regulatory body in charge of telecom services, networks, equipment, and licensing (except broadcasting and television) in Singapore. It has been an active contributor to the country's economic development as well (see, e.g., Chen and Kuo 1985, pp. 240–44).

In 1982 the value of incorporating the postal services, then under the Ministry of Transportation, with Telecoms was recognized. The merger was deemed economically feasible and necessary to improve efficiency and avoid unproductive competition between postal mail and electronic mail. They are run as separate profit centers with significant administrative and financial autonomy (see Toh and Low 1989).

Singapore Telecom is structurally a statutory board under the Ministry of Communication. The chairman and five to seven members are appointed by the minister. Daily operations are the responsibility of a president and chief executive officer who reports to the minister and the board. Telecoms enjoys greater autonomy than government departments and can thus respond to changing conditions more effectively than other civil service structures.

Since the early 1970s Telecoms has formulated strategies and plans to develop a modern telecommunication infrastructure in line with Singapore's efforts to enhance and consolidate its position as a center of trade, commerce, finance, and communications. Acting as a major catalyst in social and economic development, particularly in the field of information communication, Telecoms has gone beyond providing basic services like telephone, telex, and telegraph. The increasing intensity and complexity of social and business activities in Singapore have generated a corresponding increase in demand for information transfer, which can be satisfied only by a much more sophisticated and efficient telecommunications system.

14.2 Basic Telecom Service

Singapore Telecom provides a range of technologically advanced services and facilities that form a comprehensive and mutually supportive telecommunications system. This section discusses the major basic services: telephone, message, data, private network message switching, radio, and postal services, and the infrastructure for international communications.

14.2.1 Telephone Services

Telephone services were first introduced to Singapore in 1879, and were completely automatic by 1930. The number of lines and the quality and kinds of services provided have grown rapidly. As befits a relatively rich urban area, Singapore has more telephones per person than any ASEAN country. There were thirty-seven lines per 100 persons in 1991, the second highest density in Asia after Japan and about the same as Hong Kong. Over half of telephone revenue has come from international calling. Table 14.1 provides overall data.

Singapore has had trunk service to peninsular Malaysia since 1931. In 1962 subscriber trunk dialing (STD) was introduced. In 1965 STD was extended to 40,000 Singapore subscribers, enabling them to direct dial Kuala Lumpur, Ipoh, and Malaaca. By 1977, all lines in Singapore had STD capacity. A year later, all subscribers could direct dial 145 towns in western Malaysia.

In 1979 Telecoms launched a ''Push-Button Telephones For All'' program to replace all rotary-dial instruments with tone-type ones. Within four years,

Table 14.1. National Telephone Service, 1968–88

Year[a]	Phones Per 1,000 Population	Cumulative Phones Installed	Increase from Prior Year (%)
1968	59.2	119,184	—
1969	66.7	136,267	14.3
1970	77.8	161,310	18.4
1971	89.5	189,847	17.7
1973	102.9	225,684	18.9
1974	115.2	256,973	13.9
1975	128.1	289,784	12.8
1976	143.4	328,749	13.4
1977	170.0	395,200	20.2
1978	201.9	475,167	20.2
1979	235.9	562,358	18.3
1980	267.2	645,028	14.7
1981	294.6	719,817	11.6
1982	321.9	795,737	10.5
1983	347.9	870,503	9.4
1984	372.3	941,489	8.2
1985	399.5	1,022,009	8.6
1986	419.3	1,084,259	6.1
1987	432.0	1,128,749	4.1
1988	447.4	1,170,000	3.7

Sources: Telecommunication Authority of Singapore *Annual Report*, various years; and *Yearbook of Statistics*, various years.

Data after 1988 are not available because of the July 1989 liberalization of first-phone ownership.

[a]Calendar yearend through 1971, then March 31 of year shown. Singapore's fiscal year ends March 31, so data here for 1988 are as of the end of fiscal 1987–1988, and so on.

Singapore had become the only country with a 100 percent push-button network, albeit a relatively small one.

International telephone services to Indonesia, Thailand, and the Philippines were established as early as 1934, and international station-to-station service was introduced in 1973. By 1978, international direct dialing (IDD) was available throughout the nation with service to thirty-five destinations. By 1990 Singapore maintained telephone links with more than 200 overseas destinations, of which over 170 could be reached by IDD. In the same year, the number of international phone calls was 77 million, compared to 16.8 million in 1986.

To cope with increasing traffic Singapore Telecom installed a second digital international telephone exchange (ITE) in 1989 in a joint venture with AT&T (as equipment supplier). The two ITE centers, each with 4,000 lines of international circuits, were expected to meet the growing demand for only a few years, and indeed by 1990 a third gateway was installed. The new systems have brought Singapore a step closer to complete digitalization of its communications network and will allow the network to evolve toward ISDN.

Having installed a network to meet basic voice needs, Telecoms initiated a number of value-added services. The Phone-Plus Service comprising six facilities—abbreviated dialing, absentee message, autoredial, call transfer, call waiting, and three-way calling—has been very popular since it was implemented in 1985. Carphones and IDD/STD Password facilities have proven popular since the late 1980s.

14.2.2 Message Services

Telegraph, telex, telexmail, telefax, and telebox are message services. Telegraph use has been on the decline since the early 1980s because faster, cheaper alternatives for overseas communications are now available. The number of telegrams sent in 1990 was less than one-fourth the 896,000 in 1976–1977. Still, telegraph remains a useful link to destinations lacking more sophisticated facilities such as telex. Such areas include China and India, both of which are important trading partners.

International telex service was introduced in 1959. By 1986–1987 telex messages from Singapore could reach 205 international destinations, 185 on a fully automatic basis. Telex traffic stood at more than 16.5 million messages, while the number of telex connections had grown to 17,895. In 1986, a new S$21.5 million telex exchange was commissioned to provide telex services for local international users. Singapore, with 6.9 lines per 1,000 population, has the world's highest telex penetration (*The Link* 1988, Mar./May). Facsimile and pagers spread quickly on introduction in the mid-1980s. Telebox, an electronic mailbox, was introduced in June 1984. Besides cutting paperwork, it is considered a cost-effective way to automate office communications without heavy investment.

14.2.3 Data Services

Data communications are very important to businesses in Singapore. The tremendous increase in data-leased circuits since the early 1980s reflects the importance of Singapore as a regional communications center. In response, Telecoms introduced a dedicated digital network in 1988 intended to meet the needs of its most sophisticated customers. The network allows bankers, brokers, and companies in Singapore to make overseas transactions via computer at 140 Mbps. This further boosts Singapore's position as a major financial center. The new domestic service complements digital international leased circuits (also known as Intelsat Business Service or IBS), which have been provided by Telecoms since 1987.

Telepac and Datel, the two major data services provided by Telecoms, allow subscribers access to various data bases and computer banks. Telepac subscribers can use their personal computers to send telexes. This complements Telex Dial-in, which allows telex users to communicate with a computer connected to Telepac for information retrieval, electronic mail, and other services. Datel provides transmission of data over national and international switched telephone networks. It caters to individuals and businesses whose traffic requirements do not warrant leasing dedicated lines.

Computer Access Service, linking terminals and computers both locally and internationally, was introduced in 1978. Subscribers can gain access to computing services in the United States via the packet-switched value-added data networks of Tymenet and Telenet. Access has been extended to vendors in Australia, Canada, France, Japan, and the United Kingdom.

14.2.4 Private Network Message Switching Services

With the increase in size and complexity of private communication networks in the business community, especially evident among multinationals operating in Singapore, effective control and increased operating efficiency are vitally important. Telecoms offers a spectrum of private network message switching services through PRISNET, commissioned in 1979. It is a computer-controlled system enabling large corporations to set up their own private international networks.

Telecoms provides three types of PRISNET services: point-to-point leased circuit, private user networks, and common interest groups networks. Two popular common-interest networks are Flight Information Distribution Service and Travelnet. These provide instant computerized travel reservations and connect travel agents to the reservation computers of participating airlines.

14.2.5 Radio Services

Three types of radio services provided by Telecoms are of particular significance: Harbourcraft Telephone Service, radio paging, and landmobile telephone. Introduced in 1960, Harbourcraft allows ships to communicate with each other and to and from the shore within harbor areas.

Automatic radio paging, commissioned in 1983, was upgraded in 1987 to include a memo service that can display messages in English, Malay, or any other language using romanized letters. The service has become highly popular. As of mid-1991 there were twelve pagers for every 100 people, a total of over 330,000, compared to just over five pagers per 100 in November 1987.

Landmobile service was introduced in the 1950s, but it was initially limited to special services such as the police and fire brigade. Operator assistance was required. By 1974, the automatic mobile telephone system (AMTS) was introduced and base stations were set up to provide island-wide coverage. AMTS was then restructured into a landmobile service and a Harbourcraft Telephone Service.

The cellular mobile telephone system (CMTS), costing S$34 million, was introduced in December 1982. It can take up to 6,000 subscribers and makes a range of facilities such as IDD, STD, and other Phone-Plus functions available to its users. By late 1991 there were over 50,000 mobile telephones connected to the network, compared with about 3,800 at the end of 1987. The number of subscribers continues to increase rapidly.

Telecoms introduced a portable cellular phone system in August 1988 that offers a range of features rivaling those available for both conventional and car telephones. Since April 1989 calls from these telephones can be made or received inside Singapore's underground Mass Rapid Transit stations and trains. According to Telecoms, Singapore was the first country in the world to have a telephone system providing this facility. A CT2 service was announced in 1990 and a contract was signed with Motorola that May; testing of the service, known as "Callzone," involving 1,000 base stations, began in January 1991. Handsets must meet type-approval and be registered with Telecoms, but they can be supplied by any manufacturer.

Another development in Telecoms' radio service involves an October 1987 agreement by Telecoms, British Telecom International, and the Norwegian Telecommunications Administration to install the first worldwide aeronautical satellite telecom service; it enables air passengers and crews to make air-to-ground telephone calls and send telex messages and data to any destination in the world. Singapore's ground earth station on Sentosa will coordinate with similar stations in Goonhilly (United Kingdom) and Eik (Norway) to provide global coverage, using Inmarsat satellites above the Pacific, Atlantic, and Indian Oceans. The network can also be used to enable ground-control points to continuously monitor a plane's position, altitude, engine, fuel, and other operating data.

14.2.6 Postal Services

Despite the advent of telex, telephones, and computers offering instant communications, the postal system remains an important channel for sending messages. About 2.4 million mail items were processed each day by Singapore's postal services in 1990, not quite one item per person per day.

Following the merger of the Postal Services Department and Telecoms in October 1982, Telecoms progressively introduced a series of measures to counter the traditionally labor-intensive nature of the postal service. Computerization of post office counter services was immediately instituted. In January 1986 Telecoms embarked on a plan to install a frontline services computer system (FSCS) with point-of-sales terminals connected to a host.

In 1991, sixty-seven post offices and fifty-four suboffices provided full postal facilities and many agency services. The latter included collection of Telecoms bills, payment of Central Provident Fund and Medisave contributions, payment of civil and police pensions, renewals of licenses for dogs, radio and television, and driving, and sale of revenue stamps and bus concession stamps. Counter services at all post offices were fully computerized by 1990.

14.2.7 International Communications

To maintain and enhance its role as a global telecommunication center, Singapore has built an extensive international communication infrastructure, and has been investing heavily in upgrading and expanding it. Satellites are an important component. The nation's first earth station was set up in 1971 at Sentosa Island. By 1988 the number of satellite circuits had leaped from 25 to 1,200, and links to other countries increased from five to fifty. The Sentosa station has two antennas carrying traffic between Singapore and the Indian and Pacific Ocean regions. In May 1987 a second, S$65 million, station in Bukit Timah was officially opened. The station can accommodate as many as five antennas. Its initial 32-m antenna carries traffic between Singapore and twenty-seven countries in the Indian Ocean region through Intelsat. A 21-m antenna installed in 1988 connects with an Intelsat satellite above the Pacific.

In addition to satellite links, submarine cables are also important. The first was established in 1859 between Singapore and Jakarta and other places in Indonesia. By the early 1890s, a worldwide telegraph network was developed. Singapore has been an important submarine cable hub for the region for decades.

An ASEAN cable network links Singapore to Indonesia, the Philippines, Thailand, peninsular Malaysia, and Kuantan–Kuching. The cable to the Philippines connects to systems providing circuits to Taiwan, Japan, and the United States. At the international level Singapore is served by three interregional submarine networks:

1. The Southeast Asia–Middle East–Western Europe (SEA–ME–WE) Submarine Cable Network. Completed in June 1986, it links Singapore to France through landing points in Medan, Indonesia; Colombo, Sri Lanka; Djibouti; Jeddah, Saudi Arabia; Suez and Alexandria, Egypt; Palermo, Italy; and La Seyne-sur-mer, France.
2. The Singapore–Hong Kong–Taiwan (SIN–HON–TAI) Submarine Network. Completed in August 1986. It has a capacity of 1,380 circuits.
3. The Australia–Indonesia–Singapore (A–I–S) Submarine Cable Network. A two-segment system commissioned in July 1984 with a capacity of 1,380 channels; links Perth with Jakarta and Singapore.

Four new submarine cable systems are planned:

1. Singapore–Brunei. This U.S. $50 million fiberoptic system with a capacity of 1,900 simultaneous conversations will stretch across the South China Sea. When completed, it will form part of the ASEAN–Pacific Optical Fiber Submarine Cable Network.
2. Guam–Philippines–Taiwan (G–P–T). Telecoms will invest $8.5 million to carry telephone, telex, high-speed data, and video transmission.
3. Hong Kong–Japan–Korea (East Asia) System. Singapore signed an agreement in January 1988 to invest S$5 million (out of a total cost of S$324 million) to build and maintain the first fiberoptic submarine cable system in East Asia. The link will stretch 4,500 km and Singapore can hook up to it in Hong Kong through the existing SIN–HON–TAI network.
4. Indian Ocean–Commonwealth (IOCOM). Telecoms' investment in this

helps enhance Singapore's international communication capability and capacity.

Singapore cooperates with and participates in the deliberations of the global telecommunications community. At the international level, this is through membership in the International Telecommunication Union (ITU), International Telecommunication Satellite Organization (Intelsat), International Maritime Satellite Organization (Inmarsat), and Universal Postal Union (UPU). At the regional level, Singapore is a member of the Asia–Pacific Telecommunity (APT), ASEAN Cable Management Committee (ACMC), and the ASEAN Committee on Shipping and Transport and Communications. By 1987, Singapore had become the thirteenth largest user of Intelsat and one of its largest investors as well as the sixth largest investor in Inmarsat.

In early 1990 Telecoms broke tradition by abandoning revenue sharing calls with Brunei. Under a new agreement the originating telco keeps all the revenue. Later in the year a similar agreement was reached with Hong Kong. In both cases the calling balance has been about equal. Other such agreements may follow.

14.3 New Information Technology for an Information Society

Telecoms has long been committed to developing a modern telecommunications infrastructure in line with Singapore's efforts to enhance and consolidate its position as a center of trade, finance, and communications.

Over the years various telecommunications networks have mushroomed, each developed as a dedicated network specifically for a single service and thus each with unique characteristics. Attempts have been made to anticipate integration of future services into existing networks. Since the 1970s and especially during the 1980s, Telecoms has experimented with and developed several new technologies and networks with the potential to augment various existing ones. Significant among these are fiberoptics, Teleview and ISDN.

14.3.1 Fiberoptics

In 1977 Telecoms in cooperation with Fujitsu (its sole supplier of domestic-network digital switches until 1991) launched one of the first fiberoptic trials to demonstrate the practicability, compatibility, and reliability of fiberoptic transmission. Using feedback and experience gained from this field trial, a nationwide multipurpose fiberoptic cable network was planned. It will accommodate growth of junction telephone traffic and serve as a backbone for ISDN.

Installation was carried out in two phases. Starting in mid-1983 the first phase aimed to provide high-capacity links among busy exchanges in Singapore. Telecoms provided the labor to lay, splice, and test the cable while Fujitsu, as cable supplier, handled overall supervision. The project was completed in 1984. With this experience Telecoms proceeded in May 1984 to implement the second phase without external assistance. This phase, linking remaining exchanges, was completed in July 1987.

Singapore boasts one of the world's highest densities of fiberoptic utilization—over 21.5 km/km^2 in 1990, when there were an estimated 13,000 km of cable, up from over 10,000 km in July 1988.

Fiberoptics has not only equipped Telecoms with the ability to meet tremendous traffic exchange demand but has also helped relieve the aggravations of expanding underground infrastructure in an urban environment. Moreover, estimated direct cost-savings of S$3 million was obtained from implementing fiberoptic systems instead of conventional copper cable. Wideband ISDN can be developed only when the copper-cable network is converted to fiber.

14.3.2 Videotex

Teleview is the name given the videotex system. It is an electronic information system allowing interactive information retrieval. While working on the videotex principle, Teleview uses a television channel for transmission instead of using only the telephone network, as is done in other countries, including Teletel in France, Prestel in the United Kingdom, Telidon in Canada, and Captain in Japan. A user needs an adapter attached to a color television monitor and an antenna connected to both the monitor and the telephone line to access Teleview. Messages and instructions are carried over the telephone network, but requested information is returned either through a UHF television channel or the telephone network. The adapter picks up the information from the antenna, then decodes and displays it. This offers two advantages over other videotex systems: (1) using radio waves for transmission gives much clearer screens, and (2) the telephone line is not necessarily tied up once a selection has been made.

Telecoms first developed the concept through a joint research and development effort with an experienced partner, Marconi of the United Kingdom, under a February 1985 contract. This allowed technology transfer through Telecoms staff sent to Britain. It was also expected to generate spinoffs to boost the local information and software industries. With the introduction of Teleview, the new market for electronic information is being opened. Companies dealing with information dissemination will gradually emerge to exploit the market. One example is Integrated Information Pty Ltd. (formerly the Yellow Pages Company), which distributes its telephone and yellow pages directories through Teleview.

In mid-1988, Telecoms began a two-year field trial of Teleview involving about 1,000 business and residential users. On completion of the trial and a determination that the project's benefits are worthwhile, Singapore Telecom launched the service on commercial basis in 1990.

The project has attracted much public attention. Reasons include its ability to provide massive amounts of information and facilitate transactions of various sorts. The fact that it is a national network (the first of its kind) means Teleview could have an impact on both businesses and individuals. Business transactions and organization, family life and community activities, and even the political process of the city-state may be affected. This is an area that warrants close monitoring and careful research by social scientists.

14.3.3 Integrated Services Digital Network (ISDN)

Under the direction of ITU, the world's major suppliers have been cooperating in a worldwide effort to create an ISDN that can interactively transmit voice, high-speed data, and even video signals over one route to a single terminal. The first milestone was an official recommendation from CCITT and ITU in 1980; this was subsequently revised and adopted in October 1984. The idea of merging existing technologies to form integrated systems in Singapore was first explored by Telecoms in 1976.

As the name ISDN implies, the integration of telecom services presupposes the existence of a digital network, and that requires the entire copper-cable network be converted to fiberoptic cable at the subscriber level. In Singapore, digital switching was introduced in 1981, making the network capable of servicing both voice and nonvoice applications.

In December 1984 Telecoms proposed a series of field trials to assess ISDN feasibility. The trials consisted of four stages. During Stage I, which commenced in April 1985, Telecoms supported digital telephone with simultaneous facsimile and other data terminal communications. Stage II, started in October 1985, added nonpacket data terminal communication from ISDN subscribers to subscribers in the packet-switched data network. Stage III, which commenced in January 1986, saw provision of telex/teletex communication from ISDN subscribers to subscribers in the telex circuit switched data networks. The last stage, begun in July 1986, insured provision of packet communication with packet terminals from ISDN subscribers to subscribers in the packet switched data network.

Phone-Plus, introduced in 1979, was one of the new services considered a revenue source for funding ISDN development. Other value-added services included Telepac, Telebox, Flight Information Display Network, and Travelnet. Telecoms' new-business policy places greater emphasis on specialized communication needs of the business sector.

In line with this, Telecoms developed plans for integrated telemetering (automated and remote reading of public utility meters), telealarm (remote fire and burglar alarm surveillance), and telecontrol (remote monitoring of lifts in high-rise buildings and remote control of electrical appliances at home).

14.4 Policy Issues and New Policy Directions

Responding to changing economic and technological conditions, Telecoms has adopted changes to its policies. This section discusses several major issues: revenues and pricing, the National Information Technology Plan (NITP), new policy orientations, and liberalization and privatization.

14.4.1 Revenues and Pricing Policies

The consensus in Singapore is that it is not economically feasible to maintain more than one telecommunication enterprise in the country's small domestic market: The ultimate result would be a natural monopoly after some wasteful competition. Moreover, telecommunications being a basic service, it has been judged necessary to entrust it to a state-run monopoly. The considerations are not exclusively economic. Thus two local economists observe that,

> Besides the public good and externality arguments, security and assurance in supply, and avoidance of wasteful competition in terms of scarce resources, especially of land, equity and even profitability may be further rationales. It is also the typical interventionist style of the government to establish public enterprises wherever and whenever private initiatives are deemed lacking (Toh and Low 1989, p. 9).

Singapore Telecom has enjoyed growing operating revenue and generated an increasing surplus over the years (Table 14.2). Thus, in the decade through 1990–1991 operating revenue increased more than 129 percent and total surplus grew over 174 percent. This was achieved in spite of a series of rate reductions. It is obvious Singapore Telecom enjoys a strong financial position, which enables it to be totally self-reliant in upgrading infrastructure and funding R&D.

Telecoms initiated its first major rate reduction in April 1979. In explaining the rationale for the reduction, Ong Teng Cheong, then minister for communications, spoke of reductions in operating costs and highlighted the fact that lowering rates would enable Singapore to remain a competitive provider of

Table 14.2. Telecoms Operating Data, 1981–1990

Fiscal Year[a]	Operating Revenue	Operating Expenditure	Operating Surplus	Total Surplus[b]
1981	917.8	592.1	325.7	396.9
1982	1014.7	718.0	296.7	370.3
1983	1077.1	856.1	221.0	299.3
1984	1215.5	950.4	265.1	369.1
1985	1237.8	996.5	241.3	317.5
1986	1336.2	1045.3	290.9	367.7
1987	1508.7	1160.7	348.0	446.1
1988	1746.0	1252.8	493.2	620.2
1989	1944.8	1230.8	714.0	869.8
1990	2100.3	1271.5	828.8	1086.4

Source: Telecommunication Authority of Singapore *Annual Report,* p. 59.

All data include the postal service, which was merged in on Oct. 1, 1982. Amounts are given in S$ thousand.

[a]Fiscal years ending Mar. 31 of following year.

[b]Total surplus is before exceptional items and contributions to the Consolidated Fund. In U.S. terminology, this is total income.

telecom services, thereby helping maintain its attractiveness as a place for investment as well as strengthen its position as a communication center.

It should also be noted that 1979 was the year a new economic policy—known as the "second industrial revolution"—was launched. Reducing rates was part of the new policy, which emphasizes technology upgrading and value-added products and services.

There were ten major rounds of rate reductions between 1979 and 1988, involving tariffs for IDD, telex, postal services, pagers, telephone, modems, and leased circuits. Of these, seven involved telex and IDD. As of 1988, international telephone and telex rates in Singapore were among the lowest in the world. For instance, a standard call to the United States was S$3.00 per minute compared to the corresponding incoming rate of S$4.04; for telex, it was S$3.80 per minute outgoing, compared to S$5.03 incoming. According to Telecoms, its rate reductions reflected the adoption of principles of equitable charging without cross subsidies between services. Telecoms also reported that the nine major revisions between 1979 and 1988 gave customers accumulated savings of about S$2.4 billion.

The revision in 1979 saw reductions ranging from 7 to 43 percent for telexes sent to overseas destinations. The charging system for international telex was revised to blocks of six seconds instead of one-minute intervals effective November 1980. In 1988 telex rates to all destinations were an estimated one-fourth of those before April 1979.

IDD rate revisions were slightly more complicated because the structure was converted to a multitier system during the revision of April 1985. Prior to this, rates only varied according to destination, which led to underutilization of services during off-peak hours and weekends. IDD standard rates dropped slightly more than 50 percent between 1979 and 1988, while economy rates went down approximately 70 percent. For instance, an IDD call to Japan cost S$7.50 a minute before April 1979, but only S$3.00 at peak hours and S$2.00 at off-peak hours in January 1988.

Rental rates for items like pagers, telephone extensions, and modems were reduced for the first time in 1985. Rental rates for extensions dropped 17–53 percent. A July 1988 rate revision reduced the price of telephones by up to 70 percent. Rental rates for modems were reduced 5–83 percent while those for pagers dropped 23–52 percent.

Telecoms rate reductions, it should be noted, have not led to reductions of total revenues; instead, they have stimulated traffic. This is especially evident in the case of IDD services; traffic has consistently increased, and revenues grew from S$188 million in 1980–1981 to S$365 million in 1986–1987.

Telecoms pricing principles appear to be based on cost recovery, whereby prices are set to insure net revenues cover operating expenditures and development of and improvements to the infrastructure. It has been hypothesized that a "slack pricing policy" is practiced because Telecoms does not necessarily charge the cheapest rates possible but sets them instead according to "what the market can bear, so long as they are among the most competitive" by international standards (Toh and Low 1989, pp. 9–10). Even though it enjoys a monopolistic position, the authority has to take note of potential competitors in view of national development goals, while at the same time maximizing its surplus.

Until December 1991 domestic calls had no per unit cost. To a chorus of complaints, Telecoms then instituted time-based charges of S$0.014 per minute (about 0.8 U.S. cents) for peak hours and half that other times. To keep the change revenue-neutral dial-tone charges were cut almost in half: A business line went from S$290 to S$150 a year, and residential lines went from S$190 to S$100. Telecoms maintains that 74 percent of traffic is generated by the top 30 percent of users, while the bottom 30 percent account for just 2.2 percent.

A residential customer can talk 107 hours a year at peak rates and still save money under the new scheme while for a business it is 166 hours; therefore, two-thirds of the customers are expected to see an overall decline in their bills. The company went as far as to promise to abandon the system if this was not true. The new pricing had been announced earlier in the year and was considered a factor in the decline in PAP's share of the vote in the September 1991 elections. (Opposition delayed implementation several months and led to a reduction in the originally planned S$0.015 per peak minute rate.)

14.4.2 National Information Technology Plan (NITP)

To understand recent telecommunications policy developments in Singapore, it is important to view the formulation and implementation of the NITP in 1986 as a component of the national economic development program.

Singapore's NITP began as a modest scheme of national computerization. Having identified the computer services industry as a key to growth in the late 1970s, the government appointed a ministerial level Committee for National Computerization (CMC) in early 1980 to formulate a strategic master plan to build up computer expertise and to lay the foundation for an export-oriented software industry. In September 1981, at the recommendation of the CMC, the National Computer Board (NCB) was established under the Ministry of Finance. Its board of directors consisted of representatives from the National University of Singapore, the Economic Development Board (EDB), the Ministry of Education, the Ministry of Finance, the Singapore Computer Society, and the Singapore Federation of Computer Industry. Telecoms, however, was unrepresented, reflecting its limited concern with the computerization plan in its initial stage. Telecoms and NCB began to have cross-representation on their boards only in 1983.

While a national computerization plan was being implemented by the NCB, computer manufacturing still came under the purview of the EDB and development of the telecommunication infrastructure was under Telecoms. It was soon realized that the focus on computerization and the computer industry was too narrow and would be inadequate for fueling growth. Sometime before 1985 there was apparently a conscious policy shift from an emphasis on computerization to a more comprehensive and ambitious integrative plan to promote information technology (IT) as an integral component of economic development. IT is defined as "the use of computer technology, telecommunications and office systems in all aspects of information flows—from collection and processing, to storage, packaging and dissemination" (MTI 1986). Telecommunications, therefore, became an integral part of the new push for IT development.

Recognizing the increasing importance of IT, a working committee initiated by the NCB that consisted of representatives from Telecoms, the EDB, and the National University of Singapore was set up in late 1984 "to examine and make recommendations for appropriate new policies and strategies on IT to support the economic and social development of Singapore in the next 10 years" (NCB 1985, p. vi; see also Kuo 1988).

NITP was formally announced in December 1986. It centered on two objectives (*Hello* 1986 Sep):

1. Encouragement of creative and strategic exploitation of IT as a tool for increasing productivity in all economic sectors. Companies should computerize and employ IT creatively in securing an edge on their competitors.
2. Promotion of IT as a new growth industry by training and upgrading local staff to a level of expertise sufficient to venture into areas such as applications and systems software services, value-added networks, and product development.

According to the NITP, information communication infrastructure is one of seven "building blocks" collectively forming an integrative IT strategy to achieve the goals of the NITP. (The other six are: IT manpower, IT culture, IT application, IT industry, climate for creativity and entrepreneurship, and coordination and collaboration.) Under NITP, NCB plays the coordinating role, and the plan itself becomes an integral component of the economic and industrial development plan, as spelled out in the report of the Economic Committee (MTI 1986).

While Telecoms was actively involved in the process of defining the NITP, it also went through a process of self evaluation and planning. It was urged to continue providing new services and developing and improving facilities to enhance the information infrastructure. In its five-year plan for 1987–1988 to 1992–1993 Telecoms confirms its support of NITP as one of its corporate strategies.

14.4.3 New Policy Orientations

In 1987–1988 Telecoms formulated development and management strategies and set new goals for development, including objectives and strategies, which have been compiled in its latest five-year plan (April 1987 through March 1993).

The plan begins by reconfirming a number of corporate objectives, including:

1. Make Singapore a communications, information, and knowledge center in the region and in the world.

2. Offer a comprehensive range of information communication services to meet the needs of customers and the public, and at a level of service comparable to the best in the world.
3. Develop telecommunications infrastructure and services to enhance the competitiveness of the business community in Singapore.

To achieve these objectives in a changing social, economic, and technological environment, some new emphases in terms of corporate strategy were identified.

The first significant strategic shift is from a production orientation to a market orientation. This was formally inaugurated in April 1989 with a restructuring (and renaming) of the company. Telecoms' division manager for telephone traffic-public message services marketing, Lim Shyong, traced the shift in marketing focus by pointing out that the "huge demand for basic services like telephone, telegram and telex in the early 1970s necessitated a production orientation. This was replaced by a market orientation [when] enhanced and value-added services came into demand. Simultaneously, a customer-oriented approach became the emphasis" (*Hello* June 1988). This includes plans to bridge the gap between technology and people. Thus, Singapore Telecom is building a demonstration center to give the public hands-on experience with yet-to-be marketed services and state-of-the-art prototypes (*Hello* July 1988).

Aiming to be more market driven, Telecoms is strengthening its international marketing activities to gain access to global markets. This is both desirable and necessary, since the long-run challenge for Singapore Telecom is to look beyond the domestic market for opportunities. One aspect of this is an increasing emphasis on consulting. Since January 1987 Telecoms has had a consultancy department to provide technical and management services. The main target market is overseas administrations and telcos. This is in line with the national aim to promote the services sector through exports of professional services to overseas organizations.

Consultancy services provided by Telecoms are of two main types, planning, and implementation and operations. The former includes corporate, network, equipment exchange, and outside-plant planning, as well as tender proposal and evaluation, engineering, and traffic studies. Implementation and operations includes systems implementation, operations, and maintenance as well as training.

Since consulting began operations in 1987, Telecoms has been approached by Fujitsu International Engineering to train the staff of its client, the People's Republic of China's Telecommunication Authority. Telecoms' training facilities and the Chinese language proficiency of its staff gave it an edge over others. Telecoms also negotiated and signed contracts with AT&T World Services to provide training consultancy services to their clients on specialized telephone switching equipment.

To cope with increased activities and demand for consulting, Telecoms in April 1988 also set up a wholly owned subsidiary, Singapore Telecom International (STI), to offer services to its overseas counterparts and telcos. Having a group of multilingual professionals is an advantage, as Telecoms can service projects in various parts of the world. STI will conduct feasibility studies, prepare tender specifications, plan networks, implement, operate, and maintain equipment, and offer training.

Immediate plans are to market selectively in order to gain experience and build up a reputation of having served in less-advanced countries in order to qualify for international projects such as those of the World Bank and Asian Development Bank. Singapore's development experience is part of what Telecoms is selling.

14.4.4 Liberalization and Privatization

When Singapore experienced a recession in 1984–1985, a committee was formed to conduct a comprehensive overview of the economy and make recommendations for future development. Among the recommendations in its final report (MTI 1986) was a renewed endorsement of the philosophy that the private sector should be the main engine of growth.

The report made specific reference to Telecoms and highlighted the potential for the private sector to play a greater role in the sales and maintenance of terminal equipment (p. 185). The report stressed "Telecoms should continue to provide the basic telecommunication infrastructure," while saying "wherever possible, we should privatize the provision of telecommunication services" (p. 178). This privatization and liberalization policy, as it currently stands, does not extend to the management of basic infrastructure, which is to continue in government hands. As a unilateral policy-making body and sole regulator for telecommunication, Telecoms has had to take initiatives to provide momentum for liberalization.

Lam Chuan Leong, former general manager of Telecoms, outlined four key points of the liberalization policy in early 1987 (*Hello* Mar. 1987). These are: (1) customers may own terminal equipment, (2) customers may choose their maintenance contractor, (3) Telecoms will sell selected equipment in competition with other suppliers, and (4) Telecoms will offer services at economical and competitive rates. The policy was noticeable as early as 1986 when Telecoms allowed private ownership of pagers and fax equipment and the active participation of private firms in these markets.

Liberalization means Telecoms has to prove its efficiency by competing with the private sector. This will stimulate innovative provision of a variety of goods and boost growth of the product industry. It is obvious, however, that competition cannot be totally equal and fair because Telecoms still enjoys advantages based on economies of scale and its greater range of existing services and networks.

Effective March 1987 Telecoms allowed owners of PABX, key telephone systems (KTS) and multiline systems (MLS) who had purchased approved equipment from various suppliers to use third-party maintenance contractors.

PABX subscribers can continue to buy approved systems from various suppliers and choose either Telecoms or a third party to service the systems. They can also buy existing extension telephones from Telecoms at depreciated prices or continue to rent them and buy new extension telephones from other suppliers.

Rolm, Northern Telecom, and Philips responded by offering maintenance to their buyers. Northern Telecom built a S$2 million service center as part of its "commitment to grow as an integral part of Singapore's business community" (*Business Times* May 19, 1987).

According to Telecoms' former general manager Lam Chuan Leong, allowing private maintenance of PABX, KTS, and MLS would mean a first-year loss of about S$3 million for Telecoms and two or three times that in 1990. Maintenance of these systems generated S$16.8 million in revenues in 1986.

In a further step toward an open-market environment, Telecoms ceased to rent pagers to new subscribers after February 1987 and introduced a three-year lease to own scheme for fax machines in March 1987. Faxes previously had to be rented from Telecoms; faxes can now also be bought privately.

A policy of liberalization and privatization in telecommunications is being implemented. In March 1987, a report by the Public Sector Divestment Committee (MOF 1987) recommended studies on the feasibility of divesting various operations of four statutory boards: the Public Utility Board (PUB), the Port of Singapore Authority (PSA), the Civil Aviation Authority of Singapore (CAAS), and Telecoms. Telecoms commissioned a foreign consultant to do a feasibility study.

Coopers and Lybrand's report suggested privatization was attractive because of increasing demand for international services. In November 1989 the government announced privatization would occur by 1992. It did not, but the process was underway. The principal incentive seems to be more a desire to increase the activity and capitalization of Singapore's stock market than to cut Telecoms loose from the government, and market doldrums have thus been an element in the delay. Another reason, stressed by Telecoms, is that the company's growth is limited within Singapore and as a private firm it would be in a better position to expand overseas, including joint ventures.

A second set of consultants was retained in 1990 to recommend the best approach to privatization. In spirited competition, Morgen-Grenfell Asia and Touche Ross won. Finally, in April 1992, the proposed structure was made public (Table 14.3).

The new companies will have at least fifteen years of monopoly. The Telecommunication Authority of Singapore Board will be the regulatory authority. It will be headed initially by a permanent secretary of the Communications Ministry and will include union and private-sector members. Nothing definite was said about how the sell-off will be conducted. There will probably be a series of stock offerings with foreigners able to participate. The government is expected to retain majority control.

Table 14.3. Singapore Telecom's Proposed Structure as a Private Company, 1992

MinCom Holdings (initially 100 percent owned by the Ministry of Finance)
will own 100 percent of
Singapore Telecommunications,
which will have the following six subsidiaries (all 100 percent owned except as noted):
Singapore Post (postal services)
Singapore Telecom Investments (portfolio investments)
Telecom Equipment (sales and maintenance of equipment)
Integrated Information (telephone directories and electronic information)
Singapore Telecom International (consulting)
Sembawang Cable Depot (already 40 percent owned by others; storage facilities for submarine cables

14.5 Conclusion

It is evident that Singapore Telecom perceives its major objective to be to provide users with advanced telecom services that are also reliable and inexpensive. Measured by any reasonable yardstick, it has achieved this objective to the satisfaction of most of its clients. It has not for all, however: In late 1991, with Telecoms' belated support, a users group of some of the company's largest customers was formed to air complaints about customer service—generally centered on concerns with the attitude of Telecom employees and procedures.

Singapore is highly internationalized and the government makes a conscious effort to attract direct foreign investment. In response to the 1984 recession, the Economic Committee set up to review the economic situation and to recommend development strategies reconfirmed in its final report (MTI 1986) the need to develop Singapore as a regional and international center of trade, banking, and services. The government adheres to principles of free market and competition, yet at the same time is interventionist, justifying this on the basis of the nation's small, open, and vulnerable economy.

Telecoms has committed itself to playing a key role in the NITP, which is an integral part of Singapore's economic and industrial policy. This calls for central planning and effective coordination, which are in part possible due to Singapore's small size and the existence of a strong and efficient government.

An example of central planning and effective coordination is the formulation and implementation of the NITP under the National Computer Board, which falls under the purview of the Ministry of Finance. The Economic Development Board, which is responsible for the computer industry, comes under the Ministry of Trade and Industry, while Telecoms is responsible to the Ministry of Communication. To the extent that their defined domains and responsibilities overlap, there are areas of potential conflict and competition stemming from contradictory interest or genuine differences in policy positions. The situation could get complicated if the different ministries and statutory boards involved strongly push policies that narrowly serve their own interest, possibly as a result of pressure from different economic sectors or interest groups, and if they ignore the interests of other government bodies or even broader national interests—economic, political, and social—as has been the case in many other countries.

In Singapore, the danger of potential clashes has been minimized because development plans are typically formulated centrally through participation of relevant ministries and statutory boards, as was the case with the NITP and the Economic Committee report. Moreover, implementation of such plans is closely monitored. Cross representation by a small number of ministers and chief executive officers on related committees and boards makes communication and coordination easy and effective. Major differences are settled at higher levels as necessary, with the cabinet being the highest of all. The government's role in central planning and policy implementation in the fields of telecommunications and IT development, as in other areas, is quite transparent.

With the liberalization trend in telecommunications, the private sector is certain to respond to expanding market opportunities. Equally important are the social and political implications of a more open and liberalized telecommunications system. These and other emergent issues should be closely monitored and researched while Singapore prepares itself for the twenty-first century and as the telecommunications sector in Singapore undergoes further structural and policy changes.

Acknowledgments

This study was supported by a research grant (GR37004E) from the National University of Singapore. The assistance of Ms. Lim Geok Choo is gratefully acknowledged.

Bibliography

Chen, H. T., and Eddie C. Y. Kuo. 1985. "Telecommunications and Economic Development in Singapore." *Telecommunications Policy,* pp. 240–44 (Sep.).

Fong, Pang Eng. 1991. "Singapore: Market-Led Adjustment in an Interventionist State." In Hugh Patrick with Larry Meissner, eds., *Pacific Basin Industries in Distress: Structural Adjustment and Trade Policy in the Nine Industrialized Economies.* New York: Columbia University Press.

Krause, Lawrence B., Koh Ai Tee, and Lee (Tsao) Yuan, eds. 1987. *The Singapore Economy Reconsidered,* pp. 1–20. Singapore: Institute of Southeast Asian Studies.

Kuo, Eddie C. Y. 1988. "Policies of Information Technology Development in Singapore." In D. I. Riddle, ed., *Toward and International Service and Information Economy: A New Challenge for the Third World.* New York: The Freidreich Ebert Foundation.

Ministry of Communication and Information. 1988. *Singapore: Facts and Pictures.*

Singapore Ministry of Finance. 1987. *Report of the Public Sector Divestment Committee*. Singapore: Author.

Singapore Ministry of Trade and Industry. 1986. *The Singapore Economy: New Directions* (Report of the Economic Committee), September. Singapore: Author.

National Computer Board. 1985. *National IT Plan: A Strategic Framework*. Singapore: Author.

Telecommunication Authority of Singapore. [no date, but c. 1980]. *A Hundred Years of Dedicated Telephone Service in Singapore 1879–1979*. Singapore: Author.

Toh, Mun-Heng, and Low, Linda. 1989. "Privatisation of Telecommunication Services in Singapore." European Institute of Public Administration and Institute of Southeast Asian Studies, [conference on] Privatization: Lessons from Europe and ASEAN, Singapore, Feb. 16–18.

In addition, a number of Singapore periodicals have been used that cover Singapore telecommunications on an ongoing basis. These include:

Business Times (newspaper)
Hello (newsletter from Telecoms, published monthly since 1973)
The Link (newsletter of the Union of Telecoms Employees of Singapore, published quarterly since 1982)
The Mirror (from the Ministry of Communication)
Straits Times (newspaper)

15

South Korea: Structure and Changes

KWANG-YUNG CHOO AND MYUNG-KOO KANG

Since the first century B.C., the Korean Peninsula has been invaded by a series of forces, including those of the Chinese Empire up until the late ninteenth century and the Japanese in 1910. The current Republic of Korea was founded in 1948. During and after World War II, Korea was once again occupied. At the Potsdam Conference, the Peninsula was divided at the 38th parallel into Soviet and U.S. spheres referred to as South and North Korea.

South Korea, or Korea, occupies the southern 29,000 km^2 of the Peninsula. It has a current population of 44 million people. It is one of the more remarkable growth stories in the Pacific with its current GDP at $US 240 billion.

Its remarkable economic growth has been translated into progressive policies in telecommunications. Largely open markets have carried the day at the level of intentions and announced goals, and it has called Korea "the fourth country in the world to introduce competition in basic services" (*Telephony,* July 23, 1990, p. 28). However, actual implementation of such policies has moved slowly. By the mid-1980s the government was taking steps toward privatization of public telecom services by allowing the private sector to participate in the market, which previously had been tightly regulated and controlled by the Ministry of Communications (MOC). A key issue was whether the government-controlled common carrier—the Korean Telecommunications Authority (KTA, now Korea Telecom)—should be privatized and deregulated. The major proponents of change have been large corporations eager to have access to advanced information services.

At the same time, South Korean telecommunications have been virtually revolutionized since 1978 by the launching of a national automatic telephone switching network supporting subscriber demands for voice and data communication services. Total subscriber lines exceeded 15 million in 1990, compared with 10 million in 1987 and just 2 million in 1979.

This chapter examines development of South Korea's telecommunications and information industries and their regulatory structures, with some consideration given the issue of privatization. Sung, in the Chapter 16, analyzes long-term policies.

15.1 Development of Telecommunications

Telegraph service between Soul and Inchon, its port some 35 km away, was begun in 1885, introducing Korea to modern telecommunications. The royal palace and other governmental bodies in and around Seoul and Inchon were linked by telephone service in 1902. After 1910 the Japanese colonial government extended the telegraph and, later, the telephone to previously remote outposts as a way of facilitating trade, production, security, administrative, and political activities. Most of this infrastructure had been destroyed by the end of the Korean War.

Neither political nor practical conditions during the thirty-five years of colonial rule allowed Koreans much participation in the industry—or even access to its services. After the Korean War, there was no make major investment in telecommunications until 1962, when the first five-year economic development plan was launched.

15.1.1 The 1960s and 1970s

Three developments transformed the telecommunications system and industry in the 1960s. The first was construction of basic networks (using imported switches, initially Strowger and then, in 1968, EMD automatic switches). The start of the Korean Broadcasting System (KBS) in 1961 and Mun Wha Broadcasting (MBC) in 1966 was the second. Establishment of the Korean Institute of Science and Technology (KIST) and the Korean Electronics and Telecommunications Research Institute (ETRI) was the third. Along with these developments the government selected telecommunications as one of the strategic sectors for economic development, and subsidized it by providing lower taxes and other financial support.

During the late 1970s, the government began using electronic switches instead of electromagnetic or Strowger switches, and direct distance dialing was introduced in most urban areas.

Beginning in 1969 the ratio of total subscribers to total capacity of switches began to increase, starting at 85.6 percent and reaching 95.3 percent by 1980. The level of unsatisfied demand began increasing during the 1970s as the Korean economy moved into a high growth period. The number of applicants on the official waiting list peaked at 619,000 in 1980. Including potential demand, the number would easily have exceeded 1 million. Not surprisingly, a black market developed. The premium for telephone installation in 1980 was some 1,800,000 won ($3,000) compared to the official installation charge of 420,000 won ($700). (Average monthly wages that year were 176,000 won.) People commonly waited a year or more for service, and delayed installation became a serious social issue.

As can be seen in Table 15.1, the number of telephone lines increased at a moderate but steady rate after 1962 (when the first five-year plan was implemented). The growth continued as the second and third economic plans were launched, and in 1982, when the fifth plan began, about 1 million new lines were added—about equal to the total number of lines in service in mid-1975.

15.1.2 The 1980s

The government launched an ambitious telecommunications investment plan in the early 1980s. The primary motive was to solve the persistent backlog of demand for telephone facilities, with secondary goals of supporting development of the electronics industry and of preparing for the anticipated information society. At the end of the 1970s the Korean government was redirecting its industrial policy focus from heavy industry to electronics, with telecommunications as a closely related concern. In 1980 the government supported the launching of color television broadcasting. In 1981 it also allowed and encouraged subscriber purchase of telephone sets from third-party suppliers for both the first telephone and extensions. There was no requirement to lease or buy from the telephone company.

These policies had their difficulties, however. Typical of many developing countries, Korea had financial problems. Foreign currency, which was desperately needed to purchase foreign switching and transmission facilities, was in short supply. In 1980 Korea experienced poor harvests and, following the second oil crisis, shrinking worldwide demand for its exports, as well as an unstable domestic political climate. Under these circumstances, the government relied on foreign resources to continue its investment in economic restructuring, accumulating external debt at a very rapid pace. It is worth noting that the government maintained its commitment to expand telecommunications facilities, unlike most developing countries, where public telecommunications was a low priority (Saunders, Warford, and Wellenius 1988, pp. 12–18).

Korea also experienced problems with internal financing. To fund construc-

Table 15.1. Number of Subscribers and Unfilled Orders

Year	Lines (thousands)	Percent Utilized[a]	Year	Lines (thousands)	Percent Utilized[a]	Unfilled Orders
1956	64	59.9	1976	1389	91.5	144
1961	123	—	1981	3491	93.4	498
1966	313	87.8	1986	8905	84.5	160
1971	624	89.0	1989	13,354	86.3	0.7
			1990	15,293	—	—

Source: KTA *Statistical Yearbook of Telecommunications* and MOC 1985, vol. 2.

The number of unfilled orders peaked at well over 600,000 in 1979, and was still above that level at the end of 1980.

[a] Subscriber lines as a percentage of switch capacity.

— Not available.

tion and equipment procurement internally, the government had to consider increasing prices or some other form of raising capital, such as 1980, and increasing telephone rates would certainly contribute to an even greater increase in the consumer price index. Nonetheless, the government felt that rates were artificially low and decided users should pay for network expansion, so the rates were raised.

Beginning in January 1980 free calling was abolished, local call prices went from 8 to 20 won, and users were required to purchase a 200,000 won KTA bond at the time of installation. The bond matured in five years and could not be sold for the first three years. (The requirement was dropped in 1990.) A 25-percent "special luxury consumption tax" was added to total telephone charges; revenue went directly to the Ministry of Finance and was not necessarily used for telecommunications. In normal situations this kind of rapid price increase would have met strong resistance from users. Such was not the case—perhaps because people recognized the need to finance new services or were simply willing to take service at any price, without complaint.

During the fifth five-year plan, 1982–1986, telephone service made significant progress. Telecommunications' share of gross fixed capital formation increased from less than 3 percent during the 1970s to 7.5 percent during the plan period. This investment resulted in the implementation of essentially universal service, inauguration of a national automatic switching network, and establishment of networks capable of meeting future sophisticated requirements. The number of telephone lines increased to about 8.8 million in 1986, triple the 1980 number; household penetration went from 68 to 90%. The number of public coin telephones increased from 58,000 in 1980 to 138,000 in 1986. This is 3.3 per 1,000 population. Subscriber penetration rates are shown in Table 15.2.

15.2 Telecommunications Equipment

In the early 1970s telecommunications entered a period of accelerating expansion, particularly the equipment industry, largely as a result of developments in the industry's technology. By the end of the 1970s, KIST had produced a PABX and was experimenting with 44 Mbps optical fiber that could be slipped into existing coaxial cable ducts for low infrastructure costs. Two earth stations for satellite communications were built in 1977, and they resulted in lower transmission costs than those with a marine cable.

Production of wireline and radio telecom equipment grew at an annual rate

Table 15.2. Telephone Density*

1980	7.2	1984	13.5	1988	24.6
1982	10.4	1986	18.1	1990	31.0

Source: KTA *Statistical Yearbook of Telecommunications,* 1991 edition.

*Subscribers per 100 people.

of almost 33 percent over the period 1970–1975, as total output increased from $14.8 million in 1970 to $60.8 million in 1975. Growth then accelerated to almost 35 percent a year despite the larger base, with production reaching $363 million in 1981. Since 1981 the growth rate has declined noticeably due to the limitations of labor-intensive production; production was $1,831 million in 1988. Wireline equipment, such as telephone instruments and switching systems, is the major component. From $14.1 million in 1970 the industry had grown to $1,351 million in 1988 (Data are from the *Korean Electronics Yearbook*). Using a broader definition, production of telecom equipment in about one-third exported (*Far Eastern Economic Review*, Mar. 7, 1991, p. 43).

Exports of equipment and services have increased steadily since 1981, except for 1984, and imports have declined since 1983. The export drop in 1984 can be explained by changes in the U.S. market, including changes in technical standards and exclusion of equipment from the preferential tariff list. There has been a positive trade balance since 1984.

Government encouragement of domestic production of equipment and R&D activities to develop digital switches and optical fiber transmission equipment was, by the mid-1980s, significantly reducing the burden of foreign borrowing (which had provided over 20 percent of investment funds in the early 1980s). By 1987 direct import of switches financed by foreign loans had disappeared.

The government intends to provide producers of facsimile equipment with 14.3 billion won in subsidies during 1989–1994 in a bid to increase local content. The domestic makers are generally allied with a Japanese technology source: Samsung–Toshiba, Goldstar–NEC–Matsushita, Hyundai–Fujitsu, Daewoo–Sanyo, and so on.

15.2.1 The TDX Switch

Switches are a particularly important component of the equipment industry for two reasons. First, they account for a very large part of value added. Second, switching technology has been a major barrier to entry; it involves understanding large-scale network behavior, as well as the production and installation of international networks and operations in other countries.

Switching technology began to change rapidly in the mid-1960s as stored program control (SPC) computers replaced wired circuits and relays. SPC permitted vast increases in flexibility. In the early 1970s digital time-division switching replaced analog electromechanical switches. Korea had to import the switches.

To attain self-reliance in network development, standardize the national network, and prepare for ISDN, Korea considered it crucial to develop a switch that could be used domestically by the public authority and enable the country to compete in the world market.

In collaboration with private businesses, ETRI successfully developed the first exchange, dubbed the TDX-1 (for Time Division eXchange), in 1983. In April 1984 a 2,400-line experimental model was installed as the main central exchange system in the West Taejon office and a 480-line unit went into the

Yusung office as a remote branch exchange system. After the success of these field tests, four TDX-1 6,000-line switches were installed at Kapyung, Jungok, Koryong, and Muju in 1985.

TDXs have been installed primarily in rural areas where large-capacity switches are not needed. Installations include TDX 1As, which have 10,000-line capacity, and TDX 1Bs, with 23,000-line capacity. Korea's domestically developed digital switches, though limited in capacity, enabled the country to become less dependent on foreign technology, and as a result foreign suppliers cut prices for their large-capacity switches. By early 1990 1.4 million TDX lines were in operation.

Given the magnitude of R&D involved, TDX-1 manufacturers Samsung Electronics, Goldstar Semiconductor, Daewoo Telecom and Oriental Telecom (Otelco) have, as intended, sought to enter international markets. While established suppliers based in OECD countries dominate the world market, Korean exchange equipment firms are trying to penetrate Asian and African markets with small exchanges. Sales have also been made to Poland and Hungary. To keep the four producers from undercutting each other, KTA formed KTA International (KTAI) to coordinate marketing and provide system consulting.

The government has provided subsidized financing through Economic Development Cooperation Fund (EDCF) projects. For example, MOC asked that $1 billion be made available for telecommunication loans to LDCs in 1992 at 3.5 percent.

15.3 Telecommunications Regulation and Its Institutional Structure

The Telecommunications Act of 1961 was the basic law governing the industry until the early 1980s. It rested on a tradition of heavy government regulation of equipment, service offerings, rates, and conditions. Policy has since moved under the control of the MOC. A regulatory framework of minimum competition and public monopoly was adopted and implemented by MOC.

In 1983 the Basic Telecommunications Act and the Public Telecommunications Business Act brought about dramatic changes in this regulatory structure. The purpose of these acts was twofold: to separate the functions of policy formulation and business operation and to take a step toward privatization. In addition, these measures were supposed to support:

1. Integrated and efficient control and regulation of the network as a national infrastructure.
2. Consistent policy formulation, industrial development, and technological advancement.
3. A greater degree of competition between manufacturers in the private market.
4. Entry by competitors into telecom services provision.
5. The direction and financing of research and development.

KTA historically had exclusive authority to permit construction of new and additional facilities by Korean carriers, as well as licensing and standard-setting power over other entities seeking to provide basic telecom services over existing or new facilities.

When the Business Act was promulgated in 1983 KTA was restructured as a public corporation with all the stock owned by the government. The Act authorized the creation of a private specialized carrier, Dacom (Data Communications Corporation of Korea). Dacom's main task was to construct a public switched data network (PSDN) that interfaces with the public switched telephone network (PSTN). Dacom had established twenty-one access nodes by 1983 and begun an international service called Dacom-net.

As Table 15.3 shows, various public entities are involved in the development of telecommunication networks before the 1989 reforms, with partial penetration by the private sector.

In addition to these organizations, the Network Coordination Committee in MOC is in charge of computer network projects. The Korean Computerization Agency has also been established for supervising and validating the National Administrative Information System Network, which is in partial use. The Korean Computer and Communications Promotion Association (KCCPA), a private entity, was founded in 1987 to enhance information exchanges between industries and to make recommendations to the government.

In summary, regulatory policy has operated under the principle that common carriers provide public telecommunication services. KTA, as a public corporation, was in charge of basic network services. Dacom was responsible for enhanced network services. Information services such as data processing and information retrieval became more widely open to the private sector.

The 1983 Business Act provided a statutory basis for market liberalization, outlining the activities, obligations, and regulatory structures. Even though KTA was made a corporation and some private companies like Lucky Goldstar, Samsung, and Daewoo began to participate in new services such as VANs, there was no decided principle of competition. Still, the government planned to develop competitive markets.

15.3.1 The 1989–1991 Restructuring

The telecommunications laws were revised again in July 1989 and July 1991. The first primarily involved VANs (discussed later). The second was part of a broader process formalized in March 1989 when the Korea Information Society Development Institute (KISDI), an MOC affiliate, assembled an advisory group from the government, industry, academia, and research institutes to look at the information and telecommunications industries. It reported six months later. Competition, but not all at once in any specific area and not immediately in every area, was the consensus. The areas producing trade friction with the United States—in particular, enhanced services—were recommended for opening first. KTA would have no restraints on its business activities, although it had to practice "fair competition."

Table 15.3. Major Korean Telecommunications Organizations

Supervising Government Body

Ministry of Communications (MOC). Formulates and implements telecommunications policies. Direct and coordinates common carriers. Promotes and supports R&D activities.

General Service Providers[a]

Korea Telecommunications Authority (KTA). Provides telecommunications services. Constructs, operates, and maintains public telecommunications facilities. Renamed Korea Telecom in 1991. 100 percent government. Jan. 1982.

Data Communications Corporation of Korea (Dacom). Originally established to construct and operate public data communications networks as a monopoly. Subsequently designated to compete with Korea Telecom in international calling. 34 percent Korea Telecom, 66 percent owned by twenty-three private companies. 1984 Sep.

Specific Service Providers[a]

Korea Mobile Telecommunication Co. (KMTC). Provides cellular and paging services. 68 percent Korea Telecom, 32 percent private (trades on Korean Stock Exchange). Apr. 1988.

Korea Port Telephone Co (KTP). Provides communication services in harbor areas. 49 percent Korea Telecom; 51 percent private. Jan. 1988.

R&D Institutes

Electronics and Telecommunications Research Institute (ETRI). Carries out R&D in the field of telecommunications, semiconductors, and computers. Promotes the astronomical and aerospace sciences.

Korean Information Society Development Institute (KISDI). Undertakes long-term research for telecommunications development. Work outs progress indicators for the information society.

National Computerization Agency (NCA) Standardizes technologies relating to computer networks. Undertakes feasibility studies and supervises computerization projects run by government or public organizations.

Engineering Services

Korea Telecommunication Authority International (KTAI). Provides telecommunications engineering services. Manages overseas telecommunications projects.

Korea Information Telesis Incorporate (KITI). Maintains and repair telex facilities and equipments.

Public Relations and Education

Information Culture Center (ICC). Publicizes the information society to the public.

Unless noted, the organizations are government-related.

[a]These were common carriers prior to the 1991 restructuring. The date given is when MOC designated the entity a common carrier under the previous regulatory structure. There was a fifth common carrier (designated Feb. 1988): Korea Travel Information Service Company Ltd. (KOTIS), which is now classified as a Value-Added Specific Provider. It provides air travel and tourist information and is owned by Dacom and private interests.

MOC accepted the recommendations, and set out to prepare specific implementing policies. In putting together proposed legislation and regulations, public hearings were held—the last, in June 1990, attracted 300 people, mostly from entities directly affected, as distinguished from the general public. The decisions in the July 1990 Structural Reform Plan included directives that the liberalization of VANs was to accelerate, and that there was to be a duopoly in international and mobile calling and in data. However, domestic long distance remains a KTA monopoly until the mid-1990s. Telecommunications is divided into three categories, as shown in Table 15.4.

KTA was privatized effective January 1, 1990. Renamed Korea Telecom, part of the stock was to be sold to the public, but this was postponed—although apparently more because of a generally weak stock market than anything else. Korea Telecom was not happy, calling the Korean telecom services market too small to support competition.

Under this liberalized regime, Korea Telecom will face Dacom, which is 34 percent owned by Korea Telecom (the rest is owned by major corporations) and had 1990 revenues equal to just 5 percent of Korea Telecom's. Dacom loses its monopoly on data transmission but does compete with Korea Telecom

Table 15.4. Classification of Service Providers

Type of Service Provider	Network		
	General	Specific	Value-Added
Scope	Telephone Telegraph Telex Data Leased and dedicated circuits Various voice, nonvoice, and mixed transmission service Extension service for telephone network	Paging Mobile phones Wireless phones and data Trunked radio service Port communications Aeronautical communications	(examples) Data base Data processing Data accumulation, processing and transmission E-mail EDI MHS CRS Video conferencing
Entry requirement	designated by MOC	approval by MOC	registration with MOC
Ownership limits	Less than 10% by one individual[a]	Less than one third by one individual[a]	No limitations
Foreign investment	Not allowed	Allowed if does not exceed one third	In stages: 50% in 1991, 100% by 1994
Duties	Establish an efficient national communication system Supply regular and universal service	Appropriate to the type of enterprise	No duties imposed

Source: Adapted from MOC *Annual Report on Telecommunications* 1990, English edition, p. 35.
[a]Or by one corporation or related group.

in international calling (from December 1991). Dacom initially competed just in the Japan, Hong Kong, and U.S. calling markets—but these are 70 percent of outbound traffic. In anticipation of the competition, Korea Telecom reduced international rates, but Dacom priced calls 5 percent under Korea Telecom. It is that expected Dacom will be allowed into domestic long-distance later.

Dacom has been barely profitable since it was created, and this is considered a factor in its selection as the second major carrier. Another reason is a government reluctance (which mirrors popular sentiment) to give the major *chaebol* (business groups) a big piece of any more pies. Dacom has indicated that once it is allowed into the domestic long-distance business it will first concentrate on building an all-digital network for major business customers, and consider a 10-percent market share a reasonable goal (*Telephony,* July 23, 1990, p. 32). In the past, Dacom's network facilities have been leased primarily from Korea Telecom.

15.4 Telecommunication Facilities and Services

In 1983 a fiberoptic system with 45 Mbps shortwave, multimode transmission was installed between Seoul and Inchon (35 km); this system was designed and manufactured under the supervision of ETRI. A similar system was constructed between the ETRI office building and the Taejon Toll Office, a distance of 17.3 km, without a repeater. Based on these experiments, a 90 Mbps system has been established for commercial use. It is capable of simultaneously transmitting 1,300 voice signals. Preparation for the Seoul Olympics was a major motivation for developing this capacity. The long-distance switching system has been fully digital since 1984. In 1990 AT&T was retained by Korea Telecom to help digitize and otherwise upgrade the international service.

Semiconductor development is important in electronic switching systems and other telecom equipment. In 1985 KTA contributed the necessary funds for ETRI laboratory facilities. ETRI subsequently successfully developed custom VLSI chips used by the TDX-1 switching system and other applications.

There is an increasing demand for data communication between computers and between remote terminals and their processors. Though this demand can be met through shipment of tapes or disks, data transfer through telecommunications is much faster and allows interaction between the sender and the receiver. The number of leased data circuits at the end of 1988 was 126,496 channels, a 23 percent increase over 1987. High-speed digital channels were made available for public lease in Seoul, Pusan, and Taegu in 1985, and digital service was extended to eleven cities for use of inter- and intracity data transport. Domestic leased line service has grown rapidly. According to a 1986 survey by the Korean Information Industry Association (KIIA) users were banks and financial institutions (53 percent), general business establishments (31 percent), and administrative and public agencies (15 percent).

15.4.1 Value Added Networks

Several major corporations have their own VANs. An example is Daewoo Motor, which started in 1984 with computer links between five sales offices and its headquarters that by 1989 had developed and expanded the system into a full-fledged VAN that included thirty suppliers. The company has expressed interest in cooperating with other automakers on standards for an industry VAN.

There has been significant resistance to foreign involvement, and the issue has been a major one in Korea–United States trade talks. It is candidly admitted that Korean resistance stems from the feeling U.S. and Japanese technology is so advanced that allowing it in would freeze out most Korean firms (see, e.g., *Korea Business World,* Apr. 1989, p. 46). Joint ventures are actively being sought.

Under July 1989 legislation the registration system for new VANs was theoretically made less stringent. In practice, because the law was silent on what market segments were to be open to competition, not much changed until more than a year later when MOC provided specifics. Data base and data processing were open to international competition in July 1991 and all other areas were to be opened by 1995. Korea Telecom has plans to actively enter the market.

In September 1991, there were more than 130 companies providing value added service (VAS), mostly data base and data processing. This compares to just twenty VAS providers in early 1990, only ten of which were licensed to deal with the general public (the others were intrabusiness-group), even though the KCCPA reported that 143 had licenses.

15.4.2 Data Communication

Improvements in microprocessor technology have increased the capacity and reduced the cost of data communications equipment, including data terminals, modems, and multiplexers. The range of terminals available has expanded to meet the demand of users (e.g., high-speed terminals for high-volume but relatively simple operations; lightweight portable terminals for field communications; and display terminals with graphic capabilities for scientific applications).

The number of facsimile subscribers reached 13,000 in 1986 and 115,000 in 1990. Facsimile uses the same lines as voice service in most cases, thus network charges are the same. Telex subscribers increased from 7,539 in 1983 to 10,304 in 1987. However, the diffusion of telex is stagnating because of incompatible standards in transmission format. It has been predicted that communicating word processors capable of storing, retrieving, and editing text will compete with telex, if not completely replace it.

Videotex as an interactive data communication service enables widespread consumer access and retrieval of computerized information. It is still in the development phase in Korea. Serious experiments are underway using specially adapted television receivers to display information accessed through telecommunication networks. Videotex combines elements of mass and point-to-point communication: The information available from the central data banks is in many cases identical to that normally distributed through newspapers, magazines, and books. The use of the point-to-point network to access this information, however, allows great selectivity in its retrieval.

ETRI developed a standardized model for Korean-language videotex based on North American Presentation Level Protocol Syntax (NAPLPS) in 1984. ETRI also developed a teletex model and its specifications were transferred to Korean manufacturers in 1986.

In addition to these services, AP Telerate (November 1983) and Reuter Monitor (May 1984) have provided on-line data base services covering interest rates and other financial statistics. These specialized providers of otherwise unavailable data are the only foreigners who have been permitted in Korea's market.

15.4.3 Cellular Service

Carphone service began in 1973 and cellular joined it in May 1984, but the market did not take off until prices were cut July 1988. Initially in Seoul and Pusan, cellular had been extended to seventy metropolitan areas by mid-1991. Korean Mobile Telecommunications Corp. (KMTC), which trades on the Korean stock exchange but is controlled by Korea Telecom, has been the monopoly provider of both paging and cellular service. Korea Telecom has supervised long-term investment and facilities planning, while Mobile Telephone Service Corp. (MTS) has handled installation and maintenance of the cellular system.

Under a July 1990 decision, MOC will select a second carrier by July 1992. To their consternation, equipment producers will be limited to 10 percent ownership of the system; as a group, foreigners can own 33 percent, but may not take part in management. The new entrant will need several years to get a system built and operating, so observers do not see it as much of a threat to KMTC any time soon.

Five local firms produce cellular equipment, including Samsung Electronics, Goldstar Telecom, and Hyundai Electronics. Motorola is also in the market, with an estimated 30-percent share (1989). Service does not come cheap. To make the first call, using the least expensive telephone, cost 1.7 million won (over $2,500) in early 1989. This included the telephone, a number of permits and other set-up fees, plus twenty-five won for the first message unit of calling.

In May 1991 KMTC reported 113,000 mobile telephones and 591,000 pagers in use. The paging system was expected to be nationwide at yearend 1991; the cellular network is restricted from some areas because of military installations. In January 1992 the Korean National Police ordered mobile equipment from an Ericsson–General Electric joint venture, using EDACS, a system not directly compatible with the general public system.

15.4.4 The Koreasat Project

A Korean communications satellite was considered in the mid-1980s, but it was rejected because of uncertainty over profitability. In 1991 Korea Telecom was soliciting requests for proposals for a satellite, and General Electric was chosen prime contractor (with Goldstar as its local coordinator) that December. During the bidding process, it was made clear that foreign supplies were expected to joint venture with (and supply technology to) domestic firms. However, the launching will be contracted to a foreign entity because no one with the technology has indicated any interest in sharing it with Korea. Five new ground stations will be built to supplement the four currently in operation. With an estimated total project cost of $400 million, launching is planned for April 1995 (*Business Korea,* Dec. 1991, p. 38 and Jan. 1992, p. 61).

Fiberoptic marine cable capacity is also being expanded. Under a February 1992 contract among AT&T, KDD, Dacom, and Korea Telecom, Korea will get its first direct link to Guam, where the new cable will link with the TPC-5 network on to Hawaii and the U.S. mainland. This leg will have 7,560 circuits. In addition, there will be a 15,120-circuit cable between Korea and Japan. Both are to be in operation by 1995.

15.4.5 Related Measures

MOC's policies extended beyond traditional telecommunications to related areas, including electronics and computers. For example, MOC—together with KTA, Dacom, and ETRI—actively participated in 4M DRAM and superminicomputer development projects. Computer technologies obtained during research for digital switches were used in the development of computer networks for the Asian Games in 1986 and the Summer Olympics in 1988.

Some liberalization measures have been instituted. For example, many types of customer premises equipment—including telephone sets, modems, and facsimile machines—can be freely sold and attached to the network without restrictions, subject only to technical-standards approval. In addition, networks for television, telephone, telex, and military and public security have been integrated wherever possible to improve efficiency.

15.5 Cable and Broadcast Television

These areas have been very contentious. In 1990 when the government announced it would license a new over-the-air television broadcaster and CATV operators, it received scores of applications. Seoul Broadcasting System (SBS), the entrant of a medium-size construction company, got the nod to become the third broadcaster (and fifth station). Two of the stations are part of KBS, the third is operated by MBC, and the fourth is educational; all are government owned.

Regarding cable, Korea Telecom awarded a contract in October 1990 to Korea Communications Engineers (KCE) to build pilot projects in two Seoul apartment projects. It began operation in July 1991 by offering seven channels (including four broadcast ones) to 8,400 households. None of the channels operate twenty-four hours a day.

More extensive CATV and broadcast service is set for 1995 when a Korean satellite is launched. Between 1993 and 1995 the government, through MOC and the Ministry of Information, plans to install interactive twenty-five–channel cable systems in fifty cities. Korea Telecom will oversee building them. The interaction feature initially involves connection to police boxes and fire stations.

Although the government will tightly control the system, there will be competition to supply hardware and programming. Seoul Telecom has been among the more aggressive of the cable firms—signing contracts with CNN and ESPN (a sports network) in 1990, more than a year before service could even begin.

15.6 The Information Industry

In Korea the term *information industry* is used to collectively cover telecom services and equipment manufacturing, data processing, and sometimes even the electronics industry generally. Before 1989 there were no systematic statistics for the information industry. Here the term covers producers of the physical equipment and the systems that drive it (software and firmware) rather than the actual information content, so print media and broadcasting are excluded—although of course much of the equipment they buy is covered. KTA began classifying data for the industry in 1989, specifically breaking out telecommunications and electronics as subgroups.

Korea's electronics industry was inaugurated in March 1959 when Goldstar brought out vacuum tube radio receivers. For the next decade the level of investment and technical development were limited to assembly of imported parts—screwdriver factories. In the mid-1960s, the government launched a policy to attract foreign investment and technology transfer. Foreign investors sought the advantages of low-wage labor in Korea, with American Signetics and Fairchild building manufacturing plants for black and white television sets, and Motorola locally manufacturing transistors and integrated circuits. As a result, total domestic production of electronics increased from 125 billion won ($522 million) in 1962 to 1,640 billion won ($5,115 million) in 1969.

The government's initiation of an eight-year development plan for the electronics industry (1969–1976) was a turning point. Five goals were established by MOC: expansion of export sales, maximum use of domestically produced equipment, rationalization of production and distribution systems, construction of electronics plants, and opening the domestic market to foreign investment. In 1976, as part of the third basic development plan, several special laws were enacted establishing national investment funds, reducing trade and sales tax, and reorganizing small electronic companies by specializing their product lines.

At the start of 1977 the structure of manufacturing industries was reorganized by the fourth basic development plan (1977–1981). The principal objective was to balance overall imports and exports. Whereas production of industrial equipment was intended to supply the domestic market (displacing imports or at least reducing their increase), home electronic items were designated for export.

Proclaiming 1983 as inaugurating "the period of the development of the information industry,'' the government selected three strategic fields: computers, integrated circuits (ICs), and telecom equipment. Until the mid-1980s expansion in export markets was led by microcomputers, electronic systems, ICs, and color television sets. Due to competition and increasing pressure from importing countries, several Korean companies such as Lucky Goldstar and Samsung built factories in the United States and Portugal.

15.6.1 The Rise of the Information Industry

The Korean information industry has focused on international markets because of the limited domestic market. Most imported information equipment has been for assembly manufacturers who are oriented toward the international market (i.e., it was re-exported). There was a positive trade balance by 1981. Investment by foreign firms was a key factor in this. Japanese and U.S. multinationals have been the principal sources.

In 1983 the government initiated a policy to promote the flow of foreign capital and investment into the country, changing to a positive list system from a negative list one. (While the former explicitly lists the sectors foreign companies are able to invest in, the latter allows foreign capital in any sector not explicitly listed as off limits). With this measure, investment by foreign information companies started to grow rapidly. For example, IBM became one of the top twenty companies in terms of sales in 1986.

Technology transfer has been regarded as a crucial factor for indigenous development because it influences both the sociocultural characteristics of the importing country as well as domestic production factors and national scientific and technological potentials. During the period 1982–1989 some $3,315 million was paid in royalties for technology transfer such as patents, technical information, and services—a third of it by the electrical machinery, electronics, and telecommunications industries. Technology originating from the United States and Japan has accounted for more than 90 percent of the total since 1977.

15.7 Public Policy for Industrial Development

Five types of broad measures taken by the government are generally considered significant in bringing about the rapid expansion of the information industry since the early 1970s. These include:

1. Encouraging investment in new information technologies,
2. Local manufacturing of products and parts,
3. Establishing research and development sectors and a system for technological innovation,
4. Encouraging new techniques and products,
5. Fostering cooperation of universities and industries.

To encourage investment in information industries, the government decided in 1982 to establish a development fund for electronics with 240 billion won

($300 million at the time). The fund was used for technical innovation and automation of production lines, as well as long-term, low-interest loans. In addition, the government encouraged new businesses, including investment of venture capital, to contribute to innovation in information technologies and services.

The Ministry of Commerce determined a list of electronics products and parts that were desired for indigenization to increase Korea's competitive power in the international market. With this measure, relationships between electronics businesses were streamlined to avoid overlapping investment and overcompetition.

To promote R&D and technological advancement, engineering centers were established abroad that were funded by the government, research institutions, and private enterprises. These centers played an important role in collecting information about recent technological development, training engineers, and recruiting foreign scientists and engineers.

In early 1990 Korea Telecom announced it would invest 3 trillion won ($4.5 billion) in R&D by 2001, starting with 133 billion won in 1991 (about 3 percent of expected revenue), moving up to 6 percent of revenue in 2001. This is to include funding of four research institutes, placement of 28,000 personal computers in 1,230 schools, and distribution of 10,000 monitors to individuals for experimental videotex services.

A committee on technological innovation provided universities and businesses with research projects to develop new products and knowledge. The committee also organized an annual electronics exhibition to display new innovations. To strengthen education and training of engineers and workers, the Ministry of Science and Technology began a program of inviting Korean scientists working abroad and foreign retired engineers to teach and train in Korea. Training centers for electronics engineering also provided a variety of courses to engineers from small factories.

The United States has made an issue of access to the Korean telecom services and equipment market. To avoid retaliation under U.S. trade laws, in February 1992 an agreement was reached to make the market more open. Among other things, registration in value added services is being streamlined and restrictions on investment are being phased out by 1994. Tariffs have been cut, and U.S. firms are given more access to the standards-setting process and government markets.

15.8 Conclusion

After the Korean government launched "the period of development of the information industry" in 1983, telecom equipment makers, computer manufacturers, and other electronics-related firms have enjoyed tremendous growth in revenues and value added services. With stringent, although now falling, barriers to import, and the government as a major buyer of domestic production, the trade balance has improved. Still, research must pay attention to the intricate formal and informal procedures that pervade the decision-making process in South Korea within and between formal telecommunications institutions.

During the 1980s the monopoly structure of KTA and the lack of competition in most aspects of telecommunications were the subjects of debate among administrators, industry, users, and other interested parties. It was pointed out that the bureaucratic administrative structure of KTA hindered provision of services. As a result of these debates, KTA's control over such areas as specification of technical standards for services and equipment has been reduced and the government has confirmed a willingness to liberalize the sector by moving toward duolopoly in long distance and relatively open markets in VAN and VAS.

Bibliography

Korea Telecommunications Authority. *Statistical Yearbook of Telecommunications.* 1980–1991. Seoul: Author.

Ministry of Communications. 1985. *100 Years of Korean Telecommunications,* vol 2. Seoul: Author.

Saunders, Robert J., Jeremy J. Warford, and Bjorn Wellenius. 1988. *Telecommunications and Economic Development.* Baltimore: The Johns Hopkins University Press.

16

South Korea: Telecommunications Policies into the 1990s

KEUK JE SUNG

Until the late 1970s the Korean telecom sector had problems typical of developing countries. A backlog of unfilled orders persisted and financial resources were allocated inefficiently. Entering the 1980s, however, the Korean government made major investments in telecommunications, financed mainly through price increases and foreign borrowing. By 1987 the waiting list for telephone installation had disappeared, and a nationwide subscriber dialing system had been implemented. As the investment in facilities stabilized, the Ministry of Communications (MOC) proposed a qualitative policy that included flat-rate calling for telephone service (i.e., distance-insensitive rates) and implementation of universal information services. In a sense, these are the ultimate form of universal service.

Other forces were at work in Korea as well. By the late 1980s users of telecommunications networks were mounting pressure for liberalization. Overall, the economy had become more sizable and complex, and the government invited more private sector initiative by privatizing certain public corporations. At the same time trading partners, including the United States, have pressed for more liberalization—indirectly through joint ventures in Korea and directly through U.S. government demands for opening of the domestic market.

These developments raise two key questions. What issues are involved in flat-rate pricing and universal information service policies? What effect would privatization have on these policies? These are discussed in the context of the debate and actual restructuring of Korean telecommunications, but they are issues in other countries as well.

The focus here is primarily on network services, including information services. Major structural changes were legislated in 1989, and in July 1991 the two basic telecommunications laws were further revised. Korea has predominantly adopted American equipment and technical standards, so although equipment, technical standards, and equipment approval procedures are important, liberalization of the service area is more controversial. The equipment industry, as well as the historical development of telecom services and the content of the 1989–1991 reforms are covered by Choo and Kang in Chapter 15.

16.1 Changes in the Legal Environment Prior to 1989

Major legal and institutional changes affected the telecommunications environment between 1982 and 1985 as a new set of laws was enacted. These included the Telecommunication Basic Law, the Public Telecommunications Business Law, and the Law on the Establishment of the Korean Telecommunications Authority (KTA). These laws provide that all telecommunications business was to be carried out by common carriers designated by MOC. MOC designated several (see Table 15.3) whose business areas did not overlap significantly. In cases of overlap, MOC was responsible for coordination.

In 1982 KTA, a 100 percent government-owned public corporation, was established and given responsibilities for telecommunications management and operation previously held by the MOC. This separation of policy and management functions was designed to facilitate fast decision making and less-restrictive supervision, enabling telecommunications facilities to be provided on a massive scale. However, the transition was not easy: Korea is a country where being a governmental employee is a long-honored tradition, even though the separation raised the salaries of those transferred to KTA and created job opportunities.

Also in 1982 the Data Communications Corp. of Korea (Dacom) was established as a private company and given the status of a common carrier, mainly providing new telecom services such as videotex, packet switching, and electronic mail. As of 1988, two more common carriers assumed the business for marine and mobile communications.

In 1985 restrictions on the use of the telecommunications network, including leased lines, were relaxed. Private companies began to provide data base and data processing services. As of December 1988, eighty-five companies were in operation, but most had sales below 100 million won ($150 thousand). Lack of domestic data bases, customer lack of understanding of the services, and remaining restrictions on use of the networks were the most likely reasons for the low sales volume. Further relaxation of restrictions on use of the public network and of leased lines came in 1987 and 1988. Closed user groups (not defined in the laws, but characterized in practice as groups having close business relationships) could construct private networks using leased lines, with no service restrictions. By mid-1991 virtually anyone could share lease lines, satisfying what had been one of the major complaints of users and potential VAN entrants.

A shift toward phasing out monopoly came in a late 1989 revision to the Telecommunications Business Act. By reducing limitations on private carriers it was expected to help expansion of the VAN market, which is discussed in Chapter 15. Because entrants are not expected to own actual physical networks, value added service (VAS) provider is the common term in Korea.

Also, in 1990, KTA was privatized and renamed Korea Telecom. Its stock was to be sold to the public gradually during the early 1990s, as market conditions dictated. Concomittantly, Dacom was to be permitted to enter the international long-distance business and eventually the domestic long distance market. The Korean market was on the path to greater openess.

16.2 Ministry of Communication Policy Targets

During the rapid expansion of the 1980s, MOC proposed a nationwide subscriber dialing system and elimination of delayed installation as targets (often called the ''one telephone per household'' policy). Both goals were accomplished in 1987. Korean subscribers could dial directly, without operator assistance, to any other subscriber and, indeed, to almost any country in the world. Order backlogs and waiting lists disappeared and the total number of telephone lines surpassed 10 million.

Having achieved these goals, MOC turned its attention to a new agenda, much of which had been developed during a two-year study of long-term policy targets aimed toward the year 2000, which the ministry had begun in 1983. General proposals included development of new technologies and active participation in international organizations.

Two particularly interesting targets were under the heading ''popular use of information'' (MOC 1985, vol. 2, pp. 1163–65): flat-rate calling (making telephone rates independent of distance) and provision of universal information service. These targets indicate a significant policy shift from growth to quality improvement. MOC has demonstrated its commitment to these goals through interim measures including the reduction of the number of local rate zones.

Three other major factors have emerged that, with these policy targets, will shape Korean telecommunications: strong demand from users, domestic and foreign interests in the Korean market, and the privatization plans for public corporations including KTA, mainly initiated by the Economic Planning Board (EPB).

16.2.1 User Demand

To date, user views about telephone service has not been explicit or represented in an organized way, beyond various complaints about delayed installation and overcharging. However, organized user demand for data communications was expressed in December 1987. The Computer and Communication Promotion Association (CCPA), established by eighty-five companies providing data bases and data processing services, asked MOC to relax restrictions on the use of leased lines. The principal request centered around free attachment of multiplexers and computers to leased lines and requested a broader definition of closed user groups. (This section is largely excerpted from Lee 1989, pp. 37–39).

MOC instead maintained a policy that encouraged use of the public network, although closed user groups were allowed to construct private networks over leased lines. MOC also kept some restrictions on the attachment of multiplexers and computers to the leased lines in order to control growth of private networks. The Ministry contended that unrestricted development of private networks would allow Korea's *chaebol* (big-business groups), which have enough technology and money, to become information giants. The worry was that information dominance could add to the groups' armory. MOC also expressed concern about security and economies of scale. Dacom took essentially the same stance, anxious about the erosion of its business opportunities.

In response to user requests, MOC announced relaxation measures in December 1988 that allowed users to attach multiplexers and expanded the definition of closed user groups to include a wider range of companies. This suggests user demand will help shape the future of the telecom sector, and as the economy grows and becomes even more complex, private sector initiatives will play a more important role.

16.2.2 EPB Privatization Plan for Public Corporations

The EPB has been at the center of economic development planning in Korea since the early 1960s. In 1987 it announced a privatization plan for public corporations. This is intended primarily to invite private sector participation, but it will also boost the stock market, absorb funds from the money supply, and ease the government's budgetary burden.

Beginning in 1986 Korea had an overall favorable trade balance and the economy boomed. Selling stock in public corporations was seen as one way to help prevent excess liquidity and keep inflation, which had been maintained at below 3 percent since 1984, low. The plan was for 15 percent of KTA to be sold to the public in 1989 and additional shares sold annually until 1992, at which point the government would hold 51 percent of KTA. Instead, no shares had been sold by the end of 1991 primarily because the entire Korean stock market has been weak.

The initiative for selling KTA stock seems to have come from EPB, perhaps indicating MOC and KTA were not prepared for, or had not considered privatization. There have also been discussions about reforming KTA, including separation of its international business and the division of KTA into regional companies. KTA's labor union has strongly opposed such ideas, contending that KTA should have remained as it was and suggesting that the introduction of competition would be preferable to divestiture.

16.2.3 International Pressure

Foreign interests have played a big role in demands for liberalization. The major members of CCPA include STM and Samsung Data Systems (SDS), which are joint ventures of Lucky Goldstar with Electronic Data Systems (EDS) and Samsung with IBM, respectively. Other foreign businesses, such as Reuters, the Associated Press, and Citicorp have urged more market access. Although MOC claims the Korean telecom sector is being liberalized at a fast rate, not all foreign interests have been satisfied.

In January 1987 the United States started Market Access Fact Finding (MAFF) talks with Korea regarding telecommunications. These continued until the middle of 1988, with the United States indirectly requesting the opening of Korean markets. When the U.S. Congress passed the Omnibus Trade Act of 1988, telecommunications consultation between the two countries entered a new stage. Part 4 of the act is very specific in directing the U.S. telecommunications industry's actions regarding trade negotiations with other countries. In accordance with this act, in February 1989 the U.S. Trade Representative (USTR) designated the European Community and Korea as priority foreign countries for negotiation. It is impossible to predict the precise course of subsequent negotiations but so far many liberalization measures have been implemented or agreed to, and more are likely.

16.2.4 Other Issues

Universal service is central to the new policy targets proposed by MOC, and the interests of the major players generally lean toward liberalization. Are these respective viewpoints compatible? More specifically, there are four key questions. How will introduction of competition in telephone service in Korea affect rates? How sustainable is flat-rate calling in a competitive environment? What issues are involved in the provision of universal information service? How do flat-rate telephone rates affect the pricing of information services provided over the telephone network?

16.3 Rates and the Move to Flat Rate Calling

Table 16.1 shows changes in the pricing structure for local and long-distance rates since 1962. The sharp increase in local rates in the late 1970s reflects a general inflation rate that averaged over 18 percent a year.

In 1984 MOC announced a long-term pricing schedule that projected flat-rate calling by the year 2001. Before implementation began, however, long-distance rates were increased in 1985 as part of a change in rate zones. This was the peak, with a call to the farthest band costing ninety times as much as a local call, reflecting a pricing structure mainly designed to finance massive provision of facilities.

Since 1985, generally following the original schedule, long-distance rates have been dramatically reduced, so by 1991 a call covering the longest distance cost 900 won. However, local rates were increased in 1986 to 25 won. From July 1, 1989, local calls have been charged according to duration, whereas previously they had not. Thus, the interim steps toward flat-rate calling have seen long-distance prices decrease and local call prices increase.

Table 16.1. Local and Long Distance Rates*

	Local		Long Distance[b]		
Date Effective	Base[a]	Per Call	<100 km	<300 km	<500 km
Jun 1962	83.3	2	30	60	70
Mar 1964	83	2	—	—	—
Jan 1966	300	3	60	120	140
Jan 1968	300	4	100	190	230
Sep 1974	390	6	150	285	345
Jul 1975	520	8.7	200	380	460
Aug 1975	520	8	—	—	—
Jan 1977	1740[c]	—	—	380	520
Jan 1980	1740[d]	12	300	560	770
Jun 1981	2200	15	350	650	900
Dec 1981	3000	20	440	750	1000
Aug 1985	—	—	900	1440	1800
Feb 1986	—	25	—	1286	1286
Mar 1990	—	25[e]	650	900[f]	

Source: Adapted from MOC 1985, vol 2, pp. 1403–5, and updated by the author.

*All rates are in won.

[a]Basic monthly rate for dial tone. The telephone company does not lease telephone sets.

[b]Station rates are for first three minutes. Prior to 1985 there were eight long-distance bands. This had been reduced to three by September 1991.

[c]First 100 units free each month.

[d]No free units.

[e]Duration charges for local calls were introduced in July 1991.

[f]Calls over 100 km.

— No change.

16.3.1 Interim Steps toward Flat-Rate Calling

If competing long-distance providers connect with the public network to serve ordinary customers, they would have to pay access charges. Unless those charges were less than the flat-rate for a call placed entirely over the public network, the long-distance provider could not cover the costs of its network, let alone make a profit.

If discounts are given to alternative carriers, it is likely that those entering the market will serve only large-volume users and concentrate on high-density routes. Ordinary customers might then be paying more for a local call than a large-volume user pays for what was previously long distance. It is possible that this situation could be socially or politically unacceptable—there are certainly political elements who would seek to paint it as another example of favoritism for the *chaebol* at the expense of "ordinary citizens."

As seen in the United States, pricing and costing do not necessarily go hand in hand. They can be separated, allowing flat-rate calling to accommodate competition. Long-distance competitors must be able to charge prices equal to or slightly lower than the former monopoly. Competitors and the local loop monopolist must develop some revenue allocation mechanism by agreeing on the total revenue requirement and on the portion competitors would have to provide for use of the monopoly's network. This kind of arrangement has been used in the United States under the station-to-station philosophy. Under this approach, flat-rate calling does not necessarily preempt competition.

If competition is introduced before flat-rate calling, the transition to flat-rate calling is more difficult because new cost allocation and revenue apportionment methods would have to be painstakingly worked out.

A goal of flat-rate calling is to pursue a type of equality through uniform pricing and economic efficiency by means of a centralized network. One can argue that competition conflicts directly with this goal. However, one must also consider whether true equality is achieved if large-volume users are treated the same as users who make infrequent long-distance calls. Flat-rate calling without competition helps guarantee control by the monopoly, which often lacks incentives for efficiency and innovation. It is necessary to weigh the cost of duplicated investments under competition against the cost of inefficient management at the monopoly.

The inefficiency of monopoly has long been a concern for most countries. In Korea, Korea Telecom is regulated by a mixture of rate-of-return and performance indexes. Breaking up the monopoly into regional companies, as has been suggested, may not be a bad idea. The main argument is that such a breakup provides the regulatory agency some basis for measuring the performance of telephone companies by creating a comparative environment.

16.3.2 Privatization and Stakeholders

Essentially everyone has a stake of some sort in the concept of flat-rate calling—including MOC, Korea Telecom, EPB, large users, and the general public. Korea Telecom has no reason to oppose flat-rate calling because it would maintain its position, but it strongly opposes the idea of breakup. EPB is concerned about possible increases in local rates triggering inflation; however, decreases in long-distance rates could compensate for this, leaving an uncertain net effect.

As seen in the United States, large users were a major force behind the drive toward competition. In Korea large users have not yet publicly complained about the high prices of long-distance calls. However, large business groups may do so as they expand internationally and become more dependent on telecommunication networks. The top thirty *chaebol* produce 15 percent of Korea's GNP and 108 companies (0.6 percent of exporting companies) generate 63 percent of total exports, giving them a prominent role in the Korean economy. If they see inefficiencies, they could lead a call for competition.

It is unclear how the Korean public would react to flat-rate calling. They would surely oppose local rate increases, but their response to decreased long-distance costs is not clear. Citing the U.S. experience, lower prices for long-distance calls and higher end-user charges—both due to competition—actually reduced the overall bills of the average customer. There therefore seem to be no obvious reasons for public opposition.

16.4 Universal Information Service

With "popular use of information" a policy target for the year 2000, MOC proposed "one terminal per household" as a specific strategy. This has been explained as "realization of equal access to, and utilization of, information service over telecommunications network through low-cost terminals" (MOC 1985, vol. 2, pp. 1163–65).

It was some time before any specific schedules or plans for reaching the targets were announced. Based on statements in the press by the minister of communications during 1988 and 1989, the impression was that the terminals would be as intelligent as PCs and free to users. Finally, in mid-1991 a pilot program involving several thousand free dumb terminals was begun in Seoul. The plan is to distribute of 3 million terminals by the end of 1996. Whether these will also be free and dumb is unsettled. There has been no explicit foreign pressure regarding standards for the terminals. It is also unclear who will provide network and information services.

The phrase *universal information service* came into use only in the late 1980s, and is still without a generally accepted definition. Information service is sometimes defined as a service that provides, processes, stores, and transmits information. This broad definition includes electronic mail, protocol and code conversion, telephone service, postal service, and book publishing, to name only a few options. In this section the scope is narrowed to those services included in MOC's universal information service policy: Services carried over telecommunication networks, including most services described as VAN service or VAS. (Basically, this is what is called telecommunication networks, including most services described as VAN service or network-based service, TNS, in the context of trade, but TNS is not used here because we are not discussing trade issues.)

By excluding services that do not make use of telecommunications networks, some important information services that may develop in the future are left out, such as services over cable television networks. Dramatic cost reductions in fiberoptic cable installation in recent years indicate that the boundaries of the telephone, information service, and video entertainment industries are blurred in principle (see, e.g., Pepper 1988, pp. 5–11). Cable television networks today in Korea, however, generally carry only regular television channels to areas with poor reception and are restricted from carrying broadcast entertainment programming. In fact, cable television systems are illegal in urban areas where regular television signals are well received.

In this analysis, "universal" information services are those offered to the general public, including data base and data processing, electronic mail and bulletin boards, reservation systems, home shopping, home banking, and telemetering services. These services require information providers to develop the information, network providers to carry the information, and terminals for input and output of the information.

Since 1985, some companies in Korea with private leased-line networks have been allowed to provide data base and data processing services to the public. Providers were severely restricted from mediating or providing third-party communication. Amendments effective in May 1987 allowed these companies to provide information services over private networks without restriction as long as users were within closed user groups that shared business interests; interconnection to public networks was still prohibited. The definition of user groups was expanded in December 1988, but it still did not allow public network interconnection.

Starting in the early 1980s the Korean government has encouraged the use of telecommunications networks for data communications, emphasizing use of public networks, including Dacom's public switched packet data network. Currently, private industries are requesting even further relaxation of restrictions on the use of leased lines.

With about 1 million new telephones installed each year in Korea, a level expected to continue until the year 2000, electronic directories are a good candidate for an information service.

Educational information services may also be a major market in Korea. Extracurricular study through terminals could become very popular among families with students—many of whom demonstrate a zealous interest in their children's educations. The public television channel broadcasts educational program, but tutoring by television was found too difficult for some and too easy for others, a situation not easily dealt with in a one-way medium. Terminals in student homes could provide a workable solution: Interactive and personally tailored programs could be a much more effective mode for extracurricular education.

Another possible application is transaction service. At present, Korea is still a cash economy, although use of credit cards is spreading. (In 1987, 40 percent of transactions were cash, compared to approximately 25 percent in Japan and Taiwan. The pattern for most transactions is for people to deposit their income in banks, withdraw cash as needed, and use credit cards when out of cash. In this environment, answering balance inquiries would be a desirable service, and some banks already provide this to customers via telephone. However, there remain many legal and security problems that would have to be cleared up before expanded services such as money transfers could be offered.

Koreans are less likely to support the provision of bulletin boards and games services, however, as chatting and playing games with computers are unlikely to gain social approval in the Korean culture. The chance for their introduction is even slimmer if the terminals or services are financed by the government.

Compared to the general public, businesses are more inclined to make use of information services, at least if the applications make or save money. Business users are often the first to seek such services even in the absence of outside encouragement.

In summary, electronic directories, business applications, and educational services appear to be good candidates for services over telecommunication networks—if prices and terminals satisfy certain requirements.

16.4.1 Terminals

How intelligent should information service terminals be and what is the method of distributing them? In many on-line and text processing situations a dumb terminal can perform adequately. If there is much downloading of data, however, even conventional PCs may be unable to compete with traditional print offerings in convenience of use. It appears MOC is considering free distribution of intelligent terminals, such as personal computers. This could provide an impetus to the Korean hardware and related software industries.

Unless the service is inexpensive and easy to use, experience in most countries suggests that the general population would rather turn to such traditional information sources as books, directories, brochures, and telephones. Free distribution of easy-to-use terminals provides customers with equipment with no financial burden, and information providers and advertisers with a large potential market.

A free terminal distribution policy, however, has possible drawbacks. While massive distribution of terminals may establish de facto technical standards and may stimulate industry growth, it could also discourage innovations and creativity. Moreover, if domestically developed standards are not compatible with overseas standards, then the export market could suffer, and communications with other countries would be problematic. To achieve one terminal per household in Korea, approximately 10 million terminals would be needed at an estimated cost of $3–5 billion, not including development costs (NTIA 1988, p. 97). If technological development renders terminals obsolete, then continuing investment in replacement could become an unbearable burden.

16.4.2 Stakes of the Players

MOC is the major proponent of a universal information service policy. It claims such a policy would elevate the general public's computer literacy. Most of the general population simply has not had an opportunity to handle keyboards. General distribution of terminals, if accompanied by easy-to-use software and applications, could greatly help people to familiarize themselves with computers. Concerned that the domestic information industry (except for the hardware side) is still in its infancy and that the domestic telecommunications infrastructure is inadequate, MOC's priority is for Korean industry to become competitive as quickly as possible.

Because MOC's strategy has not been announced in detail, discussion of other players' stakes is only speculative. The general trend in Ministry of Trade and Industry and EPB policies has been to favor less government intervention and more private sector initiative, indicating these two powerful ministries may oppose MOC policy. If the trade surplus and the threat of inflation continue, however, the MOC policy of terminal distribution could offer a good opportunity to expand domestic demand.

Several ministries besides MOC have shown interest in distribution of PCs, although it has been confined to computer education for students. For example, the Ministry of Science and Technology provided three years of support (1987–1990) for a program to develop software for extracurricular tutoring (*Computer Vision* July 1988, pp. 90–96). In December 1987 the Ministry of Education announced a program to promote computer education, including computer labs in the regular curriculum and distributing 280,000 educational PCs to schools through 1996 (*Computer Vision* Jan. 1988, pp. 130–31).

Korea Telecom and Dacom, as the two major common carriers, both have major stakes in MOC policy. As long as the financial burden can be passed on to subscribers, Korea Telecom may not be seriously concerned. If the proposed universal information service policy is to be implemented under the present regulatory system, there is a possibility both Korea Telecom and Dacom may claim jurisdiction, as they both can be network providers. However, the dispute will likely be resolved in favor of Korea Telecom because it owns most of the physical plant.

The private sector has generally advocated liberalization of VANS, a request that will be even stronger if a universal information policy is implemented. Firms see many potential business opportunities for networks, terminals, and information provision and naturally wants to participate.

Foreign interests also have a stake in MOC policy. They would strongly oppose any restrictions on their participation or technical standards that preclude them. During trade negotiations between Korea and the United States in May 1989 both sides agreed restrictions on foreign investment should be almost completely lifted, including in telecommunications. From Korea's perspective, technical compatibility of Korean products with foreign goods is important because of the desire to export. It is safe to assume that foreign concerns about domestic technical standards will be a trade issue.

16.5 Conclusion

Korea is changing rapidly in many ways. The changes have sometimes happened so fast that people have not been ready for them. Politically and economically Korea is a very different place in 1991 than it was in 1987. Then, people believed democracy was a distant ideal, that relationships with countries governed by communists were impossible, that their country would never enjoy a trade surplus, and, indeed, might never get out of foreign debt. By 1991 the Soviet Union and China had agreed to open trade relations with Korea, North Korean products were imported directly, the trade surplus surpassed $10 billion, the summer Olympiad was successfully carried out, and Korea had become a creditor country in 1989. Domestically, dramatic democratic measures have been taken and more are likely to come.

The same momentum can be predicted for telecommunications. The penetration rate for telephones is still about half that of advanced countries, and more facilities are needed. While many people before the mid-1980s had to wait over a year for a telephone, however, there is no longer a backlog. This may explain why there has not been any explicit domestic movement to support competition or liberalization. Once set in motion, however, these trends may accelerate as fast as the liberalization seen in other sectors.

Flat-rate calling and the distribution of intelligent terminals to every home are radical changes. Few countries have ventured to implement such policies. Most people are unable to believe implementation will ever take place and may consider any discussion just government propaganda. But looking back, in the early 1980s when MOC announced its policies of "one telephone per household" and a "nationwide automatic dialing system" before 1990, people had the same kind of doubt. Those goals were actually achieved by 1987. By implementing the policies discussed in this chapter, Korea has established an unprecedented model for telecommunications development.

Acknowledgments

The author wishes to thank this volume's editors for their assistance in revising and updating an earlier paper of mine for inclusion here. That paper was originally prepared for the Program on Information Resources Policy, Center for Information Policy Research, Harvard University.

Bibliography

Collins, Susan M., and Won Am Park. 1988. *External Debt and Macroeconomic Performance in South Korea.* National Bureau of Economic Research Working Paper 2596. May.

The Computer Vision. 1988. "Second Educational PC Boom" (Jan.).

———. 1988. "Educational Software to Students" (July).

The Economist. 1988. "Pricing the Privatized," July 30, p. 65.

The Korea Economic Daily. 1988. "On the 7th Anniversary of Korea Telecommunication Authority: Rapid Growth . . . Contribution to the Enhancement of Telecommunications," December 10.

———. 1989. "Self Sufficiency of Digital Switches for Small to Medium Sized Switches," Jan. 11, p. 8.

Korea Telecommunications Authority. *Statistical Yearbook of Telecommunications.* 1980–1991. Seoul: Author.

Lee, Dong W. 1989. "The Challenges for Value-Added Services in Korea." Program on Information Resources Policy, Harvard University, May.

Ministry of Communications. 1985. *100 Years of Korean Telecommunications,* vol 2. Seoul: Author.

Pepper, Robert M. 1988. "Through the Looking Glass: Integrated Broadband Networks, Regulatory Policy and Institutional Change." U.S. Federal Communications Commission, Office of Plans and Policy, Working Paper 24 (Nov.).

U.S. Commerce Department, National Telecommunications and Information Administration. 1988. *NTIA Information Services Report 88-235* (Aug.).

17
Taiwan

TSENG FAN-TUNG AND MAO CHI-KUO

Telephone service in Taiwan has always been good by world standards—extremely good by developing economic standards when that was the category applicable to the island. Taiwan's remarkable post–World War II growth was achieved by many small businesses and industrious people. The government helped this by making a widespread telephone network a priority, although such facilitation was not a goal in itself. Despite being authoritarian, the government did not feel compelled to preclude a public network or usurp resources to build one primarily for its own control purposes—although it did have access to the military's system.

By the 1980s it was clear that technological and other changes required a rethinking of goals and structure. A decade has been spent studying these topics, as discussed later, and in the 1990s something will be done. Taiwan has the need and the ability to utilize the latest in value added network (VAN) services and transmission technologies, but has been slow to implement the former as the decision-making process on the method to be used, has dragged on.

This chapter begins with a review of telecommunications on the mainland from the nineteenth century to 1949. Many of the institutions were moved to Taiwan at that time. Succeeding sections look more closely at historical development in Taiwan, the organizational framework of telecom services, and the services offered. The transition process is discussed in the fifth section, while the sixth looks at some of the specific proposals for structural reform.

17.1 History, Mainland to 1949

The Ch'ing Empire adopted a policy known as "state oversight with private run" to develop telegraph and telephone systems in the late nineteenth century. The government was simply too poor to build them itself. As a result, the earliest telecommunications system in China was constructed by a Danish company—Great Northern Telegraph—in 1871. It ran along the seashore between

Hong Kong and Shanghai (about 1,800 km). From 1873 to 1881 telegraph systems were run primarily by foreigners, and were thus overseen by the Customs General Tariff Office of the Ministry of Foreign Affairs.

The first system constructed by Chinese was a telegraph line completed in 1877 between Tainan and Kaohsiung (about 54 km) in southern Taiwan. In 1879 the governor of Jilin province (southern Manchuria) proposed a line connecting China proper to where talks were taking place with Russia, which was aggressively seeking to seize areas of Asia that had been under Ch'ing suzerainty. A start was made, but the line was not finished. Parts of it were used in China's second domesticly built line, which ran some 1,200 km between Tienjin and Shanghai. It was completed in December 1881. That year the directorate general of telegraph was established in Tienjin as the headquarters of the state-run telegraph company. It had operations in Shanghai and other cities along the route. Within ten years there were approximately 36,000 km of telegraph lines and over 500 new branch offices throughout China.

The first telephone system was introduced by a British trading company in Shanghai in 1881. Companies from Germany, Denmark, France, the United Kingdom, Russia, and Japan were involved in the construction of virtually all major installations through the 1890s. Systems were built in port cities, such as Xiamen, as well as in inland ones, such as Chongqing. Foreign companies operated most of the systems, so, again, administrative responsibility resided in the Customs General Tariff Office.

In 1906, a team of high-ranking officials returning from a foreign fact-finding trip recommended to the Ch'ing Government that it follow the European example and establish a Ministry of Post and Transportation (MPT), nationalize all communications and transportation services—including posts, telecommunications, railroads, and ocean transportation—and establish technical training schools. By 1910 the newly created MPT had successfully converted all private and provincial telegraph operators into government institutions through reimbursement (from subsequent operating revenues) or loan programs, thus laying the foundations for government monopoly. (The railroads also were nationalized in 1910, but ocean transport was not.)

A total of 70,000 km of telegraph line and 239 telegraph offices were taken over by the MPT in 1910. However, foreign companies still held cable landing rights and equipment manufacture patents in some coastal cities, and these companies continued to provide international wire and wireless services.

17.1.1 Legal Foundations

After the founding of the Republic in 1911, the MPT was renamed the Ministry of Communications (MOC). Because large parts of China were occupied by independent military strongmen during the postrevolutionary period, each province had full autonomy over its own telecommunications. When the turmoil came to an end and the nation was reunified in 1927, MOC set up three divisions to integrate all national administrative matters: railroad, telecommunications, and post.

The first Telecommunications Law had been drafted in 1915; with minor exceptions, it reflected the European model of government monopoly. Revisions made in 1929 prescribed that all long-distance and international services be operated directly by MOC. Thus, MOC began to merge telegraph and telephone operations in the 1930s. However, with permission from the central government, local telephone systems could be capitalized and operated by local governments or private enterprises. The 1929 Law also required government permits and payment of a license fee to MOC before installing broadcasting stations or operating radio receiver sets. MOC had a large number of committees working on technical standards, system specification, equipment manufacturing, and research and development. Many university students majored in electrical engineering and an entrance examination system was established to recruit technical staff.

In 1935 MOC formed the directorate general of post (DGP) to operate the postal services. Unfortunately, China fell into war against Japan in 1937 and facilities were seriously disrupted. During the war MOC moved to the western part of the country and began plans to create a unified post, telegraph, and telephone policy and supervisory structure. In 1943, it set up the Post and Telecommunications Department to oversee public policy and supervision. A directorate general of telecommunications (DGT) was established to be responsible for development, operations, and management of telecommunications. Regulation and provision of service have been separated ever since. Unfortunately, the nation fell into more turmoil after V-J day. In 1949 the government moved to Taiwan.

17.2 History, Taiwan

The first telegraph line on Taiwan was constructed in 1877 between Tainan and Kaohsiung. Because of the increasing importance of military defense, efficient communication channels to Taiwan from the mainland and within the island were vital. In the mid-1880s Taiwan was ruled as a prefecture of Fujian province, across from it on the mainland. Under Liu Ming-Ch'uan, governor of Fujian from 1884 to 1891, a number of projects were undertaken to create infrastructure. In 1886 a DGT was charged with laying marine cable along the west coast from Keelung via Tamsui and Tainan to Kaohsiung, and extending land lines to Taipei, Hsinchu, Changhua, Taichung, and Chiayi. Two submarine cables were installed, one from Tamsui to Fuzhou (on the mainland) and the other from Anping to Penghu (an island in the Taiwan Straits). More than 2,200 km of line were in service by 1888.

Taiwan was under Japanese occupation for about fifty years from 1896. Post, telegraph, and telephone systems were constructed and operated under the supervision of the Japanese Ministry of Communications. Facilities were seriously damaged by Allied bombing during World War II. After V-J Day, supervision and operations of Taiwan's post and telecommunications returned to the Chinese government.

After moving to Taipei in 1949, MOC's first step was to divide the operations of the post and telecom services into two separate units: the DGP for post and the DGT for telecommunications. MOC delegated supervisory power to DGT to establish a Taiwan Telecommunications Administration (TTA) for operation and development of regional services and a Chinese Government Radio Administration (CGRA) for international services. This administrative arrangement proved to be effective for recovery and reconstruction; telecommunications in Taiwan was restored to its prewar level in 1952, when there were 39,363 telephones (about 0.49 per 100 people).

17.2.1 Local Telephone Development

Since the 1950s, there has been a high demand for telephone service. Under the government monopoly system, development capital could come either from the government budget or from public loans. The government was in financial difficulty, however, and there was no place for the TTA to get a loan. Given this situation, the government adopted ''self-sustainment and self-development'' guidelines for all state-run businesses.

In the early 1950s, a study team found that it would cost NT$14,000 to install one telephone line, including NT$6,800 for subscriber line plant and NT$7,200 for switching and transmission equipment. In 1956, with the support of the Taipei City Government, City Council, and Chamber of Commerce, the telephone installation fee in Taipei was adjusted from NT$7,200 to NT$14,000. The higher fee was soon applied to other cities (see Lee 1981). Only business firms and wealthy individuals—the fee was several years' earnings for most people—could afford to apply at this price. Nevertheless, it was a supplier's market and backlog kept building. The fees were part of a series of four-year construction plans TTA launched in the 1950s and 1960s that involved end users sharing construction costs. As incomes increased in the 1970s, the number of residential phones increased.

The first electronic switching system was put in service in 1980. Subscriber toll dialing (STD) services reached 100 percent penetration in 1981. The years 1969–1980 saw annual growth of more than 20 percent for subscriber stations, and density exceeded twenty in 1981. Data are in Table 17.1.

17.2.2 Toll and International Telecommunications

The first microwave backbone toll system was completed in 1963 along the west coast (where most people live). A second backbone system was put in service along the east coast a few years later. Microwave links were soon added and branched to every toll switching office. Carrier cable systems were commonly installed between tandem switching offices. The first digital coaxial cable backbone system, with a capacity of 5,600 channels, was put in service in 1981. Of the total 70,000 long-distance circuits in place at that time, 70 percent were digital. Long-distance construction also enhanced rural development; since 1975, all 231 townships have had telephones, and service reached all 7,239

Table 17.1. Telephone Growth in Taiwan

Year	Stations[a]	Percentage Up[b]	Density[c]	Year	Lines[a]	Percentage Up[b]	Density[c]
1950	33.4	10	0.4	1982	4,356	14	22.5
1960	67.9	10	0.6	1983	4,854	11	26.0
1963	99.5	14	0.8	1984	5,278	8.7	27.9
1966	153	17	1.2	1985	5,653	7.1	29.5
1969	265	26	1.9	1986	6,077	7.5	31.3
1972	596	24	3.9	1987	6,548	7.8	33.4
1975	1,117	23	7.0	1988	7,159	9.3	36.1
1978	2,099	23	12.2	1989	7,834	9.4	39.1
1981	3,820	22	21.2	1990	8,431	7.6	41.6

[a] Cumulative yearend installed base, in thousands.
[b] Percentage increase from previous year shown.
[c] Stations per 100 people.

villages in 1980. By the mid-1980s even people in mountainous areas or on remote islands could easily call each other (see Liang 1981).

The first over-the-horizon troposcatter microwave system—to Hong Kong—was completed in 1967. A Sino–Philippine troposcatter followed shortly thereafter. Telephone traffic grew rapidly and soon reached capacity. The facilities were replaced by satellites and submarine cable communication systems in 1981. That year an AT&T 4-ESS digital switch was installed to serve as an international gateway, providing 700 telephone circuits.

Taiwan's telecommunications entered the space age in 1969 when the Taipei Satellite Communication Earth Station was established in a suburb of Taipei to provide commercial services via the Intelsat Pacific satellite. An antenna was constructed in 1974 to link with the Indian Ocean Region satellite. A second antenna was later built to link with the Pacific satellite. These provide 1,600 circuits to forty locations. International traffic has grown 20–25 percent each year since then.

In 1979 a cable landing station was built at Toucheng, about 60 km east of Taipei, to provide additional capacity. Four coaxial submarine cable systems subsequently have been set up through joint ventures: OKITAI (to Okinawa, July 1979), TAILU (Luzon in the Philippines, Mar. 1980), TAIGU (Guam, May 1981), and SINHONTAI (Singapore, Hong Kong, Oct. 1985). In addition, the International Telecommunications Authority (ITA, CGRA's name from 1981) has actively participated in planning and constructing ten fiberoptic submarine cable systems in the Asia–Pacific region (including two that continue into Europe) and is purchasing indefeasible right of user (IRU) on three Atlantic fiberoptic cables (TAT-9, TAT-10, TAT-11).

A second cable landing station opened at Fangshan, about 60 km south of Kaohsiung, in 1989 as part of long-range planning for replacement of the coaxial cables. Two fiber links are in place: GPT (Guam, Philippines; 1989 Dec) and HONTAI-2 (Hong Kong; 1990 Jul). A third—APC (Asia–Pacific Region, which will run from Singapore to Japan and to Guam)—will open in 1993. With APC, Taiwan will have seven cable landings. It has IRU circuits on thirteen other cables.

As the international fiber network in the Pacific Basin continues to expand, carriers may cross-invest to ensure route diversity and spread costs. Taiwan will have more cables than Hong Kong, Singapore, or Japan, an important factor in making Taiwan a network center for Asia. As part of this goal, ITA decided to build an earth station at Fangshan in 1990. This will further diversify route options and increase service reliability. A second international gateway switching system has been installed in Kaohsiung. Taiwan now has fiber and microwave backbone loops linking two satellite earth stations and two submarine cable landing stations to two gateway switching systems. This is unique in the region (see Tseng 1991).

17.3 The Organizational Framework

A new Telecommunications Act was drafted and enacted in 1958. At the time the nation was under Martial Law because it faced a military threat. The nature of the government monopoly in the 1929 Law remained unchanged. In fact, the 1958 Act also regulated radio and television services. The Act was revised in 1977 to reflect technical progress in satellite and data communications. Although MOC/DGT proposed the concept of a public corporation and relaxed data regulations, neither proposal was accepted by the Legislative Yuan.

DGT's organizational structure has experience two major changes since 1949. The first was in 1969 when DGT had about 10,000 employees and several big projects—including crossbar telephone switching, microwave backbone systems, and a satellite earth station—were being constructed and put in service. DGT decided to expand the TTA and the CGRA organizations to manage the expanding services. Two new institutions—the Telecommunications Laboratories (TL) and the Telecommunications Training Institute (TTI)—were established for technological development and manpower training.

DGT was responsible for policy guidelines and supervision, while its four subordinate institutions oversaw routine operations. It is interesting to note that from 1968 to 1974 the director-general of DGT was also appointed one of the two vice ministers of MOC as a way to emphasize the importance of the DGT post. At the same time, a senior DGT staffer served as director of the Post and Telecommunications Department in MOC. This arrangement proved to be very efficient.

MOC/DGT submitted a draft proposal for the revision of the Telecommunications Act that involved changes in DGT's name and organizational structure, including making DGT a public corporation. The plan was rejected by legislators. MOC nonetheless ordered DGT to make some structural changes, beginning in May 1981, based on the postal service's 1979 reorganization model. For the next three years, DGT's stability suffered from this transition.

TTA (with about 26,000 employees before reorganization) was divided into three geographic units—northern, central, and southern—and the Long-Distance Telecommunications Administration (LDTA). CGRA (with about 1,500 employees at the time) changed its name to the International Telecommunications Administration (ITA). These five divisions were responsible for planning, constructing, operating, and maintaining the systems and services. In addition, the Data Communications Institute (DCI) was created to provide the support of data communications services. The resulting organization structure is shown in Table 17.2.

17.3.1 Budget Approval and Procurement Procedure

As a government institution, the DGT budget is submitted first to the Executive Yuan via MOC and then to the Legislative Yuan for approval. Because DGT's director-general is not an official of ministerial level, he cannot be present at the legislature to defend budget requests or recommend policy. This tends to slow implementation of evolving technology.

Procurement is regulated by the government; it is complicated, particularly when foreign firms are involved. Many procedures can be traced back to the 1930s when an audit system was created to cope with major public investment for construction projects. The Ministry of Audit (MOA), which reports to the Control Yuan, standardized practices in 1939. The Statute on Audit and Procurement was revised in 1950, 1955, 1967, and 1972. Open bids are generally specified for procurement.

The International Trade Bureau issues policy guidelines on the status of countries supplying equipment. Although there are no quotas, the supplier list

Table 17.2. Directorate-General of Telecommunications (DGT) Structure*

Abbreviation	Employees June 1990	Operation (major cities served)
Taiwan Telecommunications Administration (TTA)		
DGT	662	Headquarters staff
NTTA	12,971	Northern Region (Taipei)
CTTA	8,028	Central Region (Taichung, Chia-yi)
STTA	8,738	Southern Region (Tainan, Kaohsiung)
LDTA	2,852	Long Distance
ITA	1,851	International (formerly Chinese Government Radio Administration, CGRA)
Other Components		
TL	896	R&D Lab
TTI	372	Training
DCI	309	Data Communication
	36,679	Total

*This structure became effective in May 1981.

is normally favorable to European and North American countries; Japan is excluded in many cases. In general, an international procurement takes twelve to fifteen months from initial request to delivery. It may drag on longer if disagreement emerges among the various regulatory parties.

The procedure for domestic procurement is simpler. To maintain systems and services quality, DGT has ruled that a survey of manufacturing ability must be conducted to see if there is any potential local supplier (DGT 1981). To ensure compatible equipment standards, DGT has centralized procurement procedures for local and long distance telephone facilities. However, in many cases this is very time consuming. If fewer than three suppliers bid, price negotiation with one or two bidders may take place.

Existing procurement policies have caused particular problems for the highly technology oriented telecom services industry. Specifications for procurement put more emphasis on overall systems performance than on individual units. Acquisitions of equipment suffer from substantial delays resulting from government bureaucratic intervention. Proper bidding procedures should focus on lifetime costs rather than on first costs. However, complicated cost analysis, with detailed breakdowns of hardware and software items, is difficult for laymen to understand. Thus, it is usually very difficult for nontechnical auditors to agree on a base price for negotiation. These problems were evident in three years of negotiations with AT&T Taiwan, ITT Taisel, and GTE Taicom on the purchase of switching equipment.

The cumbersome purchasing procedures extend to land acquisition. The market price is almost always much higher than the "official announced land pricing" (assessed) value set by the government. It very hard to convince auditors that a price acceptable to the seller is reasonable. DGT proposed changes to the procurement policy, which would affect transportation projects as well. Finally, the government set out to review the land appropriation system with the intention of instituting uniform standards that "avoid confusion and difficulties between the Government and landowners." The changes are to include revision of assessments to reflect prevailing market prices.

17.3.2 Development

DGT established the TL in 1969 to promote industry development. TL reports to DGT and receives about 2 percent of DGT revenue for R&D. Since the 1950s DGT and TTA have used their purchases to aid development of a domestic equipment industry. In response to a series of four-year national economic development plans, which began in 1953, the TTA signed long-term procurement contracts with a dozen local manufacturers. These included Pacific Cable, Hwashin-Lihwa Cable, Universal Cable and Hawzun Cupro (for cable); Taicom and Far East (crossbar switches); Universal Electric and Tongya (telephone sets); and Vidar (PCM equipment) (see Lee 1989).

While short of natural resources, Taiwan has an abundance of well-educated workers, which is essential to development of the electronics and information industry. The integrated circuit and computing equipment industries began to emerge in the early 1980s. Many new venture capital companies came to the Hsinchu Science-Based Industrial Park. The complex, opened in December 1980, was an early key piece in government policies to encourage high-tech R&D. Others included such programs as duty free import of instruments, tax deductions on imported material for export products, and loans.

In the early 1970s, the government began to view the United States as a major technology source. In 1973 ITT Telecommunications and GTE International, respectively, created ITT Taisel and GTE Taicom as joint ventures to produce electronic telephone switching equipments. In 1984 AT&T established AT&T Taiwan to produce digital switching equipment. As a major customer, DGT was requested to become a shareholder in all three companies, which it did, as shown in Table 17.3.

When the Executive Yuan told the DGT that the network was to be divided into three regions, it was also specified that each of these three digital switching equipment manufacturers would get the contract for one region. This was intended in part to reduce maintenance costs for each region. With the sixth tender for equipment in 1990 the market was open to competition.

17.3.3 Employee Staffing and Labor Union

Employment practices were defined in the Statute on Communication Business Staffing of 1957. It stemmed from the entrance examination system established in 1934, which was similar to the British civil service system. Two parallel routes are available for employee ranks and job grades. There are six ranks for the two professional ladders—technical and business. New employees have to pass an open examination held by the Ministry of Examination in order to get certification for ranks R2, R3, and R4—depending on their education background. Most TL employees have postgraduate degrees and are hired without passing through the open examination.

The number of DGT employees increased only 10 percent during 1981–1987 despite a 85 percent increase in installed lines. In March 1988 there were 30,082 employees—including 12,763 (42 percent) technical and 4,130 (14 percent)

Table 17.3. DGT Investment in Its Major Suppliers, 1990

Company	Year Founded	DGT Investment NT$[a]	DGT Investment Percentage[b]	Region Supplied[c]
ITT Taisel	1973	164.0	40	Central
GTE Taicom	1974[d]	63.0	15	Northern
AT&T Taiwan	1985	233.7	15	Southern and toll network

[a]Includes subsequent capital increases.

[b]Of total. DGT paid for its shares on an equal basis.

[c]During first five tenders for switches. See text discussion.

[d]DGT investment made in 1978. Siemens acquired GTE's equipment business in 1989.

temporary. DGT traditionally has provided a life-long, stable working environment. Thus, more than 90 percent of DGT employees have stayed at DGT until compulsory retirement at age 65.

The TTI was established in 1969 for career training. The training program consists of orientation for new employees, on-the-job training and advanced studies at various levels. Seminars on operations, management, and special technologies are also often held in the TTI.

As Taiwan moves into the Information Age, DGT is facing several staffing issues: How to recruit employees with advanced degrees? How to assess manpower needs in light of organizational structure expansion? How to differentiate the role of a government employee from that of a business employee (see Chang 1985)? Any attempt at major revision of the 1957 Statute may be complicated. However, a special measure proposed by DGT to recruit research personnel on the basis of higher educational degrees for the TL recently has been approved by the Executive Yuan.

Following the establishment of the government's Council for Labor Affairs in 1987, DGT's labor union was reorganized and became more active. It has asked for certification of temporary employees who were unable to pass the open examination and for a reasonable promotion path and job allowances.

17.4 Facilities and Services

Under the 1977 Telecommunications Act, basic network circuits and services are a DGT monopoly. Development plans initially emphasized installation of local telephones to ease the burden of a long waiting list. During the 1980s, the main goal had been transition from analog to digital switching systems. Between 1979 and 1988 DGT invested the equivalent of 1.06 percent of GNP to improve services. Long-term plans have been formulated for construction of basic networks leading to establishment of ISDN by the year 2000 in metropolitan Taipei, Taichung, and Kaohsiung.

In the toll network, analog microwave links have been replaced by digital systems, and coaxial cable by fiberoptical. There have been three primary centers—Taipei, Taichung, and Kaohsiung—and thirteen toll centers since July 1990. Toll transmission was 92 percent digital and toll switching was 76 percent digital in 1990.

As of June 1990 the number of telephone stations exceeded 8.15 million (40 per 100 people). Almost two-thirds of households had telephones. There were well over five coin telephones per 1,000 people. Total local switching capacity was 7.84 million lines—20 percent digital, 24 percent electronic, 56 percent still crossbar. All local telephone networks are expected to be digital by the year 2000.

The growth rate of telephone density was 18–21 percent during 1965–1980. It then gradually slowed to 7.5 percent, reflecting the larger base and density. Still, compared to many advanced countries that had penetration levels over 50 percent before 1980, the slowdown reveals two problems. The first is the bureaucratic procurement procedure, which caused some delays on major construction projects. The second is that restrictions on nontelephone applications limited demand for second lines.

Cellular service had a capacity of some 145,000 lines in 1991, which was to be tripled by mid-1993 under a contract with Ericsson reported in January 1992. Ericsson also provided switches and radio-base stations for the existing capacity, all of which is analog. Handheld and car telephones were banned until July 1989. That same summer DGT drafted plans for full competition in mobile telephones in 1993.

17.4.1 Data Communications and VANs

Data communications began in 1971 when leased circuits were first opened to the public. Since the Data Communication Institute (DCI) was established in 1981, additional services have been put into operation. Available services are listed in Table 17.4.

Data Communications Rules were drafted in 1981, in accordance with the Telecommunications Act; they were subsequently revised in 1983 and 1986. Domestic data communication services are monopolized by DGT/DCI. Private entities cannot operate or resell leased circuits. There have been numerous complaints about this from both foreign and domestic firms. However, MOC

Table 17.4. Types of Services Available

Date Began	Subscribers 1985	Subscribers 1990	Service
Nov. 1971	4,452	24,556	Domestic leased line data
Jan. 1975	41	222	International leased line data
Dec. 1979	63	212	International Universal Data Access Service (UDAS)
Jun. 1982	115	112	Circuit switching data (CIRNET)
Jan. 1984	103	730	Dial-up data
Jan. 1984	103	92	PIPS[a]
Oct. 1984	531	2,578	Packet switching data (PACNET)[b]
Aug. 1987	—	11,307	Chinese Videotex Service (CVS)[c]
Aug. 1987	—	175	Public message handling (MHS). E-mail.
Sep. 1989	—	177	MVDNET[d]
Nov. 1989	—	326	MARNET[e]

Source: Data Communications Institute.

[a] Public Information Processing System. Subscribers use their personal computers or terminals to access DCI mainframe for information processing.

[b] Can link to international packet networks such as Datapac, Telenet, Tyment, and so on.

[c] Subscribers can use television sets or personal computers to access 165,000 screen pages of data from twelve providers.

[d] Motor Vehical Driver Network. For fleet administration and tracking.

[e] Multi-Access Reservation Network, for airline ticketing.

is concerned about the effect of transborder data flow on national security. In 1988 the government granted special approval for shared usage of data services for international networks such as Citibank, Sita, AP, Reuters, GE, and Swift.

VAN services were opened to competition in July 1989, and thirty-eight private firms had registered as of March 1991. Because the United States and Japan have well-established data base services, eleven of the firms provide links to one or the other of these two countries; twelve have other overseas links, including five to Hong Kong—a reflection of its role as a region telecom center. Many of the systems are dial-up rather than dedicated-line, reflecting low initial traffic volume. Almost every VAN provides information storage and retrieval service. Different names are used, but they are basically videotex. Only one VAN was providing electronic document storage and delivery (a form of E-mail). For the services to expand, we feel further revision of the regulations are required—particularly as the relate to leased circuits (see Tseng 1991). There has been a proposal to allow foreign firms to own 30 percent of VANs, but failure to implement it has led to friction with the United States in particular.

17.4.2 Customer Premises Equipment

Traditionally, all subscriber telephone sets were the property of DGT/TTA; rental charges were included in the monthly bill. Taiwan has become a major exporter of push button telephones since the early 1980s. The Local Telephone Rules were revised in August 1987 to permit subscribers to own telephone sets, thereby opening the CPE market to the public.

Similarly, all data communication modems had been provided by DGT/DCI. In November 1987, this constraint was relaxed to allow customers to provide their own modems for speeds below 2,400 bps. Further relaxation is expected in the early 1990s to allow intelligent communications terminals and modems with speeds up to 9,600 bps.

Private companies are allowed to own PBX equipment and connect it to the public network. No specific rules govern installation and operations of LANs, so the regulation on PBXs are applied. Private companies may design and construct their own LAN; following inspection and approval, the LAN may be connected to the public network.

17.4.3 Rate Structure and Financial Health

DGT policy traditionally combined high installation fees with a low service rate structure. In 1956 installation fees were set to recover virtually all of the cost (including switching equipment) immediately. This provided DGT/TTA with a sound financial foundation from which to launch a series of development plans, with end-users sharing initial construction costs. Even though installation fees were beyond what most people could afford, and the number of installed lines was increasing at double-digit rates each year, the installation backlog grew.

The rate of return for pricing, which must approved by the Legislative Yuan, is between 8.5 and 11.5 percent—with VAN and other new services, such as paging and cellular, allowed a return at the higher end of that range. Any major rate adjustment on basic services must be approved by the Executive Yuan. However, DGT can adjust rates on data communication services or other minor items simply by giving notice to MOC. Rates have been adjusted three times since 1949, based on a ''zero effect'' principle (i.e., revisions were structured so that the change in overall revenue from the adjustment was 0).

In the immediate postcolonial period, when all calls were connected manually, there was a flat monthly fee—tracking the number of calls made from any given phone was impractical. Automatic switching allowed charging by volume, and this led to the first rate adjustment, which took effect in January 1957. A rate of NT$0.70 was set for calls beyond a base number. A public coin phone call cost NT$1.00 for three minutes.

The second change came in 1976, increasingly the basic call rate to NT$1.00. Monthly rates are dependent on the number of lines in the calling area; the number of divisions has been reduced from four in 1957 to three and then to two. Similarly, long-distance regions (rate bands) have gone from eleven to six to three. These changes resulted in a price increase for local services and a price decrease for long distance.

A third adjustment was completed between November 1987 and January 1988. The basic call rate went to NT$1.10, and installation was reduced to NT$12,000 (U.S.$418), equivalent to a few months of the average wage. Rates for international direct dial calls decreased from 20 to 30 percent. In 1991 basic monthly dial-tone charges were reduced about 14 percent for nonresidential lines and just over 20 percent for residential lines. The additional message unit charge dropped back to NT$1. An urban business line is about U.S.$11.

Domestic long distance is billed in NT$1 increments, with the number of seconds varying with time of day and distance. Under the July 1989 tariff, the highest rate is about US$0.37 per minute (for a call that would cost 21 cents in the United States). The farthest band begins at 140 km, less than half the length of the island.

Financially, DGT and its subsidiaries are dealt with as one unit: international revenue subsidizes domestic service, long-distance subsidizes local, and basic services subsidize data services. Results for fiscal year 1987 showed that less than 2.0 percent of total revenue was from data services. Data on DGT finances are in Table 17.5.

17.4.4 Regional Collaboration

In the 1950s and 1960s, CGRA had exchange visit programs with KDD, RCA, WUI, and ITT for international telecommunications development. The construction of a troposcattering communications system between Taiwan and Japan was a direct result of CGRA–KDD collaboration in the early 1960s. The TTA also had an exchange visit program with NTT for telephone engineering, construction, and operations.

Table 17.5. Directorate-General of Telecommunications (DGT) Operations

Year[a]	Total Revenue	Owner's Equity	Liabilities	Employees	Revenue Per Worker (in NT$ thousands)
	(in NT$ billions)				
1975	7.1	23.2	3.3	18,604	381.6
1980	24.0	64.4	16.0	24,602	975.5
1985	45.6	143.2	15.4	29,798	1530.5
1990	87.2	223.7	31.0	36,679	2377.2

[a] Fiscal year ended June 30 of following year.

The DGT established the International Telecommunications Development Corporation (ITDC) in 1974 to facilitate regional collaboration after Taiwan withdrew from the ITU and Intelsat. With assistance from Comsat, ITDC–ITA maintained international satellite communication operations and services, as well as cable services.

DGT has collaborated with Comsat since 1977, sending research engineers to Comsat to study satellite transmission technology. Advisors from Comsat were invited to assist in the construction of the Taipei Satellite Communication Center. DGT began to lease Intelsat satellite transponders for domestic service in 1989.

The Ato Rengikai was founded in Tokyo when diplomatic ties between Taiwan and Japan were severed in the mid-1970s. Ato has acted as a gateway to maintain DGT–NTT collaboration on systems and technology. (Since privatization of NTT in 1985, Ato has been replaced by NTT International.) DGT has collaborated with many other telecommunications entities such as MCI, BellSouth, Nynex, and Hawaiian Telephone Company.

17.5 The System in Transition

Although MOC/DGT proposed corporatization in 1977, the proposal was not accepted by the Legislative Yuan. Under this constraint, DGT reorganized in May 1981 and continued as a government monopoly. DGT has been hampered by slow budget approval, procurement procedures, and employee staffing systems, resulting in deterioration in DGT's ability to expand and upgrade services.

17.5.1 Studying the Issues

During the 1980s policy makers and parties interested in Taiwan continued to pursue the issues of competition and privatization in a Taiwan context, encouraged by the shift toward competition in telecommunications underway in the United States, United Kingdom, and Japan. When the statute on DGT Organization was revised in October 1984, the Executive Yuan instructed MOC to restudy the possibility of corporatization. MOC again treated telecommunications like postal services. However, after a two-year study, it concluded the two services have different characteristics. The study group proposed that DGP remain a government department, while DGT should be reorganized as a government-owned or privatized corporation.

In the wake of this and various other studies that indicated an "appropriate" level of competition was beneficial to all telecommunications providers and users, the Council for Economic Planning and Development (CEPD) of the Executive Yuan agreed to study the relevant legal issues. More specifically, the proposal was to consider what policies and laws would support restructuring telecommunications into a competitive industry and to analyze the feasibility of privatization.

The research contract was awarded in 1986 to the China Interdisciplinary Association (CIDA). The team consisted of fifteen members, including consumers, professors, scholars and experts in law, business, mass media, telecommunications, computers, and electronics. After twelve months of work, a final report was issued in July 1987. (One of the authors was secretary general of CIDA at the time and served on the team. His is the lead name on its report: Tseng et al. 1987.)

CEPD studied socioeconomic and legal aspects of competition, and recommended separation of government regulatory and system operations functions through a new legal framework. It also drafted a conclusive recommendation on the legalization of policies.

Since that study was completed, various government agencies have undertaken additional research. MOC studied regulatory administration, and also recommended that DGT be reorganized into corporate organizations. DGT studied managerial and organizational issues of competition, and recommended priorities and procedures to be considered to ensure a smooth transition. Consequently, an ad hoc committee was established in MOC to work out the processes of change, and a DGT seminar was held in September 1987 with the theme, "Liberalization, Rationalization, and Privatization."

17.5.2 Re-evaluation of Policies

In the late 1970s people in Taiwan still studied Chinese keyboard input methods and debated the standardization of Chinese computer code books. In 1980, when the first Information Show was held in Taipei, people began to recognize the computers' ability to stimulate economic growth. With the advent of personal computers, there was a realization that computers could be used to process Chinese characters.

Networks are necessary channels for information flow to every business. In the past, land, energy, and capital served as Taiwan's resources. Information is now the major resource and the information industry is Taiwan's growth engine. The Information Age is the result of the growing interrelation between information processing and telecommunications.

These two industries developed in very different environments. As in other advanced countries, the information-processing industry in Taiwan has always been competitive, while the telecommunications industry has been a regulated monopoly. Introduction of competition in telecommunications may rapidly produce positive results in terms of lower prices, improved services, wider choice and acceleration in the availability of advanced products. The CEPD has argued that the transition to competition would benefit users, enabling them to choose from a wide menu of transmission services and interconnect these as their needs require. A variety of competitive VANs and information service providers would be able to provide network management and data base access. Users would also be able to choose from a variety of cost-effective CPE. Long-distance rates would most likely decrease due to a surge in traffic.

The CEPD also argued that service-oriented businesses could become potential information providers. Computer software houses could expand their scope to provide VAN services as new Type 2 carriers, enabling these newcomers to become the fastest growing segment of the telecommunications marketplace. Their use of network capacity would provide services that would fuel the economy in the 1990s. CEDP pointed out that Taiwan's information industry producers have depended mainly on exports. All 5+ million subscribers in Taiwan, however, are potential consumers.

17.5.3 A Study for Cable Television Development

Taiwan has had three commercial television networks—TTV, CTV,and CTS—since the 1960s. The first community antenna system was approved in 1970 to provide services in rural areas with poor reception. As of 1990, there were over 120 systems with 800 thousand subscribers—a 15 percent television—household penetration. There were also so-called Fourth Channel systems in urban areas to provide programming from video tapes. These often involve obscene and pirated material and in any case are illegal.

CATV is a new direction for television development. In August 1983 an ad hoc working group of the Executive Yuan for cable television development was established in MOC to study CATV systems and services. After eighteen months, the group presented recommendations to the Executive Yuan indicating that it was technically feasible and economically viable to develop CATV in Taiwan. However, the practical issues of program supply, legal restrictions, and administrative workload would present a different set of concerns.

The group recommended selected pilot projects be undertaken to determine market needs and to gain operational experience (see Tseng 1985). When a private entity pursues a CATV pilot project, it may choose a proven coaxial cable technology and aim at a market niche. But when DGT undertakes a project, it may construct fiberoptic subscriber loops and integrate with broadband ISDN development. The study reports and recommendations were submitted to the Executive Yuan in June 1985 via MOC; however, CATV policy was still pending in the early 1990s.

17.5.4 Survey Study on Operation and Organization

During 1986–1987 a CEPD/CIDA team studied three areas—basic network services, CPE markets, and VAN services. To distinguish basic and VAN services, they adopted the international convention of distinguishing a Type 2 carrier as one that provides value added services over circuits leased from a Type 1 carrier (ie, DGT or its subsidiaries). In most cases, Type 2s provide VAN, share-use, resell, and information-processing services.

The study team interviewed about 120 people in November 1986, including those in the industry, scholars, and consumers, asking about perceptions of the optimal organizational structure. Questions addressed both Type 1 and Type 2 service and whether each should operate as a regulated monopoly or a deregulated competitive structure. It broke this down further to analyze whether they should be provided by one national organization, government or private, or by private corporations.

Almost no one felt DGT should retain its existing monopolistic status as a government institution. However, a public monopoly was considered best for Type 1—62 percent gave this response regarding the domestic market, 59 percent for international. (It should be noted the Type 1 questions did not distinguish local and long-distance services.) Almost all interviewees preferred introduction of competition in Type 2 telecommunications (96 percent for domestic, 95 percent international) with private organizations (59 percent, 52 percent) providing services.

17.6 Legal Framework and Organizational Issues

The telecommunications market environment consists of three elements: Type 1, Type 2, and customer premises equipment. In many cases, Type 1 carriers provide both basic circuits and VAN services. Type 2s lease basic circuits from Type 1s to provide VAN services. Type 2s are strictly defined as having no license to construct any wired or wireless transmission lines. Therefore, a Type 2 needs neither a channel frequency allocation from the national radio frequency registration authority or right-of-way permits from the local government. Since a Type 2 does not require public resources, it has less of a social obligation than a Type 1. Customers can purchase their CPE on open markets.

The CEPD/CIDA study team indicated that a new law is necessary to ensure fair play between the monopolist and new competitors. A conceptual framework has been proposed. There are four parts to the proposed business environment:

Part 1: A new "Telecommunications Authority" (possibly still called DGT, but organizationally reformed) is responsible for ensuring fair competition between Type 1 and Type 2 carriers. The objectives are to establish proper operations of public properties, ensure quality services, and protect users' benefits.

Part 2: The basic circuit network is a national infrastructure. It is both an integral part of the total service industry and the underlying pathway for information flow. The network has a high level of public obligation—including serving senior citizens, remote villages, and low-income households as well as wealthy urban markets. Type 1 operators thus to some extent should be monopolies and be required to pay close attention to new technologies in order to remain cost effective.

Part 3: Type 2 carriers should provide various VAN services at their own risk. Like transportation carriers running on highways, Type 2s offer multiple services in a fully competitive market. The small size of the geographical area in Taiwan may not require Type 2s to divide into domestic VAN and international VAN.

Part 4: Users are in the center of the model and should be given ample opportunities to actively express their requirements. To meet user demand, there are three factors to consider. First, regulation should be kept to a minimum. Next, the basic network should be widespread and of high quality. Third, there should be a large and fair competitive environment.

This legal framework concept implies that the new law should strictly regulate Type 1 carriers and give official sanction to Type 2 carriers. Because it is expected that Type 1s will also provide VAS, an interesting competitive relationship exists between Type 1s and Type 2s. To ensure fair competition, there are two safeguards: a separate accounting system to prevent cross subsidization and fair lease line interconnection.

As of mid-1991 telecommunications was still under DGT monopoly. To implement the new concepts, it is necessary to separate the regulatory and operational functions—that is, to clearly define a ''referee'' and the ''players.'' It has been proposed to spin-in certain DGT departments to a government body called the Telecommunications Authority or DGT. Other parts of DGT would be spun-off and reorganized as corporations that continue monopolistic operations of various Type 1 networks. An equipment approval institute and a technical standard board have also been proposed within the authority.

17.6.1 Priorities

The first plan of action is developing marketing techniques for telecom services. Data services are a small part of DGT revenue and many households do not have a telephone. Clearly, telephone growth could be increased by encouraging more households to get residential lines and by marketing various data communication services to business users. Under present regulations, the tele-information processing service market is not effectively open to nongovernmental entities. This is limiting growth.

Conventional services were provided under a concept of ''supply orientation'' during the monopolistic period. However, data services require providers to use a ''market orientation'' strategy. In response, a new marketing division was organized within DGT to be responsible for marketing strategic planning, with end-users as the focus.

To open the market for competition means liberalization or deregulation of telecom services. The DGT recommended a sequence of priorities in a 1987 seminar. First, take immediate action to liberalize services, including parts of the CPE market. Next, revise the relevant regulations to support liberalization of some Type 2 services. Third, create a long-term strategy to revise the 1977 Telecommunications Act so DGT's organization structure can be corporatized or privatized, allowing for the time required to pass these revisions through the Legislative Yuan. The first two of these had been done by 1989.

DGT is not only a rule maker but also a player. To separate regulatory and operational functions, the CEPD has urged MOC/DGT to use a spin-in and spun-off. Two regulatory bodies might be created within DGT: Telecommunications Equipment Inspection and Approval Institute and Telecommunications Standard Review Board are proposed names that reflect their function. However, an organizational alternative is to include these functions in the proposed Telecommunications Authority or a restructured DGT.

17.7 Conclusion

Telecommunications policy in Taiwan usually refers to policy made and executed by MOC/DGT. Conventional policy objectives have been relatively simple. Roughly speaking, the main social goal has been to provide basic telephone services. Objectives focused on engineering or construction, with eliminating the backlog of orders a typical goal. It had originally been a supplier's market. In 1981, when every village had automatic dialing and density was over twenty stations, universal service could be said to have been successfully achieved. Taiwan was thus ready to enter a new age of telecommunications, pursuing qualitative improvement and expansion.

The industry changed to a users' market in the 1980s. The decade was spent considering structural changes to how service was provided—with an eye on possible lessons from the spread of privatization and deregulation in Japan, the United Kingdom, and the United States. Many proposals have been made, some of which have been rejected at least once, but which may again be considered.

An ad hoc committee was established in May 1988 in MOC to study all the recommendations. In addition to MOC and DGT officials, people from the electronics and computer industries and other relevant areas are again being asked for input. The ad hoc committee is to propose solid plans for the liberation of telecom services and the organization of the DGT, and to prepare drafts of various legislation (including the revision of existing acts) that will provide the foundation for the creation of a new telecommunications era in Taiwan. It is impossible to know the speed of change or the extent privatization of services and competition will be allowed. The consensus, however, is that this time plans along these lines will be implemented, and Taiwan's regulatory structure and telecommunications markets will be transformed in the 1990s.

Acknowledgments

The authors wish to thank the Accounting, Commercial, and Plant Departments of DGT for providing data. Special recognition is also due Ms. Hu Chang-Hung, Hsu Yu-Rong, and Syloia SH Hsiao for their efforts in providing statistical and other material for turning our conference paper into its present form.

Bibliography

All these items are published in Taipei and are in Chinese, except Tseng 1991, which is in English.

Chang, W. C., et al. 1985. "A Study on the Promotion Path for the Telecommunications Employees." In *DGT R&D Annual Report*. Taipei.

Chen C. C., et al. 1985. "A Study on Telecommunications Operations and Organization Structure." In *DGT R&D Annual Report*. Taipei.

Diretorate General of Telecommunications. *Statistical Abstract of Telecommunications*. [Annual].

———. 1971. "ROC Telecommunications Records."

———. 1981. "DGT Domestic Procurement Practice."

———. 1987. "Records on the 1987 Seminar on DGT Strategy," October.

———. 1988. "Telecommunications Development for the Next Decade." In *Presentation for the STAG Meeting*.

Lee, Wan. 1981. "Story to Be Remembered in the 1950s." In *Centennial Excerpt of ROC Telecommunications*. Taipei: DGT.

Lee, P. Y. 1989. "Telecommunications in R&D." *China Times* (May 4).

Liang, Ken-Peng. 1981. "Address from the Director-General." In *Centennial Excerpt of ROC Telecommunications*. Taipei.

Ministry of Communications. *Communications Year Book*. [Annual].

———. 1987. "Report on the Study of the Corporatization of Post and Telecommunications."

Tseng, Fan-Tung. 1985 May. "Final Reports on the Development of Cable Television Systems and Services." Taipei: Ad Hoc Working Group on CATV.

———. 1991. "The Development of Taiwan Telecommunications as a Telecom Network Center in Asia." Keynote Report to the 12th Science and Technology Advisory Board Meeting, May.

———. et al. 1987. "Modernization of Telecommunications and Information—A Study of Related Regulations." Taipei: CPED.

18

Taiwan: Changes in the Environment for Development

CHINTAY SHIH AND YEO LIN

Deregulation has been pursued by Taiwan's government since 1980, but it has been singularly elusive. Although the worldwide trend toward liberalization of telecommunications has been a strong influence favoring reform, only limited progress has been made. Until 1987 the directorate general of telecommunications (DGT) provided both services and equipment on a monopolistic basis. Customers may now purchase many (but not all) types of equipment from multiple vendors, but for an economy as dynamic as Taiwan's, value added network (VAN) services are conspicuous by their rarity. Liberalization of administrative matters, privatization of DGT, and adjustment of service rates have been seriously discussed, but a sufficiently broad consensus as to what to do has meant proposals have languished in limbo.

This chapter focuses mainly on changes in the environment and on the development of information-related industries in Taiwan. Related policy issues are dealt with in Chapter 17.

18.1 Policy Overview, 1945–1989

When Taiwan reverted after World War II, its telecommunications facilities were extremely poor. In Taipei, for example, there were only some 700 telephone subscribers. Lacking sufficient construction funds, yet desiring to meet basic communication requirements, the government adopted policies known as "Self-Sustainable and Self-Developing" and "Popularization of Telephone Services." Users contributed to construction costs as a part of their installation fees.

Throughout the 1970s, Taiwan experienced very rapid economic development. Economic growth forced the government to select more effectively and explore more aggressively the telecommunications technologies available throughout the world. As a result, a six-year *Telecommunication Development*

Plan was implemented in 1976. It focused on increasing the popularity of telephones and modernizing long-distance and international services. Modernization was understood to mean that people could directly dial domestic and international calls.

During the 1980s, booming international trade created a huge demand for services. The expanding business sector required information on both domestic and international markets. Not only did the demand for regular telephone and telegraph services increase, so did demand for new services. For example, videotex was introduced to transmit stock market information, commodity prices, and weather information. Recognizing the significance of such an evolution, the government designated both the information and the telecommunications industries as strategic industries in the hope their development would insure continued growth of the economy.

Paralleling the national ten-year (1980–1989) Long-Range Economic Planning Program, which predicted even more pressing future demand for telecom services both in terms of quantity and quality, a Ten-Year Telecommunication Development Plan was set forth. It addressed the necessity of volume expansion so that tensions from imbalances between demand and supply could be alleviated, and the nationwide digitalization of telecommunications.

In 1985 the government announced fourteen major construction projects, totaling some NT$900 billion. Modernization of telecommunications is one of the fourteen, and it was listed for NT$65 billion. Goals included viable ISDN services in Taipei, Kaohsiung, and Taichung and a density of fifty by the year 2000 in all metropolitan areas, popularization of rural phone service, and upgrade of networks. By 1987 nationwide density had already reached thirty-one and was more than forty-one in mid-1991.

Taiwan's Council for Economic Planning and Development commissioned a study, led by Tseng Fan-tung, to analyze the causes and effects of reforms in the United States, United Kingdom, and Japan along three dimensions: the motivating forces behind the reforms, reform activities themselves, and the impact of reforms. The result appeared in 1987 (Tseng 1987). The three primary driving forces for reform are seen to be the need for convenient and reasonably priced information demanded by businesses seeking help in achieving higher productivity, advanced by-pass technology, and achievement of universal service.

These same forces exist in Taiwan. Pressure from them continues to build, and proposals and plans for liberalization are emerging. In 1990 the government initiated a six-year plan in which privatization of DGT and liberalization of Type 2 service have been clearly indicated as major policy goals.

18.2 Regulatory and Operating Bodies

Based on the Telecommunications Act and its related by-laws, which were formulated in 1958 and revised in 1977, the Ministry of Communications (MOC) regulates the industry. The law indicates clearly that MOC is the primary telecommunications policymaker within the government. However, this has not been fully realized in the actual policymaking process. This is because MOC gave its highest priority to planning and developing transportation, which diverted time and energy away from telecommunications matters. Responsibility for policymaking thus fell on the DGT, where it has primarily remained.

On many occasions, DGT has initiated policy proposals, while MOC has only played the role of friendly watchdog. For example, DGT drafted the formulas governing rates of return for state-owned telecommunications providers (i.e., for itself). MOC reviewed the proposal, which was then submitted to the Executive Yuan and passed by the Legislative Yuan to become effective. Any revisions would most likely follow the same procedure.

18.3 Policy Evolution

Under its monopolistic structure, DGT has been both the supervisor and the operator of Taiwan's telecommunication. It has drafted and executed regulations for equipment manufacturing and service provision. At the same time, it monopolizes the networks and telecommunications business. DGT is, for all intents and purposes, running a self-supervised business. Popularization of telecom services has always been DGT's primary goal.

Liberalization in Taiwan has been following the same sequence as in the United States, United Kingdom, and Japan: first the CPE market, next competition in VAN services, and, finally, open competition with common carriers. Thus, since August 1987 users have been able to purchase their own telephones. Most of what DGT provided were pulse; users flocked to buy touchtone from other providers.

In 1988, several types of peripheral equipment were deregulated. End-users could own fax machines, modems, and telex and videotex terminals. Mechanical telex terminals provided by DGT have been replaced by electronics ones. In 1989 DGT deregulated VANS. Such services can be offered by DGT or any private company that has a service contract with DGT.

18.4 Five Major Telecommunications Interests

There are five major interest groups relevant to this discussion: the government, telecommunications institutions, end users, labor, and equipment manufacturers. Each group's primary focus is discussed in the following, except equipment makers, which are covered later.

Effective and efficient telecommunications are important to the development and maintenance of modern national economies, including national security and social progress. The Executive Yuan, when announcing fourteen important construction projects in September 1984, set an agenda for the nation referred to as the Telecommunication Modernization Plan. It called for investments totalling more than NT$70 billion (about U.S.$ 1.75 billion at the time) over the six years, 1985–1990, to digitalize urban telecom services. Since by law a portion of DGT profits must be remitted to the Treasury, these investments will be at least partially recovered by revenues to the government. Although martial law in Taiwan province was lifted in July 1987, antagonism between Taiwan and mainland China still officially exists. Thus, the Garrison Command under the Ministry of National Defense retains its authority to sample mail and telegraph transmissions.

DGT, which operates as a monopoly, has enjoyed substantial growth over the years. Its average annual profit as a percentage of total revenue was 37 percent for 1977–1988. In principle, DGT observes what the government mandates. All planning and objectives are in line with national master plans. However, stated goals may sometimes be in conflict with the government's intentions. For instance, in 1987, before resolution of debates over price negotiations with its privileged suppliers (the Big Three, discussed later), decisions affecting the procurement budget for switching systems were stalled in the Legislative Yuan and awaited administrative instruction.

Users want quantitative expansion and qualitative improvement of services. Expansion of service to rural and remote areas is always welcome by those reached. Stepwise liberalization of the CPE market has won general applause, despite slow progress and often underestimated demand forecasts. However, with ongoing rapid urban development and economic growth, it is sometimes difficult for users to get new services or sufficiently expand original service capacities in some business districts. Even after several rate reductions, users complain about rates on outbound overseas calls being higher than on inbound ones. Considering Taiwan's heavy reliance on foreign trade (imports and exports combined account for more than 95 percent of GNP), the high rates are seen as a burden on business users.

18.4.1 Labor

A nationwide telecommunication qualification test is held regularly under the auspices of the Ministry of Examination to recruit new employees for DGT. After passing the test, new recruits obtain civil service qualifications and a lifelong job. Employees are entitled to all the fringe benefits of, and subject to all the restrictions pertaining to, civil servants. Career development patterns are fairly rigidly defined. Some employees have complained about the lack of mobility. A person assigned to a job in a certain area is unlikely to be transferred unless a colleague elsewhere is willing to exchange posts.

Although DGT is a government institution, it is committed to operating as a businesslike entity. A union—the Taiwan Telecommunication Workers Union (TTWU)—has been organized by the employees to promote their interests. Owing to the large number of members and a relatively strict organizational structure, the union has greater political strength than other labor unions, including the railway workers and the postmen. According to Taiwan's constitution, seats in both the National Assembly and the Legislative Yuan are reserved for candidates from labor unions. Since the TTWU is the most powerful union in

Taiwan, its candidates are most likely to win these seats. The government and the ruling Kuomintang Party (KMT) pay close attention to the labor movement.

The TTWU functions primarily as a benevolent association. Its main activities have been protecting or advocating member personal benefits—aid for medical expenses, children's education, birthday gifts, group travel, and employee study. The union, despite the technological fluency of its members, was not involved in the policy-making process. The issue of most concern to the union in the early 1990s is probably the proposal to convert DGT into a privately owned corporation, which would greatly affect employee rights and benefits.

The union's role vis-à-vis the ruling party has changed since the establishment of an opposition party, the Democratic Progress Party (DPP), in 1986. That year a DGT employee ran for a legislative seat as a DPP candidate, and upset the KMT incumbent, who also happened to be the union's chairman. Another employee was elected a DPP member of the National Assembly. DPP members have also been elected to TTWU offices. Although the DPP has not proclaimed any explicit objections or alternate proposals to current telecommunications policy, the underlying force of an opposition voice in the union cannot be neglected. In short, the TTWU will no longer support KMT positions as unquestioningly as it once did.

18.5 System Development

Utilization of intrafirm networks is still in its infancy, although it has enjoyed significant growth since the mid-1980s. These include electronic key telephone systems (EKTS), PABX, and computer network systems. By 1987 there were 113,216 EKTS installations, 85 percent with capacities below ten lines, and only 3.6 percent handling twenty or more lines. This implies that most users are small or medium businesses, as are most firms in Taiwan. About 25,000 systems per year were being installed in the late 1980s, half of them replacements. The PABX market has expanded tremendously since the mid-1980s, from 1,513 installations in 1986 to 4,200 in 1990. Locally made PABX captured 25 percent of the market in 1986. By capacity, 86 percent were under 100 lines, 4.3 percent were over 200.

Computer networks are mostly for short distances between a company's main computer and its department terminals. A few large companies have island-wide in-house systems, with some even including overseas branches. Formosa Plastic Group, for example, has on-line factory operation. Franchise service businesses frequently dependent on computer network systems for communications with headquarters. It is common for a franchiser to join computer system vendors to develop software packages as part of what is provided to franchisees.

18.5.1 Data Communications

Data communications revenue accounted for a very small portion of total telecommunications revenues, but its share is increasing. DGT began a data circuit leasing service in November 1971. In June 1988, a breakdown of customers showed 9,360 leased domestic data circuits, and eighty-four international; 1,051 had packet switching, 4,600 were subscribers to dial-up services, 139 used circuit switching, 2,282 used public information processing, and 187 used universal data base access service (UDAS). Private businesses were 44 percent of users, 32 percent were state-owned businesses, and 19 percent were government organizations. Information industry and education and research institutes accounted for the remainder. Of the eighty-four international circuits, fifty-one ended in Hong Kong, and thirteen in the United States. Private businesses used sixty-one circuits, and the information industry used seventeen. Most customers were local branches of foreign companies.

Circuit switched public data communication service has been offered since June 1982. In June 1988, the total number of lines was 139, 59 percent used by state-owned businesses, 21 percent by government organizations, and 26 percent by private businesses. By the end of 1990 the number of lines had reached 23,000, and were mostly in the private sector.

UDAS service started in December 1979. In June 1988 there were 187 circuits reaching data banks in more than thirty-eight countries, although utilization concentrated in the United States. Private businesses had 47 percent of the circuits, government organizations had 25 percent. Most used UDAS for internal purposes, but a few of them provide information retrieval services to the public.

18.6 Information-Related Industries

Over the period 1953–1990, Taiwan has generally enjoyed rapid economic growth. In terms of GDP composition, as one would expect, industry has increased its share at the expense of agriculture (which has gone from 35 to 5 percent). The service sector fluctuated in a 40–45 percent band until the early 1970s. Since 1978 the share services have been drifting up, reaching almost 51 percent in 1990.

Although information-intensive service sectors (finance, insurance, data processing, etc.) do not account for a high percentage of GDP, the sophistication of such activities has increased since the mid-1980s as many specialized services have arisen and foreign companies have rushed into the local market. These newcomers, with their ever-increasing demand for information services, help the telecom service industry grow. Another major part of the information industry is the computer industry. By the end of 1987, the total number of microcomputers installed in Taiwan approached 350,000. The total of all other kinds of computer systems was 6,367, an increase of 29 percent from the preceding year. At the end of 1990 there were some 950,000 microcomputers and over 16,600 minis and mainframes.

18.6.1 Electronics

Taiwan's electronics industry came to life in the 1960s with the advent of television manufacturing. It still mainly turns out consumer and other mass products and components. In the 1960s, foreign investors were attracted by government-sponsored incentive programs as well as by a diligent and abundant labor force with comparatively low wage scales. Foreign technology and professional training propelled rapid development.

In the 1970s Taiwan displaced Japan as the largest supplier of video and audio products such as television sets, radios, and tape recorders to the U.S. market. Since 1980 the structure of the domestic industry has shifted away from being highly labor intensive toward becoming more technology and capital intensive. Likewise, consumer items gradually gave way to industrial and information ones. A milestone was reached in 1984 when electronic products became Taiwan's number one export; the industry has remained the nation's key export industry. Dramatic growth of the information hardware industry, primarily personal computers and related peripherals, has made a significant contribution. Production of such hardware was U.S.$2.13 billion in 1986 and U.S.$5.17 billion in 1988, and U.S.$6.15 billion in 1990. Taiwan's global rank has risen from ninth in 1985 to sixth in 1990. This is the foundation for Taiwan's move into the computer and communication age.

During the initial stages of the electronics industry in the 1960s, foreigners and overseas Chinese dominated investment. In 1977 electronics exports were 79 percent by nondomestic companies. (This falls to 67 percent if a pro-rata share of exports is attributed to Taiwanese minority interests.) As native companies have grown and started business, the domestic share has increased. In 1987, total electronics exports were U.S.$8.1 billion, only 41 percent by companies owned by foreigners or overseas Chinese. Exports in 1990 were U.S.$16.4 billion.

The combination of a developing computer industry and the government's determination to develop indigenous manufacturing capability mean Taiwan has created a good environment for development of its telecommunications industry.

18.6.2 The Big Three

To stimulate telecommunications technology development, the government decided to solicit world-class companies. As a result, joint ventures were established with GTE (the operation is now owned by Siemens) and an ITT subsidiary (now part of Alcatel) in 1973 and in 1984 with AT&T. These are referred to as the Big Three. Each involves a locally owned minority stake. All three produce a range of products, including central office switching systems for the domestic market. The government hoped to attract the best technology available globally to establish an internationally competitive equipment industry. Thus, for example, the agreement with AT&T calls for technology transfer in eight areas—although the exact nature is rather broad and AT&T has a good deal of discretion. To solidify the results of the 1973 agreements, the government instituted a policy of "Local Purchase if Locally Available." Central office switching systems must be brought from the Big Three.

Unfortunately, the original objectives of the government's technology transfer initiatives have not been achieved. Needed switching systems are nominally supplied locally, and DGT is in compliance with the domestic purchase policy. In reality, however, key components and parts are imported; only a small portion of the assembly and modification work is actually done in Taiwan.

18.7 Promoting Development

To stimulate development of telecommunications the government has designated it a strategic industry and offers various measures such as financing facilities, technological support, R&D incentive programs, and assistance for new product development.

Two major institutions manage the financial aspects. One is the Bank of Communications, a government bank for development and reconstruction. It provides mid- to long-term loans and equity investment to strategic industries. By the end of 1987 it had invested NT$1.8 billion in forty-four new ventures, nearly half of which were telecommunications and information companies. The Bank has played an important role in helping many entrepreneurs during startup. However, as the nature of each business varies, it lacks the expertise to guide or assist management. Moreover, because it is only a minority holder (legally restricted to 25 percent of any company's equity), it often wields only limited influence.

The Development Fund of the Executive Yuan is the other institution. It was formed in 1973 to provide financial support to domestic industry. It can assist strategic industries or ones deemed important to the country that are financially unattractive to the private sector. Its equity investments and loan facilities amounted to NT$7.9 billion by the end of 1988.

On the R&D front, the Electronics Research and Service Organization (ERSO) of the Industrial Technology Research Institute (ITRI) has played an important role. It specializes in electronics, has helped build up the domestic industry, and has made important contributions to the integrated circuits and personal computer industries. Its pilot plant has been an important source of telephone ICs for local makers competing in the U.S. market. Many of its R&D results have been transferred to local manufacturers for commercialization, including technology for LANs and X.25 packet switching data networks. To continue its strength and sharpen its R&D capability, a project titled "Development of Integrated Service Telecommunication Technology (1989–1992)" was undertaken.

ERSO was reorganized in mid-1990, with a computer and communication research lab (CCL) spun off. The "new" ERSO focuses on semiconductors and electronic components. In a bit over a year after its founding, CCL has developed a number of ISDN products, including an interface card, a telephone set, a terminal adapter, and a PBX. Other developments include several types of communication ICs (e.g., video compression), and work has been done on personal communication networks (PCN) and the deployment of fiber optics to residential users.

The DGT has coordinated joint-development efforts by local equipment producers and computer companies. However, with increasing demands from local industry for government support, DGT has adopted a more aggressive attitude in technology development. The Telecommunication Laboratory (TL) used to do planning for DGT, and, as the R&D arm of DGT, it has been dedicated to introducing foreign technologies essential for telecommunications construction work. In 1988, TL took a more active role by cooperating with ERSO to develop integrated communication technology. TL broadened its staff role to participate in actual technology development work. Some 80 percent of its work load relates to servicing DGT requirements, and the remaining 20 percent is targeted at developing technologies or products and smoothly transferring the results to local firms.

TL has developed technology for electronic telephone sets, microwave communication systems, intelligent Chinese terminals, M13 digital MUX, D4 PCM terminals, touchtone converters, automatic message accounting equipment, modems, facsimile terminals, data ports, and 12 GHz DBS transponders.

18.7.1 Equipment Industry

Local development and production of telecom equipment is mainly determined by the demand from DGT and the general public. DGT takes the lion's share. In 1990 the procurement of equipment and facilities for the Taiwan network was U.S.$1,018 million, 91 percent obtained locally. Switches involve the most complicated technology. Using their manufacturer as an index, the development of Taiwan's equipment industry can be divided into two major periods: 1957–1981, when traditional systems predominated, and since 1981, when manufacturing of electronic switches has predominated.

During the earlier period there were only two producers, each with a relationship with a Japanese firm: Far Eastern Electric Industry Corp., established in 1957, and Taiwan Telecommunication Industrial Corp., formed in 1958. The former transferred technology from Oki Denki, the latter from NEC. Major products were relay systems (1957–1963), step-by-step systems (1957–1969), and crossbar systems (1969–1981). After the entry of GTE and Taisel (Alcatel's local subsidiary) in 1973, production of electronic switches began.

In addition to switches, local manufacturers supply transmission equipment such as PCM and FDM carriers, repeaters, modems, and M13 multiplexers for fiberoptic systems. Many types of terminal equipment are made, including telephone handsets and fax machines.

Total sales of locally produced equipment exceeded NT$23 billion in 1985, with wired telephonic apparatus such as handsets and switches accounting for 60 percent. In the wake of the opening of the U.S. market and capitalizing on integrated circuits supplied locally by ERSO and the United Microelectronics Corporation (UMC), Taiwan has become a world leader in telephone exports.

Production and export of hand sets is the most active sector of the local industry. With exports of U.S.$268 million in 1986, and U.S.$250 million in 1988, Taiwan has become a major supplier to the United States. Nevertheless, for more sophisticated equipment, Taiwan is still basically a technology recipient and most manufacturers maintain foreign contacts for technology inputs. To cultivate domestic R&D capability, some companies have initiated R&D projects by themselves or in cooperation with local research institutions.

Domestic demand for PBX and electronic key telephone systems (EKTS) almost doubled from 1988 to 1990, when it reached U.S.$54 million. Because domestic demand consists mainly of high-end products, 81 percent of them were imported in 1990. At the same time, local firms produced U.S.$67 million of mainly low-end PBX and EKTS, 80 percent of which were exported.

More than twenty local firms make modems, mainly 1,200 bps and 2,400 bps, exporting about U.S.$28 million in 1990. DGT relinquished its control over modems below 1,200 bps in 1986 and modems below 2,400 bps in 1987; therefore domestic demand grew in the late 1980s as prices fell.

Foreign capital and technology have played important roles. In general, there are three forms of cooperation: technical cooperation such as United Optical Fiber licenses from AT&T, off-shore manufacturing plants, such as 3M (Taiwan), and joint ventures, such as by GTE Taiwan, Taisel, and Taiwan Telecommunication.

18.7.2 Developing Domestic Technology

To cultivate indigenous telecommunications technology capability, the government has sponsored R&D projects and transferred technology from abroad. However, due to the lack of systematic policy planning, the good intentions of the government often failed to yield the desired results. For example, Taiwan had some experience manufacturing analog switches and fiberoptics, but the scale is too small to be efficient. Often, users and manufacturers simply purchased products or parts abroad for assembly in Taiwan. Even with government efforts, local manufacturing capability was difficult to cultivate. The entry of the Big Three is another example. The government had reserved the domestic market for their CO switches, but technology transfer has not met expectations.

The Ministry of Economic Affairs, whose primarily responsibility is to develop local industry, has also been very active in promoting technology transfer and R&D. It sponsored several high-priority, large-scale research projects carried out by ERSO and CCL: submicron ICs, VLSICs, and HDTV, distributed computing. In the late 1980s there was also project on superminicomputers. The two laboratories have also participated in projects to develop ISDN, CPE, PBX, large-scale software, and related chip sets.

The absence of adequate supporting industries has handicapped the development of domestic telecommunications technology. In fact, with their preemptive technological advantage, financial strength, and market power, the Big Three pose formidable entry barriers to would-be competitors. Increasing awareness of intellectual property rights and national measures abroad to protect against high-technology disclosure overseas have further marred the prospects for unilateral technology transfer.

In 1987, opposition parties raised a fierce debate on this issue in the Legislative Yuan. DGT finally agreed to abandon its procurement approach with the

Big Three and adopt open bidding for CO switches. Traditionally, when DGT planned to purchase equipment from the Big Three, it would decide a purchasing price using certain officially regulated formulas. DGT then negotiated with the Big Three. Negotiations occasionally failed due to the differences in price.

18.7.3 Procurement from Overseas Suppliers

If equipment is not available domestically, DGT is allowed to buy it overseas, provided the purchase is processed through the Central Trust of China, a state-run agency. This regulation can be traced back to the decades when Taiwan was short of foreign exchange and had to resort to centralized planning and control of limited resources. With an extraordinary U.S.$70 billion in foreign exchange reserves in the late 1980s and early 1990s, the situation is totally different. Continuation of the rule introduces delays and lowers efficiency.

Beginning in the late 1980s, the accumulation of foreign reserves and a continuous trade surplus with the United States imposed intense pressure on the government to open the domestic market. The government has made efforts to simplify the procurement process and to increase purchases from abroad, especially from the United States.

18.8 Conclusion

Early regulation of Taiwan's telecommunications industry was the result of assumed natural monopoly conditions as well as the economic and political environment after World War II. Only the government was considered to have the ability to initiate development. This is an interesting contrast to the situation in the nineteenth century, when the Ch'ing felt it was better to leave such expensive undertakings to others. Private parties responded by constructing a network. In the twentieth century, private parties were not deemed capable of such initiative. Even if they were, security reasons led the government to want to maintain tight control of telecommunications.

Since 1980, however, economic development in Taiwan has made information transmission and achievement of universal service more important than ever. Advances in microwave, satellite, and fiberoptic technology have meant the network need not to be restricted by limited space. Meanwhile, the merging of computers and telecommunications, and the decline of manufacturing costs for microcomputers, have stimulated the emergence of VANs and diversification of customer terminal equipment. These developments have all shaken the logical foundation of a "natural monopoly" in telecommunications. The worldwide trend toward liberalization has further influenced public opinion and government policies.

Deregulation became the major policy concern during the 1980s. New telecommunications laws are expected during the early 1990s. In anticipation, DGT reached an agreement with Cable & Wireless in December 1991 under which C&W will assist DGT "ensure optimum benefit" from the new laws. This includes staff training, R&D, technology transfer, and other aid in expanding DGT's network and services. Telecom users on Taiwan should clearly benefit from DGT's reaction to potential competition, and the competition itself.

Acknowledgments

Research for this chapter was supported by the Industrial Technology Research Institute. The authors appreciatively acknowledge research assistance by Terry Taihor Huang and Chien-Tsai Hung.

Bibliography

Although titles are given in English, all these items are in Chinese.

Council for Economic Planning. *Monthly Report of Electric Industry & Development*. Taipei: Author.

Directorate General of Budget. *Quarterly National Economic Trends China-Taipei*. Taipei: Author.

Directorate General of Telecommunications. *DGT Statistics* [Annual]. Taipei: Author.

Inspectorate General of Customs. *Monthly Statistics of Import & Export*. Taipei: Author.

Institute for Information Industry. *Information Industry Yearbook*. Taipei: Author.

Tseng, Fan-tung, et al. 1987 Jul. *Modernization of Telecommunications and Information–A Study of Related Regulations*. Taipei: Council for Economic Planning and Development. Taipei: Author.

IV
BEYOND UNIVERSAL SERVICE

19
Canada

HUDSON N. JANISCH AND BOHDAN S. ROMANIUK

By almost any conventional measure, be it ownership, regulation, internal organization, or industry structure, the Canadian telecommunications industry must appear puzzling if not utterly enigmatic to the foreign observer. If it is any consolation, it is no less confusing to many a Canadian observer as well. What other country in the industrial, and perhaps the entire, world can "boast"—if that is the word—of as tangled and complex a patchwork of foreign and domestic, public, private, and mixed ownership of its telecommunications infrastructure? In how many other countries did the regulatory regime consist (until the 1990s) of three different levels of government acting for the most part entirely independently of each other, yet collectively still leaving large gaps and with no single level of authority responsible for the industry as a whole? Still, Canada's eclectic mix works and has produced one of the finest telecommunications systems in the world.

19.1 The Past

The origins of telecommunications in Canada are found in the privately owned telegraph companies. The first official telegram was sent in 1846 from Toronto to Hamilton. There followed a period of rapid formation and eventual consolidation of companies. By 1915 there were three main ones, all associated with railroads: Canadian Northern, Grand Trunk Pacific, and Canadian Pacific. By the end of World War I the railway companies found themselves in serious financial difficulties. In 1920 the federal government took over the Canadian Northern and Grand Trunk, and in 1921 Canadian National Telegraphs was created to provide the communication necessary for the newly formed Canadian National Railway system, as well as a public telegraph service. In 1928 the Grand Trunk Pacific Telegraph, previously operated independently, merged with Canadian National. The following year, the federal government acquired all the Canadian land mileage of the U.S. giant Western Union, which had operated in the Maritime provinces. By the 1930s, Canada was essentially served by two systems, both operated by railway companies. One, Canadian National, was government owned; the other, Canadian Pacific, was private. (For more on the telegraph, see ComCan 1988).

The introduction of telephony was somewhat sporadic. Alexander Graham Bell's father, Melville Bell, who lived in Ontario, was assigned Canadian patent rights, but he acted with little energy. By 1880 a number of independent companies had sprung up, and, as there appeared to be great risk of fragmentation, American Bell acquired Melville Bell's interests and sent Charles Fleetford Sise to Montreal to establish Bell Telephone Company of Canada (later Bell Canada). Sise proved to be a man of remarkable vigor and competence and is the true father of the Canadian telephone system in the same sense as Theodore Vail in the United States.

There has been little serious research into the early history of the Canadian telephone industry. Armstrong and Nelles (1986) is a welcome exception, but it covers all public utilities and thus lacks real depth. Two popular histories are Collins (1977) and Ogle (1979). Fetherstonhaugh (1944) is a biography of Sise (see, as well, Babe 1990).

The most important early decision was to use the same organizational structure employed in establishing what became the Bell System in the United States. Far from being the monolithic monopoly it was subsequently portrayed to be, that early organization was, in effect, a patent franchising operation (see Garnet 1985). Each operating company was required to raise much of its capital locally. This was applied to Bell Canada despite urgent requests for more direct American investment (see Taylor 1982). Ironically, this corporate policy—rather than any government initiative—was to insure Canadian ownership of Bell Canada.

Bell Canada, federally incorporated from the outset, hoped to provide service nationally, a task that proved beyond the capability of a single company. In the Maritime provinces, Bell emulated American developments by spinning off three provincial companies in which it retained minority interests. Things did not go as smoothly in the west, where unrealistic expectations for a rapid spread of service in rural areas, combined with strong antagonism toward the eastern monopoly, triggered political demands for government ownership. By 1909 Bell Canada had sold its local holdings to the provincial governments in Manitoba, Saskatchewan, and Alberta, trading territorial dominance for security in its lucrative central Canadian market.

Overall, there can be little doubt as to the pervasive influence of American practices. Consider, for example, the policy of leasing and not selling telephones, value of service pricing, flat-rate local calling but measured long-distance rates, and vertical integration in equipment manufacturing.

However, the flow of ideas was not one way. For instance, it was initially assumed that the very survival of the Bell companies depended on Bell's original patents, which were to expire in the United States in 1894. However, in 1884 they were voided in Canada for lack of local manufacture. After the initial shock, Sise realized that the viability of a telephone system depended on "occupying the field," a lesson later of the utmost importance to AT&T. The

Canadian company similarly proved the critical importance of access to long distance as the primary means of binding a single system together, and the earliest reassuring experience with regulation, especially at a federal level, occurred in Canada.

Outside the Prairie provinces, the British move to a government-owned monopoly had little influence. For a brief moment it appeared that this might not be so. The postmaster general, Sir William Mulock, in opening a wide ranging parliamentary investigation into the telephone industry in 1905, declared himself in favor of government ownership of long-distance lines and municipal ownership of local exchanges. Despite a good deal of complaint about Bell's predatory practices with respect to the remaining independents, and strong urging from Ontario municipalities, especially Toronto, the Select Committee on Telephone Systems fizzled out without even making a final report. Government ownership was rejected as too extreme, although it was also apparent that an appropriate government response to private monopoly had yet to be developed.

19.1.1 Institutional History of Telecommunications

Although it remains a mystery why the established Canadian and U.S. telegraph carriers did not make a determined entry into the telephone business, Canadian telegraph companies later sought to provide a broad range of telecom services. A crucial issue for the 1990s has been whether the reorganized descendent of the telegraph companies, Unitel, should be allowed into all aspects of modern telecommunications. In June 1992 the government granted Unitel's application to provide competitive long-distance services. Thus, the future of Canadian telecommunications appears to be one of more open competition.

The first coast-to-coast transmission of a commercial radio broadcast was over Canadian National lines in 1925. In 1932 Canadian National and Canadian Pacific jointly secured the national network contract of the Canadian Radio Commission, the forerunner of the Canadian Broadcasting Corporation system. They also inaugurated the first nationwide weather information gathering and dissemination service in 1939, and provided a Canada-wide voice communications system for air traffic control during World War II.

In 1946 Canadian National entered directly into provision of telephone service when it took over the Northwest Communications System, a government wartime trunk line established from Alberta to Alaska. This grew into NorthwesTel, a Canadian National subsidiary providing a full range of telecom services in northern British Columbia, the Yukon, and the western portion of the Northwest Territories. In 1949, when Newfoundland became Canada's tenth province, Canadian National became further involved in telephone service by assuming responsibility for much of the rural service previously provided by Newfoundland Post and Telegraphs.

In 1947 Canadian National Telegraphs and Canadian Pacific Telegraphs began joint operations to provide private wire services. This was the first step toward the formation of CNCP Telecommunications. In 1956 CNCP introduced

Telex to North America, and in 1964 completed a microwave network across Canada.

The most significant achievement in the interwar years was the full national interconnection of the separate regional telephone systems. Unlike other national links such as the railways and later the Trans Canada Highway, this was achieved without government subsidy. It was also done without there being a single, separate-long distance company. Each company retained exclusive responsibility for its own territory, but agreed to extend long-distance lines to its boundaries to exchange traffic, to enter interconnecting agreements with neighbors, and to share revenues for calls from or to nonadjacent companies.

In 1921 the Telephone Association of Canada (TAC) was organized. Its technical committees began to explore the possibility of developing a national system. At that time, many long-distance calls between Canadian cities went via the United States because of a lack of cross-country circuits. During the late 1920s TAC decided to construct an all-Canadian network from coast to coast. The link between Montreal and Winnipeg was completed in 1928, In 1931 the Trans-Canada Telephone System (TCTS), renamed Telecom Canada in 1983, was formed to develop and maintain a Canadian transcontinental long-distance telephone network. The network was completed before the end of 1931 and inaugurated in January 1932 (an epic story well told in Ogle 1979).

Although the legal structure of this arrangement as an unincorporated association has remained unchanged, two new major technological innovations have greatly increased its transmission capabilities. In 1958 the member companies built a 139-station microwave route. At the time the world's longest, it extended 5,400 km from Sydney, Nova Scotia, to Victoria, British Columbia.

The next important step was the introduction of communications satellite technology and yet another independent organization with mixed ownership. Telesat Canada was incorporated in 1969 to introduce satellite technology to domestic telecom systems. Jointly owned by the federal government and the major carriers (provision for direct public participation has never been acted on), Telesat became a member of TCTS in 1977.

Regarding international services, connections to U.S. points have been provided since the earliest days by means of interconnection agreements between Canadian and U.S. firms. By the 1980s, these private arrangements had been supplemented by international agreements and treaties.

U.S.–Canada traffic accounts for some 85 percent of international calling. The remainder, termed "overseas," was handled by the Canadian Overseas Telecommunications Corporation (COTC), renamed Teleglobe Canada in 1975. This government corporation was formed in 1949 to comply with the 1948 Commonwealth Telegraphs Agreement whereby each signatory government agreed that external telecom operations would be acquired by a government department or corporation that could then represent its government as a "national body" at meetings of the Commonwealth Telecommunications Board. Prior to 1949, Canadian overseas telecom services had been provided by privately owned telegraph companies including Western Union.

As early as 1882, Bell Canada started manufacturing equipment through a local subsidiary, The Northern Electric Manufacturing Company, forerunner of Northern Telecom, was incorporated in 1895. Canadian manufacture was a major irritant in relations between American Bell and Bell Canada, as it had been assumed from the outset that the Canadian company would purchase all its equipment from the Bell system's vertically integrated supplier, Western Electric. This issue came to a head in 1901 when Bell Canada proposed to expand its cable manufacturing capacity and to end its dependence on Western Electric. Western Electric demanded, and was eventually granted, a half share in Northern Electric. It then gave Northern full access to all its new technology by way of a series of very broad service agreements (see Taylor 1982, pp. 26–28).

These arrangements meant that until the 1950s, Northern Electric was a branch operation entirely dependent on Western Electric for its technology. It was weaned from this reliance by AT&T's 1956 Consent Decree. In 1957 Western's share in Northern was reduced to 10 percent, and that was divested in 1962. The terms of the Consent Decree regarding the disclosure of technical information led to a concern at AT&T and Western that they might have to extend to all U.S. manufacturers the same information Northern was obtaining. Accordingly, beginning in 1959, the Western Electric–Northern Electric Technical Information Agreements became progressively more costly and restricted. By 1972 this flow of information had essentially stopped, and in 1975 the last AT&T-Bell Canada agreement expired. As a result, Northern was on its own at a most fortuitous moment: the shift from electromechanical to electronic switching.

By the 1970s Canada had achieved an extraordinarily advantageous situation in equipment manufacturing: vertical integration at home and competitive entry abroad, especially in the United States (see Waverman 1989). After a long investigation culminating in the early 1980s, the Restrictive Trade Practices Commission, then the nation's competition policy watchdog, concluded that the benefits of vertical integration far outweighed any disadvantages. The commission was particularly impressed by Northern's success in penetrating the U.S. market and concluded that risky product development would have been undertaken only with the assurance of a large share of the Canadian market (RTPC 1983).

19.1.2 Major Legal Foundations

Bell Canada's primary legal foundation, its initial patent rights, proved to be of little lasting value. Under the 1870 Patent Act, patents could be declared void if after two years there had been no manufacture in Canada. In 1884 Bell's patent, granted in August 1877, was canceled for this reason.

Unlike the Bell companies in the United States, which sought charters at the state level, Bell Canada was incorporated federally from the outset. The Bell Telephone Company of Canada Act of 1880 gave the company extensive rights of way to the apparent exclusion of any residual provincial or municipal control. This was successfully challenged in 1881 in a Quebec court on the ground that telephone service was only local at that time and did not cross any provincial boundaries. Bell Canada in 1882 asked the federal Parliament to declare conclusively that it was under federal jurisdiction. As it turned out, this legislation was not vital. When the City of Toronto sought to challenge the company's right to enter its streets to lay cables without municipal consent in 1904, the Privy Council, then Canada's highest court, held that the scope of the business contemplated in its act of incorporation was sufficient in and of itself to exclude Bell Canada from being a local undertaking subject to municipal or provincial jurisdiction.

Bell Canada's legal victory in the Privy Council was a major blow to the municipal ownership movement in Ontario and greatly weakened the impact of municipal intervention before the Select Committee on Telephone Systems in 1905. Sise was very aware of the crucial importance of the federally guaranteed right of way granted Bell Canada.

While Bell Canada had not been granted a legal monopoly to provide telephone service, its federally guaranteed right of way gave it a very important advantage over any would-be competitors. The latter would face very significant costs in obtaining municipal rights of way, assuming any municipality was prepared to have two sets of poles on its streets. Bell Canada also had a great advantage in that Parliament would be very reluctant to grant such rights of way to another company given the criticism it already faced for having been so generous to Bell Canada.

The other crucial element in early legislation was the absence of any requirement for interconnection. Before the 1905 Select Committee and elsewhere, Bell Canada stoutly resisted interconnection on the grounds that a long-distance competitor could use Bell's local exchanges to take long-distance traffic away from it.

There was no provision in the original Bell Telephone Act for regulation. However, Bell Canada did have to return to Parliament whenever it wished to raise its level of capitalization, which provided an opportunity for those who felt there should be some form of public control. A power with reference to rates was slipped into the Bell Telephone Act of 1892, apparently more by accident than design, and its ill-conceived nature soon became evident.

At the turn of the century, debate was shifting from the right-of-way issue to the issue of monopoly. It was widely agreed that the nature of telephone service required it be provided on a monopoly basis, which justified some form of rate regulation. In 1902 a public service dimension was added to Bell Canada's legal obligations, and all rates were subject to regulation by the Governor in Council, elected politicians serving as the Cabinet.

What had not been resolved was the institutional competence of the governor in council to deal effectively with rate matters. Under both the 1892 and 1902 Acts, the politicians found it impossible to devote adequate time and energy to this complex task, in which there would often be no clear political winners, so this early experience in regulation satisfied no one.

At this time, telephone rates were a minor matter compared with the highly contentious issue of railway rates. Much thought had been devoted to the institutional design issue with respect to those rates, and the Galt Commission in 1888 and the McLean Reports in 1899 and 1902 advocated rate regulation be undertaken by independent, expert commissions. This would remove it from the immediacy of politics. At the same time, the U.S. model of a fully independent commission was rejected as inimical to the political accountability essential to parliamentary government. This led to a compromise in which independence and expertise would be reconciled with political accountability by means of a broad "cabinet appeal" power whereby politicians had the final say. By seeking the best of both worlds, the Railway Act of 1903—applied to Bell Canada in 1905—built a troubling ambiguity that remains unresolved into the regulatory scheme.

In the Maritime provinces, the shareholder-owned provincial spinoffs of Bell Canada were brought under provincial boards. Unlike the Board of Railway Commissioners, these were more closely modeled on the more fully independent state public utility commissions in the United States. It is interesting that government ownership and regulation were not seen as alternatives, for the prairie telcos were also made subject to provincial public utility commissions. This came about in no small measure as a result of early mismanagement and the desire of politicians to be shielded from the embarrassment of having rates rise above those previously charged by the much maligned eastern monopolist, Bell Canada.

Extensive regulation came earlier in Canada than it did in the United States, and Canadian experience taught Vail that regulation could be accepted in return for monopoly. Bell Canada did not actively seek regulation; rather, it fought hard for the regulatory regime likely to be the least threatening to its well-being. Detached expertise, measured legal process, and acceptance of prior industry structure were the attributes that would be most sympathetic to the interests of a telephone company. As Vail was to remark some years later, there could be no objection to "independent, intelligent, considerate, thorough" regulation that recognized that ". . . capital [was] entitled to its fair return, and good management or enterprise to its just reward" (AT&T *Annual Report* 1907, p. 32). In no small measure his private-ownership monopoly-regulation philosophy grew out of the benign form of regulatory regime adopted in Canada.

This structure was put to its first real test in the years immediately following World War I when the telcos were placed under immense financial strain. Overall, the Board of Railway Commissioners and the various provincial public utility boards proved to be most understanding—too much so, according to Armstrong and Nelles (1986, p. 282). While their view might be seen as going too far toward a full-blown capture thesis, it is true that regulation legitimated telcos activities that would otherwise have been denounced as contrary to the public interest. Only in the 1980s has it been asked whether this type of broad "regulated conduct exemption" from the normal rules governing competition should be maintained.

19.1.3 Telecommunications Industry to 1980

The first century of telephony in Canada was one of striking achievement. Penetration rates for basic service (POTS) were comparable to those in the United States and Scandinavia; satellite and microwave services had been introduced in a timely fashion—indeed, Canada pioneered developments in domestic satellite telecommunications; and a number of innovative business services had been made available. Three overarching considerations explain this success.

First, territorial exclusivity: Each carrier operated as a monopoly within its territory. Second, regulatory simplicity: Each carrier was subject to only one regulator. Third, common philosophy: Despite wide variations in ownership, all carriers shared common objectives. This is particularly important with respect to the provincially owned telcos that, after a short period of aloofness, brought their policies in line with those of Bell Canada.

The role of rules and practices in creating discrete territorial monopolies remains an open question because of the somewhat limited ambit of legal restrictions and the relatively isolated nature of the recorded instances of predatory behavior. If Canadian telephony in its early development had been considered suitable for competitive entry, one might have expected more attempts at entry and a far greater array of defensive measures.

For example, look at Sise's success in ''occupying the field'' in Ontario and Quebec for Bell Canada. While it is true the 1920s still saw many small independents in operation, they were actually feeders that conveniently relieved Bell of some of the burden of providing rural service. There had been some genuine competitors such as the Montreal Merchants' Telephone Exchange and the Canadian Pacific Railway-backed Montreal Federal Telephone Company, but they never lasted long.

Bell Canada had never been granted a legal monopoly as such, but certain rules placed it in an advantageous position. It supplemented its federally guaranteed right of way by entering into exclusive service contracts with major municipalities in the 1890s and by arranging with the railway companies to have only Bell service at their depots, the center of commercial activities in small towns at the time. Most importantly, until 1906 it was under no legal obligation to provide interconnection; when interconnection was required, the regulators were persuaded to rule it only had to be provided on terms very favorable to Bell.

It also appears Bell Canada was vigorous in its dealings with potential competitors, on occasion combining predatory pricing with strategic acquisition of independents that might mature into serious rivals. Only well into this century was an aggressive policy of acquisition abandoned and noncompetitive interconnection allowed.

Many economists agree that local service is most likely a natural monopoly and that it is therefore not surprising competitors seldom sought to enter that market. Regarding long-distance, there is evidence it was provided at a loss in the interwar years. For example, in a 1926 proceeding Bell argued successfully that it would be harmed by interconnection because local service was subsidizing unprofitable long-distance service. Only after World War II, when microwave and satellites brought transmission costs down dramatically, did entry became attractive and new entrants in long distance emerged.

The major foreign involvement has been in British Columbia Telephones (BC Tel). Despite Bell Canada's ambitious national aspirations in 1880, it was considered totally unrealistic to think of providing service on the other side of the Rocky Mountains. As a result, a number of independent companies sprang up in the 1880s, but BC Tel, originally organized in 1898, had emerged as the dominant carrier by 1920. In the mid-1920s BC Tel approached Bell Canada seeking to be acquired, but Bell was not interested, apparently believing as long as its western counterpart continued to buy Northern Telecom equipment, there was no need to acquire it.

This turned out to be a miscalculation. BC Tel sold out to an American public utilities consortium headed by Theodore Gary, which happened to own Automatic Electric, the largest manufacturer of telephone equipment in the United States after Western Electric. The Gary interests were sold in 1955 to GTE, the largest independent telco in the United States. In 1975 a federal study confirmed the most serious charge against this foreign ownership: In the move to electronic switching, BC Tel lagged other carriers because its planning was determined by the technology available from Automatic Electric (DOC 1975).

In the 1970s Northern Telecom began to develop and apply electronic switching on a massive scale and to enter the liberalized U.S. market with considerable success. With the establishment of Bell Northern Research (BNR), the tricorporate synergism of Northern–Bell Canada–BNR meant that while only 10 percent of Northern's manufacturing sales were of Northern proprietary design in 1970, it had risen to 75 percent in 1977 and to 82 percent in 1980.

19.1.4 Special Circumstances

19.1.4.1 Demographics

Canada is said to have too much geography and not enough history. It certainly has a small population compared to its land mass. For example, its territory is some twenty-five times that of Japan, while its population is only one fifth as large. Some 80 percent of Canada's people live within 300 km of the U.S. border, creating—from a demographic standpoint—a country 5,000 km long by 300 km wide. This creates an immense challenge to maintain east–west traffic flows in the face of powerful north–south attractions.

There are wide differences in density within the population strip. With 25 percent of the nation's population, Toronto and Montreal create a Tokyo–Osaka type business concentration, while the Prairie provinces have a thinly dispersed rural population.

19.1.4.2 Contiguity to the United States

Two phenomena arise from U.S. proximity. First is the demonstration effect of liberalization, especially for the business community. The "why can't *we* have it *here?*" complaint, grounded in the high degree of business mobility and

American ownership in the general economy, has a strong impact on Canadian telecommunications policy. At the same time it must not be thought that all U.S. ideas, particularly disruptive ones such as the divestiture of AT&T, will be automatically imported into Canada.

Second, there continues to be concern that if Canadian long-distance rates are not brought down closer to those in the United States—in the early 1990s the differential may have been as high as 40 percent—trans-Canada traffic will be diverted to the United States. While the attractiveness of this has been somewhat reduced by substantial increases in short-haul private-line rates to the border, the threat remains an important factor in Canadian policy-making. As a senior official at Telecom Canada succinctly put it, "The biggest deregulator in Canada is a private line to Buffalo" (Harvey 1983, p. 67).

19.1.4.3 A Weak Central Government

When compared with other federal states, the central government in Canada has been strikingly deferential to provincial interests, especially during the 1980s. Canada is a prisoner of its history; once provinces were allowed exclusive authority over telcos within their borders, and with only Bell Canada and BC Tel under exclusive federal control, it became very difficult to ask them to give it up. What may have had little value in the past has acquired great symbolic importance in the ongoing federal–provincial maneuvering so characteristic of Canadian federalism. More than a decade of fruitless negotiations since the late 1970s bears witness to the intractable nature of the jurisdictional issue (see Buchan et al. 1982).

The federal government has no preemptive power, as it does in the United States, which means it has been impossible to introduce truly national policies on competition in long distance or for new services such as cellular (see Dalfen and Dunbar 1986, pp. 139–202). The federal government has not sought national authority with any degree of vigor, but has coyly held back in the hope of not offending the provinces. However, it seems that the federal government may be prepared to assert itself on the necessity of national telecommunications policies in the 1990s.

19.2 The Present

19.2.1 Industry Structure

Although there are still more than 100 common carriers in Canada, most are extremely small. The five largest—Bell Canada, BC Tel, Alberta Government Telephones (AGT), Manitoba Tel, and Saskatchewan Tel—account for over three-quarters of industry revenues, with Bell alone generating over half. These five, along with New Brunswick Tel, Maritime Tel & Tel, Island Tel, and Newfoundland Tel, provide the basic terrestrial infrastructure of Canada's domestic telecommunications network.

Three other common carriers merit attention, CNCP Telecommunications, Telesat Canada, and Teleglobe Canada. CNCP was, until 1988, a partnership of the telecom divisions of Canada's two major railways—Canadian National Railways (CNR), a crown (government) corporation, and Canadian Pacific Ltd. (CP), an investor-owned company. CNR sold its half interest to CP in the fall of 1988. A few months later Rogers Communications Inc. (RCI), flush with cash from the sale of its U.S. cable television assets for U.S.$1.365 billion, purchased a 40 percent stake in CNCP for an estimated $250 to $275 million. The renamed company, Unitel, applied to federal regulators (the Canadian Radio-television and Telecommunications Commission, CRTC) to enter the public voice long-distance market.

Unitel is the only company in Canada owning coast-to-coast microwave facilities. Using the right of way of the railways that preceded it, Unitel has rapidly laid a transcontinental fiberoptic line. The irony is that Unitel, Canada's only truly "national" telecom carrier, is severely restricted in the types of service it is allowed to provide. Although it competes in varying degrees with several of the regional telcos in data and other business services, including private-line voice, it does not provide an alternative to either local or toll basic voice services. This is because it owns no local distribution loops linking subscribers to local switching offices. Moreover, it has not been allowed to interconnect with the telephone companies to provide switched voice message services in competition with them.

With a massive infusion of capital and entrepreneurial vigor from RCI, Unitel promises to create an entirely new universe of competitive opportunities. Led by Ted Rogers, its energetic and visionary chairman and CEO, RCI not only owns Canada's largest CATV company, Rogers Cablesystems, with nearly 1.4 million subscribers representing a 22 percent share of the Canadian market (in 1987), it also owns 97 percent of Cantel Inc., the country's largest (and only national) cellular telephone company.

In 1989 Cantel was supplying service to over half the country's nearly quarter million cellular subscribers, and was positioning itself for further gains with plans to invest U.S.$1.3 billion in cellular and cable service by 1994. By 1991 Cantel's service corridor extended 7,500 km from coast to coast, the longest cellular network in the world. Plans to replace the coaxial cables wiring homes of existing cable subscribers with fiberoptic technology have also been outlined by Rogers.

The combination of RCI's extensive cable and cellular operations with Unitel's microwave and fiberoptic long-distance networks, as well as Unitel's leased satellite offerings, not to mention CP's 30 percent interest in Telesat Mobile Inc., a supplier of national mobile satellite telephone service scheduled to begin operation in 1993, poses the most serious challenge ever to face Canada's traditional telephone carriers.

Telesat Canada was established in 1969 as Canada's national satellite communications carrier. The federal government owns half; the rest is held by Canada's major terrestrial carriers including Unitel. Prior to 1986, Telesat's role was essentially restricted to that of a carrier's carrier; most of its revenue was from providing long-distance transmission for common carriers and broadcasters. This restriction was relaxed in mid-1986 with the introduction of a new federal policy permitting broadcasters, business users, and others to independently own and operate transmit/receive earth stations. (For more on this policy, see ComCan 1988, pp. 49–51.)

By the late 1980s Telesat offered a broad range of competitive voice, data, and video services to business customers. In addition, it offered custom designed networks for highly specialized needs or, for users with private networks already in place, leased satellite channel capacity in increments of 1 percent. Telesat continues to provide voice and data communications to isolated communities, as well as oil and gas exploration camps, mining and forestry centers, and other remote work locations, primarily in the north. It is announced government policy that Telesat be fully privatized in the "near future."

Teleglobe Canada, meanwhile, is the exclusive provider of overseas telecom services. Canada–U.S. traffic, 85 percent of all Canadian international traffic, falls outside Teleglobe's reach, being supplied instead by carriers belonging to Telecom Canada. Prior to 1987, Teleglobe was a federally owned crown corporation. It was sold to private interests, however, in the spring of 1987 as part of the government's privatization program.

The winning bidder was Memotec Data Inc., a small and until then little-known company providing data services. The price was $563 million, to be raised in part by a public offering by Memotec once the sale had been completed. In May, to a great deal of surprise, BCE (Bell Canada's parent) acquired one-third of Memotec's shares, gaining de facto control. This cast considerable doubt on the government's announced policy of seeking to insure the independence of Teleglobe from Canada's other major carriers. Bell Canada continues its push to acquire full control of Teleglobe, arguing that international services should be provided on an integrated basis.

The sheer simplicity and elegance, others might say audacity, of these moves was reminiscent of Bell Canada's corporate reorganization in 1982 when, in a move designed to thwart what it perceived to be excessive regulation, Bell Canada created its own unregulated parent company, BCE, and then transferred to it most of its subsidiaries. Although legislation restored some of the regulatory powers thus lost by the CRTC, Bell's move is still seen by many as having been largely successful.

One feature distinguishing the Canadian telecommunications industry from that in other countries is the absence of any single common carrier offering fully integrated services nationwide. It has instead, dozens of small and nine relatively large full-service terrestrial carriers operating essentially within the boundaries of single provinces. Thus, aside from long-distance traffic crossing provincial boundaries within Bell Canada's operating territory—which spans much of Ontario, Quebec, and the eastern Northwest Territories—all interprovincial telecommunications must pass through the facilities of two or more separately owned and operated common carriers. Interprovincial communications are therefore largely dependent on the existence of interconnect agreements between the various service providers. The most important of these is the master contract between Canada's nine major telcos and Telesat Canada.

The Telecom Canada Connecting Agreement covers nonadjacent member companies, leaving arrangements between adjacent companies to bilateral negotiations, a revenue sharing plan, and a commitment to cooperate in the development and implementation of uniform standards and operating procedures, the adoption of new technologies, and in the marketing of new services. It also provides for coordination with U.S. carriers to facilitate handling North American traffic and with Teleglobe for exchange of overseas traffic (see McManus 1973).

The precise legal status of Telecom Canada, and TCTS before it, has always been somewhat unclear. Telecom Canada remains an unincorporated enterprise best described as a voluntary association of independent carriers bound by a common purpose as defined in a set of multilateral agreements. From an economic perspective it exhibits many of the classic traits of a cartel.

Telecom Canada's organizational structure has necessitated a number of functional compromises to facilitate smooth operation. For example, because each member has one vote and all decisions must be unanimous, the smaller members wield disproportionate influence. Though decisions relating to budgets, construction plans, marketing, the introduction of new technologies, and so on, must be jointly made, Telecom Canada itself owns neither plant nor equipment. Its function is to plan, administer, and coordinate; it is not to operate. Telecom Canada's headquarters are located in Ottawa, with all its personnel, premises, and facilities on loan from member companies. Administrative costs and services are shared among members as are profits in accordance with the revenue sharing plan (Janisch 1984).

An important consequence of the Connecting Agreement, or more precisely, its revenue sharing plan, has been the ability of the Telecom Canada Board of Management to effectively set rates on all interprovincial traffic utilizing the facilities of three or more members (Brait 1981, p. 56). Indeed, Telecom Canada members have been remarkably successful in creating and sustaining a rate structure based on system-wide average pricing and toll-to-local cross subsidization, the benefits of which have flowed primarily to the local subscribers in the smaller, less-urbanized provinces.

Also characteristic of Telecom Canada has been its engineering—as opposed to marketing—approach to service provision. Member companies have historically been much more committed to providing reliable, high-quality service on a universal basis than they have been to marketing a diverse mix of services or rapidly introducing new ones. This arguably reflects genuinely shared values and beliefs of the members, especially at a time when monopoly was the rule and declining costs allowed considerable leverage in experimenting with income redistribution between regions and classes of subscriber, while at the same time facilitating one of the highest service penetration levels in the world.

A second factor explaining the relative success of Telecom Canada is the fact that it has never been subject to direct regulation (see McManus 1973, p. 424). The CRTC has, at most, only indirectly affected Telecom through its regulation of Bell Canada, BC Tel, and Telesat Canada. Provincial regulatory agencies, by comparison, have shied away from meddling in Telecom affairs despite the growing reliance of provincial carriers on Telecom toll revenues. Fear of upsetting system-wide uniform pricing is the principal reason.

19.2.2 Current Regulation and Ownership

To see why Canada has had the richest and most eclectic mix of ownership patterns and divided regulatory jurisdictions in telecommunications of any country in the world, one need only scan Tables 19.1 and 19.2.

The complexity of these arrangements has been somewhat reduced in the early 1990s, at least with respect to jurisdiction to regulate. The constitutional basis for provincial regulation had long been suspect, but the federal government, out of deference to long-entrenched provincial interests, had not been prepared to launch a legal challenge. In the mid-1980s, however, CNCP (now Unitel), frustrated at its inability to obtain interconnection with the provincially regulated companies for data and private-line voice services, sought to have the validity of provincial regulation tested in the courts. In August 1989 the Supreme Court of Canada ruled unanimously that members of Telecom Canada are subject to federal jurisdiction *(AGT* v *CRTC).* It is possible that *all* carriers interconnecting to the interprovincial network will be swept into the federal fold. As jurisdiction in Canada is determined on an all-or-nothing basis, there is no room for shared jurisdiction as in the United States. The ruling has raised concern that the pendulum has swung too far in favor of exclusive federal authority (Janisch and Schultz 1991).

A move to privatization remains on the ownership front. In October 1990 AGT was privatized as a subsidiary of a new holding company, TELUS Corp. The initial public offering was the largest in Canadian history, raising $896 million from more than 139,000 Canadian investors. AGT Ltd. is now the

Table 19.1. Specialized Carriers Regulated by the Federal Government*

Carrier	Service Provided	Owners
Telesat Canada	Satellite carrier	Government of Canada 50%, others 50%[a]
Teleglobe Canada	Overseas (i.e., international other than to United States)	Memotec 100%[b]
Unitel	Specialized transcontinental carrier	Canadian Pacific Ltd. 60%, Rogers Communications 40%[c]

*The federal regulator is the CRTC (Canadian Radio-television and Telecommunications Commission)

[a]The others, which include Unitel, are approved common carriers listed in Schedule I of the Telesat Canada Act of 1969.

[b]Memotec is 33 percent owned by BCE (Bell Canada Enterprises). The other 67 percent is widely held.

[c]CP and Rogers are widely held publicly traded companies.

regulated entity. Cumulatively, the withdrawal of government-owned CN from telephone service, the sale of Teleglobe and AGT, and the announced plans to fully privatize Telesat mean that, of the major carriers, only Sasktel and Manitoba Tel remain under government ownership.

Simply stated, in the last half of the 1980s Canada went from government and private ownership with federal and provincial regulation to a system that is, with minor exceptions, federally regulated and investor owned.

19.2.3 Regulatory Oversight

19.2.3.1 Federal Jurisdiction

Although the Canadian telecommunications industry has undergone a profound transformation since the mid-1970s, the steady shift from monopoly to competition has been accomplished entirely without benefit of new legislation. Indeed, the Railway Act, which governs telecommunications regulation at the federal level, dates back to the first decade of this century. Sections 334–41 are easily the most crucial provisions to telecommunications regulation. They set out the jurisdiction, duties, and most of the powers of the CRTC and outline a number of the responsibilities, duties, and obligations of service providers coming within the commission's jurisdiction.

The CRTC's principal responsibility is to insure rates are "just and reasonable" and "not unjustly discriminatory or unduly preferential." Given that neither the Railway Act nor any other federal statute anywhere contains a general policy statement outlining the aims and objectives of telecommunications regulation, it is no exaggeration to suggest that the statutory mandate of the CRTC both begins and ends with the responsibility of insuring these two requirements are met. The Commission interprets and realizes how these goals are to be met.

As the CRTC pointed out in its first public statement after assuming jurisdiction over federally regulated carriers from the Canadian Transport Commission in 1976:

> [T]he principle of "just and reasonable" rates is neither a narrow nor a static concept. . . . Indeed, the Commission views this principle in the widest possible terms, and considers itself obliged to continually review the level and structure of carrier rates to ensure that telecommunications services are fully responsive to the public interest. (CRTC 1976.)

The Railway Act limits the jurisdiction of the Commission to companies rather than markets or market activities. The definition of companies unfortunately employs hopelessly antiquated terminology. As new technologies proliferate, the resolution of jurisdictional questions must become more arbitrary and subject to dispute, as shown by two CRTC decisions issued within three months of each other. In one, the commission held cellular radio providers to be "companies" within the meaning of the act, while the other found nontelephone company suppliers of enhanced services not to be "companies" and, hence, outside commission jurisdiction. The definition in the act could quite legiti-

Table 19.2. Area-Specific Telephone Companies, as of Mid-1989

Carrier	Area Served*	Regulator[a]	Owners
British Columbia Tel (BC Tel)	British Columbia	CRTC	GTE 51%[b]
Alberta Govt Tel (AGT)	Alberta	Alberta Public Utility Board	province
edmonton Tel (Edtel)[c]	Edmonton, Alberta	unregulated	city of Edmonton
Sask Telecom (Sasktel)	Saskatchewan	unregulated	province
Manitoba Telephone System (MTS)	Manitoba	Manitoba Public Utility Board	province
Thunder Bay Tel	Thunder Bay, Ontario	Ontario Telephone Services Commission	city of Thunder Bay
Ontario Northland Tel	parts of northern Ontario	unregulated	province
Northern Telephone	parts of northern Ontario	Ontario Telephone Services Commission	BCE† 98%
Bell Canada	the most-populated parts of Ontario, Quebec, and eastern NW territories	CRTC	BCE
Telebec	parts of northern Quebec	Quebec Regie des Services Publics	BCE
Quebec Tel	parts of eastern Quebec	Quebec Regie des Services Publics	GTE 55%[d]
New Brunswick Tel (NBTel)	New Brunswick	New Brunswick Public Utility Commission	BCE 31%[e]

(*continued*)

mately support opposite conclusions, which may be why no supporting rationale was offered in either case.

A second problem with a company definition of jurisdiction is that it brings all the activities of a given "company" within the commission's reach, whether warranted on other grounds or not. This, in fact, was the principal rationale for Bell Canada's reorganization in 1982. By creating a parent corporation that fell outside the definition of "company" and then transferring to it a number of subsidiaries, Bell was able to significantly lessen the scope of the CRTC's reach.

Insofar as the CRTC's actual powers are concerned, the act expressly authorizes it to regulate: (1) pricing, (2) the terms and conditions of network interconnection, (3) all working agreements to be entered into between a "com-

Table 19.2. *(continued)*

Island Tel	Prince Edward Island	Prince Edward Island Public Utility Commission	BCE 55%, Maritime Tel & Tel 38%[f]
Maritime Tel & Tel (MTT)	Nova Scotia	Nova Scotia Public Utility Commission	BCE 33%[g]
Newfoundland Tel (Nfld Tel)	Newfoundland	Newfoundland Public Utility Commission	BCE 55%[h,i]
Terra Nova Tel	Newfoundland		Newfoundland Tel
NorthwesTel	Yukon and Northwest Territories	CRTC	BCE[i] —

*The Provinces from west to east followed by the Territories.

†BCE (Bell Canada Enterprises) is a widely held publicly traded company.

[a]The CRTC (Canadian Radio-television and Telecommunications Commission) is the federal regulator. All the others are provincial.

[b]BC Tel is 51 percent owned by Anglo–Canadian Tel, which is 100 percent owned by GTE, a widely held publicly traded U.S. company. The remaining 49 percent of BC Tel trades on Canadian stock exchanges.

[c]Company spells its name uncapitalized.

[d]Quebec Tel is 55 percent owned by Anglo–Canadian Tel, which is 100 percent owned by GTE, a widely held publicly traded U.S. company. The remaining 45 percent trades on Canadian stock exchanges.

[e]NBT is 100 percent owned by Bruncor Inc., which is 31 percent owned by BCE. The remaining 61 percent of Bruncor trades on Canadian stock exchanges.

[f]Maritime Tel & Tel is 33 percent owned by BCE. The remaining 7 percent of Island trades on Canadian stock exchanges.

[g]The remaining 67 percent trades on Canadian stock exchanges. Shareholders, including BCE, are restricted to voting just 1,000 shares under 1966 Nova Scotia legislation.

[h]Newfoundland Tel is 100 percent owned by Newtel Enterprises Ltd., which is 55 percent owned by BCE. The remaining 45 percent of Newtel trades on Canadian stock exchanges.

[i]Nfld Tel and NorthwesTel were owned by Canadian National Railways until 1988. (CN is federal government owned.)

pany'' and other providers of telecom services, whether or not the latter are subject to the commission's jurisdiction, and (4) the terms and conditions under which traffic may be carried by the company.

Those powers and duties created by a number of Special Acts of Parliament should also be added to this list. The most important of these are the Bell Canada Act, the British Columbia Telephone Company Act, and the Teleglobe Canada Reorganization and Divestiture Act. Among the restrictions imposed is that it is the commission's responsibility to enforce are: (1) the obligation to serve all customers under specified circumstances, (2) the terms and conditions that such carriers may control attachment to the network of customer-owned equipment, (3) an outright ban on entry into certain types of markets, (4) limits on foreign ownership of the voting shares (20 percent in the case of Teleglobe), and (5) a host of restrictions on the size, dimension, location, appearance, and so on, of physical plant and equipment erected on public property.

Comprehensive as this list may appear, there are a number of important elements of market conduct that the commission lacks authority to regulate. These include corporate policies related to the nature, level, and quality of service provision, marketing and sales promotion excluding pricing, investment in new capacity, research and development expenditures, and, to some extent, horizontal and vertical integration. This lack of express statutory authority, however, has not prevented the CRTC from regulating every one of these on the grounds intervention is necessary to insure that rates are just and reasonable and not unduly discriminatory!

The commission's reliance on the broad discretion afforded by its enabling legislation has been most conspicuously demonstrated in recent years in its approach to two very important issues: competition and regulatory forbearance. (On this topic generally, see Janisch and Romaniuk 1985 and Romaniuk and Janisch 1986.) Federal legislation is essentially industry-structure neutral as far as competition is concerned. Neither the Railway Act nor any other federal statute dealing with telecommunications has ever expressly conferred or ruled out (except in limited circumstances) a monopoly franchise on any carrier.

It is precisely this statutory ambivalence between competitive and monopolistic market structures that has allowed the commission to gradually introduce more competition into different segments of the industry without the need for new legislation. Since the late 1970s, for example, the commission has allowed terminal interconnection, facilities-based competition in data and private-line voice services, and liberalization of the rules with respect to enhanced services competition and resale and sharing.

This is only half the picture, however. While the CRTC has used its discretion to introduce new players into telecommunication markets, it has simultaneously withdrawn from intensive regulation of these same activities through a process of regulatory forbearance. The commission, for example, has determined in the course of a number of decisions that the following markets would be better served with less regulatory intervention on its part: cellular radio, enhanced services, data and multiline business equipment provision, satellite earth station services, and public mobile satellite communications services.

The CRTC has justified its policy of forbearance on two closely related grounds: (1) in appropriate circumstances market forces alone may be sufficient to assure "just and reasonable" rates, and (2) the costs of regulation to the industry, the regulator, and ultimately to consumers and taxpayers are not always warranted.

The difficulty faced by the commission in seeking to selectively withdraw from tariff regulation has been its questionable statutory authority to do so. The issue of the CRTC's power to suspend the tariff filing and approval process finally came to a head with the appeal by the Canadian Telecommunications Workers' Union (TWU) of a 1987 commission decision relieving CNCP of the obligation to file tariffs for its competitively provided services. The Federal Court of Appeal, although apparently sympathetic to the rationale underlying the commission's forbearance policy, held that the relevant statutory provision was mandatory: Tariffs must be filed and approved before any tolls for services may be charged. In June 1989 the Supreme Court declined to hear an appeal.

As a result, much of the commission's efforts to reduce the regulatory burden faced by the industry has been placed in jeopardy, although certain options remain. In particular, (1) the CRTC may, in selective cases, interpret the definition of "company" under the Railway Act more narrowly and determine that certain service providers fall outside its jurisdiction in any event; (2) the commission may develop more streamlined tariff approval mechanisms; or (3) Parliament itself may enact amending legislation modeled on §16 of the Teleglobe Canada Reorganization and Divestiture Act, which expressly grants the commission broad powers of forbearance regarding Teleglobe.

The CRTC's efforts to cut regulatory costs within its jurisdiction, which includes all broadcasting, cable television, and federal telecommunication matters, appear to have been relatively successful. The CRTC's total operating budget has increased only very modestly, once inflation is taken into account, since the mid-1970s. Staff has been slashed from 492 in 1978 to 388 in March 1987. In the same period, applications processed by the Broadcasting Directorate have approximately doubled from 1,653 to 3,079. Although the number of hearings and decisions rendered per year on the telecommunications side have not increased significantly, their average length and complexity has, due in no small part to the increased participation of intervenors. In 1986 the CRTC was authorized to impose fees, levies, or charges on carriers under its jurisdiction in order to recover its regulatory costs. While a welcome source of revenue, this power carries with it the potential for conflict of interest.

Although the CRTC is nominally independent of the federal government, its decisions are not immune from government interference. The principal avenue is through "cabinet appeal" (§67 of the National Telecommunications Powers and Procedures Act). While the cabinet has rarely relied on this power of its own motion, various interested parties have increasingly resorted to §67 to challenge commission decisions—most spectacularly in the Call-Net case in 1987, discussed later. The extent of the cabinet's discretion under §67 has been held to be virtually unlimited by the Supreme Court, although this view has been questioned. For example, Romaniuk and Janisch argue that the power is not unlimited (1986, pp. 626–28).

Decisions may also be affected by cabinet policy directives (see Bureau 1988), informal consultations with government ministers and their staffs, indirect pressure—principally through ministerial speeches and statements of government policy made in Parliament or other public forums—and ultimately, through the enactment of new legislation.

The minister of communications also plays an important role in regulating the industry, principally through the authority granted under the Radiocommunications Act to control entry into telecommunications markets and to make regulations prescribing service and equipment standards. The power to control entry, however, is limited to services in which the transmitted signals are propagated through open space without benefit of a tangible, physical medium of carriage. Such services include cellular radio, radio paging, and all local and long-distance services relying on satellite or microwave technology. No federal license appears to be required for carriers providing end-to-end service exclusively by unbroken physical medium, be it copper wire, coaxial cable, or fibe-roptic.

The minister, together with the Department of Communications (DOC), also plays an important policy development role, usually in consultation with the provinces, and is ultimately responsible for all new federal legislation. As might be expected, the minister also carries considerable weight in determining the outcome of appeals to the cabinet from CRTC decisions, although in the late 1980s there has been some successful bypassing of ministerial authority.

19.2.3.2 Provincial Jurisdictions

Although now largely of historical interest in the wake of the Supreme Court's 1989 decision, each province has had one or more pieces of legislation in place governing the regulation of telecom service providers either in their own right or as one of several provincially regulated public utilities. With the exception of Manitoba and Saskatchewan, where the major provincial carriers are statutorily created and publicly owned monopolies, monopoly franchises were generally not created for individual carriers operating within provincial boundaries. Nevertheless, provincial carriers have been treated as de facto monopolies within their operating territories and have either been subject to simple rate regulation or the more complex process of capital base, rate of return regulation.

The governing regulatory principles in most provinces are essentially the same as at the federal level. The major difference has been the relative lack of speed and enthusiasm with which competition has been allowed to develop at the provincial level. The most opposed to competition, even in matters as basic as terminal interconnection, have been Saskatchewan and Manitoba. These are two of Canada's most rural, generally less well-off, and most heavily toll-revenue dependent provinces (Schultz and Alexandroff 1985, p. 73). Any incursion, however small, into the provincial monopoly is treated as a direct threat to the rural subscriber, whose political importance is considerable.

The Atlantic provinces, although poorer still, have no similar tradition of government involvement, not as great a disparity in toll versus local revenues, and no comparable political stake in preventing competition. As a result, many of the CRTC's decisions relating to competition have been adopted to a greater or lesser extent in most of these jurisdictions, albeit after a lag of some years (FP 1986a, pp. 37–51).

19.2.4 Interests the System Seeks to Protect

Canadian confederation dates to 1867, but the process of nation building has never really stopped. Canada's sprawling territory and scattered population have always demanded strategic counterweights to the threat of American domination. The ninteenth century response was a three-pronged policy of transcontinental railroads, high tariff barriers, and regional economic specialization to compensate for the small size of domestic markets. The western economy produced agricultural products in exchange for eastern manufactured goods, while the federal government subsidized the cost of shipping the bulky, low-value produce of western farms. The policy worked, Canadian industrialization proceeded apace, and national sovereignty and economic well-being were secured.

With the decline of the railroads, however, as well as the crumbling of tariff barriers and the growth of export opportunities, new policies had to be developed. Instead of concentrating on the coast-to-coast movement of goods, the government, particularly through the agency of federal crown corporations such as Trans-Canada Airlines (TCA, later Air Canada), the Canadian Broadcasting Corporation (CBC), and similar government-initiated ventures, began to focus on the rapid movement of people and then, to a much greater extent, on the exchange of information and ideas. Today, the wrap that bind the national fabric is communication links.

Ministers of communications in the 1980s speaking on universal availability of telephone service at affordable prices gave the impression it had always been a central priority of the federal government. However, this objective has never been codified in legislation of any kind. Canadians have universal telephone service thanks to the efforts of the telephone companies and the shared priorities of their regulatory agencies.

The most complete statement of Canadian policy in the late 1980s emerged from a meeting of federal, provincial, and territorial ministers of communications in April 1987. Their communique endorsed the following six principles intended to guide formulation of government policies and regulation in the industry (see DOC 1987, pp. 3–4).

1. The future development of the industry presents uniquely Canadian challenges requiring uniquely Canadian answers.
2. Canadians must continue to have universal access to basic telephone service at affordable prices.
3. Policies must maintain the international competitiveness of the Canadian telecom sector and the industries it serves.
4. Policies must insure that all Canadians benefit from the introduction of new technology.
5. A Canadian policy must reinforce the goal of fair and balanced regional development and respond to the interests of all concerned governments.
6. Policies should be established by governments and not by regulatory bodies or by the Courts.

These principles may be difficult to reconcile. The most inherently contentious ones may be the desire to preserve low-cost, universal basic service while maintaining international competitiveness. In 1989 two studies concluded that direct-dial rates in Canada were two or more times higher than those in the United States (Koelsch 1989 and Hoey 1989). This is so because toll revenues continue to subsidize local services. A great fear of Canadian policymakers is that rate rebalancing, however justified on grounds of economic efficiency, may imperil the principle of universality. This is discussed extensively in "Rate Rebalancing Decision" (1988, pp. 89–119).

Just how far Canadian regulators are prepared to go in protecting universal availability may be seen in a remark by former CRTC chairman, André Bureau, that even if only 1,000 subscribers "drop their telephone service because of an increase in local rates, it would be 1,000 too many" (CRTC *1987–1988 Annual Report,* p. xii). Bell Canada's local rates rose only marginally from 1983 to 1989. By contrast, from January 1987 to May 1988 alone, the cumulative reductions in long-distance rates have averaged 26 percent for calls within Bell's territory and 31 percent for calls to other provinces.

Canadians have watched U.S. rate rebalancing with intense interest. The heartening news thus far is that significant increases in the price of basic service have not been accompanied by any appreciable loss of customers. It was this fact, perhaps more than any other, that led the CRTC to conclude in its mammoth 1988 hearing into Bell Canada's application for limited rebalancing that the principle of realigning rates with costs is one that should be followed in Canada—subject to one condition. The commission has stated that if it is satisfied that a safety net can be provided for subscribers for whom access would no longer be affordable at cost-based rates, then it is prepared to approve future applications for a more efficient rate structure.

This is the source of considerable friction and hostility between the federal and provincial, especially Prairie, governments because of regional economic disparities. Most of the industries benefiting from lower costs are based in Ontario and Quebec, jurisdictions served by Bell Canada. Were the CRTC to significantly lower Bell Canada toll rates, few competitive benefits would accrue to the Prairie provinces. Should their toll revenues plunge, however, local rates for rural subscribers could soar, creating a political maelstrom.

19.2.5 The Extent of Universal Services and Challenges in Extending the Network

Canada's telecom penetration level, defined as the percentage of households with telephone service, is among the world's highest. In 1985, the Canadian average was over 98 percent, ranging between 94 percent in Newfoundland and 99 percent in Ontario. The average was just under 50 percent in 1947. (For complete data, see FP 1986a, Table 4.1, p. 205.)

Challenges to extending the national network, although great given the extent of the country and its extremely diverse topography, have become much more manageable with the advent of satellite technology and, more recently, cellular radio. It may soon be the case that there is not a single community or household in this, the world's second largest country, without access to some telecommunications facility.

19.2.6 Types of Services Offered

Most telecom services offered somewhere in the world are also available in Canada. Telecom Canada members together with Teleglobe and the nation's independent telcos offer fully integrated, transcontinental, and international public-switched telephone services as well as MTS and WATS. MTS includes a number of long-distance calling options to nonlocal destinations within Canada, as well as to points in the United States and overseas. WATS includes a variety of bulk-rated long-distance options usually used by business customers. Outward WATS is available within Canada, while 800 service is available both on a Canada-wide and U.S.–Canada basis. By early 1989, Toronto had one of the highest subscription rates for cellular phone service of any major metropolitan area in the world.

Private lines may be leased on an individual or bulk basis from the telcos or competitive suppliers such as Unitel for voice and data communications or program transmission. Private-line services include foreign exchange (FX) lines, off-premises extensions, and tie trunks. Telesat Canada has provided satellite-based private-line services directly to users since 1986. International leased circuits are provided by Teleglobe at the point of interconnection to its gateway facilities.

An entire array of nonvoice services are also provided on a competitive basis, including data, switched teleprinter, facsimile, electronic message, mail and text services, public message, audio and video program transmission. The network market was expected to grow by 30 percent in 1989 and at comparable rates thereafter.

Among the newer technologies and service types being considered, the most promising appears to be ISDN. Canada is particularly well-positioned to take advantage of ISDN because it has been moving toward a fully digitized national network at a faster rate than any other industrialized country. It was estimated that more than 70 percent of intercity circuits would be carried on digital transmission facilities and 80 percent of local and long-distance calls would be digitally switched by 1990. (A particularly valuable study, by a former CRTC vice-chairman, is Lawrence 1989.)

Typical of the services being introduced by the late 1980s was "Alex," a Bell Canada videotex information service. It allows consumers to call up more than 120 listings on a small terminal screen, including home shopping, banking, television and movie reviews, and restaurant, travel, and transportation information. The twenty-four-hour service was initially made available to some 20,000 Montreal area households in December 1989 and was eventually expanded to other cities.

19.2.7 Telecommunications Rates and Rate Structure

In January 1986 Canadian federal and provincial ministers of communications commissioned an extensive study of the pricing objectives, principles, and practices of the major telcos and the effect of these policies on the universal availability of phone service. The result, the "Federal-Provincial Report" (FP 1986b), issued some ten months later, indicated that the two most important objectives of telcos in setting service prices were the need to maximize access to and use of the public-switched network and the need to maintain adequate and stable revenues. Other goals individual companies mentioned as important include the need to keep rates simple and easy to administer, maintenance of customer satisfaction, efficiency, competitiveness, and equity among different classes of customer.

Notwithstanding the diversity in rating objectives, there has been total unanimity among telcos regarding the principles governing rates; the two most important ones are value of service pricing and company-wide price averaging (FP 1986a, p. 63).

Value of service pricing is most important at the local level. Generally, the larger the local calling area—whether measured by the total number of toll free numbers that can be reached, the number of main stations, or the number of access lines—the higher the monthly local rate. Business users are charged more than residential subscribers for essentially the same grade of service, as Table 19.3 indicates.

Company-wide price averaging means that customers obtaining similar services pay similar rates regardless of the actual cost of providing them. Thus, within a local exchange, all subscribers pay the same flat monthly rate regardless of usage or distance from the central office. In the case of long-distance service, all callers pay the same rate per mileage band. Company-wide price averaging, by definition, means low-cost uses of telephone service to subsidize higher-cost ones. Table 19.4 provides information on business day (peak period) long-distance rates by band for representative major companies.

Table 19.3. Monthly Exchange Local Telephone Rates*, Individual Line (January 1986)

Residence	Business	Residence as % of Business	City
11.70	31.85	37	Victoria, BC
9.28	23.74	39	Calgary, Alberta
8.30	20.85	40	Regina, Saskatchawan
7.50	20.00	38	Winnipeg, Manitoba
10.70	34.45	31	Ottawa
12.80	38.85	33	Rimouski, Quebec
12.05	35.45	34	Moncton, NB
13.10	37.50	35	Halifax, NS
12.60	38.30	33	Charlottetown, PEI
13.15	41.00	32	St John's, Newfoundland

Source: Federal-Provincial Examination of Telecommunications Pricing and the Universal Availability of Affordable Telephone Service *Working Papers,* Table 2.1, p. 64. Ottawa: Supply and Services Canada, 1986.

*Rates are in Canadian dollars.

Includes rental for rotary-dial telephone set and, where applicable, EAS.

These are flat rates, unlimited local calling. EAS is widely available, with the flat-rate calling area in Metropolitan Toronto and surrounding areas probably constituting the largest in the world.

Table 19.4. Long-Distance, Two-Point Service Rates* for Customer-Dialed Business Day, Three-Minute Duration Calls (December 1985)

Mileage	Canada to U.S.	Telecom Canada	Bell Canada	Nfld Tel	Manitoba Tel	BC Tel
10	.32	.60	.52	.75	.30	.77
50	.55	1.01	1.34	1.14	.72	1.34
100	.94	1.41	1.64	1.50	.96	1.70
500	2.04	2.55	2.04	2.34	1.50	2.24
1,000	2.46	3.00	2.04	2.34	1.50	2.24
2,200	2.78	3.30	2.04	2.34	1.50	2.24
3,000	2.92	3.30	2.04	2.34	1.50	2.24

Source: Federal-Provincial Examination of Telecommunications Pricing and the Universal Availability of Affordable Telephone Service *Working Papers*, Table 2.10, p 89. Ottawa: Supply and Services Canada, 1986.

*Rates are in Canadian dollars.

Telecom Canada and Canada to U.S. rates are those applicable to Bell Canada customers.

Off-peak discounts are generally 30–35 percent during the evening, and 50–67 percent from 11 P.M. to 8 A.M. daily. On Sunday discounts are 35–67 percent for all carriers; Bell Canada's weekend discount begins at noon Saturday. The structure and level of discounts usually does not vary between intraprovincial and trans-Canada calls.

19.2.7.1 The Level of Cross Subsidies

Perhaps the most serious consequence of setting prices for basic local services at levels designed to maximize accessibility and use has been the growing disparity between the cost and the revenues generated. To meet overall revenue requirements, rates for other services, especially long distance, have had to remain above costs. Long-distance transmission costs have fallen continuously in both nominal and real terms since the introduction of microwave in the 1950s, with no corresponding decline in rates for the most part until the 1980s.

Some indication of the resulting cross subsidy from long distance to local services is provided in a cost-revenue study released in the early 1980s (Bell Canada 1983). It concluded the cost to Bell of generating $1 in revenue from monopoly supplied local service was $1.93. By comparison, it cost only $0.32 for each $1 of revenue earned in providing noncompetitive toll services. The resulting total shortfall from provision of local services in 1982 was $1.2 billion, virtually all of which was made up from revenues generated in supplying noncompetitive long-distance services, primarily MTS and WATS.

In its 1987 application for rate rebalancing, Bell Canada supplied further evidence of the growing subsidy flowing from monopoly toll to local services. The company argued that the local/access service category shortfall was $1.4 billion in 1984 and would increase to $2.4 billion in 1986 without corrective action. The commission concurred in the need for rebalancing, but delayed restructuring rates until a complete plan for targeted subsidies was developed.

19.2.8 The Equipment Manufacturing Industry

The Canadian equipment market includes over 100 firms of various sizes, although this large number is very misleading in two important respects. First, the shares of all but one are virtually inconsequential. Northern Telecom accounts for approximately two-thirds of the telecom equipment market; its share is somewhat lower in the communications equipment market. The next largest competitor, Microtel, accounts for less than 10 percent. Other well-known firms are Gandalf Technologies (network processors, including PABXs), Mitel (a major PABX manufacturer), and NovAtel (cellular telephone terminals and systems). Northern Telecom, Mitel, and Gandalf all sell more in the United States than they do in Canada and have widely followed, publicly traded stocks.

The second point is that most firms are highly specialized. Very often, smaller producers are actually in the business of supplying larger firms specialized parts and components. Only two firms supply a fairly complete range of equipment—Northern Telecom and Microtel.

The more important foreign-owned firms include Plessey, Siemens, Ericsson, Phillips, Rolm, TIE, Toshiba, and AT&T. None has a significant share of the Canadian market although some successfully occupy niches. Many foreign-owned firms have plants in Canada, but some rely only on finished imports. During the 1970s these nonmanufacturing importing subsidiaries began to lose market share as telcos, especially those on the Prairies, increasingly switched to domestic producers. As a result, foreign, especially European-owned firms, began to establish manufacturing subsidiaries in Canada.

Northern Telecom has strategically placed manufacturing plants in all Canadian provinces; this fact alone produces strong incentives for all companies to consider the negative regional economic impact of buying elsewhere.

Although the Canadian telecom equipment market achieved sales in the vicinity of $3 billion in 1987, a substantial portion of these took place in vertically integrated or captive markets. The two biggest telcos—Bell Canada and BC Tel—each have manufacturing affiliates: Northern Telecom (53 percent BCE-owned), and Microtel, respectively. BCE also has significant direct and indirect ownership links in the four Atlantic telcos, which significantly rely on Northern Telecom for their equipment. Similarly, Quebec Tel, owned by a GTE subsidiary, Anglo–Canadian Telephone, shares the same parent as BC Tel and its preferred supplier is thus also Microtel.

Given the dominance of Bell and BC Tel as purchasers, together accounting for some 70 percent of the market, and their strong tendency to buy from their manufacturing arms—on average Bell purchases some 85 percent of its requirements from Northern, while Microtel supplies BC Tel with over 50 percent of its need—the vertically integrated market probably accounts for over one half the value of all telecom equipment sold in Canada.

Both tariff and nontariff barriers exist. The height of these barriers, however, is difficult to estimate even given a schedule of the rates. This is partly because different components are often taxed at different rates. Thus, some hardware, including PBXs, is assessed at the maximum rate of 17.8 percent, wire and cable enters at 4 percent, and software crosses the border free. In addition, rates differ according to the degree of assembly on importation.

Nontariff barriers are by their very nature even more difficult to assess except when they take the form of virtual prohibitions. For example, if "native son" or "buy at home" policies are enforced, then nontariff import barriers become impenetrable. Other barriers of varying effectiveness include transportation and communications costs, various government design standards, regulations, and antidumping laws. There are also nonprice considerations, including the greater perceived risks of dealing with foreign sources of supply, especially when contracts are of a long-term nature; established trust in and goodwill of domestic producers leading to ingrained buyer preferences; greater availability of spare parts and technical expertise; and finally, what might be called a "follow the leader" approach, especially on the part of the smaller telcos (Beigie 1973, pp. 91–93).

The Canada–U.S. Free Trade Agreement (FTA), implemented at the beginning of 1989, requires elimination of tariffs over a ten-year period, and nontariff barriers may also be expected to diminish.

Evidence exists suggesting economies of scale, and possibly of scope, have historically been an important factor. As long as trade barriers confined Canadian producers to the domestic market, few firms could expect to grow to minimum efficient size. The eventual emergence of only one world-scale manufacturer of telecom equipment in Canada is therefore not surprising. The opening of the U.S. market after the AT&T divestiture reduced the significance of Canada's own small market as a determinant of market structure, as Canadian firms could now spread costs over longer production runs.

Just how important foreign, especially U.S., markets have become can be gauged by the relatively small percentage of sales accounted for by the domestic market. Northern Telecom, for example, had total revenues of U.S.$6.1 billion in 1989—some 60 percent of it in the United States and 25 percent to BCE. Mitel generated only 16 percent of 1989 sales in Canada, while for Gandalf it was 30 percent.

19.2.8.1 Procurement Policies

For the better part of this century Bell Canada has procured most of its equipment from its manufacturing affiliate, Northern Telecom. Their supply agreement designates Northern as Bell's preferred supplier and requires Northern to supply as much of Bell's equipment needs as it is able to meet at reasonable prices. Bell, however, is not required to make its purchases from Northern if better products or terms are available elsewhere. Conditions of sale are normally subject to negotiation with one important exception: Bell is accorded "most favored purchaser" status. Northern's compliance with the agreement is monitored annually by an independent audit of its sale prices.

This agreement was the object of intense scrutiny by both the federal competition law authorities and the CRTC during the 1970s and early 1980s. The principal concern, and conclusion, of the director of investigation and research, Canada's chief enforcement officer under the Competition Act and its predecessor legislation, was that Bell's vertical relationship with Northern was having an inimical effect on competition in the equipment industry. The remedy initially proposed by the director was structural separation of Bell Canada from Northern Telecom. The director was later to argue only for the introduction of a competitive bidding process in the procurement procedure followed by Bell Canada.

After an exhaustive study of vertical integration, the Restrictive Trade Practices Commission (RTPC) concluded in 1983 that, on balance, the benefits of Bell's relationship with Northern Telecom outweighed the costs. The RTPC's recommendation was essentially that the status quo should be maintained (RTPC 1983, pp. 199–211). Northern Telecom was Canada's only major success story in the high-technology sector in the early 1980s, so it is extremely unlikely the government would have allowed it to be tampered with by overzealous competition law enforcers in any event.

The CRTC's concerns in the 1970s were of a different nature. What the regulator wanted to know was how the Bell–Northern Telecom supply agreement guaranteed that Northern, even if it offered Bell the best prices available in Canada, was still not overcharging. The commission argued that Northern's position as the low-cost, dominant producer in Canada, protected by high tariff barriers, made it possible for Northern to charge monopoly prices even to its best customers. The question, therefore, was what assurance did Bell have that it was receiving the best possible prices?

If the prices were higher than those possible in a more competitive environment, the commission reasoned, then Bell, through its ownership of Northern, could be making profits in excess of its allowed rate of return. To minimize the possibility of this occurring, the commission directed Bell to provide annually detailed price information on equipment sold by other Canadian suppliers comparable to that purchased by Bell from Northern Telecom.

Early in the 1980s, the CRTC also began to express a very different concern. Observing that Northern was then going through difficult financial times—returning only a 2 percent dividend on $100 million Bell had just spent on a major Northern share issue—the CRTC expressed concern Bell subscribers were being asked to subsidize Bell's investments in nonregulated markets. To counter this possibility, the Commission adopted a somewhat arbitrary and controversial solution. It simply deemed the return on Bell's investment in Northern Telecom to be 15 percent after tax for the purposes of calculating Bell's regulated revenue requirement. This solution was later modified to make it more equitable to shareholders; it was later extended to apply to Bell's average total investment in all other subsidiaries and associated companies. However, Bell's 1982 reorganization undid much of what the commission had hoped to accomplish because Northern became a subsidiary of unregulated BCE; it was no longer a subsidiary of regulated Bell Canada.

19.2.8.2 Policies on Terminal Attachment

The liberalization of rules dealing with attachment of customer-provided equipment began in a very limited way with the Terminal Attachment Program (TAP) in the mid-1970s. Participants were the federally regulated carriers acting in cooperation with certain provincial governments and a variety of manufacturers and major users. Between 1976 and 1979, a number of carrier tariffs were revised to permit direct connection of a limited number of customer-provided, network nonaddressing (i.e., nondialing) terminal equipment as long as it had been tested and certified by the DOC. The first types of equipment permitted included answering, dictation, and two-way voice recording devices. Under Phase II, alarm reporting, graphic, facsimile, and data modems, and traffic measuring equipment were also allowed. Other categories have subsequently been included.

The major breakthrough came with the CRTC's interim and then final decisions (in 1980 and 1982, respectively), to authorize terminal interconnection of virtually all types of customer provided equipment to the network, provided certain technical standards were met. Somewhat surprisingly, it was a formal application in 1979 by Bell Canada to the CRTC to inquire into the merits of liberalized terminal attachment that triggered the entire process in the federal jurisdiction. Although most provincial jurisdictions have followed suit, Manitoba and Saskatchewan had still maintained restrictions as of 1989.

The process of testing and setting standards is handled by DOC in concert with the Canadian Standards Association, an independent body comprised of telecom service providers, equipment manufacturers, and user groups.

19.2.9 International Trade and Collaboration

Canada has quite a number of treaties and arrangements with the United States and other nations, both bilateral and multilateral. In fact, North American networks are so interconnected and interdependent that at times it is more accurate to think of them on a continental rather than national basis (Grant 1988).

The FTA with the United States that came into effect at the beginning of 1989 is the first international agreement to deal with telecom services, and it significantly precedes possible similar developments in the General Agreement on Tariffs and Trade (GATT).

In view of Canada's more cautious approach to competition, the agreement recognizes the right of either country to retain monopolies with respect to "basic" telecom services. However, with respect to "enhanced" and "computer" services there is a commitment to maintain and support further development of an open and competitive market. Measures envisaged included structural separation, although at the time the Agreement came into effect Canada had not insisted on separate subsidiaries for carriers providing enhanced services.

To take advantage of the prospective nature of the agreement, the Canadian government announced a limit of 20 percent on foreign ownership of Type I carriers (facilities based) with no restrictions on Type II (nonfacilities based) carriers in July 1987. Existing foreign ownership, especially of BC Tel, was grandfathered in.

The agreement also contains a provision very dear to the hearts of Canadian policymakers concerned that trans-Canada traffic may be carried on American facilities. Nothing in the agreement is to be construed to prevent a party from maintaining or introducing measures requiring basic services to be carried on its network within its territory. Canada has already announced that a statutory obligation will be placed on its carriers to employ Canadian facilities wherever feasible.

The prevailing conventional wisdom is that the FTA will have little direct impact on telecommunications. The agreement does not require policy changes, does not address the pivotal issue of competition in public long-distance voice, and merely confirms existing regulatory policies governing enhanced services. This may well be too narrow a perspective.

First, the mere inclusion of computer and enhanced services in the agreement is, in and of itself, an important signal of the centrality of telecom services in contemporary international trade. Second, while it is true that the agreement does not change the ground rules, it does mandate a positive duty to take effective measures to ensure further development of an open and competitive market in a crucial growth area.

Third, the agreement preserves the status quo of a remarkably open border in telecommunications and will provide barriers against visceral protectionist reflex actions in less prosperous times. For example, the agreement applies to the movement of information across borders and access to data bases; this indicates a significant commitment not to impose limits on transborder data flows in the longer run. Fourth, if multilateral agreements are to include telecom services, then they will need to respect national concerns to protect basic networks. In this, Canada is really no different from other countries. It will simply not be possible to export the American fully competitive model, but it may be possible to espouse the sort of careful delineation and segmentation to be found in the FTA. Thus this U.S.–Canadian experience should be of the utmost interest in the ongoing GATT negotiations (Janisch 1987a).

A major irritant in Canada–U.S. relations was not dealt with in the Agreement. There has been a persistent U.S. demand that Canada's vertically integrated market structure be revised along the lines adopted in America. U.S. manufacturers have argued that Northern Telecom's preferred supplier arrangement with Bell Canada is inconsistent with the spirit of free trade. Despite the relatively small size of the Canadian market, there may well be pressure on the U.S. administration to use its powers under the Omnibus Trade and Competitiveness Act of 1988. Canada, no doubt, will be very resistant to any such export of American domestic solutions to telecommunications issues, but will also be very concerned to preserve its access to U.S. markets (see Janisch 1989).

19.2.10 *The Policy Role of Trade Unions*

The fact that Canadian trade unions have had very little say in policy formation in any area, much less telecommunications, should come as no surprise to observers of Canadian politics. In the country's history not once has a union-affiliated or labor-supported political party held the reigns of power at the federal level. Provincially, there have been several New Democratic Party (NDP) governments at different times, but without much impact on national telecommunications policy.

This is not to say labor has been entirely ineffectual in putting forth its various positions on important issues. The TWU and the Canadian Federation of Communications Workers (CFCW), among other unions, have been very active as intervenors in federal and provincial regulatory hearings, before parliamentary committees considering new legislation, and elsewhere. The impact of these activities on government policy, although difficult to gauge, has probably not been all that significant given the promonopoly, pro–status quo views typically adopted by the labor union movement. In fact, the progressive shift to increased competition and less regulation in recent years, stands in direct opposition to the positions taken by Canada's major trade unions.

19.3 Process of Change

19.3.1 *Emergence of the Electronic Industry*

Although some 40 percent of Canada's GNP comes from resource-based industries, there are clear signs of a move toward an information-based economy. Symptomatic of this has been the rapid rise in the service sector compared to the relatively small industrial sector. At the same time, the "information economy"—defined to include computer manufacturing and the secondary information sector (information services produced and consumed internally, such as the market research department of a manufacturing firm)—accounted for some 35 percent of Canada's domestic GNP in 1971, and 47 percent in 1981. Of particular interest is the growth in the data communications market.

Although Canada's overall R&D effort is generally acknowledged to be weak, the information technology sector is something of an exception. The computer and telecommunications industries spent about $1 billion on R&D in 1986, about 30 percent of total industrial R&D. BCE, far and away Canada's largest R&D spender, spent some $623 million that year, 22 percent of all R&D expenditures by Canada's private sector.

Still, in 1986, apart from BCE and a handful of other major companies such as IBM Canada and Mitel, which spent $89 million and $52 million, respectively, along with $44 million spent by the federal government, the vast majority of Canadian firms were simply too small to mount sustained R&D efforts on their own. Furthermore, the overwhelming presence of BCE meant that Canada's information technology R&D efforts were far too concentrated on the needs of the telephone industry.

In March 1987, the federal government announced a national science and technology policy entitled Innov-Action. It stressed the need to improve the productivity and competitiveness by assisting industry in identifying and securing economically exploitable niches in strategic technology areas and improving the transfers and commercial application of new technologies through greater cooperation among government, firms, and universities.

19.3.2 Disagreements within Government

There are considerable tensions and differences within the federal government as well as between different levels of government, although not on the scale of the great ''Telecom Wars'' within the Japanese bureaucracy in the early 1980s (see Janisch 1988b). Some of these have been brought about by structural defects in the machinery for policymaking; others reflect genuine differences as to appropriate policy, especially with respect to greater reliance on competition.

Around the turn of the last century a compromise was reached regarding independence and political accountability in regulation. For a time it worked well enough because most disputes were seen to involve complex, technical rate issues, and the federal cabinet was understandably reluctant to intervene. By the 1980s, however, with the fragmentation of monopoly, it became apparent there were winners and losers in the regulatory game and what were previously seen as technical issues had important distributional effects for which there should be political accountability (Janisch 1979). As a result, parties in the regulatory process have begun to push their claims at a political level, and the cabinet is demanding that it, not the CRTC, make major policy decisions.

This has created something of a policy vacuum: The cabinet asserts it must make policy, but does not have the expertise or—as yet—legal authority to do so, while the CRTC—which has both—becomes progressively more reluctant to act decisively.

The most dramatic instance of a breakdown in understanding of the role of an independent regulatory agency in a parliamentary system happened in 1987 in Saskatchewan. There the Conservatives, in opposition, demanded that Sasktel be brought under independent regulation. When, on their election, this was done, they complained bitterly about the political insensitivity of some of the decisions of the new Public Utilities Review Commission (PURC). The relationship between government and regulator was not made any easier by PURC having its decisions vindicated in the Saskatchewan Court of Appeal, which was highly critical of unauthorized government interference.

Things came to a head when PURC had the audacity to propose that rural telephone rates should be cost justified. The government, entirely reliant on rural support (it did not have a single urban seat) responded by firing all the PURC commissioners (Janisch 1987b). While a somewhat extreme example, much obviously remains to be learned about the give and take required in the relationship between government and an independent regulatory agency.

The other structural issue concerns the federal DOC. While it had clearly been envisaged in 1969, when the DOC was established, that it would be the primary telecommunications policymaker within the federal government, it was widely agreed by the 1980s that DOC had failed to assert itself effectively. Reasons include its being given little actual decisionmaking authority, especially compared with the CRTC; its not having been blessed with strong ministers except in its earliest years; its high-profile responsibilities in broadcasting and the cultural industries have diverted time and energy away from telecommunications; and the absence (until the very end of the 1980s) of any clear national jurisdiction and the resultant nonstop squabbles with the provinces have sapped whatever creative abilities it might have had.

This is not to say there have not been many policy proposals over the years. In Canada there never has been a shortage of proposals; the problem has been with implementation!

The earliest of the studies was the ambitious Telecommission in 1971, which undertook a wide-ranging review of all aspects of telecommunications policy (DOC 1971). This was followed by the Green Paper (Canada 1973a), the Grey Paper (Canada 1975), and three proposed Telecommunications Acts in 1977–1978. None of these bills went beyond first reading. Despite all the study, no provision had been made to deal with competition, even though it was quite apparent by the late 1970s that this was the crucial issue.

There was a report from the Computer Communications Task Force (1972) followed by a government policy statement (Canada 1973b). Nothing concrete came of these proposals. A similar fate awaited the report of the Consultive Committee on the Implications of Telecommunications for Sovereignty (1979). The 1980s saw a large number of studies sponsored by DOC, but little actual policy implementation. (See the list in Janisch et al 1987).

The general position of the DOC has been cautious, especially with respect to the introduction of competition. However, there have been other voices within the government that have spoken more forcefully. For instance, the Economic Council of Canada (ECC), an advisory body, has urged greater reliance on competition (ECC 1981, pp. 37–50, and 1986, pp. 38–45). The director of investigation and research has been a very active intervener before the CRTC and certain of the provincial regulators.

The most important indication of emerging new players in policymaking within the federal government may be found in a long-running battle over the future of Call-Net, a would-be enhanced service competitor. What is intriguing about this saga is the evidence that it provides of the minister of communications being outmaneuvered by a cluster of ministers more favorably inclined to support competitive entry.

On the face of it, Call-Net, which provided a computerized long-distance billing service for small businesses that allows easy allocation of calling costs amongst clients (and is thus particularly popular with smaller law firms), should have been quickly knocked out of business. The CRTC ruled in early 1989 that Call-Net amounted to an unauthorized long-distance service and was not a truly "enhanced service." DOC agreed. The cabinet previously had deferred to the DOC when a matter came before it; however, Call-Net has been kept alive as a result of successful lobbying of the Departments of Regional and Industrial

Expansion, Finance, and Consumer and Corporate Affairs, as well as the Office of Privatization and Regulatory Affairs. (It is not easy to identify precisely where support is coming from because of the secrecy surrounding cabinet appeals, itself a matter currently before the courts.) In any case, the company has remained in business, and the rules governing resale and sharing were liberalized by 1990.

It appears that as concern for telecommunications issues permeates more broadly within the federal government and lobbyists employed by would-be entrants learn to exploit the different interests involved, opportunities will emerge to remove policy logjams. It remains to be seen whether these actors will be as successful as Japan's MPT and MITI in arriving at a workable compromise between stability and change (see Janisch and Kojo 1991).

19.3.3 Impact of Reforms in the United States, Japan, and Great Britain

Canadian identity all too often finds its expression in anti-American sentiment, so it is not surprising to find little overt recognition of the impact of U.S. reforms. Indeed, there has been a consistent tendency to exaggerate the disruptive effects of change to the south. For instance, a minister of communications on one occasion expressed profound condolences to the American people for the confusion caused by divestiture and competition and assured Canadians that they would not be subject to any such catastrophe.

The usual way of discounting the relevance of the American experience is by ridiculing the unalloyed free enterprise, free market ideology said to lie at the heart of the move to competition. Canada is portrayed as a more caring country that has never adopted such a philosophy and is not prepared to jeopardize universal access at affordable rates. (For an excellent review of U.S. developments and their significance for Canada, see Schultz 1989.)

Evidence from the United States that competition can be introduced without any subscriber drop-off, the success of targeted subsidies, and the staying power of MCI and Sprint in the face of prophecies of doom have gone some way to mute the criticism. Still, continuing nationalistic sensitivities make it important to distinguish between the very significant actual impact of U.S. reforms and the inadvisability of openly acknowledging this reality. This is not to suggest Canada will simply eventually "go American"; there are too many important differences for that to happen.

Canadian interest in Japanese reforms only arose in the late 1980s. It is probably fair to say that virtually nothing was known in Canada of the organization of Japanese telecommunications markets until then. Interest perked up with passage of legislation privatizing NTT, opening domestic markets to competition, and reorganizing the structure of Japan's international telecommunications. Even then, however, this curiosity was limited to a few academic investigations and studies (e.g., Janisch and Kurisaki 1985). Almost out of the blue, however, appeared the July 1987, "A Policy Framework for Telecommunications in Canada," issued by the Federal Minister of Communications (DOC 1987). It represented a commitment to introduce new legislation, substantial elements of which appeared to be based on Japanese developments. The minister specifically proposed dividing the industry into Type I (facilities-based) and Type II (nonfacilities-based) carriers, just as Japan had done, with restrictions on ownership, facilities, and the types of activities that could be undertaken in each category.

With DOC's release in January 1988 of its "Proposed Guidelines for Type I Telecommunications Carriers," however, it became clear that little more than the nomenclature had been borrowed. Whereas the Japanese had arrived at this division for the purpose of reorganizing its industry by allowing in substantial competition, Canadian use was much more restrictive.

Competition among the four subclasses of Type I carrier, itself an unnecessary complication avoided in Japan, will be severely limited by specific entry requirements for each subclass. Canadian concerns with universality, a structured (i.e., regulator rather than market-controlled) approach to rate rebalancing, and a general national penchant for caution, have all combined to minimize the practical effect of the Japanese experience for Canada. Indeed, it appears that the primary purpose of the July 1987 policy was to restrict foreign ownership to 20 percent in Type I carriers in advance of the Canada–U.S. FTA (see Janisch 1988c).

Just as Canadian policymakers reject the notion that competition can be introduced before rates are rebalanced, so, too, do they reject British and Japanese approaches to the introduction of competition. Favorable terms for interconnection, as developed by OFTEL for Mercury in Britain, or no specific access charges at all for new entrant competitors, as in Japan, have, at least prior to 1992, been summarily rejected as inapplicable to Canadian circumstances.

The pessimistic Canadian assumption is that competition cannot be introduced at this stage without harm—local Peter will have to pay long-distance Paul. The British, and the particularly optimistic Japanese, appear convinced that rapidly expanding markets mean competition can be introduced without harm. Canadians are quick to point out that competition without rate rebalancing will inevitably be contrived and orchestrated as well as being of questionable economic validity. To this, the Japanese and British reply, some competition is better than none and many of its benefits can be obtained without waiting for "perfect" competition in a fully rate-rebalanced world (see Janisch 1988b).

19.3.4 Positions of Political Parties and Major Interest Groups

There are three major political parties in Canada—Liberals, Conservatives, and New Democrats. There has been a Liberal prime minister most of the time since 1921, with the party last turned out in 1984. Traditionally viewed as the party of the (eastern) middle class, the Liberals can be socially activist, but conservative in economic and foreign policies. Based on several drafts of proposed legislation in the 1970s their telecommunications policies are neither strongly for nor against competition; if anything, they are against deregulation.

The Conservatives, in power since 1984, are more cautious fiscally, less adventurous on social policy, and generally tend to be the party of rural voters, small business interests, the working class, and the well-off. They have tended to favor less regulation and the cautious introduction of competition on telecommunications.

The New Democratic Party is the voice of social democrats, organized labor, and social reformers. It has never held power federally. Its policies are strongly promonopoly and proregulation.

These generalizations, however, have little application provincially. In Saskatchewan, for example, where the carrier is publicly owned, all three parties are uniformly against competition. The 1987 annual TWU convention amply demonstrated this point. There, addressing the delegates and assuring them "we stand united, shoulder to shoulder against competition" was none other than Saskatchewan's minister of communications, from the most conservative of Conservative governments in Canada!

Canadian interest groups, as might be imagined, are spread across the entire spectrum in their views on policy. The Consumers' Association of Canada (CAC), speaking for some 120,000 contributing members, is, of all Canadian interest groups, perhaps the most difficult to fathom. Although it strongly supported introduction of competition in provision of terminal equipment in the past, it came to support the status quo by the late 1980s. It has argued against facilities-based competition, against rate rebalancing, and generally against reductions in the regulatory burden facing monopoly carriers. It does favor regulatory forbearance for nondominant carriers in competitive markets. CAC's views are essentially governed by a single overriding objective: keeping local rates at present levels.

Canadian trade unions such as the TWU and CFCW have virtually mirrored CAC on every issue but one. Labor both opposes competition and strongly rejects deregulation, regardless of the nature of the market or the individual participants in it. Unlike CAC, however, which identifies itself with the interests of local subscribers, the labor unions have offered every reason—except that union jobs and pay scales may be imperiled—for their opposition to competition and deregulation.

Procompetitive forces are generally to be found in the business community, including equipment manufacturers and larger institutional users. These groups have consistently supported such policies as liberalized entry into facilities-based service provision, resale and sharing, enhanced services competition, and unrestricted terminal attachment. Most have also argued for structural separation of Bell Canada from its manufacturing arm and from its affiliates engaged in selling and installing customer premise equipment. In addition, they have called uniformly for adoption of a fully distributed as opposed to incremental costing approach during the CRTC's multistage hearings preceding the development of a formal costing methodology. The only major issue on which competitive service providers have tended to diverge from larger business users relates to the lessening of regulation. Many users stand to benefit from aggressive pricing policies by the major carriers. Competing suppliers in such areas as customer premises equipment, enhanced services, and resale and sharing, on the other hand, may face more vigorous competition if the regulatory burden facing major carriers is lifted.

19.3.5 Exceptions to the Telecommunications Monopoly

The provision of basic switched voice services, at both the local and toll level, remains the preserve of Canada's monopoly common carriers. Nevertheless, impressive inroads into other product and service areas have been made over the past decade.

Competition is most pronounced in the manufacture, sale, installation, servicing and repair of customer premises equipment. While Northern Telecom retains its dominance in the business office switchboard (PBX) market, the key telephone systems market is more wide open. The markets for mobile and cellular radio as well as radio paging, alarm, and similar services are quite competitive. Interconnection of radio common carriers to the networks of federally regulated telcos was authorized by the CRTC in 1984.

The introduction of cellular radio in Canada paralleled the approach taken by regulatory and government authorities in the United States. The governing principle in Canada is that entry will be limited, via federal controls over radio spectrum allocation, to two sets of carriers, a single, federally licensed national carrier and the major wire-line carriers operating within their regional territories (DOC 1984). DOC selected Cantel as Canada's national cellular carrier in December 1983, a little over a year after it first issued calls for applications. Selection was based on the assumption that Cantel would be a wholly Canadian-owned company and its promise to employ only domestically developed and manufactured cellular equipment. Yet, barely a year after the license was awarded, Ameritech Mobile Communications, a subsidiary of a Baby Bell, purchased a 20 percent equity interest. This was sold to Canadian investors in 1987. Also disappointing to the DOC was Cantel's subsequent failure to adopt Canadian technology. Instead, it entered a licensing agreement to use technology from Ericsson.

Overall, the choice of a duopoly market structure for cellular service appears to have been largely vindicated. In jurisdictions where Cantel has been allowed to interconnect with the local switched network, competition between Cantel and the telcos or their affiliated cellular divisions has been vigorous and intense.

Another promising area for increased competition has been resale and sharing of various telco-provided services. Terrestrial and satellite transmission and switching capacity can now be leased for the purpose of providing enhanced services, data services, noninterconnected long-distance voice services, local voice services except public pay telephones and, subject to stringent limitations, interconnected long-distance voice services.

The principal restrictions within federal jurisdictions are concerned with making sure each telco-provided circuit is dedicated to one user, including the "leaky PBX" problem. MTS may be resold or shared to provide MTS only (e.g., hotels reselling MTS to their customers). There are no restrictions on resale and sharing of data services, basic local voice services excluding public pay telephone service, and enhanced services (provided their primary purpose is not provision of MTS, WATS, or pay telephone service).

The primary goal of the CRTC's policies on resale and sharing is clearly to prevent the erosion of revenues from MTS, WATS, and public pay telephone services—the primary sources of the cross subsidy to local service. In practical terms, this means that the provision of data services is the only area in which competition is likely to be meaningful and vigorous in the resale market. The rapid expansion of public and private value added networks, especially in non-voice applications, appears to bear this out.

The only other major sector of the telecom services market in which competition has been introduced and continues to expand is provision of private-line voice and switched public data services. The major national competitors are the telcos, Unitel, and, as of 1986, Telesat Canada. Unitel's participation was limited to the federal jurisdiction when provinces had the power to refuse it the necessary interconnections. Telesat was similarly limited to the federal jurisdiction and Alberta.

Two other areas of potential competition merit some mention. First are the CATV companies, which are considered broadcast media in Canada and are regulated under the Broadcasting Act. Because they own no switching capacity and provide only one way, noninteractive broadcast service, they are not viewed as a major competitive threat to the telcos. This may be a dangerous misconception, however. CATV companies have attained remarkably high penetration, about 60 percent of all households, and, given switching equipment, could begin to provide a wide variety of services.

The threat goes both ways, however. With increased telco reliance on fiber-optic, not to mention the progress in deploying ISDN technology, the day may not be far off when telecom services begin to encroach on the territory of CATV providers. The only legal impediments to this are the restrictions on companies such as Bell Canada from influencing in any way the content of messages they carry and the prohibition on holding broadcasting licenses. Given the progress of technology, however, this prohibition is already being violated to the extent electronic messages are transformed by computer processing and then reconstituted at the intended destination. Some resolution to this potential conflict, including even a decision to allow the telcos and CATV providers to compete head on, will inevitably be called for. Indeed, DOC has commenced a wide-ranging inquiry into convergence (see DOC 1989).

Future competition may come from Cantel, the national cellular radio provider. Once Telesat launches the world's first mobile telecommunications satellite system in partnership with the American Mobile Satellite Consortium, as it plans to do by 1993, it will possible for Cantel to provide coast-to-coast cellular service completely bypassing the terrestrial carriers. If nothing else, this possibility increases pressure on the CRTC to introduce rate rebalancing and allow long-distance competition between Canada's two national terrestrial systems.

A third source of potential competition Canadian Satellite Communications (Cancom), which was licensed in the early 1980s to provide both DBS and satellite-to-cable service to remote northern communities. Since then, it has managed to expand its mandate to include the more populous southern market. By the mid-1980s it had become an aggressive marketer of data communications services.

19.3.6 Policy Positions on Bypass

The CRTC has been a jealous guardian of long-distance monopolies in public-switched voice services within its jurisdiction. Any attempts at bypass by supplying MTS/WATS alternatives have been systematically prohibited. The Call-Net case provides a very useful example.

In another case the CRTC received a complaint from BC Tel that its toll revenues were being threatened by discount U.S. toll services provided by Camnet and Longnet. These companies had been leasing toll circuits from BC Tel, originating in Vancouver and terminating just inside the U.S. border some thirty miles away. They then began offering BC Tel subscribers access to lower priced long-distance service to U.S. destinations. It also announced plans to later provide trans-Canada service via the United States. After a brief hearing to consider the problem, the commission announced that it would authorize BC Tel to restructure its short-haul rates to the United States to make the competitive service offerings provided by Camnet and Longnet unattractive. Little has been heard from Camnet or Longnet since.

19.3.7 The Long View

Even as pressures for change continued to build, many decisions were put off awaiting the outcome of the jurisdiction issue. With federal supremacy established, this backlog can be addressed. The issues include increasing concern for competitiveness in a global economy, the enhanced credibility of Unitel as a competitor, deepening divisions within Telecom Canada and the federal government, the future of Teleglobe Canada, a greater willingness on the part of the federal government to assert itself with respect to a national dimension in policy, and the growing strength of the procompetition business lobby.

The FTA with the United States deals specifically with the further development of a competitive marketplace in enhanced services. It is likely that the indirect impact of the agreement will extend far beyond enhanced services. The most effective arguments for change in the Canadian industry have always been those closely linked to the international competitiveness of Canadian business, especially when compared to its U.S. counterpart. In view of the ever-increasing importance of telecom services to business, arguments in favor of choice and the benefits of competition will no doubt receive a more sympathetic hearing than they did in the past.

If Unitel has not been an aggressive competitor in the past, it is because it is not a genuine new entrant bringing a fresh approach and the sort of entrepreneurial drive and opportunism of Bill McGowan and MCI. Full privatization in 1988, however, removed the restraining hand of government ownership through CN Rail, while major managerial reorganizations and the infusion of capital, coupled with the drive and verve of Ted Rogers, promise to make the new Unitel a far more effective competitor.

On June 12, 1992, the logjam in full facilities based long-distance competition was broken. The CRTC ruled in favor of "open competition." The government gave Unitel permission to provide national long-distance service. It also approved BCRL's (a joint venture of B.C. Rail, Call-Net and Lightel) regional long-distance service and greatly expanded reselling. The ruling was appealled by Bell Canada; however, it appears that the government has irrevocably moved away from protecting the Bell monopoly in favor of a more liberalized environment.

It is evident that telcos no longer share the same values. Bell Canada and BC Tel recognize the inevitability of competition and through rate rebalancing seek to position themselves to greatest advantage when it finally comes. There is no basic tenet of monopoly service that they are not prepared to seriously reconsider. By contrast, the provincially based companies cling with varying degrees of desperation and tenacity to the received truths of monopoly telecommunications. Under these circumstances, although change will be resisted, there will be less of a united front among the incumbent firms.

As the Call-Net saga indicates, the relatively conservative DOC no longer has a monopoly on policy making within the federal government. Significant pockets of new interest and expertise are found in a number of other ministries and bureaus. These agencies are seeking their own sources of information and attracting their own client base independently of the DOC. This will make it possible—in an era of greater politicization of regulation—for determined lobbyists to finesse restrictive regulation.

Critical issues with respect to Teleglobe remain to be dealt with. These include its relationship with Bell Canada, its protected status as Canada's only overseas carrier, opportunities for bypass via the United States, and the appropriate form of regulation.

The federal government may at last be prepared to take a firm public stand to implement the Supreme Court's decision on jurisdiction, as may be seen from these excerpts from a June 1989 statement by the minister of communications (Masse 1989):

> The real issue involves the refusal by certain provinces to open their markets to a level of competition equal to that in Ontario, Quebec, and British Columbia. This refusal has considerably limited the choice of services and equipment available in the Prairies and the Maritimes.
>
> At the same time it has made it difficult to sustain effective competition in networks and services on a national scale. Today anyone wishing to offer services across Canada has to obtain approval from no less than eight regulatory agencies, which is extremely time-consuming and costly.
>
> Given the unified markets of our major trading partners, which are also our

> major competitors, can we afford not to move toward a significant simplification and unification of our own?

The minister's warm endorsement of the recommendations of the 1989 Report on ISDN Implementation in Canada (Lawrence 1989)—which called for national policies and standards in order to achieve the competitive objective envisaged for ISDN—also acts as a further indication of a growing recognition of the need for a stronger federal role. Whether all this will carry over into much-needed new legislation, however, is still unclear.

Masse acknowledged that he was not making original observations, and quoted from a May 1989 public statement of the Information Technology Association of Canada (ITAC) that ". . . the current regulatory regime is a direct threat to the international competitiveness of Canadian companies." As ITAC had emphasized at the time, it is an association representing all facets of the information and telecommunications industry. This is the sort of voice no government can ignore, and it will be joined by new organizations of major users that will supplement and reinforce the role of the Canadian Business Telecommunications Alliance (CBTA). An indication of an increased role for the user community may be seen in the participation of a representative of CBTA in Melbourne at WATTC in 1988, the first time a business user group has been represented on a Canadian delegation at an international telecommunications forum.

The minister also highlighted the belief by the Europeans, Americans, and Japanese that ". . . communications must increasingly develop in an open market." What he did not go on to say, but what will be increasingly important in the Canadian debate, is that it is becoming evident this can be achieved without any threat to universal service. Experience seems to indicate that, while local rates may have to go up, network drop-off can be avoided through targeted rather than undifferentiated subsidies. Also, some of the worst attributes of regulated competition can be avoided if the regulator is determined to wean competitors away from a subsidized form of entry.

The late 1980s were a time of considerable reticence in policy development and, above all, implementation. Yet it is evident that the Canadian telecommunications industry and its regulators have managed to adjust to much change in the past and have displayed considerable ability to adopt unique institutional structures appropriate to the times. For Canada, a history of proud achievement must not be used as a justification for the status quo, but as an inspiration as to how change can be successfully accomplished.

Bibliography

AGT v CRTC. Alberta Government Telephones v Canadian Radio-television and Telecommunications Commission, 1989 2 Supreme Court Reports 225.

Armstrong, Christopher, and H. V. Nelles. 1986 *Monopoly's Moment, The Organiza-*

tion and Regulation of Canadian Utilities, 1830–1930. Philadelphia: Temple University Press.

Babe, Robert. 1990. *Telecommunications in Canada: Technology, Industry and Government*. Toronto: Toronto University Press.

Beigie, Carl. 1973. ''An Economic Framework for Policy Action in Canadian Telecommunications.'' In H. E. English, ed., *Telecommunications for Canada: An Interface of Business and Government*. Toronto: Methuen.

Bell Canada. 1983. ''Submission to the Royal Commission on the Economic Union and Development Prospects for Canada,'' Oct 14.

Brait, R. A. 1981. ''The Constitutional Jurisdiction to Regulate the Provision of Telephone Services in Canada.'' *Ottawa Law Review* 13: 56.

Buchan, R. J. et al. 1982 *Telecommunications Regulation and the Constitution*. Montreal: Institute for Research on Public Policy.

Bureau, Andre. 1988. ''Notes for an Address to the Law Society of Upper Canada's Conference on Communications Law and Policy,'' Toronto, Mar 25.

Canada, Government of. 1973a. ''Proposals for a Communications Policy for Canada'' (Green Paper). Ottawa: Information Canada.

———. 1973b. ''Computer/Communications Policy: A Position Statement by the Government of Canada.''

———. 1975. ''Communications: Some Federal Proposals'' (Grey Paper).

Collins, Robert. 1977 *A Voice From Afar*. Toronto: McGraw-Hill Ryerson.

Communications Canada. 1987. ''Communications for the Twenty-First Century: Media and Messages in the Information Age.'' Ottawa: Supply and Services Canada, 44.

———. 1988. ''Canadian Telecommunications: An Overview of the Canadian Telecommunications Carriage Industry.''

Computer Communications Task Force. 1972. ''Branching Out.'' Ottawa: Information Canada.

Consultative Committee on the Implications of Telecommunications Sovereignty. 1979. ''Telecommunications and Canada.'' Ottawa: Supply and Services Canada.

Canadian Radio-television and Telecommunications Commission (CRTC). 1976 ''Telecommunications Regulation—Procedures and Practices.'' Ottawa: Supply and Services Canada (J4.20).

Dalfen, Charles M., and Laurence J. E. Dunbar. 1986. ''Transport and Communications.'' In Mark Krasnick, ed., *Case Studies in the Division of Powers*, pp. 139–202. Toronto: University of Toronto Press.

Department of Communications. 1971. ''Instant World: A Report on Telecommunications in Canada.'' Ottawa: Information Canada. (There were some twenty-five published studies that also accompanied the Telecommission Report.)

———. 1975. ''Review of the Procurement Practices and Policies and the Intercorporate Financial Relationships of the British Columbia Telephone Company'' (Jul.).

———. 1984. ''Policy with Respect to Competing Cellular Radio Service'' (Mar. 14).

———. 1987. ''A Policy Framework for Telecommunications in Canada'' (Jul.)

———. 1989. Notice DGTP-09-89 ''Local Disribution Telecommunications Networks.'' *Canada Gazette Pt I,* Sep 2.

Economic Council of Canada. 1981. ''Reforming Regulation.'' Ottawa: Supply and Services Canada.

———. 1986. ''Minding the Public's Business.''

Fetherstonhaugh, R. C. 1944 *Charles Fleetford Sise, 1834–1918*. Montreal: Gazette Printing.

Federal/Provincial Examination of Telecommunications Pricing and the Universal Availability of Affordable Telephone Service. 1986a. ''Working Papers.'' Ottawa: Supply and Services Canada.

———. 1986b. ''Report.''

Garnet, Robert. 1985. *The Telephone Enterprise: The Evolution of the Bell System's Horizontal Structure, 1876–1909*. Baltimore: Johns Hopkins University Press.

Grant, Peter S. 1988. *Canadian Communications Law and Policy*. Toronto: The Law Society of Upper Canada, Department of Education.

Harvey, J. A. 1983. ''The Emerging Agenda: Response by the Carriers.'' In *Policy Issues in the Canadian-American Information Sector*. Montreal: McGill University, Centre for the Study of Regulated Industries.

Hoey, Eamon. 1989. ''Long Distance Calls Twice As Expensive As In The US.'' *Telemanagement,* vol. 64 (Apr 4).

Janisch, Hudson N. 1979. ''Policy Making in Regulation: Towards A New Definition of the Status of Independent Regulatory Agencies in Canada,'' *Osgoode Hall Law Journal* 17: 46.

———. 1984. ''Telecommunications Ownership and Regulation in Canada: Compatability or Confusion?'' *Canadian Regulatory Reporter* 5: 5–27.

———. 1986. ''Federal-Provincial Relations in Canadian Telecommunications.'' Fourteenth Annual Telecommunications Policy Research Conference, Airlie House, Virginia, Apr. 27–30.

———. 1987a. ''Telecommunications and the Canada–U.S. Free Trade Agreement.'' *Telecommunications Policy,* vol, 13, no. 2, p. 99 (Jun).

———. 1987b. ''Independence of Administrative Tribunals: In Praise of 'Structural Heretics'.'' *Canadian Journal of Administrative Law & Practice,* vol. 1; p. 1.

———. 1988a. ''Emerging Issues in Foreign Investment in Telecommunications.'' International Business and Trade Law Programme, *Working Paper Series,* 1988–89–(1).

———. 1988b. ''Japanese Telecommunications Developments.'' Centre for the Study of Regulated Industries, McGill University, *Working Paper Series.*

———. 1988c. ''Comments on Type I and II Telecommunications Carriers: New Policy Developments.'' A paper presented at the Law Society of Upper Canada Conference on Canadian Communications Law & Policy, Toronto, Mar 26.

———. 1989. ''Canadian Telecommunications in a Free Trade Era.'' *Columbia Journal of World Business* 17: 5 (Spring).

——— and Y. Kurisaki. 1985. ''Reform of Telecommunications in Japan and Canada.'' *Telecommunications Policy* 9: 31.

——— and B. S. Romaniuk. 1985. ''The Quest for Regulatory Forbearance in Telecommunications.'' *Ottawa Law Review* 17: 455–89.

———, S. G. Rawson, and W. T. Stanbury. 1987. *Canadian Telecommunications Regulation Bibliography*. Ottawa: Canadian Law Information Council.

——— and Makato Kojo. 1991. ''Japanese Telecommunications after the 1985 Regulatory Reforms.'' *Media of Communications Law Review* 1: 307.

——— and Richard Schultz. 1991. ''Federalism's Turn: Telecommunications and Canadian Global Competitiveness.'' *Canadian Business Law Journal* 18: 1.

Koelsch, Frank. 1989. ''Regulatory Climate Hobbles Innovation.'' [Toronto] *The Globe and Mail,* May 29, p B1.

Lawrence, John. 1989. ''Report on ISDN Implementation in Canada.'' Ottawa: Communications Canada (Mar.).

Masse, Marcel (Minister of Communications). 1985. ''Looking at Telecommunications—The Need for Review.'' Notes for an Address to the Electrical and Electronic Manufacturers' Association, Montebello, Quebec, Jun 20.

———. 1989. Speech to the Canadian Satellite Users Conference, Ottawa, June 19.

McManus, J. C. 1973. ''Federal Regulation of Telecommunications in Canada.'' In H. E. English, ed., *Telecommunications for Canada: An Interface of Business and Government,* pp. 419–23. Toronto: Methuen.

Ogle, E. B. 1979. *Long Distance Please, The Story of the TransCanada Telephone System.* Toronto: Collins.

''Rate Rebalancing Decision'' 1988. *Bell Canada—1988 Revenue Requirement, Rate Rebalancing and Revenue Settlement Issues.* Telecom Decision CRTC 88-4, Mar. 17.

Restrictive Trade Practices Commission. 1983. ''Telecommunications in Canada, Part III, The Impact of Vertical Integration on the Equipment Industry.'' Ottawa: Supply and Services Canada.

Romaniuk, B. S. and H. N. Janisch. 1986. ''Competition in Telecommunications: Who Polices the Transition?'' *Ottawa Law Review. 18: 561–661.*

Schultz, Richard J. 1989. ''United States Telecommunications Pricing Changes and Social Welfare: Causes, Consequences and Policy Alternatives.'' A study prepared for the Regulatory Affairs Branch, Bureau of Competition Policy, Department of Consumer and Corporate Affairs, Government of Canada (Feb.).

Schultz, R., and A. Alexandroff. 1985 Economic Regulation and the Federal System. Toronto: University of Toronto Press.

Taylor, Graham. 1982. ''Charles F Sise, Bell Canada and the Americans: A Study of Managerial Autonomy, 1880–1905.'' *Canadian Historical Association Papers* 23–24.

Waverman, Leonard. 1989. ''R&D and Preferred Supplier Relationships: The Growth of Northern Telecom.'' Paper presented at the ITS Regional Conference, Ottawa, Jun. 19.

20

New Zealand: The Unique Experiment in Deregulation

PATRICK G. MCCABE

Dramatic changes occurred in New Zealand telecommunications policy starting in the mid-1980s. A traditional government department with a virtual monopoly in the provision of post, telecommunication, certain banking, and ancillary services was separated into three distinct corporations, while its regulatory and policy advice functions were transferred to another government department. Almost all regulatory barriers to market entry were then removed, and the government-owned Telecom Corporation was privatized. In 1991 competition in the provision of domestic and international long-distance services emerged, and plans for competition for local calls and mobile telephony were revealed. At the beginning of the 1990s New Zealand had one of the most liberalized and modern telecommunications markets in the world.

The changes in policy are the most dramatic elements of a complete revision of New Zealand's communications legislation, which in turn reflects a larger, but comprehensive shift in direction of the government's role in New Zealand's economic development. This chapter reviews the development of telecommunications in New Zealand. It considers the corporatization and later privatization of Telecom, identifies the other network operators, analyzes the regulatory environment, reports the reforms in related markets, and concludes with an assessment of the changes. Dordick provides another view of the changes in New Zealand in Chapter 21, concentrating primarily on their context and implications.

20.1 Background

New Zealand has a population of just 3.4 million people but covers a land mass of 269,000 km^2, with literally hundreds of islands. More than 99 percent of the population lives in the two largest: North Island and South Island, which stretch across mainly hilly and mountainous terrain for a combined length of

1,600 km. One-third of the population is in the Auckland metropolitan area. While New Zealand is a highly urbanized society, the rural community has a key economic role because much of its production eventually earns a large part of total export receipts.

Lieutenant Governor William Hobson opened the first post office for the receipt and distribution of letters just a month after New Zealand became a British Crown Colony in February 1840. The government assumed a major role in the provision of communications services for the next 150 years.

Terrain and population density were such that New Zealand was late in accepting the telegraph, but by the end of the 1850s settlement had increased to the point where efficient communication was necessary, and the first experiments in telegraphy took place. A merchant firm built a line between Dunedin and Port Chalmers for its own use, but opened it to the public in 1861. In 1862 the Canterbury provincial government established service between Christchurch and its port at Lyttelton, after which other provincial governments established their own systems. These systems did not provide integrated services for the rapidly growing colony. The Electric Telegraph Act of 1865 consequently brought all telegraph services under the control of the central government and gave the newly formed Telegraph Department the sole right to construct, establish, maintain, and regulate telegraphic communications. Thus, the pattern for a government-owned communications structure throughout the six provinces was set.

Provincial authorities were unwilling to relinquish control to the weak central government, but the latter pushed ahead. A main trunk line was built from Nelson to Bluff—the length of South Island which, until the twentieth century, was the most heavily populated island. By 1866 more than 1,000 km of lines had been established and a cable was laid under Cook Strait to connect South and North Island. This was followed by a North Island main trunk from Wellington to Auckland. It took six years to complete in part because of the terrain. Most towns had the telegraph by the early 1870s. In 1876 a cable from Wellington to Australia was completed, providing a link to the world, as Australia was already linked to Great Britain.

20.1.1 Julian Vogel

Dramatic leadership was provided by Julian Vogel, whom Sinclair has called "the first politician in New Zealand whose talents were at all remarkable" (1984, p. 152). Vogel was a Londoner who had abandoned his grandfather's merchant trading business to make his mark in Australia's Victorian gold fields. He eventually found his way in 1861 to Otago on South Island where he established New Zealand's first daily newspaper and was drawn into provincial politics. It was not long before he was also drawn into national politics, leading the Parliamentary opposition during 1865–1869. In 1870, as treasurer in the new government of William Fox, he propounded "that grand go-ahead" policy with which his name has been associated. (Biographies of Vogel are Dalziel

1986 and Burdon 1948. The standard general history of New Zealand is Sinclair 1984; also useful is *The New Zealand Official Yearbook.*)

Vogel understood that further growth in the colony required much improved transport and communications. Roads were few and poor, the 80 km of railway had three different gauges, and there were only 1,100 km of telegraph. He proposed to borrow £10 million over ten years to finance a rapid extension of transport and communications and to encourage immigration to provide the needed labor. Security was 6 million acres (over 24,000 km^2, equal to about 9 percent of the islands' land area) that lay along the new railway lines and roads.

Vogel believed a strong central government was necessary for the development of the colony. He was opposed by the "provincialists" who wanted to preserve their local authority and power. In 1876, after several years of bickering with his opponents, Vogel, prime minister at the time, abolished the provincial governments. This action accelerated growth of telegraphy.

Without his "grand go-ahead" policy and victory over his provincialist opposition, Vogel could not have created the strong central government needed to provide the environment necessary for a centrally administered national telecommunications system in New Zealand. Vogel, who also served as postmaster general in 1884–1887, thus had a seminal impact on development of telecommunications in New Zealand.

20.1.2 The Telephone

The first official New Zealand trial of a telephone link took place in Blenheim in 1877, using part of the Blenheim–Christchurch telegraph line. New Zealand seemed very aware of the possibilities the telephone offered, and the government encouraged its introduction. The telephone was initially used primarily to extend the telegraph service to communities too small to justify employment of a trained telegrapher. Messages could be called into or from the nearest telegraph office over a direct line from a telephone bureau in the local community. The first telephone exchange was opened in Christchurch in 1881 and had about thirty subscribers. Exchanges were soon established in Auckland, Wellington, and Dunedin. Initially, however, only local calls were possible. Long-distance services began in 1897.

A clause inserted at a late stage during the passage of the Electric Telegraph Amendment Act 1880 provided the Telegraph Department with exclusive rights to the new telephone service. To minimize administration costs, the Act also brought together the Post Office, the Telegraph Department, and the Post Office Savings Bank (which had been started in 1867). Until the passage of the Post Office Act of 1959 the combined departments were known as the Post and Telegraph Department. The new department provided postal, banking, and telegraph services, as well as, by 1900, at least thirty agency tasks for other government departments, including the collection of fees under the Homing Pigeons Protection Act.

When radio communication was introduced in the early 1900s, it, too, was taken under the wing of the Post and Telegraph Department. The first coin-in-the-slot telephone was introduced in 1910 and automatic exchanges were introduced in 1913. The Postal Telegraph Amendment Act 1924 made provision for a radio-broadcasting service, then beginning to develop rapidly.

In 1926 a Cook Strait cable specifically for telephone use was laid. Until then telephone services were largely restricted to the island in which the caller was located. New Zealand's first overseas phone call—to Australia—was made in 1930 using a short-wave radio link; the United States was called using the same means the following year.

Subscribers faced a flat-rate tariff (i.e., a fixed charge was made for access), and no usage charges were made for calls within a local area. Until 1923 the base rate area was limited to one mile from an exchange. In subsequent years the free calling area was gradually extended. Pressure for extension was, in part, prompted by high toll tariffs for calls—especially between neighboring suburbs in metropolitan areas. (Wagner 1984 cites a case study of the integration of the Auckland metropolitan region into one free calling area, which took nearly twenty years to accomplish during the 1960s and 1970s.)

A satellite earth station at Warkworth, north of Auckland, opened in 1971, which enabled New Zealand to join Intelsat. International cable links were also regularly extended, especially to Australia and the Pacific Island nations, and, from 1963, to Canada. A trans-Tasman fiberoptic cable was commissioned in December 1991, and a fiberoptic link to Hawaii is due to be commissioned in 1993.

20.1.3 Legal Framework Before 1987

The legislative framework was periodically revised and regularly amended. The consecutive principal acts were the Post Office and Telegraph Act 1908, the Post and Telegraph Act 1928, and the Post Office Act 1959. The 1959 act provided the legislative basis for the monopoly in public switched telecom services enjoyed by the Post Office until mid-1987. It empowered the postmaster-general, as the cabinet minister responsible for the Post Office, to establish and operate telephone and telegraph services. Section 4 (1) enabled the postmaster-general to delegate any or all of these powers to the director-general as the administrative head of the Post Office.

The erection, construction, establishment, and maintenance of privately owned lines was generally prohibited, unless they belonged to local authorities or the lines were situated entirely on land owned or occupied by one person. Provisions were made, however, for licensing private lines under the telephone regulations. These regulations gave the postmaster-general the power to revoke a license if the line was used "for any purpose that might be construed as an attempt to deprive the Crown of revenue." The regulations also generally prohibited interconnection of private and public lines in more than one place, eliminating the opportunity for bypass of the long-distance toll network. A 1962 amendment defined telegraph to include "any communications transmitted to a distance by any apparatus other than a telephone."

The Post Office, however, was not the sole provider of physical networks.

Other government organizations—such as the railways, forest service, and electricity departments—operated networks, but these were strictly for their own use. Establishment during the 1960s and early 1970s of a microwave relay network for television programming by the government-owned Broadcasting Corporation faced opposition from the Post Office, which considered the activity part of its common-carrier functions. In the end, Broadcasting Corporation got its own network for provision of public broadcasting services, with any telecom services permitted only if they are incidental to the broadcasting function.

New Zealand has a unicameral parliament modeled largely on Great Britain's House of Commons. (A second chamber was abolished in 1950.) Each parliamentary term is of three years' maximum duration. Government by political party has existed since the 1880s. Since 1935 the government has been singularly formed by either the Labour Party or the National Party. Over time, both main parties have become broad coalitions with a strong degree of pragmatism, and without the characteristics of strong ideology and formal factionalism which feature elsewhere. Until 1990 cabinet ministers and ancillary officeholders in the governing party formed a majority of the government caucus. This feature, together with very strong party discipline, meant that the executive was seldom challenged in Parliament.

Government policies were traditionally directed to encouraging the primary productive sector, especially agriculture, protecting the domestic manufacturing sector, financing the development of infrastructure, and providing social assistance. Agriculture was the mainstay of New Zealand's export-dependent economy, and it still accounts for more than half of export receipts.

Great Britain's entry into the European Economic Community in 1973 forced the search for new markets, which, although initially successful, became fragile in the early 1980s. By 1983 the deteriorating economic conditions had resulted in government policies controlling wages, prices, and interest rates, as well as a burgeoning government budgetary deficit equal to 9 percent of gross domestic product, and providing substantial subsidies to export meat producers and large-scale funding of infrastructural development, especially in the energy sector.

The change to a Labour government following the July 1984 general election marks the commencement of a program of comprehensive economic, social and political reform in New Zealand. The average age of the new Cabinet was nearly twenty years younger than its predecessor. Reflecting greater educational opportunities, for the first time in New Zealand history it was predominantly university educated, including several with graduate degrees. An inner core of five ministers led by Prime Minister David Lange and Finance Minister Roger Douglas, dominated the government and dictated the rapid pace of change. (The other three were Deputy Prime Minister Geoffrey Palmer and two associate finance ministers, David Caygill and Richard Prebble.)

The economic reforms included widespread financial deregulation, the cessation of subsidies to farmers and exporters, comprehensive tax reform, including introduction of a value added tax, abolition of import licensing, a major program of tariff reduction, commercialization of government trading (commercial) activities, plus, in the second term from 1987 to 1990, reform of government administration and a major privatization program. (Publications reviewing this period include James 1986, Douglas and Callen 1987, Bollard and Buckle 1987, and Boston et al. 1991. The origins and nature of economic policy changes are addressed in Bollard and Duncan 1992, Williams 1990, and Easton 1989.)

20.2 Corporatization of the Post Office

Prior to the 1984 general election, Labour had no plans for major changes to the organization of the New Zealand Post Office—although a 1982 party internal memorandum prepared by Mr. Douglas declared that "the need to examine 'state commercial enterprises' to see whether they should be pursuing the 'same profit objectives as the private sector'." Two months earlier, the party's leader, Mr. Lange, assured the Post Office Union that Labour's policy was to maintain the enterprise as a single entity under public ownership, and to ensure it had adequate resources to ensure adoption of new technology. Three weeks after the election, in an address to the Post Office Union Annual Conference, Mr. Lange identified the Post Office as one of the most efficient public sector organizations. (See Roth 1990, pp. 260–63.)

After the 1984 general election, the newly elected government, as is standard practice, received a set of briefing papers from the Treasury on the state of the economy. The documents were more coherent and extensive than any before. They revealed that for the ten years until 1983, compared with the average situation for the OECD countries, the annual economic growth rate had been lower, inflation had been higher, the unemployment rate had risen markedly (albeit from exceptionally low levels), and the government annual budget deficit before borrowing had increased, with a consequent rise in government debt both domestically and internationally, and a sharp rise in debt servicing costs. Governments had used inappropriate interventions, relying on specific controls rather than general policy instruments, and thus had failed to address the underlying causes of New Zealand's economic malaise. These poorly targeted interventions included "unwarranted state monopolies in the communications sector" and the "underpricing of state-supplied goods and services".

In its advice, the Treasury drew attention to the activities of the state-owned commercial enterprises (called trading enterprises in the report) which then accounted for over 12 percent of gross domestic product and 20 percent of gross investment. Factors adversely affecting the performance of these enterprises included (p. 270):

1. Their lack of clear, nonconflicting objectives
2. Their operating environment—that is, the special assistance they received and restraints on competition with them
3. The incentives arising from existing arrangements for monitoring performance.

The solutions for improved efficiency lay in separating policy and regulatory functions from trading operations, providing subsidies for social objectives, and greater use of standard commercial methods of accountability and management procedures.

The Treasury's advice did not result in immediate structural changes, although the November 1984 government budget required enterprises, including the New Zealand Post Office, to pay dividends and tax. This largely formalized the previous government's policy of requiring payment in lieu of taxes.

During 1985 a number of common problems emerged as the government focused on performance of major state trading enterprises. In May 1985 the finance minister, attempted to get cabinet agreement to a "comprehensive approach" to all state trading enterprises, but the issue was deferred. Late in 1985 a dismal forecast of a rising budget deficit for the 1986–1987 financial year prompted a major expenditure review. During the review Associate Finance Minister Richard Prebble said "it became apparent that the Post Office was not being customer-driven". On September 19, 1985, the government commissioned two reviews of the Post Office. One focused on the telecommunications, postal, and agency services, while the second examined the banking activities.

On December 12, 1985, in a government economic statement to parliament, the finance minister announced five principles for government state trading enterprises. Managers would be required to focus on commercial objectives and would report to establishment boards whose directors comprised mainly business sector leaders.

20.2.1 The Post Office Review (Mason–Morris) Report

The Post Office Review report was presented to the government in February 1986 and published two months later (Mason and Morris 1986). The report recommended the Post Office be reorganized into three discrete businesses with separate support functions and chief executives so that each business—telecommunications, postal and agency, and banking—was a completely independent unit. Each chief executive would to report to a group chief executive—the director-general of the Post Office. The director-general would be assisted by a deputy, a secretariat, and a policy-regulatory advice group in a small head office. The idea was that if authority and responsibility regarding day-to-day operations was decentralized within each enterprise, they would be able to respond more quickly to market conditions. The split-up was to take place by April 1, 1987.

The report also recommended that basic network services remain a government monopoly controlled by Telecom, that customer premises equipment be deregulated approximately twelve months after the decision to reorganize the Post Office, and that, a further six to twelve months later, the areas of enhanced services be deregulated and opened to market competition. The report also recommended that the Post Office have the flexibility to determine its own pricing policies, that prices be related to current costs, financial objectives, and market conditions, that cross subsidies be progressively minimized, except in relation to urban and rural users, and that the pricing of monopoly services be subject to price surveillance. Further, the Post Office was to carefully examine the effects of its equipment leasing and procurement policies on the locally manufactured equipment industry, especially as import tariff protection was to be lifted concurrently with deregulation. Telecom was instructed to investigate the availability of alternative capital financing, keeping in mind it would be selling rather than leasing CPE in a competitive marketplace.

On May 19, 1986, the finance minister announced "a major restructuring of the Post Office . . . [under which it] will be separated into three government-owned independent business groups." As part of this, the postmaster-general announced establishment of the Post Office Steering Committee. In July 1986 the review of the Post Office Savings Bank was released, and Establishment Boards for the three proposed new companies were created the following month to replace the Post Office Steering Committee (which was formally disbanded in September).

In the latter part of 1986 it was apparent that corporatization could mean significant reduction in staff numbers. Large reductions had already been signaled for the government's coal mining and forestry corporations. In November 1986 the Post Office Union received assurances from the three chairs that all staff would be placed within the restructured organizations. Because the finance sector had already been deregulated and was very competitive, it was agreed approximately 1,300 staff would be transferred from PostBank to Telecom and NZ Post.

The State-Owned Enterprises Act 1986 provided the statutory framework for the reform of major government trading enterprises. Under the Act, Telecom's conversion to a commercial enterprise requires it to be run as such, which means it is to be as profitable and efficient as comparable businesses not owned by the Crown. Another provision states that when the government wishes Telecom to provide a service Telecom would not otherwise provide, the government will pay part or all of the cost for Telecom to provide the service in question, with the amount negotiated by the two parties. The objective is to make subsidies explicit. To ensure availability of service in rural areas (which house about 15 percent of the population) and for low income households, the use of subsidies is anticipated. (The SOE Act is discussed further by Dordick in Chapter 21.)

20.2.2 Corporatization of Telecom

Telecom Corporation of New Zealand Ltd. (Telecom) commenced business April 1, 1987. As the valuation of assets had not been completed, it operated the commercial telecom services formerly provided by the New Zealand Post Office under a special license from the postmaster-general. The regulatory and policy advice functions, principally the Radio Frequency Service responsible for managing the radio frequency spectrum, were transferred to the Department of Trade and Industry (which on December 1, 1988, became the Ministry of Commerce) to form the basis of a new communications division.

Telecom commenced service with 26,500 staff (24,472 full-time equivalents), and the first managing director was the former director-general of the Post Office. Later in 1987, an international search for a deputy managing director located Dr. Peter Troughton, formerly of British Telecom, but no appointment was made. The managing director retired instead, and Dr. Troughton was recruited as the new managing director. He took up his appointment in March 1988. Meanwhile, in December 1987, Telecom's chair, Mr. Roy Mason resigned, and was replaced by Sir Ronald Trotter, who was also Chair of New Zealand's largest corporation, Fletcher Challenge Ltd. Employment at Telecom had been reduced to under 15,000 by 1992.

The Post Office monopoly in the provision of public switched telecommunications network services was transferred to Telecom from its first day of operations. Legislative action to replace the relevant parts of the Post Office Act 1959 commenced shortly thereafter. While Telecom was concerned with the establishment of its operation, officials from the Department of Trade and Industry and the Treasury commenced a major review of the regulatory regime for telecommunications.

20.2.3 How Much Competition?

In July 1987 the management consultant arm of Touche Ross was engaged by the government to report on whether it was in the interests of economic efficiency to introduce greater competition into the network services market, the likely economic and social impact of introducing greater competition, and the possible phasing in of that competition. As part of determining these, Touche Ross examined Telecom's operations.

The Touche Ross report was subsequently published by the government and provides a good summary of the state of the business at the commencement of Telecom's operations. The results were not flattering. A survey of major users found that Telecom was "poor at providing services, communicating with customers and understanding their requirements. It was engineering rather than market driven." It went on to say:

> Telecom is not achieving a level of efficiency comparable to the best practice of overseas telephone companies. Its management systems are outdated and grossly inadequate, making efficient management very difficult. Automation of clerical functions has lagged behind investment in the network. The [efficiency of] utilization of engineering staff on necessary investment and maintenance is low by world standards (p. 99).

The report identified the existence of large cross subsidies between access and toll (long-distance) charges, and between access charges for subscribers in different circumstances. Possible price reductions in real terms of 50 percent for domestic toll calls on most routes, and of 25 percent or more for international calls, were identified. At the same time the report recognized that local access charges—there were no local usage charges—could possibly double.

Subject to interconnect arrangements, the report concluded that competition in network services was possible and sustainable; the resultant losses of economies of scale and scope would be small, and would be outweighed by dynamic gains arising from the greater pressure on Telecom to be efficient, to offer better service, and to be more innovative. Overall, the national economy could only gain.

The government welcomed the report's conclusions and on December 17, 1987, announced competition in the provision of network services would be permitted beginning in early 1989.

20.2.4 An Independent Telecom

On March 31, 1988, an agreement was signed transferring ownership of the former Post Office's telecommunications assets to Telecom. This agreement valued the business at $NZ3,200 million, with an issued and paid-up capital 2,350 million fully paid ordinary shares of $NZ1 each, and a loan agreement for $NZ850 million which Telecom was to repay in six equal installments over a three-year period.

On June 8, 1988, Dr. Troughton advised the shareholding ministers that the corporation was to be restructured and arrangements made to enable competitors to enter the market once the corporation's monopoly was removed. A number of subsidiary companies were established. The organizational structure has been modified subsequently and in 1992 comprised four regional operating companies (ROCs) providing customer services: Telecom New Zealand International Ltd. providing international services; Telecom Networks and Operations Ltd. providing network and toll services to the ROCs; and a group of new venture and joint venture subsidiaries providing such value added services as cellular telephony, mobile radio, and directory services. A small corporate office also provides policy direction.

The 1988 restructuring was designed to ensure a more commercial focus within the corporation and to prepare it for competition in 1989. At the same time Telecom confirmed its willingness to consult widely with the industry about technical and operational arrangements for interconnection. Arrangements would be fair, and charges for interconnection were to be based on costs.

The pace of change accelerated following Dr. Troughton's appointment as managing director. A major capital investment program begun in 1987 had resulted in the expenditure of $NZ2,500 million (U.S.$1,460 million) through 1991, largely for network modernization. As a result, half the exchanges were less than three years old and *more than* 90 percent of access lines were hooked to digital switches. More than $NZ100 million was expended on a new computerized billing system, and the billing cycle was reduced from two months to one month, thereby improving cash flow. Another marked improvement in service was the reduction of the waiting period for a new telephone connection from six to eight weeks, to just forty-eight hours.

In the first five years, over half of the original staff had departed the company, while many others have seen dramatic changes to their duties. Staff reductions were achieved in part by engaging subcontractors, including former employees. People with new skills—especially in finance, accounting, sales, and marketing—have been recruited. A small number of senior executives have come in from overseas—a trend that has continued since privatization. By late 1991 fewer than ten of about seventy senior management personnel remained from the old Post Office days. Remuneration rates for a significant number of staff are at least in part determined by performance bonuses.

The changes were mirrored in the Post Office Union. In the past, even senior management were voluntary members. In 1987 Telecom immediately identified about 2 percent of all positions that it negotiated as being exempt from union coverage. By 1991 the number of exempt positions had reached more than 10 percent of all positions. The much shrunken Post Office Union, which had also witnessed a 35-percent reduction in the number of postal workers, merged with the Engineers Union to form the Communications and Engineering Workers Union on April 1, 1992.

In November 1988 Telecom commenced a gradual program of tariff rebalancing with reductions in charges for domestic and international long-distance services and leased lines, while residential tariffs have increased. Business customers received a reduction in monthly access charges but faced introduction of timed usage charges for local calls. As with many of the changes, this was first done in one regional operating company, then refined before its gradual extension in the remaining ROCs. The tariff rebalancing program and consequent marketing promotions, combined with the introduction of new services (such as a cellular telephone service) has changed the relative proportions of the sources of Telecom revenue. In the financial year ended March 31, 1991, local services earned 34 percent of total revenue, domestic toll calls 22 percent, international calls 20 percent, and other services provided the remaining 26 percent. The strongest revenue growth was in international calls and other services.

The November 1988 reductions to domestic toll tariffs were the first changes since a set of price increases in 1985. This change, together with changes in April and November 1989, reduced the number of tariff steps, increased the number and extent of discounts according to the time of day, reduced the minimum charge call time from three minutes to one minute (additional time remained rounded up to the next full minute), and shifted the basis for charges away from the actual distance called toward the density of traffic volumes on each route. Telecom subsequently introduced occasional weekend special rates that represented a further reduction in the higher step tariff rates.

20.3 Privatization

In 1986–1987, the Labour government vigorously denied suggestions that corporatization was just a halfway house to privatization. Corporatization was designed to achieve considerable gains in the efficiency, financial performance, and accountability of these state trading activities. This view was readily reaffirmed in the Party's manifesto for the 1987 general election, which stated that

Labour would ensure that all three post office corporations remain in public ownership.

During this period there was a growing feeling that a privately owned company, subject to stock market scrutiny, was likely to be more efficient than a state-owned enterprise (see, e.g., Jennings and Cameron 1987). While accepted by key economic ministers, such a view was unlikely to command a majority in the government caucus. An alternative rationale was provided by the government's own debt situation. Despite early success in reducing the deficit—initially through slashing subsidies to farmers and exporters, as well as comprehensive tax and expenditure reform, significant problems remained. The main drivers of government expenditure were social services (health, education, housing, social welfare benefits, and pensions) and Government debt servicing. Since social services were inviolate to the Labour Party, asset sales targeted to reduce debt were advocated.

In December 1987 the minister for state-owned enterprises, Richard Prebble, confirmed a major assets sales program with "an objective of retiring one third of the public debt by 1992" because "the current level represents a serious impediment to economic growth" and accounted for 20 percent of government expenditure. The desirability of any sale would be assessed on a case-by-case basis, and only made when competition had been permitted.

In 1987–1988 the privatization program focused on enterprises operating in very competitive markets, but it was apparent that if the target of $NZ14 billion in debt reduction was to be achieved, some major assets, such as Telecom, would need to be sold. For telecommunications, however, the focus throughout 1988 was on implementing reforms following the Touche Ross report.

In 1989 National Economic Research Associates (NERA) was commissioned to evaluate the economic issues relating to the privatization of Telecom. The report, completed that June, assessed a number of options in an endeavor to promote competition. These options included removal of international, intercity, or radio-based (cellular, mobile radio) services from the corporation, although no restrictions on Telecom's subsequent re-entry to these markets would be made. NERA concluded that such actions would reduce the benefits of economies of scale and scope, and would also mean a delay that would affect possible timing of Telecom's privatization. NERA also recommended Telecom be obliged to disclose prices for certain services.

In October the government announced it would take action to establish regulations requiring information disclosure by Telecom. In March 1990 Parliament enacted the legislation necessary to permit a sale (essentially removing Telecom from coverage of the State-Owned Enterprises Act and associated legislation) and to allow regulation-making powers for information disclosure. The Telecommunications (Disclosure) Regulations 1990 were promulgated in May and took effect on July 1, 1990.

Sale of Telecom by tender was announced on February 23, 1990. The call for expressions of interest yielded nineteen responses. Indicative bids were invited and the list of interested parties was then reduced to five for the due diligence phase.

The government offered up to 100 percent of the corporation for sale, but was willing to accept tenders for a more limited shareholding. A key aspect was the condition that a maximum 49.9 percent "strategic" stake could be taken by any overseas party, although the government was willing to entertain the possibility that the strategic stakeholder could purchase 100 percent and then reduce that to 49.9 percent over a three-year period. The government also required that at least $500 million worth of shares be made available by public offering on the New Zealand market. The government also announced it would retain a single share (called a Kiwi, or golden share) with special voting rights to control the maximum shareholding of any single foreign party and transfers of blocks of shares among parties, as well as to ensure Telecom Corporation's compliance with its residential services pledges. (The kiwi bird is New Zealand's symbol.)

At the same time as the due diligence phase, work was undertaken to prepare for the subsequent float of any shareholding not required by the strategic shareholder, and to value that remaining parcel.

Although the identities of all the parties and the compositions of other bidding consortiums were never formally disclosed, except for the successful bidders, others known to be actively involved included Cable & Wireless, SouthWestern Bell, GTE, OTC (the Australian international carrier), and Brierley Investments Ltd. (a large New Zealand company). Another New Zealand firm, Fletcher Challenge, attempted to provide a late bid emphasizing local ownership, but this was rejected.

Public reaction to the sale was very mixed, to say the least. During the due diligence process Wellington's morning paper editorialized that reference to the Government's debt problem as a justification for the sale, was a "red herring". After noting the considerable public investment built up over 130 years, and the recent improvements in operating efficiency, the editorial concluded: "Telecom's best days are yet to be. The benefits ought to be shared by the whole community. If it were not for Mr. Prebble and his myopic colleagues, they would be" (*Dominion,* June 1, 1990).

The role of the main Opposition Party in the New Zealand Parliament is to provide a responsible outlet for criticism of the government. The National Party, reflecting the public opposition to the sale, stated that if elected to be the government, it would limit overseas ownership in Telecom to 24.9 percent and would require any successful overseas owner to reduce its shareholding to that level. It also considered that the service commitments were insufficiently protected, and its spokesperson introduced a private Member's Bill to enshrine the commitments into the statute books. The move was unsuccessful.

The precise value of each bid was never disclosed, but all bids were assessed on the basis of the total revenue achieved from the sale of a full 100 percent of shares.

On June 12, 1990, the government announced the successful bid, for 100 percent of the shareholding, was from a consortium lead by two U.S. regional Bell holding companies, Ameritech and Bell Atlantic. The bid was for $NZ4,250 million, and the sale would be effective beginning September 12, 1990. The consortium would sell 5 percent to each of Fay, Richwhite and Company Ltd and Freightways Holdings Ltd., both New Zealand companies, that had acted as advisors to Ameritech and Bell Atlantic during the sale of Telecom. Over three years Ameritech and Bell Atlantic agreed to reduce their remaining 90% to 49.9 percent.

The government agreed that $NZ250 million would be earmarked for health and education programs in an attempt to pacify some members of the public, while $NZ4,000 million would be used to retire government debt. After the 1990 general election, the new National Party government cancelled the $NZ250 million commitment and redirected it to debt reduction. The sale represented the largest sale in Labour's privatization program, and accounted for almost half of all sales, which totaled approximately $NZ9,000 million. The proportion of total funds used for actual debt reduction, however, was considerably less.

The success of the bid by Ameritech and Bell Atlantic was attributed to the parties beginning their assessments of Telecom New Zealand in the second half of 1989, well before the government formally announced its intention to sell. The two companies had other joint business arrangements elsewhere in the world.

The sale price represented $NZ1.82 for each ordinary share. Financial market reaction was positive. Most commentators observed that the Government had received a good price. Many suggested the government had achieved possibly as much as $NZ1,000 million beyond the expected realization. Mr. Prebble claimed the price even exceeded the Treasury's top estimate of $4,190 million. The sale quadrupled American foreign investment in New Zealand.

The new owners retained the existing management team. A new nine-member board of directors was appointed. Only Mr. Peter Shirtcliffe and Dr. Peter Troughton remained from the previous board, with Mr. Shirtcliffe being confirmed as Chair. Five were New Zealand citizens, including Dr. Troughton who had just taken out New Zealand citizenship, while the remaining four were appointed by Ameritech and Bell Atlantic. Telecom's articles of association (see section 16.4.1) requires at least half of the Directors be NZ citizens.

In 1991 Ameritech and Bell Atlantic undertook the initial public offering of ordinary shares. The offering was made on a worldwide basis, with particular emphasis in New Zealand, Australia, the United States and Canada. The company is listed on the New Zealand, Australian, and New York Stock Exchanges (as ADRs equal to twenty ordinary shares).

In New Zealand the shares were offered at $NZ2.00 each, a 10 percent increase on what Ameritech and Bell Atlantic had paid. About 30 percent of Telecom's total shares were sold, and in the process each made an after tax profit of NZ120 million. Soon after listing the stock soared to $NZ2.70 but has subsequently traded just above its offering price, as investors assess the impact of a new tariff structure for domestic long distance calls. A further 10 percent of shares are offered by the majority holders by September 1993, reducing their combined holdings to 49.9 percent.

A 1991 amendment to the Broadcasting Act 1989 removed the specific limits on foreign investment in the New Zealand broadcasting industry. Shortly thereafter, Ameritech and Bell Atlantic, in conjunction with subsidiaries of two of the largest U.S. cable television companies, Tele-Communications Inc. and Time Warner Cable Inc., acquired 51 percent of Sky Network Television Ltd. Sky provides subscription television services in Auckland, Hamilton, Wellington, and Christchurch, which are New Zealand's four largest cities.

In February 1992 the resignation of Dr. Peter Troughton was announced. He had played a key role in the transformation of Telecom, but there were some differences of opinion with the major shareholders. He continued to act as a consultant.

20.4 Other Network Providers

There are four other main service providers: Clear Communications Ltd., Netway Communications Ltd., BellSouth New Zealand Ltd., and Optus Pty Ltd.

20.4.1 Clear Communications Ltd.

Clear is jointly owned by Bell Canada International, MCI International, Television New Zealand (the state-owned television service), and Todd Corporation (a private New Zealand investment firm).

In addition to utilizing the digital microwave telecommunication links owned by Broadcast Communications, a subsidiary of Television New Zealand, Clear leased, then purchased, two fiberoptic cables from New Zealand Rail that run between Auckland and Wellington. It also leases fiberoptic capacity between Wellington and Christchurch from Electricorp.

New Zealand Rail (the government-owned railways service) commenced interconnection negotiations with Telecom soon after competition in network services was permitted in April 1989. Later that year it suspended these negotiations to join a consortium with MCI International and Todd Corporation to provide long-distance and international services. At the same time, a consortium of Television New Zealand and Bell Canada International was established to provide long-distance, international, and local services. Both consortia commenced interconnection negotiations with Telecom in December 1989.

The use of the telecommunications assets owned by other state-owned enterprises (SOEs) on commercial terms highlights a vital component in the early development of competition. The interests shown by some SOEs in entering telecommunications markets (and other markets as well) prompted the government to establish principles for business diversification. SOEs were obliged to confine their activities to those covered by their statement of corporate intent. A distinction was made between diversification within the core business (i.e., Telecom's earlier establishment of a cellular telephone service) and diversification using existing, but noncore, assets (i.e., NZ Rail's fiberoptic cables). This distinction ruled out new telecommunications investments by other SOEs unless it was incidental to core business new investment, such as Electricorp's link between Wellington and Christchurch during installation of new power lines, and primarily for the purpose of the corporation's own communications needs. One affected SOE was NZ Post, which was concerned that the policy would largely confine it to the more traditional segments of the communications market. Prior to the decision, NZ Post had been considering a joint venture with Cable & Wireless to test CT2 service, a radio-based telephone service pioneered in the United Kingdom.

Although both consortia were able to utilize the existing telecommunication assets of SOEs, they soon that realized the New Zealand market was not large and that their existing network facilities complemented each other. The two North American parties discussed the situation and announced a merger in May 1990, creating Clear Communications. The parties confirmed their commitment to the venture in October 1990.

Negotiations for an interconnection agreement with Telecom proceeded through 1990. A memorandum of agreement was concluded in August, and the definitive agreement for long-distance and international services was signed in March 1991. The interconnection agreement was internationally unique in that it did not involve the government, any government agencies, or any intervention by the courts: It was negotiated by the two carriers alone.

The agreement provided for seventeen points of interconnect in fifteen different localities (free calling areas) throughout the country. Clear is able to receive calls from Telecom's network in any of those localities, which contain approximately 85 percent of all telephones (and almost the same proportion of the population) in New Zealand. Clear made a request to Telecom for fifteen additional interconnection points in late 1991 representing other local exchange areas. The matter is still under negotiation, but some difficulties have been identified. In all Telecom has 110 local exchange (or free calling) areas. After conveying the call on its own trunk bypass facilities, Clear is able to deliver that call to any telephone in New Zealand by linking back into Telecom's network. Telecom charges Clear an agreed per minute rate for the tails at each end of a call, together with any long-distance fees incurred in the final delivery stage and a flat fee for the automatic number identification for each call. Clear operates its own billing service, as an earlier request to Telecom by MCI–Todd for the provision of both billing, name and address (BNA), and billing on behalf, services was rejected. At the time of that request Telecom filed proceedings in the High Court seeking a declaratory judgment supporting its right to decline provision of these services.

Clear commenced services in April 1991. Its first-year growth was remarkable. It garnered some 40,000 business and residential customers, and, more importantly, had approximately 9 percent of the domestic toll market one year later. Clear initially relied solely on Telecom for dispatch of international calls, but in 1992 it commissioned independent facilities. It now has a satellite earth receiving station in Auckland and is a member of the Tasman-2 fiberoptic cable consortium. It also joined the PacRim East consortium.

Clear also wants to provide local business access and 0800 (toll-free) services. Although these were also contemplated by the August 1990 Memorandum of Agreement on interconnection, difficulties have arisen in the negotiations. In particular Telecom is adamant that Clear should pay an access levy as a contribution toward Telecom's service obligations. Clear filed an action and a High Court hearing was set. This was the first major court action between the two parties.

Clear Communications' entry into the long-distance market has resulted in a number of changes to toll tariffs. Clear introduced six-second rounding after the first minute. It also introduced a number of discounts—including those for time of day, large volumes of calls, prompt payment, and direct payment. Altogether the discounts represented a reduction of approximately 15 percent on Telecom's standard charges. The occasional weekend specials, first introduced by Telecom, saw both companies offering the same rate, at a considerably lower level. In April 1992 Telecom announced one-second rounding after the first minute for domestic toll calls. Clear promptly matched this and extended it to international call charges—a move Telecom followed.

Clear customers use its network by dialling an access code prior to dialing the national number of the recipient. Some customers, especially businesses with PBXs, have programmed their lines, but most customers have to dial the three-digit access code each time. BellSouth's cellular customers will dial an access code of similar length and type to Telecom's cellular customers. The concerns at the possible competition implications of the numbering pattern prompted the Ministry of Commerce to conduct an investigation.

20.4.2 Netway Communications

Netway was originally owned jointly by Telecom and Freightways (which owns 5 percent of Telecom). Telecom had an option, which it exercised, to buy all of Netway during 1992. Netway offers integrated voice, image, and data network services to corporate and government customers throughout Australasia.

20.4.3 BellSouth New Zealand

BellSouth NZ is a subsidiary of BellSouth International, which in turn is the international operating subsidiary of the Atlanta, Georgia-based regional Bell holding company. In 1990 the Company successfully bid for one of the bands suitable for cellular telephony. The company's bid was $NZ85 million and, under the second-price bidding system used, the actual price was $NZ25 million.

BellSouth has chosen to use digital GSM, which has just been launched in Europe. It offers many added service features, and subscribers would be able to use their own telephones in Europe and Australia. Over the ten years subsequent to 1992 the company plans to employ 100 staff and invest more than $NZ150 million in establishing its network.

BellSouth has announced Clear Communications will provide the main network trunking services, and it expects to commence commercial services in early 1993. In March 1992 the company reached agreement in principle on the key commercial aspects of an interconnection agreement with Telecom; negotiations are proceeding on the key technical, operational, and contractual arrangements.

20.4.4 Optus Communications Pty Ltd

In late 1991, Optus (in which BellSouth Australia has a 24.5 percent shareholding) was licensed to become the second Australian telecommunications carrier. As part of the deal Optus acquired the assets of the satellite company Aussat. These include satellite earth stations in Auckland and Wellington that provide a link to one of its trans-Tasman satellites. Aussat provides trans-Tasman voice, image, and data services.

20.5 Regulatory Policy

The unique feature of New Zealand telecommunications is the regulatory environment. There is no industry-specific regulatory agency or comprehensive set of industry-specific statutes, nor are there lines of business restrictions within the sector, or between this sector and similar activities such as cable television. In addition there are no specific foreign investment restrictions, except for Telecom, for which the major shareholding must be reduced to 49.9 percent within an agreed time and any other shareholding exceeding 10 percent requires the consent of Telecom's Board and the government.

Instead, the main reliance is placed on the Commerce Act 1986 (as amended by the Commerce Act 1990), which is the general competition (antitrust) law in New Zealand. The 1986 statute reflects a keen desire to establish a robust competition framework for New Zealand to improve the performance of the economy. The Commerce Act thus formed an integral part of the Labour government's economic reforms, although its enactment was partly the result of the 1983 commitment by the then National Party government to harmonize New Zealand's competition laws with Australia under the Australia–New Zealand Closer Economic Relations and Trade Agreement (ANZCERTA). The Act is modeled on the Australian Trade Practices Act 1974, which in turn reflects American antitrust policy (see Ahdar 1991). Three parts of the Act have major relevance to the telecommunications sector.

Part II includes prohibitions on contracts, arrangements, and understandings that substantially lessen competition, exclusionary provision, price fixing, resale price maintenance and the use of a dominant position in a market for the purpose of restricting, preventing, or deterring entry or eliminating a person from a market.

Part III relates to mergers or takeovers. The objective is to prevent any acquisition that results in acquiring or strengthening a dominant position in a market, unless the acquisition can be justified in terms of public benefit—a difference with U.S. standards. The regime includes offense and remedy provisions aimed at encouraging the prior clearance or authorization of all mergers that raise competition issues.

Part IV makes provision for the imposition of price control generally, or on particular firms, or even specific products and services, in circumstances where the minister of commerce is satisfied that conditions of effective competition do not exist and control is necessary to protect users or consumers or, as the case may be, suppliers. Price controls were used extensively before 1985, but have subsequently been used very sparingly.

The Fair Trading Act 1986 (as amended by the Fair Trading Amendment Act 1990) prohibits defined unfair trading practices, and false or misleading representation. It also contains product safety requirements.

The Commerce and Fair Trading Acts are enforced by the Commerce Commission, which is an independent statutory body. The commission has five members and its own staff. To assist in enforcement, especially in inquiries involving network industries with natural monopoly elements such as telecommunications, it is able to engage specialists.

There have been three major cases involving the Commerce Act with respect to competition in telecommunications. One relates to the interest of Clear Communications in providing local access services to selected business customers. Clear has filed proceedings directly in the High Court demonstrating the process anticipated in key cases involving major parties. Another case relates to acquisition of additional cellular spectrum licenses (and is discussed later).

The third and longest running case involves a complaint to the Commerce Commission over Telecom's tariffs for short-distance 2 Mbps data transmission links. This service was originally marketed as Megaplan. Telecom raised the tariff in December 1988 and again in April 1989. The cumulative effect was an increase of 588 percent. The Information Technology Association of New Zealand (ITANZ) nearly pursued action after the first price increase, and did lodge a formal complaint with the Commission after the second. The complaint was amended to its present form by August 1989. In addressing the complaint, the Commission has utilized the full range of its investigative powers (see Lojkine 1991). In late 1991 the Commission concluded its investigation and announced it was filing proceedings in the High Court against Telecom's actions. During 1991 Telecom reduced the particular tariffs to a level lower than those prevailing in March 1989. Also in late 1991 the Commerce Commission launched its own inquiry into the competition in telecommunications markets and the factors affecting its development, including the information disclosure regime. At the start Telecom questioned the Commission's powers to hold such an inquiry and initially withheld cooperation. It eventually agreed to make a submission.

20.5.1 Telecommunications Act and Regulations

To augment this reliance on general competition law, there is a Telecommunications Act as well as two sets of regulations made under the Act. The Telecommunications Act 1987 became effective July 1, 1987. The Post Office Act

1959 was repealed on the same date. Under the Act it is not essential to be designated a network operator to provide telecom services. Telecom is deemed to be a network operator and fourteen declarations, mainly by companies proposing to establish cable television services, have been made since this legislation came into force.

Restrictions on the provision of telecommunication goods and services have been abolished progressively since 1987 with passage of the Telecommunications Act 1987 (providing, inter alia, for phased relaxation of restrictions on customer premises equipment) and with the commencement of the Telecommunications Amendment Act 1988 (removing restrictions on the supply of telecom services of all kinds on April 1, 1989). The Act's remaining provisions relate to the promotion of competitive conditions with respect to land access and regulatory powers with respect to international telecom services. The Telecommunications Amendment Act 1990 established regulation-making powers used to establish information disclosure requirements on Telecom, with the purpose of facilitating effective competition.

The Telecommunications (International Services) Regulations 1989 require registration of any person providing public switched telecom services or leased circuits between New Zealand and any overseas operator in a territory outside New Zealand. There are currently five registrations. Registered providers are required to comply with such international telecommunications agreements and conventions to which New Zealand is a party. When representing New Zealand at meetings of international organizations, Telecom is expected to comply with New Zealand's foreign policy and the government's telecommunications policy.

New Zealand is bound by the ITU Nairobi Convention of 1982, which it has ratified. Technical standards in New Zealand are generally in accordance with the recommendations issued by the ITU's International Consultative Committees on Radio (CCIR) and on Telegraph and Telephone (CCITT). The Ministry of Commerce represents the New Zealand government at the ITU. It has wide-ranging consultations with users and providers of telecommunications facilities within New Zealand. Telecom and major broadcasters participate directly in the work of the CCIR. CCITT representation is mainly provided by Telecom, with arrangements for national coordination through the Ministry of Commerce.

Telecom is New Zealand's designated signatory to the Intelsat and Inmarsat Operating Agreements, but the government retains the right to review that appointment. Telecom has established an Office of Signatory Affairs to process applications from competitors for satellite capacity on a separate basis from Telecom's own commercial interests.

The Telecommunications (Disclosure) Regulations 1990 impose certain information disclosure requirements on Telecom and its subsidiaries. These recognize that Telecom does not presently face effective competition. Under the Regulations, Telecom is required to publish information with respect to its subsidiary regional operating companies, which most closely represent the natural monopoly component (the local loop) of its operations. This includes:

1. Financial statements following generally accepted accounting principles for each ROC as if they were independent and unrelated companies, together with information on the accounting policies adopted.
2. Prices, terms, and conditions for all local loop products and services (including interconnection) including those provided to other parts of Telecom or sold under private contract.

Local loop products or services are defined as line rental, local telephone call charges, local and international toll call charges, leased circuits, and interconnection for other networks (telephone, telex, packet switched, cellular radio, mobile telephone, and radio paging). In addition to the standard or list prices for the specified products or services, Telecom is required to disclose the principles or guidelines applied in determining discounts.

Financial statements are published every six months. Information on prices, terms, and conditions was published in October 1990; there are quarterly updates on any modifications, and a comprehensive set of the information published every two years. Each publication forms a supplement to the Government Gazette.

On November 7, 1989, Telecom's Board provided the government with three pledges with respect to its residential services. The government included these in the Articles of Association of the Corporation prior to the settlement of the sale. Changes to these specific Articles, and certain others, require consent of the government as Kiwi shareholder. These commitments are described in the following from Telecom's Articles of Association.

11.4.2.1 Local Call Charging. A local free-calling option will be maintained for all residential customers. Telecom, however, may develop optional tariff packages that entail local call charges for those who elect to take them as an alternative.

11.4.2.2 Price Movement. Telecom will charge no more than the standard residential rental for ordinary residential telephone service and from November 1, 1989, the pre-GST standard residential rental will not be increased in real terms provided that overall profitability of the subsidiary regional operating companies, as evidenced by their audited accounts, is not unreasonably impaired.

11.4.2.3. Standard Prices and Availability. The line rental for residential users in rural areas will be no higher than the standard residential rental and Telecom will continue to make ordinary residential telephone service as widely available as it is at the date of adoption of these Articles.

In 1990, Telecom Corporation also agreed to a request from the minister of consumer affairs to publish quality of service indicators for residential service.

In addition, Telecom Corporation operates a permit to connect program under which anyone seeking to provide equipment for connection to the Telecom network must gain prior approval. The purpose is to check safety and compatibility features. This has gained widespread acceptance from the industry and from the public. As with all the conduct of Telecom, operation of this program is subject to the Commerce and Fair Trading Acts.

Both the present National Party government and the previous Labour Party government remain committed to the present "light handed" regulatory regime for telecommunications. The present government issued a formal statement reaffirming this commitment in December 1991. The statement also indicated that the government would introduce further regulatory measures, should that prove necessary, in the interests of facilitating the development of competition.

20.6 Reforms in Related Markets

The Telecommunications Act 1987 made provisions for the phased deregulation of CPE beginning with residential wiring and telex equipment from October 1, 1987, other wiring, and telephone handsets from May 1, 1988, and PABXs from April 1, 1989. Telecom continues to offer equipment for rental. There is a large number of equipment suppliers, and a significant range of equipment available.

Telegram service was transferred from Telecom to The Telegram Company, a division of Netway Communications, a nongovernment company, in August 1988. There is a three-digit toll-free number for access.

20.6.1 Import Tariff Policy

Since 1986 the government has been engaged in general tariff reduction as part of an overall strategy of reducing industry protection levels. The Post Office purchased only a small share of the output of the local electronics industry. Domestic manufacturers, however, received a one-off production stimulus with the conversion from rotary to touch-tone dialing handsets in 1986–1987. Between 1986 and July 1992 tariffs have fallen from 45 to 19.5 percent, and further reductions, to 13 percent, are planned through 1996. Canada enjoys a preferential rate. Under the ANZCERTA agreement, Australian telecom equipment is imported tariff-free. The level of tariff is also influenced by the level of domestic production, and some goods have a lower tariff level. The range of equipment manufactured in New Zealand is relatively limited, while the tariff descriptions are fairly general. As a consequence the list of concessions in Part II of the Customs Tariff has grown in recent years.

Imports of telecom equipment have grown steadily since the early 1980s (although there was a decline in the 1991 calendar year), and have readily exceeded $NZ200 million annually since the late 1980s. More than 60 percent of these have been for high-value goods, reflecting the considerable investment program of Telecom and, more recently, of Clear Communications. High-value goods include switching apparatus (of which NEC of Japan is the main supplier) and line systems.

20.6.2 Radiocommunications

The Department of Trade and Industry and the Treasury commissioned NERA in July 1988 to examine options for the management of the radio frequency spectrum. This was to assist the design of policies for the deregulation of the telecommunications and broadcasting industries. The main objectives were (1) to assess options that would maximize economic efficiency in the use of the spectrum; and (2) to evaluate and recommend practical and equitable options for the implementation of an allocation and management regime.

Until 1987 the Post Office was responsible for management of the radio frequency spectrum. Bands were allocated to the main government agencies—the Post Office, the Broadcasting Corporation, and the Ministry of Defence—for their respective use. The radio frequency service (RFS), which managed the spectrum, was transferred to the Department of Trade and Industry following corporatization of the Post office in 1987 to form, with the addition of a small policy group, a new Communications Division. The cost of the RFS operations is met by revenue from license holders.

The RFS planned and coordinated the use of spectrum required for private use. Assignments were determined in an administrative manner, after having due regard to government policy, and in accordance with internationally recognized practices. Those for broadcasting were licensed by a warrant system, which in practice seriously limited opportunities. As a consequence, the administrative process—a public tribunal hearing—was long and expensive. Thus, the total cost to all parties of the hearing for the allocation of the third national television license is estimated at $NZ20 million. The station was licensed in October 1987 and began broadcasting November 27, 1989, reaching about two-thirds of the population.

NERA concluded deregulation of the telecommunications and broadcasting industries and future growth in demand for spectrum will mean the RFS increasingly will have to make choices between competing uses and users. In the absence of price rationing, it was expected that excessive demand for certain services, including mobile radio and cellular telephony, could occur within the next ten years (NERA 1988, p. 1).

Provision was made for two cellular networks following the American AMPS standard. Telecom was assigned the B block of frequencies, while the A block was reserved for a possible competitive system. Telecom commenced services in Auckland in late 1987. It then extended to Wellington and Christchurch, and now covers most of the country (in terms of population).

NERA recommended the use of market-based solutions to ensure the optional and efficient allocation of spectrum resources. It specifically recommended the use of a "second-price" sealed-bid auction system to determine the successful bidder, which is a method that avoids payment of an "excessive" price. The two existing cellular blocks were identified as top priority for auctioning as this would help promote competition and maximize revenue.

The report recognized that "a small number of organizations could dominate the spectrum market, and hence stifle competition in the main downstream markets of telecommunications and broadcasting." To overcome this difficulty, NERA recommended the use of existing competition law, especially with respect to the essential facilities doctrine, or even government-imposed limitations on what could be acquired by one firm.

The Radiocommunications Act 1989 introduced fundamental reforms to the management of the New Zealand radio spectrum in order to facilitate competitive entry in telecommunication and broadcasting, as well as to promote efficiency in spectrum management. Provision is made in the Act for establishment of management rights of up to twenty years and subordinate licenses in the name of the secretary of commerce, and for transfer and subdivision of such rights. It is the Government's general policy that where demand for such rights or licenses exists, supply will be forthcoming.

In 1989, during the Parliamentary passage of the Radiocommunications Bill, sufficient concern was expressed by incumbents, mainly sound radio broadcasters, at the possible loss of existing spectrum rights under competitive tendering, that the government amended the legislative proposals to provide surety. (NERA had recommended an initial three-year grace period for incumbent users to vacate the spectrum in the event they were unsuccessful in the auction.)

The Ministry of Commerce invited expressions of interest in spectrum for cellular radio (mobile phones) and other telecommunications purposes in October 1989. Five spectrum bands were identified for consideration, and were designated as the AMPS-A and B, and TACS-A, B, and C bands. The AMPS-B band was already used by Telecom Cellular Ltd., and it enjoyed incumbency rights under section 154-161 of the Radiocommunications Act.

John Murray Associates, a U.S. consulting firm, reviewed the responses to the expressions of interest. Their 1990 report identified that the TACS-A and B bands were lightly used for point-to-point communication links while there were 124 assignments in the TACS-C band. The report recommended nationwide allocation and identified three options—to offer one, three, or four bands for tender—in addition to the band already utilized by Telecom Cellular. Of the options, offering the AMPS-A and TACS-A and B bands was recommended as this would provide "abundant spectral capacity to meet expected end-user demand through the twenty-year planning period."

Rights or licenses in cellular telephone frequencies were tendered in 1990. The successful bidders are required to obtain Commerce Commission clearance before actually receiving licenses. Telecom enjoys incumbency rights to one of the frequencies, and three further frequencies were offered for tender. As a result, up to four cellular telephone services could be established.

Telecom won the AMPS-A and BellSouth International was successful bidder for both TACS-A and TACS-B. BellSouth sought only one TACS frequency band, and preferred the TACS-A. The next highest bidder for TACS-B was the Australian international telecommunications carrier OTC. That bid was rejected as OTC wanted the New Zealand government to agree to a limit of only two cellular operators. The third highest bid for TACS-B was from Telecom's mobile radio subsidiary.

As noted, acquisitions of the bands required clearance or authorization by the Commerce Commission. The Commission cleared the acquisition of TACS-A by BellSouth International, declined Telecom's acquisition of the AMPS-A band, and cleared Telecom's acquisition of the TACS-B band, subject to an undertaking that required Telecom to dispose of TACS-B if Telecom was successful in its High Court appeal against the Commission decision on AMPS-A.

The High Court subsequently upheld the Commission's decision on AMPS-A, but the Court of Appeal granted Telecom leave to appeal the ruling. The High Court, however, rejected the Commission's decision on TACS-B because in proposing and accepting the undertaking the Commission did not give other unsuccessful bidders the opportunity to make further submissions. Telecom has also been granted leave to appeal that decision.

20.7 Conclusion

The dramatic changes in New Zealand's telecommunications are the result of the remarkable confluence of a number of factors: fiscal imperatives, which necessitated the demand for a better financial result from the state trading enterprises; political leadership, whereby a small group of cabinet ministers was able to dominate the executive, and through it the government caucus, and, in turn, Parliament; new economic ideas, especially public choice theory, contestability, and agency theory; and rapid changes in telecommunications technology.

While this confluence has resulted in dramatic changes, four distinct phases—corporatization, deregulation, privatization,and competition—can be identified. The corporatization phase between December 1985 and March 1988 was primarily concerned with establishing Telecom as a separate commercial entity, within an accountability framework consistent with the requirements of a Parliamentary democracy.

In 1988 and 1989 the policy focus shifted to deregulation, following the Touche Ross report, with the enactment of the Telecommunications Amendment Act 1988. At the same time Telecom, under new leadership, prepared itself for competition with an organizational restructuring plus commencement of both a major capital investment program designed to upgrade its network and a major tariff rebalancing program.

The third phase occurred in 1990 with privatization of Telecom. The regulatory regime was adjusted and a restatement of universal service commitments was made by Telecom.

The fourth phase began in 1991 with domestic and international long-distance (toll) services from Clear Communications. Within a year, Clear Communications accounted for approximately 9 percent of the domestic and international toll markets, and its share is expected to continue to grow. Clear established independent facilities for the transmission of international calls in early 1992, and is already planning to introduce local services to major business customers. BellSouth introduced a competing cellular service in early 1993.

Together these phases have comprehensively transformed telecommunications services to an extent well beyond the expectations of the 1987 Touche Ross report and decision makers at the time. The range and quality of services and equipment has significantly increased, while a degree of tariff rebalancing has occurred.

New Zealand's unique experience over the five years 1987–1992 was strongly influenced by its political economy of the preceding fifty years. As a result, there remains a strong commitment to the present ''light handed'' regime, but that regime is likely to be strongly tested with the intensification of competition during the remainder of the 1990s.

Acknowledgments

The views expressed in this chapter are the author's alone and do not necessarily reflect those of the New Zealand Ministry of Commerce. Parts of this chapter draw on Goodin, Saunders, and Stewart, 1988.

Bibliography

Ahdar, Rex. 1991. ''American Antitrust in New Zealand.'' *Antitrust Bulletin,* pp. 217–47 (Spring).

Bollard, Alan, and Robert Buckle, eds. 1987. *Economic Liberalisation in New Zealand.* Wellington: Allen & Unwin New Zealand in association with Port Nicholson Press.

——— and Ian Duncan. 1992. *The Corporatisation and Privatisation of State Trading Activities in New Zealand.* Auckland: Oxford University Press.

Boston, Jonathan, John Martin, June Pallot, and Pat Walsh, eds. 1991. *Reshaping the State: New Zealand's Bureaucratic Revolution.* Auckland: Oxford University Press.

Burdon, Randal M. 1948. *The Life and Times of Sir Julius Vogel.*

Dalziel, Raewyn. 1986. *Julius Vogel: Business Politician.* Auckland University Press and Oxford University Press.

Douglas, Roger, and Louise Callen. 1987. *Towards Prosperity.* Auckland: David Bateman.

Easton, Brian H., ed. 1989. *The Making of Rogernomics.* Auckland: Auckland University Press.

Goodin, Stuart C., A Max Saunders and Angela El Stewart. 1988. ''Telecommunications in New Zealand.'' Working Paper 318, Institute for Tele-Information, Columbia University, New York.

James, Colin 1986. *The Quiet Revolution: Turbulence and Transition in Contemporary New Zealand.* Allen and Unwin, New Zealand Limited, in association with Port Nicholson Press, Wellington.

Jennings, Stephen, and Rob Cameron. 1987. ''State-Owned Enterprise Reform in New Zealand.'' In Bollard, Alan, and Robert Buckle, eds., *Economic Liberalisation in New Zealand.* Wellington: Allen and Unwin New Zealand in association with Port Nicholson Press.

John Murray Assocs. 1990. *Radio Frequency Band Selection for Cellular Radio and Related Services in New Zealand, A Report to the New Zealand Ministry of Commerce.* Auckland: New Zealand Ministry of Commerce.

Lloyd-Prichard, M. F. 1970. *An Economic History of New Zealand to 1939*. Auckland: Collins.

Lojkine, Susan. 1991. ''Enforcement of the Commerce Act in the Telecommunications Sector.'' Proceedings of the 1991 Annual Conference of the Telecommunications Users Association of New Zealand (TUANZ), Wellington, pp. 213–32.

Mason, R. N. and M. S. Morris. 1986. *Post Office Review*. Wellington: Government Printer.

Mueller, Milton. 1991. ''Reform of Spectrum Management: Lessons from New Zealand.'' Policy Insight, No. 135, The Reason Foundation (40 pages).

National Economic Research Associates. 1988. *Management of the Radio Frequency Spectrum in New Zealand, A Report to the New Zealand Ministry of Commerce*. Wellington: Ministry of Commerce.

———. 1989. *Economic Issues Relating to Privatisation of Telecom Corporation of New Zealand Ltd, A Report to the New Zealand Department of Trade and Industry and the Treasury*. Wellington: Department of Trade and Industry and the Treasury.

Robinson, Howard. 1964. *A History of the Post Office in New Zealand*. Wellington: Government Printer.

Roth, Bert. 1990. *Along the Lines: 100 Years of Post Office Unionism*. Wellington: New Zealand Post Office Union.

Touche Ross. 1987. *Competition in Telecommunication Networks*. Auckland: Author.

Treasury [New Zealand]. 1984. *Economic Management*.

Wagner, Graham A. 1984. ''Telecommunications Policy-making in New Zealand During the Last Two Decades.'' *Telecommunications Policy*, pp. 107–26 (Jun.).

Williams, Michael. 1990. ''The Political Economy of Privatization.'' In Jonathan Boston and Martin Holland, eds., *The Fourth Labour Government: Politics and Policy in New Zealand*, 2d ed. Auckland: Oxford University Press.

21

New Zealand: Testing the Limits of Nonregulation

HERBERT S. DORDICK

Of all the countries that have announced telecommunications reforms during the 1980s only New Zealand can claim to have achieved deregulation. The United States, United Kingdom, and Japan have deregulated portions of their structures, but they have also shifted regulatory burdens rather than lifting them. Cynics often claim that reregulation rather than deregulation has taken place. Not so in New Zealand.

The origins of reform in New Zealand are diverse. First, the monopoly services provider was perceived, in time, to be unsuitable and inappropriate for the domestic economy and international dynamics New Zealand has found itself dealing with in the last decades of the twentieth century. Second, these dynamics had become highly competitive. Third, strong challenges to the state monopoly were mounted by important interest groups and stakeholders—including the computer and information industries, innovative industrialists and financiers willing to break down the walls of protectionism and take competitive risks in world markets, and economists who believe a freewheeling competitive economy less burdened by social goals is best. Finally, there was a Labour Party that realized its traditional policies for advancing the well-being of New Zealanders were not working and could not work. To the surprise (and outrage) of many, it sought solutions that would work, including deregulation.

The Post Office in the twentieth century closely followed the Postal Telephone and Telegraph (PTT) model of the German Bundespost. Successive governments enlarged the social role of the Post Office by enabling and mandating that it provide a wide range of public services. The telephone, the telegraph, and the post provided social services, and banking at the Post Office was convenient, safe, and proper. No politician could successfully challenge the Post Office as the provider of the nation's basic infrastructure for trade and industry, nor its role in social integration. No one, that is, until a Labour government came to power in 1984.

The process of privatizing and selling off Telecom are discussed in Chapter

20. My concern here is with the broader sociopolitical context and what I see as some of the implications and consequences of New Zealand's liberalization.

21.1 The Challenge of Computerization

New Zealand has experienced computerization comparable to that in the United States and the major industrialized nations in Europe. There were about 140 digital computers in New Zealand, all imported, the decade after the Treasury purchased its first IBM 650 in 1960. Import licensing was already a fact of New Zealand politics, and those wishing to import computers and associated equipment had to apply to a government committee for licensing. There were over 2,000 mainframes and 13,000 personal computers in the country by 1985 (Beardon 1985). The cost of computerization to New Zealand has been higher than in many other industrialized nations because it had to use its limited foreign exchange at a time when the country was undergoing serious balance of payments problems.

21.1.1 Barriers to Increasing Productivity with Information Technology

The expectations for savings and improved productivity through computer technology were not realized as quickly or as easily during the 1980s as had been anticipated. Many firms found that they had to expand their data processing staffs, increase training, and suffer long and tedious periods of debugging. As in many other industrialized nations, software costs rose more rapidly than expected.

In a 1986 survey (Dordick 1987, pp. 94–99) the single biggest difficulty encountered in networking was given as unavailability of Post Office lines (38 percent of respondents). The Post Office did not provide alternative transmission means such as satellites, cable, or microwave, and type approval for equipment took months.

Firms also had difficulty finding qualified personnel. Almost half of the firms reported difficulty finding staff or consultants. A troublesome finding was that 62 percent of the respondents did not believe the universities could assist them in managing their information resources. Further, 79 percent did not believe researchers and consultants in the government's Department of Scientific and Industrial Research (DSIR) could assist them. While these results are similar to other industrialized nations, New Zealand appears to have suffered more because of the very rapid application of information technology in the face of very slow growth in the availability of adequate technical assistance and the inability to network distributed computers and terminals effectively.

Post Office telecommunications were among the earliest and largest users of information technology, initiating purchases of computers for billing, order taking, repair scheduling, and other operating functions in 1975, and in stored program controlled switches in the early 1980s. To increase revenues, it offered services such as videotex in the early 1980s and provided microprocessor-based customer equipment in addition to purchasing such equipment for its own system. To provide world class services, the Post Office argued it would have to purchase this equipment overseas. The domestic industry viewed this as a threat and sought a policy to encourage development of the local industry.

Some New Zealanders were becoming concerned about dependency on multinational organizations for their best opportunities to re-establish a position in world markets. They felt if they did not develop a policy for adoption of computers, they would be dictated to at the expense of their own economic and social interests. The primary consumer of information technology, however, the Post Office, continued to go offshore. To further exacerbate matters, networking of computers among distributed locations remained difficult because the Post Office followed its own internal planning cycles and adapted only slowly to the growing demands for data communications.

21.1.2 Communications Advisory Council Studies

To resolve this issue, in 1976 the government established a commission to advise it—including broadcasting and other technical fields involved in modern communications. Following its initial report to the government in 1977, the cabinet agreed to establish a Communications Advisory Council (CAC) to serve the broad functions of formulating, coordinating, reviewing, and recommending long-term national telecommunications policies, monitoring their implications and insuring "balanced consideration of each telecommunication sector as an entity and in comparison with all others." CAC was established in 1977 and fully staffed in 1978. Members of the Council included representatives of the Post Office, the Post Office Employees Union, Air New Zealand, Bank of New Zealand, the manufacturing, agriculture, and horticultural sectors, and academics.

Beginning in 1980, CAC undertook a series of studies aimed at resolving issues raised by challengers of the existing telecommunications structure. The inquiries addressed barriers to efficient telecommunications that, if lifted, could allow manufacturers, financial institutions, and the information industry to operate more productivety and thereby compete more effectively in world markets. CAC recommendations were advisory rather than regulatory; regulatory functions for telecommunications remained with the postmaster-general acting on behalf of Parliament. CAC was disbanded in 1987.

21.1.2.1 Network Transparency

In 1980 CAC issued a report on data transfer networks and the introduction of packet switching. This established the basis for Post Office provision of data communications services and development of the packet network. It also dealt with one of the more important, perhaps the most important, question arising as a result of the convergence of telecommunications and computing technologies—how to establish the boundary between terminal equipment and the transmission network. CAC raised the issue of competitive services and the necessity for network transparency. (This is the ability of any piece of terminal equipment to easily communicate with any other piece. A well-designed public switched telephone network is transparent.)

The council concluded that a packet-switched data transfer system was needed. The Post Office argued that it should retain its traditional role and thus provide the processing necessary to enable terminals and computers to communicate on the new network. It also favored adoption of Consultation Committee on Telephone and Telegraph (CCITT) standards, and suggested that a packet-switched data transfer system should include the conversion of digital to analog as well as analog to digital signals and that this be performed by the Post Office as the network services provider. In addition, the form of the data for presentation at the network interface was to be specified by the Post Office. The Post Office wanted to provide the data communications equipment but was willing for terminal equipment to be provided by its manufacturers, subject to type approval by the Post Office. CAC agreed with the Post Office's position; however, this required a precise delineation of the point of interface.

In keeping with international standards, an interface was defined that was dependent on the service provided. By adopting international standards X.25 and X.28 the Post Office proposed a moveable interface. By preventing any fixed definition, it preserved for itself the opportunity to provide equipment and software for network communications on both the terminal side and the network side. The Post Office also protected its historic role as sole provider of telecommunications services, as mandated by the government.

As the boundaries between computing and communicating blurred and the economics of new technologies opened the market to new entrants, questions were raised about whether monopoly provision of telecommunications was efficient and fair to resident consumers. Further, the very definition of telecommunications services was questioned. Whether in analog or digital form, it had become difficult to differentiate between communicating and computing in the transmission of information. As the cost of microprocessor-based equipment had fallen, firms traditionally not in the telecommunications business sought to enter that business, and telecommunications firms sought to enter the computer business. However, such moves are not easily accomplished.

In a submission to CAC in 1980, the Post Office announced its intent to provide ISDN services. It maintained that the Post Office was in the best position to provide the diversity of services and terminal equipment required. However, all equipment would need to pass a type approval procedure. There were immediate cries of distress from the computer and information industries. Post Office approval procedures were perceived as barriers by competitive providers. The matter came to a head in 1980 when CAC was asked to comment on the provision of a viewdata type service.

Because viewdata has usage patterns quite different from those for which the voice telephone network was designed, heavy use could overload the public switched telephone network. Videotex services, CAC recommended, would be delivered on the Post Office packet network, thereby denying access to competitive providers of enhanced networks.

CAC did not question the Post Office's monopoly transmission of videotex.

It did, however, break with traditional policies by allowing for a special tariff based on the cost of the information accessed as well as the duration of access. Further, CAC supported a strong private industry role in providing viewdata services and terminals. To this end, CAC recommended special import license and sales tax considerations for equipment (including television broadcast receivers for teletex), computers, and modems. The Post Office was to coordinate development of viewdata to insure adequate standards and efficient implementation.

CAC did not support monopoly, either public or private, in provision or control of database facilities, terminals, or information. It held that the Post Office could offer its own information services and equipment in competition with other providers of enhanced services on the public switched telephone network. This would conceivably require some form of oversight authority to insure that charges were fair.

CAC suggested that the Post Office insure the network be as transparent as possible, and that interconnect standards be required to insure consumer satisfaction. Type approval was not to be used to limit access to Post Office equipment. Long delays in obtaining approval were common still. There was consequently a relatively small equipment market. A 1985 DSIR study of the domestic electronics industry reported that communications equipment for both radio and telecommunications accounted for only 15 percent of all electronic equipment output (which also includes computers, data processing terminals, and measurement and test equipment).

The Post Office was both provider of services and the agency insuring fairness in the marketplace. The postmaster-general was expected to make sure the Post Office did not abuse this status. It was evident to almost everyone else that this was unworkable. Although several proposals were made throughout the years 1983–1986 for establishing some alternative form of regulation and a regulatory body, they were not seriously discussed.

21.1.2.2 Competitive Transmission Alternatives

In 1984 CAC examined the issue of using satellites for communications and remote sensing and touched on cable television. The council concluded that there would be major benefits in using satellites for the domestic network, especially trunk routes, network expansion, and replacement of terrestrial equipment at the end of its economic life. Further, satellites could provide services in areas inaccessible to the terrestrial network and be used to establish emergency communications. They could also be used for broadcast television. The council concluded, however, that a domestic satellite was not warranted in view of existing terrestrial services and Intelsat. It noted that, with appropriate negotiations, New Zealand could utilize Australia's AUSSAT II after its launch in the early 1990s.

CAC recommended that the government accept proposals for future satellite use and that the Post Office should provide all satellite transmission services, including television distribution. However, the Post Office would have the obligation of allowing access for small and even part-time applications.

CAC examined coaxial cable primarily as an alternative distribution system for television, but also recognized cable can be an important transmission technology for business services. It noted that fiberoptics cable would ultimately provide more economical broadband local telecommunications than twisted pair copper cable. The council concluded that it would waste national resources to have two separate networks with broadband capabilities and that the greatest economies and benefits would be achieved by integration through a single network. It therefore recommended that the Post Office construct and own any cable television network in the short term. It argued for an integrated transmission network for both telecommunication and video transmission to be provided by the Post Office, essentially supporting the Post Office's plan for ISDN.

21.2 The New Zealand Political Economy

The rapid worldwide economic growth that followed World War II especially benefited New Zealand. Strong demand for wool products helped produce a boom, and New Zealanders enjoyed one of the highest living standards in the world. Automobiles and telephones per thousand, common measures of living standards, approaching U.S. levels. GDP grew steadily at about 4 percent per year through 1973. In the words of Sinclair, "This prosperity was the all-pervasive fact in New Zealand life for the first two postwar decades; it was the dominant influence on social attitudes and on politics alike" (1984, p. 288). This was also the period during which the New Zealand welfare economy essentially achieved its goal of an egalitarian state, even though the "conservative" National Party was in power for all but three years of the 1949–1972 period.

Throughout its history New Zealand had sought to achieve a middle class standard of living without the inequities of the British colonial-era class system so many of its inhabitants had wished to leave behind. Income distribution during the 1950s and into the early 1960s showed relatively few people in the lowest income brackets and, similarly, relatively few people at the highest level. In addition, New Zealand was unique among industrialized nations in that land ownership was widely distributed and largely occurred on an owner-occupied basis. New Zealand provided womb to the grave health and welfare services, including attractive pensions.

Labour, under Norman Kirk, came to power in 1972 and set out to expand social welfare programs. Kirk was an international and social idealist whose professed goals were to re-establish the nation's position in the world, to fight what he saw as anti-Maori racism in the country, and to continue toward the goal of a social democratic egalitarian nation. New Zealanders valued their social security programs; these were among their "most treasured possessions, one of the last things they would give up" (Sinclair 1984, p. 271); however, they were losing them anyway, as the system deteriorated. Quality health services became scarcer as doctors and nurses left for higher paying positions in Australia, Canada, and the United States. The pension program was also short of money. Kirk was not helped when the United Kingdom joined the common market in 1973. When oil prices quadrupled shortly thereafter, New Zealand's boom collapsed.

While there was a considerable diversification of export products and trading partners, the terms of trade dropped 46 percent between June 1973 and March 1974, the steepest decline since the depression of 1938. Trade surpluses turned to deficits in 1973. The current account deficit mushroomed to $NZ1.3 billion (U.S.$1.8 billion) in 1974, almost 15 percent of GNP. By 1975 external debt stood at $NZ863 million. It had almost tripled by 1978, reaching $NZ2,447 million.

To a considerable degree this debt was the result of an extraordinary trade imbalance brought on by protectionist import licensing, which had begun in 1938 during the first Labour government when New Zealand had almost exhausted its foreign currency reserves. Import controls continued during World War II and remained a major instrument of economic policy. While this coincided with the prosperity experienced from the 1950s through the early 1970s, it also contributed to a lack of incentive to innovate. Consequently during a period when foreign companies were modernizing or building new facilities, incorporating new information and telecommunications technologies, protected New Zealand manufacturers were content to continue along less innovative paths.

New Zealand lost its major export market when the United Kingdom was admitted to the European Economic Community (EEC) despite arrangements that had been negotiated to give it access to the British market for reduced exports of butter and cheese up to 1977. New Zealand was already making efforts to diversify products and markets and was not caught entirely at a disadvantage. Export sales lost to the United Kingdom were replaced by increased sales to other world markets.

In 1974 Kirk died and Labour was swept away by a National Party landslide in the 1975 elections, making Robert D. Muldoon prime minister.

In the mid- and late 1970s annual inflation averaged over 10 percent and generally was above the rates in other industrial nations. The standard of living declined as real income per capita fell by over 11 percent from 1973 to 1977. To make matters worse, many of the skilled workers needed to extricate the nation from its economic difficulties emigrated.

Because the economic climate had not encouraged innovation, there were few attractive, highly valued export products except in agriculture. To compete with less-developed countries New Zealand's farmers had to reduce their selling prices, a difficult task in a climate of high taxes and rising living costs. While productivity in agriculture is among the highest of any sector, and despite the fact farmers and horticulturists had developed new products such as the kiwi (derived from a plant introduced from China about 1906), this was not sufficient to lift the country out of its depression. No government policies had been directed toward increasing competitiveness through improvements in productivity across other sectors. Manufacturers had gotten use to being protected by high import tariffs, so they did not recognize that they had serious quality problems and were unprepared to search for those niche markets in which New Zealand could compete.

Muldoon muddled along, and was able to win re-election in 1978 and 1981 despite continued trade deficits, inflation, and unemployment. In the late 1970s New Zealand believed it was being increasingly isolated by its traditional allies. During the Vietnam War the United States had sought support in the South Pacific but now no longer needed it to the same extent. The United Kingdom also lost interest in the region as it moved out of Singapore, Malaysia, and the Mideast, and focused on Europe. Politically and economically, New Zealand was adrift. After nine years of things not getting better, voters again turned to Labour.

21.2.1 Economic Reforms

Labour came to power in 1984 under David Lange on a platform of drastic reforms aimed at creating a more dynamic economy and a ''fairer'' distribution of income and wealth. ''The tangled mess of economic policies we inherited had inhibited economic initiative and growth. It had created an environment of controls which left the more privileged in our society largely unaffected while penalizing the less well off,'' stated Finance Minister Roger Douglas. He promised his free-market oriented reforms, soon dubbed Rogernomics, would deliver higher quality services without the current high tax burden that helped weaken the economy.

In a radical departure from traditional economic management, in July 1984 the new government—in an attempt to make New Zealand products more attractive overseas—devalued the New Zealand dollar about 20 percent against the U.S. dollar, and announced on Christmas Eve that it would be allowed to float. Other dramatic shifts in the economy were initiated, including major tax revisions, increased competition, and less regulation for the banking sector, the airline system, electric utilities, and the state-owned coal mines.

Reforming the status of the many state organizations, including the Post Office, would make these organizations more accountable to Parliament and the taxpayer, Douglas asserted, and new policies on publicly supplied goods and services would make the government more responsive to the needs of those using the goods and services. The government expected business to adapt to the new economic environment, and was prepared to do so itself.

Pointing to the massive amounts of taxpayer money poured into government departments such as the Airways System, Lands and Survey Department, Forest Service, and the Post Office, Douglas argued these activities could and should be made to perform more efficiently in order to restore sustainable economic growth. In many of these enterprises roles were unclear and conflicting objectives abounded; there were burdensome controls, lack of commercial freedom, and inadequate incentive structures. Commercializing them could provide clearer objectives and incentives for growth. A vital ingredient in the commercializing process was reform of the regulatory environment in order to permit real competition between state and private enterprises.

21.3 The Post Office in 1985

The Post Office Act of 1959 firmly established the Post Office as a department of state responsible under the direction of the postmaster-general for the administration of the Act and regulations pursuant to it. Through the addition of further functions dictated by changing communications technology and economic factors, the Post Office was organized under a head and district office structure. In 1985 the Post Office was the largest organization in New Zealand and carried out a central role in providing communications services domestically and internationally. Its operations were distributed among three functions: banking, postal, and telecommunications services.

Post Office operations were carried out by twenty-two postal districts and were controlled by a chief postmaster. The exception was telecommunications engineering activities, which were divided into seventeen engineering districts under the control of district and regional engineers.

Over the years the Post Office Employees Union became the largest and perhaps most powerful union in the nation. The largest number of its members were in the Telecom operation. Post Office personnel policies reflected long-standing attitudes and agreements between management and the union. Indeed, these had evolved into traditions that were difficult to change. Similar wage rates were established across the three operations. Rewards for higher level skills were difficult to obtain, and advancement was were severely limited by promotional criteria that gave more weight to seniority than skills and labor market conditions. It was not at all unusual to find two employees starting work on the same day, one in the Savings Bank and one in Telecom, receiving equal rates of pay even though, as one worker put it, ''the Savings Bank fellow was still using a pencil and paper while I learned to use a personal computer and test optical fiber networks.'' As a result, telecommunications employees were being attracted to less secure but higher paying jobs in the information industries. Despite this slow but nevertheless significant exit of skilled workers, the union made it difficult to recruit staff specialists from outside the Post Office.

There was also a long-standing joint consultative process between management and the union that was charged with resolving disputes and obtaining agreements on the introduction of new technologies. It resulted in long delays not only for the introduction of new technologies needed for improved services to customers but also for the introduction of office technologies and modern procedures in the Post Office itself. The union and middle managers were both concerned that no jobs be lost or downgraded, and if there was to be little productivity gain, senior management had little incentive to push.

Post Office telecommunications were vertically integrated. When available, equipment was purchased from local suppliers; however, it was usually purchased from foreign sources. The Post Office primarily got new technology in the form of imported equipment, although some software was procured domestically. To a much lesser degree than overseas suppliers, the DSIR, a government department, and the domestic electronic manufacturing industry, were also sources of R&D. Close working relationships between the Post Office and its foreign suppliers enabled it to integrate its network operations and facilities.

21.3.1 The New Zealand Computer Society Position

Driven largely by the structural changes taking place in the United Kingdom, United States, and Japan that freed computer users from what they felt was the "tyranny of the telephone company," the New Zealand Computer Society (NZCS) undertook a comprehensive review of telecommunications in New Zealand in 1985. NZCS represents independent software developers, software and computer engineers in major financial institutions, firms that utilized computer assisted design and manufacturing (CAD and CAM), and computer and terminal equipment manufacturers and importers. Parallel to this effort, and encouraged by NZCS, a Telecommunications Users Group was organized. Impetus for formation came from the major computer suppliers: IBM, Unisys, Burroughs and NEC, and almost all of Telecom's major business and industry customers became members. The NZCS position paper was reviewed at its first meeting in 1986.

NZCS concluded that "with the convergence of technologies and the rapid growth in acceptance of computers, the existing policies of the New Zealand Post Office will require change in order to better suit the changing and emerging needs of new classes of users." Further, NZCS flatly stated that the regulatory and monopoly powers given the Post Office could inhibit valid and effective application of computer-related technologies. Although NZCS argued for a wide range of reforms, it did not seek to stop the Post Office from competing with private enterprise and recognized the Post Office's duty to serve the national interest. How and on what terms competition was to be achieved were the questions the society posed.

The NZCS paper raised the issue of the appropriate interface point for terminals that provide value added services and suggested a well defined point be located as far as possible into the network. On the customer premises side, NZCS argued in favor of free and open competition in provision of terminals and software designed to facilitate computer-to-computer communications. This was to preclude the Post Office from involvement in setting of specifications for network protocols, which NZCS felt should be left to terminal, computer, and other equipment manufacturers and software designers.

Other recommendations were that users have more freedom of choice in equipment and protocols required to interconnect equipment, and that they be permitted to interconnect in geographically dispersed offices as well as with suppliers and consultants with whom they do business. Further, users should be allowed to encrypt and be permitted to provide value added services on their own communications and on the public switched network with payment of access fees to the Post Office.

The paper went on to argue that the Post Office should continue to provide a full range of high-level transmission services and should not compete unfairly with independent providers of value added services, customer premises equipment, and specialized transmission services. Finally, NZCS recommended that the government insure equity in the provision of basic services and that its decisions concerning rates and service levels be open to public discussion.

Many top executives of the Post Office criticized the NZCS paper. Favorable references to liberalization in the United States came under particularly vehement attack because of a purported lack of concern for universal service and other social values. In fact, NZCS had accepted that the New Zealand policy of wide free-calling areas and rural areas served by party lines had become part of the New Zealand way of life. Where subsidies occurred for social reasons, NZCS held that they should be made explicit and that there should be public representation in such decisions.

21.4 Rogernomics and Public Sector Reform

The New Zealand Planning Council had begun to examine the public sector in 1983. Their reports questioned the monopoly status of its enterprises, so there was already a broad feeling that it was time to re-examine original assumptions and earlier decisions when Labour took up the call.

The Council noted that despite widespread desire to maintain strong government monopolies, restructuring in the public sector had been underway for some time. Government corporations with some private board members that operated in a commercial fashion already existed. For example, New Zealand Railways had been corporatized in 1979; Air New Zealand was competing internationally and domestically; and the Hotel Corporation of New Zealand had aggressively competed with privately owned hotels to become a leader in upgrading the tourist industry—all before the Planning Council began to examine the public sector in 1983. The climate for a radical revision of the government enterprise structure was already established—such as asking why the government had to be aggressively competing in the hotel business.

Four options for competition in telecommunications were available to the government before deregulation began.

1. Remove the legislative barriers to competition in all network services and permit a free market subject to the constraints of the Commerce Act
2. Permit a free market in all network services based on interconnection arrangements favoring new entrants for an initial period, much as the FCC dealt with new entrants in the U.S. long-distance market
3. Allow Telecom to retain its monopoly of all facilities, but permit the resale of leased lines in competition with Telecom and permit attachment of switching and other electronics to these lines so competitors can build public networks using Telecom's main facilities; bypass and free

competition in enhanced services would be allowed as in the first two options
4. Limit open competition solely to enhanced services and local bypass

The government elected the first option because this was clearly in keeping with its policies towards state-owned enterprises. Further, the demands of stakeholders calling for access to what they believed to be a rapidly growing telecommunications market were met.

The dramatic regulatory reforms in the United States significantly affected policymakers in New Zealand, denials by Post Office executives involved not withstanding. New Zealand initially did not wish to emulate the reforms of the United States model, especially while instituting dramatic reforms in the domestic economy as a whole.

In late 1985 the Postmaster-General on behalf of Parliament requested an evaluation of the Post Office organization. The result, called the Mason–Morris report, is discussed in Chapter 20. Two legislative acts with important provisions for Telecom and the environment in which it operates were passed during 1986: the Commerce Act and the State-Owned Enterprises Act.

21.4.1 The Commerce Act 1986

The Commerce Act was a fundamental shift away from patterning legislation on a British model. In particular, it is considered by New Zealanders to be tougher on anticompetitive activity. Its importance to Telecom is that it prohibits a firm with the dominant position in a market from using that position to restrict entry, prevent or deter competitive conduct, or eliminate competitors. Telecom has and will continue to have, even under market liberalization, a dominant position.

Under the Act, a firm found to be exploiting its dominant position for anticompetitive purposes is exposed to certain penalties for three years thereafter. The Commerce Commission, the body enforcing the act, can ask a court to fine the firm up to $NZ300,000. More important, the firm has unlimited liability to anyone suffering a loss or damage caused by its actions. Officers of an offending company can be imprisoned. The Commission or any other person can apply to the courts for an injunction restraining the firm from using its dominant position for anticompetitive purposes.

Telecom was initially exempt from these provisions with respect to network services, but, with liberalization, Telecom's competitive behavior has also become fully subject to the provisions of the Act.

The Act allows the government to impose price controls on services supplied in a market where competition is limited if such price controls are in the public interest. However, there are no specific procedures for instituting these controls nor, in the case of Telecom, are there any provisions that explicitly call for the upgrading or advancement of services not yet on the network.

How the courts interpret the provisions of the Act is uncertain; consequently, the effectiveness of these provisions in regulating anticompetitive behavior and encouraging the development of competition remains to be seen. Litigants might find it difficult to prove Telecom (or any other firm) has acted anticompetitively. The threat of litigation, however, and the consequences of losing might inhibit Telecom from acting in a manner that could bring suits—which is what is intended. A potential entrant might seek a preemptive injunction against Telecom; then, if the injunction is granted, and Telecom ignores it, Telecom is exposed to the risk of unlimited fines and damages.

21.4.2 The State-Owned Enterprises Act 1986

The SOE Act provided for the creation of nine new state-owned enterprises (SOEs) (including the three from the Post Office) from six government departments on April 1, 1987. Each SOE was to be established as a limited liability company under the Companies Act 1955, with capitalization based on an agreed valuation of the assets, and an appropriate debt to equity ratio. A board of directors would be appointed by the government and the shares would be held by two cabinet ministers. The Act set the following objectives for each SOE:

1. To operate a successful business, but being as profitable and efficient as comparable private sector businesses.
2. To be a good employer.
3. To exercise a sense of social responsibility.

The Act imposed an accountability framework on each SOE to the ministers, and on the ministers to Parliament. The board and the shareholding ministers must establish a statement of corporate intent each year that specifies the information to be disclosed by the corporation—including its objectives, nature and scope of activities—and financial performance targets for the current and next two years. In addition, SOEs remain subject to auditing by the controller and auditor-general, the requirements of the Ombudsmen Act 1975, and the Official Information Act 1982.

SOEs such as the newly formed Telecom should enjoy "competitive neutrality" (e.g., as far as possible, they should be subject to the same regulatory and tax regime as comparable companies in the private sector). Some specific exceptions from competitive neutrality were permitted for Telecom. The most important is the monopoly right to operate a telecommunications network, as provided by the Post Office Act of 1959. This right was transferred to Telecom for the period April 1 to December 31, 1987. Telecom's monopoly was not universal, as a number of private networks existed—including ones for banks, insurance companies, department stores, libraries, and Fletcher Challenge (one of the country's largest conglomerates).

These networks were legislatively permitted because many of the organizations concerned were Crown corporations and thus not bound by the Post Office Act; their lines are situated on their own land or were negotiated as private licensed lines with the Post Office.

21.4.3 The Path to Deregulation

The Telecommunications Act (July) 1987 effected most of the Mason–Morris recommendations. Postal services, telecommunications, and banking each became independent entities. Regulatory activities such as spectrum management was assigned to the Radio Regulatory Service, which was transferred from the Post Office to the Department of Trade and Industry. Finance Minister Roger Douglas noted that there was a tremendous demand for a skilled telecommunications staff and that they were being bid away by the private sector. A Post Office with commercial objectives and services in a competitive market would remedy this situation.

The Act maintains the prohibition against competition with Telecom in network services, except for previously licensed private or leased networks and transmission facilities on private property and established a timetable for competition that envisioned completion of the process by April 1989.

Following re-election of the Labour government in 1987, the consulting firm of Touche Ross was engaged to study telecommunications in New Zealand, working closely with Telecom, users, and potential competitors. Touche Ross examined Telecom's operations, its cost and pricing structure, the cost structure of the industry, and the markets for network services. Among other things, Touche Ross concluded that competition with Telecom's network is feasible and desirable. In light of this, SOE Minister Richard Prebble announced that the government was reviewing Telecom's monopoly and that Telecom's board had developed "goals and implementation strategies which it believes can only be attained in a fully deregulated environment."

In late 1987 Prebble announced that legislation would be introduced in early 1988 to permit competition in network services by late 1988 or early 1989. At the same time, ministerial responsibility for Telecom was transferred and placed with the other SOEs under Prebble, who had long argued for a more open and competitive industrial climate for New Zealand. Prebble's announcement reflected government policies toward SOEs: They were to be placed in a competitively neutral position, should not be hindered by government-imposed disadvantages such as employment rules, and should not benefit from special advantages or privileges given by the government. The Telecommunications Amendment Act (December) 1988 was the result.

Premise wiring, telex equipment, telephone sets, cabling for PABX equipment, the telex, and the line side of telephone network were deregulated by April 1989. The trunk side of the network, including mobile telephone and paging, were deregulated in July 1989. Deregulation was rapid, on schedule, and proceeded without delays.

21.4.4 Privatization of Telecommunications

By 1989 the government had sold the Post Office Savings Bank, Air New Zealand, the Forestery Corporation, the national oil company, a synthetic fuel plant, and a steel mill. The sell-off reduced overseas debt, and the government maintained that long-run benefits would accrue to the economy from the more efficient operation of these concerns as private firms.

The initial steps toward telecommunications deregulation were taken in April 1987 and the final steps came in April 1989; therefore, the entire process took some twenty-four months. In that time, telecommunications in New Zealand were transformed from a traditional government PTT monopoly to the most deregulated system in any developed nation. The government's objective was to sell Telecom for the highest possible price; little or no thought was given other goals, such as promoting wide shareholding or selling shares to employees.

21.5 Social Costs of Economic Success

From the government of Vogel, who instituted compulsory free education for Maoris and whites in 1876, through Dick Seedon's 1890s social legislation, down to Muldoon's 1975 promise to raise pensions, New Zealand had been known as a model egalitarian society with extensive government-provided services. As a result of the associated costs, the country was near bankruptcy by 1983 and the old joke that capitalist countries share the wealth unequally while communist ones share the poverty equally looked like it might need to be modified to fit New Zealand.

The Labour government that came to power in 1984 had a clear mandate to take drastic steps to reverse the falling fortunes of the nation. The resulting economic success has had social costs. The rapidity of the reforms led to what were called "speed wobbles." Interest rates over 20 percent (the high was 26 percent) and an inflation rate well over 10 percent persisted from 1987 through the early months of 1988 despite the floating of the New Zealand dollar and reductions in the trade deficit. Professional and skilled workers continued to emigrate to Australia, the United States, and the United Kingdom.

Toward the end of 1988 prime minister Lange found himself under pressure to slow the rate of economic change. Inflation had been brought down to 4–5 percent, but at a cost in unemployment of more than 10 percent, the highest level in the nation's history. Interest rates had also fallen, but still hovered around 13 percent. The budget was balanced and there was a small trade surplus. Crime, however, was increasing and unemployment was about 25 percent among minority groups, primarily Maoris and Polynesians. Maori unemployment led to increasing demands from them for reexamination of provisions of the 1840 Treaty of Waitangi (whereby New Zealand became a British colony) which the Maori and their supporters argue are not being respected.

While Lange and Douglas agreed on the restructuring of the economy, they could not agree on social policy. The prime minister was unwilling to further reduce expenditures on health, education, and welfare, while Douglas believed that not doing so would slow down the rate of change and that, once momentum was lost, it could not be recaptured. Prebble agreed with Douglas, and Lange first dismissed Prebble and then Douglas from his cabinet in late 1988.

David Caygill became the new finance minister and promised to continue the economic policies of his predecessor. Nevertheless, it became clear that economic reform was going to slow while the social consequences were considered (see Pinfield 1987 and Royal Commission on Social Policy 1988).

It was fairly clear that simply exploiting New Zealand's natural resources could no longer subsidize high living standards even for a population of 3.3 million; therefore, people were simply going to have to work harder. A large part of Labour's traditional constituencies were finding out the personal short-term costs of a competitive society were likely to be large long-run costs, and they did not like it. Labour Party infighting led to Lange's replacement by Geoffrey Palmer as prime minister in August 1989.

21.6 Testing the Limits

New Zealand wants to achieve its goals in a manner that is unique to New Zealand, in a cooperative rather than an adversarial manner. Unlike other nations, it is experimenting with a nonregulatory process. New Zealand wants to avoid a system of regulated competition and thus seeks to let the marketplace set fair prices, control market entry and exit, and otherwise perform all of the tasks traditionally undertaken by regulation in the United States and United Kingdom.

If the market does not work, regulation will be added back, preferably in very small doses. This is in marked contrast with liberalization in the United States, the United Kingdom, and Japan, where regulations are removed if it appears the market for a specific service has emerged as a competitive one. A mix of regulatory authorities has been overseeing this process. In the United States, these are the Federal Communications Commission, state regulatory commissions, the Antitrust Division of the Justice Department, and the courts. The United Kingdom has the Office of Telecommunications (OFTEL), the British version of the Commerce Act, and one or more parliamentary offices. In Japan the Ministry of Posts and Telegraph and the Ministry of International Trade and Industry (MITI), as well as the Diet, have a role.

Policymakers in New Zealand concluded that the U.S. model is expensive and inflammatory and does not suit the country's parliamentary structure, while OFTEL, even with a staff of just 150, is expensive and does not have sufficient experience to be taken seriously. They believe New Zealand courts, acting through the Commerce Act, will insure competition through the ''essential facilities'' legal principle, which asserts that owners of essential facilities cannot deny new entrants fair and reasonable access on grounds of competitive advantage.

A precedent was set when the High Court ruled that the Auckland Regional Authority could not deny a rental car firm access to the airport. The judge noted that the essential facilities doctrine had been widely applied to telecommunications in the United States, clearly referring to the common carrier principle that was the basis of telecommunications regulation. This was a clear warning to New Zealand Telecom.

The telecommunications market is a complex one both in terms of technology as well as in how costs of services are determined, how costs are allocated between local and long-distance services, and how these services are priced. Whether access to the network by competitive service providers is to be priced on the basis of incremental costs or on the basis of fully distributed costs must be determined. The Commerce Act and the courts alone may not be sufficient to keep the cost of regulatory oversight low, to avoid engaging in the adversarial activities New Zealanders see as unseemly in the United States or in the United Kingdom, and to achieve the competitive market in telecommunications that New Zealand wants.

Backstopping faith in the market and the courts is the Commerce Commission. As now constituted, however, the Commission is not geared to deal with the complexities of telecommunications regulation; its members are concerned with all aspects of fairness in all the newly liberalized markets. These include electricity, domestic air travel, railway service, and petroleum distribution. There are no carefully crafted procedures for determining the extent of competition that exists in certain markets and, inversely, fully determining if anticompetitive practices are occurring. There appears to be no formal oversight proceedings currently in place. Thus, how well the New Zealand approach works depends in part on the Commission's staffing.

21.6.1 Whither Telecom?

The threat of competitive entry could force Telecom to bring its prices closer to costs for those segments of its business where commercial entry is feasible. This is initially likely to be in long-distance, international calling and high-speed data networks. Touche Ross estimated that in a liberalized market over a two- to three-year period toll prices could be expected to fall by about 50 percent in real terms, leased lines by about 40 percent, and international calls by 25 percent. Lower prices are expected to increase demand. Data just after deregulation showed the number of long-distance calls rising sufficiently that revenue increases can be expected. As in the United States, the share of the long-distance market for new entrants will not be large and Telecom is likely to maintain a dominant position.

Finland had a reasonably comparable economy but twice the productivity of Telecom, measured in access lines per full-time equivalent employee, in 1985. There is thus presumably some room for increased productivity and consequent significant decreases in Telecom's costs.

There is concern over the fate of the cross subsidy from long-distance to local services. Based on Touche Ross' allocation of Telecom's network costs between long-distance and local calls, the price of local calls is far below the full costs of providing the services, just as toll costs are above full costs. Telecom does not agree with this, claiming that cost recovery from local calls is not far below full cost, but toll costs may still have to provide a subsidy to residences in any case, especially to rural subscribers. Residential prices have not been increased in real terms (although they have in nominal terms), but residential customers, both urban and rural, may not be sharing in the productivity improvements and lower costs created by the digitalization of the network and staff reductions at Telecom.

21.6.2 Prospects

New competitors have found it difficult to enter the United States, United Kingdom, and Japanese markets without explicit regulatory assistance by government. There have been no efforts by the New Zealand government to stimulate competition, as has been done in the United States and United Kingdom. Indeed, Telecom's revenue is likely to be greater than estimated by Telecom's analysts, thereby making entry even more difficult. Telecom will remain the dominant provider. As has been learned in the United Kingdom and in the United States, a monopoly provider of essential services is not easily displaced by new competitors.

Difficulty in plugging into the Telecom network is the main anticompetitive activity new entrants could face. There are two approaches in dealing with this problem: one is to rely on the Commerce Act to force Telecom to interconnect; the other is to draft comprehensive regulations so all entrants know just what interconnection conditions and terms apply. The ideal might be a mixture of the two, as it is difficult to believe Telecom will allow new entrants unobstructed access without some sort of resistance. Telecom might volunteer interconnection standards, and the Commerce Act would provide legal redress. If Telecom were to obstruct a new entrant by either unfair or unreasonable pricing or by restricting interconnection information, the Commerce Act and the courts could, as a last resort, set prices and other conditions for competitive entry.

Telecom's interconnection policy appears reasonable. For example, should a competing firm wish to offer long-distance toll service between Wellington and Auckland and use Telecom's local network in both cities for origination and termination, Telecom will charge the competing company the same interconnect charges as it would charge any business phone or PBX to connect to the system. The competing company would pay a low monthly fixed charge at each end of its circuit and a charge per minute for each call terminated or originated on Telecom's local networks in the two cities. There is a small fee for extra programming charges, but no other up-front charges.

Although these technical matters are important, competition will emerge only if there is sufficient growth in demand for its products and services. This depends on significant growth in other sectors of the economy.

Telecom promised to provide equal or nondiscriminatory installation and repair services to competing networks, a matter now being tested in the courts, as is the ability of the Commerce Act to deal with such issues and, further, if New Zealand can continue a policy of nonregulation.

The arguments for liberalization that have been set forth by consultants and by government economists draw heavily on the experiences in the United Kingdom and the United States. However, such comparisons must be made with extreme care. It is not simply that New Zealand is much smaller, but also that its economy has not been an expanding or particularly vibrant one. It has only been since the early 1980s that the nation has begun to confront the fact it can no longer be the egalitarian society it was and that it cannot provide a high standard of living with a protectionist economic policy. Although "speed wobbles" toppled the reform wing of the Labour government in 1989 and helped bring the National Party to power in October 1990, the need for society-wide reforms has come to be accepted.

There is evidence that Telecom's operations can be more efficient through better management, changes in labor practices, and continued application of new technologies. Its costs can thus remain essentially the same or even fall in the face of sustainable competition. A liberalized market where competitive entry is possible clearly allows the limits of natural monopoly to be tested continually and in ways that regulators may not be able to foresee.

Natural monopoly has been dismissed as a basis for communications policy. The question of how to regulate the new communication infrastructure has been answered with *minimally*—indeed, with a policy on nonregulation. The Labour government that began the process said it was "testing the limits of nonregulation" and that has been the case.

Bibliography

Beardon, Colin. 1985. *Computer Culture; The Information Revolution in New Zealand.* Auckland: Reed Methuen.

Communications Advisory Council. 1980. "The Question of the Provision of a Viewdata Type Service in New Zealand." Wellington.

———. 1980. "Report on Data Transmission Networks and Introduction of Packet Switching to New Zealand." Wellington.

———. 1984. "Satellites and Alternative Transmission Services for New Zealand." Wellington.

———. 1985. "Submission to the Royal Commission on Broadcasting and Related Telecommunications." Wellington.

Dordick, Herbert S. 1987. *Information Technology and Economic Growth in New Zealand.* Wellington: Victoria Press.

Douglas, Roger O. 1986. "Statement on Government Expendieture Reform." Wellington: Government Printer.

Department of Scientific and Industrial Research. 1985. "Electronics in New Zealand; Progress in the Manufacturing Industry." Auckland: Auckland Industrial Development Division.

Ministry of Trade and Industry. 1986. "State Owned Enterprises Review of 1986." Wellington: Government of New Zealand.

New Zealand Computer Society. 1985. "Position Paper: The Provision of Telecommunications Services in New Zealand." Wellington: Victoria University.

Pinfield, Christopher. 1987. *Equity and Economic Policy.* Wellington: The Treasury. Processed.

Prebble, Richard W., Minister of Trade and Industry, ''Deregulation of the Telecommunications Industry.'' 1987. Wellington: Government Printer.

Royal Commission on Social Policy. 1988. *Towards a Fair and Just Society*. Wellington: Government Printer.

Sinclair, Keith. 1984. *A History of New Zealand*. London: Penguin Books.

22

Japan: Creating the Domestic and International Network

YOUICHI ITO AND ATSUSHI IWATA

Making a unified nation of islands that for over 200 years had intentionally restricted contact not only with the outside world but internally among its constituent parts was a daunting task confronting mid-nineteenth-century Japan. Communications has always been a key element of state building. It is thus not surprising that the telegraph was introduced into Japan in 1869, just a year after the Meiji Restoration, which is conventionally considered the beginning of the country's modern era.

This chapter traces the development of the telecommunications industry in Japan from these beginnings, concluding with a discussion of factors affecting policy change, including changes and goals and the means available to achieve them. The next chapter will cover the domestic deregulation process that began in the late 1970s as well as the data processing services that were part of the technological impetus for deregulation.

22.1 Establishing a Domestic Network

When Matthew Perry and his fleet visited Japan for the second time, in 1854, they brought some fifty items as presents from the president of the United States. Among them were two telegraph sets, the first ever in Japan; many others quickly followed. They were used experimentally and studied in many parts of Japan. In 1857 Nabeshima Naomasa, daimyo of Saga in Kyushu, had a Dutch set copied. The result was the first telegraph instrument made in Japan. Production soon was underway in several places (see KDT 1968, pp. 15–28).

A decade later two merchants petitioned for permission to construct a telegraph line between Tokyo and Yokohama "to contribute to the trading business" (see KDT 1968, pp. 80). The plan was approved by the central government the next year, but was not realized because of the government's collapse later in 1868—an event known as the Meiji Restoration.

The new government received several petitions in 1868 for private telegraph lines, including one from a Frenchman for service between Osaka and Kobe. Acting on the recommendations of Terashima Munenori, who had studied telegraphy since its arrival, the government decided it should operate the telegraph itself. Terashima's report emphasized the importance of the technology for national integration and administration, effective use of the military, and economic development. One of the government's slogans was "prosperous country strong military" *(fukoku kyohei)*, and telegraphy was considered critical in achieving these goals.

Telegraph service began December 25, 1869, as a direct enterprise of the Ministry of Technology with a line between Tokyo and Yokohama. Service rapidly expanded all over Japan. When a rebellion against the central government broke out in 1878 in southern Kyushu (the Seinan Civil War) the nationwide telegraph trunk network, which had been completed in 1876, was a key communications channel in mobilizing soldiers and supplies and in executing the government's military operations. This incident reinforced government leaders' recognition of the telegraph as vital to their ability to govern all of Japan. As a result, the government invested heavily in the construction of a nationwide telegraph network. Thus, despite a relatively late introduction, the network covered the country as effectively as those in continental Europe by the 1920s.

Records from the nineteenth-century show a much higher percentage of telegrams were of a commercial nature than was the case in Europe or the United States, where many messages were congratulations or condolences. In 1881, when the population was around 36 million, over 2 million telegrams were sent (see KDT 1968, p. 126).

When news of the telephone reached Japan in 1877 (the year after Bell's patent), a number of private companies expressed interest in entering the field. They were opposed, however by advocates of a government monopoly, who stressed the importance of preserving communications secrecy for the military, police, and other government agencies. There was also a fear that investment would be made only in urban areas under private management, so diffusion to rural areas would be delayed. In addition, foreign countries were mostly opting for government monopolies—the United States and Canada being notable exceptions. Others in the government—such as the Ministry of Finance, which was concerned about the huge investment required for this new enterprise—supported private initiatives.

The controversy over the appropriate structure of telephony lasted several years. During this time the government rejected all requests from private parties seeking to establish public service. Finally, in 1889 it was decided telephone service should be run by the government. This was codified in the Telegraph Law of 1900, which prescribed that all wired telegraph and telephone services be directly operated by the Ministry of Communication (MOC) and managed as a complete government monopoly.

At the time, many in and out of government felt that the telephone was less important than the telegraph, in part because its quality was such that messages often were misunderstood if not unintelligible. The government was thus pushing expansion of the telegraph network, while not becoming very enthusiastic about telephony. Still, by the time of the decision a number of government agencies—including the police and military—and private companies—including railroads—had established systems for their own use. However, there was no interconnection among them until 1890 when an exchange service was established. It started with 344 subscribers. This had expanded to 2,672 in 1893.

Due to the Sino–Japanese War, which erupted in 1894, government investment in telephony shrank, shutting down expansion. As a result, in 1896, one year after the end of the war, the number of unmet service requests had reached 6,500 more than double the 3,200 total subscribers served at the time. The size of the backlog increased yearly—and the problem continued into the late 1970s. In other words, it took eighty years for supply to catch up with demand.

The demand for telephone service was so strong that the business was highly profitable from its introduction. In contrast, the telegraph ran cash flow deficits every year but three from its inception until 1895. That year MOC changed its bookkeeping system in such a way that it was impossible to tell. The deficits were primarily the result of capital expenditures for network expansion—on an operating basis the system was profitable. In any case, profits in telephony covered the telegraph deficit and became an important source of revenue for the government.

Because of the backlog problem, individuals and companies whose needs remained unmet began to buy telephone service from existing subscribers. MOC could not ignore the high black market prices and so in 1901 established a priority system based on willingness to pay. The basic idea was similar to an auction. Those in a hurry could leap the queue by bearing a higher proportion of installation costs. This differentiated cost-burden, which was unique to Japan, meant the network could be expanded more quickly than might otherwise have been the case.

The Great Kanto Earthquake in 1923 resulted in serious damage to the system. In spite of this, the number of subscribers steadily increased, reaching 1 million in 1938.

22.2 The Early Postwar Period

As a result of bombing raids during World War II about half the telecommunications facilities in Japan were destroyed. The peak number of subscribers was 1,080,000 in 1943, but this had dropped to only 470,000 by war's end. For the urgent need to rebuild facilities, it was decided to create a ministry specific to the task. In June 1949 the MOC was divided into the Ministry of Postal Services and the Ministry of Telecommunications.

At the same time, the government established the Advisory Council for the

Reconstruction of Telegraph and Telephone to consider postwar policies. It published its final report on March 31, 1950. The Council considered direct government operation, private management, operation by a public corporation, and a dual system of direct government operation and private management. It insisted independence and flexibility were critical to quick recovery from war damage. The report concluded that although private management provides the greatest independence and flexibility, the public nature of telecom service, the need for technical standardization, and a natural monopoly structure meant a public corporation acting as much as possible as though it had private management was most appropriate. In addition, to provide funds for reconstruction, the government was waiving taxes on the telecom provider, which was easier to do if it was a public corporation rather than a private venture.

After the government-monopoly recommendation was sent to the prime minister, trading companies in Japan adamantly advocated that international service be separated and operated by private management. Government leaders were willing to consider this. The argument in favor was that trade was of vital importance, and international telecommunications are critical to international trade. To the trading companies, recovery of the overseas system was quite simply more urgent than rebuilding the domestic system. It was feared that if both services were operated by the same organization, recovery of international services would be delayed due to difficulties providing domestic services. It was expected that the recovery of international telecommunications would be more quickly realized if managers could concentrate on international services alone.

Separation prevailed. The NTT Law for domestic service, and the KDD Law covering international service, were drafted at the same time; both passed the National Diet on July 31, 1952. Although NTT started operations August 1, KDD did not begin until the start of the next fiscal year on April 1 1953. From August 1 through March 31 NTT operated international services. (NTT's formal Japanese name is Nippon Denshin Denwa—telegraph and telephone—and the company is referred to as Denden. The English acronym is used in this chapter. KDD is Kokusai [international] Denshin Denwa. The company is referred to as KDD even in Japan.)

Large parts of the Ministry of Telecommunications were transferred to the newly created NTT and KDD, and the remaining regulatory sections were combined with the Ministry of Postal Services as the Ministry of Posts and Telecommunications (MPT).

NTT was a public corporation *(kosha),* one of three (with the railways and the salt and tobacco monopoly). As recommended, it was intended to act like a private corporation in most ways, but it was entirely owned by the government. In contrast, KDD was quickly turned into a private company. In May 1953 about half the shares in KDD were sold to private parties by the Ministry of Finance. Further shares over to NTT. A few months later NTT sold half of those. NTT still owns about 10 percent of KDD. Other large holders include MPT's Mutual Aid Assoc. (11 percent), NTT's Mutual Aid Assoc. (2 percent), and, as one would expect of a large utility, insurance companies. In September 1991 MPT announced it would end its opposition to any foreign ownership of KDD and NTT shares. Non-Japanese would be collectively limited to 20 percent of each company.

22.2.1 Funding

One other argument in favor of government ownership had been that profits from the telephone business would be a good revenue for the government. Restoring and expanding basic services was a major mission of NTT, a task that required considerable capital investment. Thus, the Diet decided profits should be retained and reinvested in the construction of network facilities and the improvement of telephone quality. NTT was allowed to charge high rates and thus earn substantial profits in the early years, but rates moved up very slowly.

As an additional source of funds, in 1953 NTT began requiring subscribers to pay an installation charge and to purchase telephone bonds. From 1960 until 1982, when the requirement was dropped, the stipulated amount was 150,000 yen (a bit more than two weeks cash earnings for regular employees in 1982). These bonds provided over 25 percent of funds available to NTT each year between 1960 and 1974; their contribution had declined to about 10 percent in 1982. (As a fringe benefit to the whole economy, these bonds could be traded, and a market developed for them that provided some indication of a ''free market'' interest rate during a period when other deposit and borrowing rates were tightly controlled.)

NTT used a series of five-year projects, beginning in 1953, to achieve its goals. As an example of how financing was obtained, during the third project period (1963–1967) internal funds provided 52 percent, subscriber bonds 32 percent, installation charges 4 percent, and government investment and loan programs 12 percent.

Rate increases came slowly after 1953. The unit call charge remained 7 yen from 1953 (when it had been increased from 5 yen) until 1976, when a 10 yen rate was implemented. Exchange line charges went nine years with no increases in major cities, sixteen years in the rest of the country. Increased usage, however, meant revenue growth, and NTT was profitable. Then, in the wake of the inflation induced by the 1973 oil crisis, the company incurred a deficit. In 1976–1977 it was allowed two increases that together doubled monthly rates.

Itemized bills were not introduced until 1986, and then as an extra-cost option—100 yen per page—although a customer can inspect the detailed billing at an NTT office. In 1991 some areas still could not provide the information. Most Japanese have forgone itemization. However, in 1990 NTT announced it had found 4.8 billion yen in accidental overcharges during the previous five years. International call billings have always been itemized.

22.3 International Service

When Japan initiated domestic telegraphy as a direct governmental enterprise it did not have either the financial or technological resources to initiate international telegraphy. Therefore, the government licensed a Danish company, Great Northern Telegraph (GNTC, called Taihoku Denshin Kaisha in Japan) on August 25, 1870, to install and operate submarine telegraph cable between Japan and other countries. A cable between Nagasaki and Shanghai began service in June 1871, reaching Europe through connecting lines. GNTC also installed cable between Nagasaki and Vladivostok, which was put in service in November 1871. Japan signed the International Telegraph Treaty in 1879.

In 1906 the Japanese government and an American private company jointly installed a submarine telegraph cable between Tokyo and Guam (some 2500 km south of Tokyo) and connected it to the telegraph cable between Guam and the U.S. mainland, ending GNTC's monopoly. The facilities were leased to the government, which operated the service. The cable itself, however, was mainly controlled by the American partner. Japan still had to rely on foreign companies for know-how and equipment.

22.3.1 Wireless and Telephone

Radio telephone communication using long waves became commercially viable at the beginning of this century, and Japan opened its first wireless radio communication circuits to Petropavlovsk from Ochi Ishi, near Nemuro City in Hokkaido, in June 1915. In the same year, a Wireless Telegraph Law was enacted. Although long wave is suited to international telecommunications, much electricity was needed for transmission, making it quite costly. Moreover, the number of available frequencies was limited to about 120, resulting in severe competition among governments to secure them. Japan thus sought to construct transmission stations as quickly as possible.

However, as the government had inadequate financial resources for this venture, it was decided facilities for international telecommunications should be owned by private businesses. As a result, a new company, Japan Radio Telegraph (Nihon Musen Denshin KK), was established in 1925. It owned facilities for international communications and leased them to the government. Because of various government subsidies, it was essentially a semi-government operation.

In the late 1920s short-wave radio technology developed. Because short wave requires less energy, and is thus less costly, intensive effort was made to use it for international communications. Short-wave service to the United States started in 1928.

The first international telephone service was between the United Kingdom and the United States in 1927. Japan had experienced a major crisis in its banking and financial system that year and the government had insufficient funds to finance international service. In December 1932 a new private company, International Telephone (Kokusai Denwa Kaisha), was established. It constructed and owned radio transmission facilities for international telephone service and leased its facilities to the government.

Thus, International Telephone and Japan Radio Telegraph owned the international facilities for telephone and telegraph, respectively, leasing them to the government, which operated the services. The companies merged in 1938 as International Telecommunications (Kokusai Denki Tsushin).

As Japan expanded militarily into Manchuria and China, international telecommunications facilities also expanded. Just before the outbreak of the Pacific War in 1941, Japan operated thirty-three international telegraph circuits, thirteen international telephone circuits, and four international telephoto circuits, more than any country except Great Britain and the United States. GNTC's license expired in 1943. Almost all the international telecommunications facilities installed before the war were completely destroyed during the war.

22.3.2 After the War

International Telecommunications was dissolved by GHQ order soon after the war, and its facilities were absorbed by MOC. Thus, in 1947 MOC owned and operated both domestic and international telecom services as a monopoly for the first time. However, GHQ renewed Great Northern's license in September 1947 and the company resumed service on November 15 using its Nagasaki–Vladivostok cable. In 1954 GNTC and KDD agreed to jointly own and operate the two cables installed by GNTC before the war. By 1957 they had deteriorated to the point the two companies agreed to construct a new broad-band cable. The new company was half-owned by each firm. It was completed in July 1969 and called JASC.

22.3.3 KDD Monopoly

KDD was established as a monopolistic, regulated joint stock company to give it freedom in management and in the adoption of new technologies, while at the same time having it serve the public interest. It was rather strictly regulated, but less so than NTT. KDD was controlled mainly by the MPT, while in many situations NTT had to obtain approval from the Diet.

In the early 1980s a scandal involving top executives of KDD and high-ranking bureaucrats at MPT made headlines. The origins go back to the mid-1960s when continued very high calling charges began producing substantial profits for KDD. Rather than reduce rates, KDD hid part of its profits by spending lavishly on entertaining "business partners"—a valid expense up to a point, except that many of the guests were politicians and MPT officials. This incident was a factor in the opening of international calling to competition.

The trend in the 1950s, particularly in Europe, was socialization. The shift from private to public management was often seen in public utilities like telecommunications. The reverse shift in Japan—from government operation to private management—was a surprise to many. Viewed from Japanese business tradition, however, it was a natural shift. There is only one instance in modern

Japanese history where existing private companies were nationalized—the major railroads in 1906–1909—and that was for military reasons in the wake of the Russo–Japanese War (1904–1905). Japan National Railway was split up into six regional companies and privatized in 1987.

KDD grew steadily as the Japanese economy recovered. International telecommunication before the 1960s depended on submarine cables and short wave. KDD initiated telex service in 1956 and started to provide leased telegraph circuits to trading companies in 1962. The number of frequencies allocated to KDD increased from 95 in 1953 to 360 in 1963, while the number of international circuits increased from 58 to 292. KDD's sales increased from 4.5 billion yen in 1953 to 10.8 billion in 1963.

Short wave has serious defects due to its susceptibility to natural conditions such as change in the ionosphere. Also, scarcity of spectrum means it is difficult to secure a large number of circuits. KDD followed installation of a newly developed submarine coaxial cable between Europe and North America in 1956 (TAT-1) with interest. The success of TAT-1 stimulated KDD to plan installation of the same type of cable between Japan and North America. The cable, TPC-1 with 138 circuits, was completed in 1964.

Production and installation of TPC-1 were totally dependent on U.S. technology. However, KDD learned enough to develop its own know-how in cooperation with NTT. In 1972 the two firms together developed a submarine cable technology—dubbed CS-12M—that provided a tremendous increase in circuits per cable. KDD then installed submarine coaxial cables on seven routes, five using technologies developed in Japan. In 1989 KDD installed a trans-Pacific coaxial cable (TPC-3) using optical fiber.

The first communications satellite was launched by the United States in 1962. KDD constructed its first earth station shortly thereafter. In 1964 the first international relay broadcasting was conducted by the Japan Broadcasting Corporation (Nippon Hoso KyokaiNHK) using the U.S. satellite and KDD's earth station. In 1964 Intelsat was provisionally established. It launched its first satellite in 1965 and began commercial operation. In 1967 KDD started using Intelsat satellites for communication between Japan and the United States, and in 1969 constructed another earth station to establish circuits to Europe and the Mideast.

The dramatic growth of Japan's international telecommunications, as shown in Table 22.1, paralleled Japan's rapid economic expansion since the Korean War. This has produced a strong demand for international telecom services, and this in turn has facilitated further expansion of trade. The number of international circuits operated by KDD increased from 71 in 1953 to 1,240 in 1970 to 8,279 in 1985 (two-thirds by satellite and one-third by submarine cables). There were more than 20,000 circuits in 1990.

22.3.4 International Public Services

Before telex service started in 1956, telegraph accounted for 80 percent of KDD's annual income and 94 percent of all international traffic through public

Table 22.1. Expansion of International Telecommunications[a]

Year	Circuits	Telegrams[b] (thousand times)	Telex[b] (thousand times)	Telephone[b] (thousand times)	Public Data Transmission (no. of Contract)	TV Transmission (frequency)
1953	71	3,430	—	1,900	—	—
1955	92	3,670	—	1,600	—	—
1960	217	4,140	400	1,900	—	—
1965	463	5,020	1,100	5,200	—	—
1970	1,240	5,820	4,360	2,180	—	—
1975	2,425	5,820	16,230	8,570	—	—
1980	4,553	3,340	37,980	23,430	406	2,559
1985	8,279	1,530	50,170	95,630	5,890	4,832
1990	20,397	600	17,530	322,750[c]	17,489	16,355

[a]This table was made based on data in *Tsushin Hakusho* (White Paper on Communications), 1973, 1978, 1983, 1988, and 1991 editions (Ministry of Posts and Telecommunications, 1973, 1978, 1983, 1988, 1991) and *Denki Tsushin Tohkei, 1991* (Ministry of Posts and Telecommunications, Telecommunications Policy Bureau, 1991).

[b]The figures for international telegraph, telex, and telephone calls include those of transmission, reception, and relay.

[c]The figure for international telephone calls in 1990 covers only calls made through KDD and does not cover calls by other new international carriers, ITJ and IDC.

circuits. The number of international telegrams peaked in 1969 at over 6 million; by 1990, the number had fallen to 600,000. The use of telex, in turn, reached its peak in 1984 with over 52 million messages and had fallen to around 17 million in 1990. International telephone volume has been increasing remarkably as users shift to fax. Another reason for this growth is the expansion of computer communication via phone circuits. Since 1970 the number of international telephone calls has increased even more rapidly than the number of circuits, exceeding 320,000 in 1990.

NTT started a data communication service called MARK I in November 1971 for domestic users. KDD began providing International Computer Access Service (ICAS) in 1980. (In Japan, there were several "ten-year gaps" between domestic and international services. This is a typical example.) The service was upgraded using a packet exchange called Valuable and Efficient Network Utility Service-Packet (VENUS-P) in 1982. The two were integrated in 1983 under VENUS-P. Another rapidly expanding area of KDD's services has been international television transmission.

22.3.5 International Private Leased Circuits

KDD leases two types of international circuits. The first is a telegraph class and the second is voice (telephone) class. These were originally to be used for telex and telephone, respectively. Voice class usage has increased dramatically due to facsimile transmission and development of modems that allow computerized data transmission. Telegraph class circuits peaked in 1981 at 625. In

1987 there were 458 telegraph class and 851 voice class leased international circuits.

There are several regulations restricting use of international leased circuits. Under the Telecommunications Business Law and the ITU Convention, use of international leased circuits is limited to three cases. These are (1) intracompany communication (inhouse use), (2) communication within the same group of companies (subscriber holds more than 50 percent of the stock of the partner company), and (3) communication between companies having close business relations (20 percent of subscriber's sales must be to the other company).

A major purpose of these rules is to prohibit the use of international leased circuits in the same way as international public circuits, including resale. Therefore, use is permitted for simple data processing, including data base service and financial market information, even if the communicators do not fit in any of the three categories.

22.4 Early Deregulation and Introduction of Competition

Using microwave systems, electric power companies and the Self Defence Forces (military) established their own communication networks. In rural areas where construction of facilities for NTT's telephone system was delayed, wired radio and telephone systems expanded rapidly.

The Wire Telecommunication Law and Public Telecommunication Law were enacted in 1953. Before then, those who wished to install private lines had to get permission from MOC. Under the new law anyone could install private lines provided that the system was reported to MPT and did not affect public circuits. This was the first instance in Japan of telecommunications "deregulation" caused by technological development.

During the 1950s and 1960s, policy goals for telegraph and telephone were very simple. They were the reconstruction of facilities destroyed during the war, technical standardization, expansion of telephone networks, and completion of automatic switching on a nationwide scale. Because these were technical problems, they were left to NTT and KDD.

In the late 1960s various new media such as data and facsimile communications began to emerge and, together with the growth of the computer industry, rapid expansion of various forms of telecommunication and information services was projected. The development of cable television in the United States and new ideas like "wired cities" caught the attention of Japanese, and the first CATV expansion occurred in the early 1970s as a part of the *johoka shakai* (information society) boom.

To cope with such a rapidly changing situation, MPT created a special section named the Telecommunications Policy Division in 1972 to plan and execute nonbroadcast telecommunications policies. During the 1970s the division completed a number of well-known projects, such as the Information Flow Census (see Ito 1981, pp. 671–98), the Tama Community Cable Information Service (CCIS) experiment, and the Character And Pattern Telephone Access

Information Network System (CAPTAIN) videotext experiment (on these last two, see Araki 1984, Ito and Oishi 1987, and Ikeda 1985). The division was later expanded and strengthened; it became the Telecommunications Policy Bureau in 1980.

Though restrictions on privately installed circuits were eliminated by the Public Telecommunication Law in 1953, many other restrictions continued, such as those regarding what equipment could be connected to NTT lines, sharing of private leased lines, message exchange using private leased lines or cable tv circuits, sublease of private leased lines, and interconnection with NTT lines. These restrictions were justified on the basis of the security of public networks, technical standardization, protection of communication secrecy, and protection of NTT's monopoly. Complaints about restrictions were rare when services were limited to the telegraph, telephone, and telex. They became loud, particularly among business circles, when data communication rapidly spread and various information and communication services became viable in the late 1960s.

Users and information service providers began to feel the restrictions were a serious hindrance to free and efficient information flow. MPT understood the problem and announced a "Basic Policies for the Liberalization of Telecommunication Circuits for Data Communications" in September 1969. The Public Telecommunication Law was amended in May 1971 based on these basic policies. This amendment is called the "First Liberalization of Telecommunication Circuits." Connection of computers, information receiving equipment (terminals), and fax equipment to NTT public telephone lines became legal. At the same time, NTT and KDD were allowed to engage in the data communication business using their computers.

Thus, after 1971 many private information service companies emerged. They provided stock market quotations to brokerage firms, data processing and storage services for small companies, literature information for researchers, and so on. Sharing private leased lines was permitted on condition that the parties involved have business transactions on a regular basis. Message exchange through private leased lines also became permissible on the condition that the messages have something to do with the data being exchanged.

NTT's two major goals for the postwar telephone system had been immediate installation and immediate connection. After eighty years, the backlog of requests for phone service disappeared in the late 1970s, and with it went the chief justifications for NTT's monopoly. Furthermore, because of developments in transmission system technology, the economies of scale that had justified monopoly were also lost. Thus, by the late 1970s and early 1980s, most experts in this area came to agree that fundamental rethinking and possibly drastic changes in the traditional public telecommunications order were needed. Nagai continues this discussion in the next chapter.

22.5 From Monopoly to Competition

Under the "Second Liberation of Telecommunications Circuits" in 1982, usage restrictions on private leased circuits were relaxed and more private companies entered this area. For example, Dentsu Kokusai Jobo Sabisu (Dentsu International Information Service) started a calculation service and a sales and inventory management service in 1982 using General Electric Information Service Co. (GEISCO) computers in Cleveland, Ohio. As a result of a major reform in 1984 it became possible for private companies to enter all areas of telecommunications, international or domestic. This allowed Shikyo Joho Senta (Market Information Center) to use private lines leased from KDD to provide stock market quotations from Reuters in London starting in 1987.

As deregulation in the domestic system proceeded, it became clear that international agreements and ITU's policies were not as progressive as those for the domestic system. Japanese government and business leaders perceived existing international rules and regulations as obstacles to free entry by private enterprise. However, those who supported the existing international framework argued that Japanese firms, together with U.S. interests, were trying to skim the cream and pursue selfish interests. These issues will be discussed with regard to common carriers and international VANs.

22.5.1 International Common Carriers

According to the Telecommunications Business Law enacted in 1984, common (Type 1) carriers are companies owning transmission facilities such as cables, microwave networks, satellites, earth stations, antennas, and the like. In July 1986 five trading companies, an electronics manufacturer, and a bank formed International Telecom Japan (ITJ) (Nihon Kokusai Tsushin) to study the feasibility of entering the international telecommunications business. (The companies were Mitsubishi, Mitsui, Sumitomo, Marubeni, and Nissho Iwai, plus Matsushita and Bank of Tokyo.)

Intelsat initially leased transponders exclusively to recognized private operating agencies (RPOA). However, that policy has changed to permit private businesses to lease international circuits. Government approval is required to hold Intelsat harmless if the private company causes any international trouble.

ITJ began regular service on April 1, 1989, by leasing transponders on Intelsat satellites (ITJ owns ground stations), part of a submarine cable between Japan and Hawaii (TPC-3) from KDD, and part of a submarine cable between Hawaii and U.S. mainland (HAW-4) from AT&T. The resulting network connects the major countries in Asia, North America, and Western Europe.

Kokusai Dejitaru Tsushin (International Digital Communications, IDC) was established on November 7, 1986. Major investors are C. Itoh & Co., (20 percent), Cable & Wireless (C&W) (20 percent), Pacific Telesis Intl. (10 percent), Toyota Motor Ltd. (3 percent), Merrill Lynch (3 percent), two electronics companies, and three banks. A notable feature of this company is that it is 33 percent owned by Japanese firms. IDC was the first case in an advanced country in which so much of an international carrier—albeit a fringe one—is held by a large foreign international carrier.

Under the 1984 Telecommunications Business Law, foreigners and foreign firms were not allowed to hold shares of NTT or KDD, and ownership of new Type 1 carriers is restricted to 33 percent. Other areas, including Special Type

2 (large-scale VANs), are open to foreign entry. Therefore, from a regulatory viewpoint, there should have been no problem with IDC's ownership. Nonetheless, on the grounds foreign companies are not permitted to control major carriers in any advanced country, MPT expressed dislike for IDC's plan.

MPT strongly suggested IDC and ITJ be unified and shares of each foreign company in the combined company be kept at less than 5%. On a more positive note, MPT also argued a combined company could better compete against the much larger, well-established KDD. Both C&W and Pacific Telesis International, a subsidiary of a Baby Bell company, strongly protested. Their protests were backed by the US Trade Representative, the US Secretary of Commerce, and British Prime Minister Thatcher, and became an international issue.

C&W planned to make IDC one of the cornerstones, and possibly the center, of its business activities outside England. Approximately 60 percent of C&W's revenue are from the East Asian market. C&W's main Asian office is in Hong Kong, but management, uneasy about 1997, has expressed interest in moving it to Tokyo. Japanese liberalization would appear to provide a perfect opportunity for this. Furthermore, C&W has an ambitious plan called the "Global Digital Highway"—the connection of all the major industrial and financial centers in the world. It is now constructing optical fiber and digital cable networks across the Atlantic and North America, and between Hong Kong, South Korea, Japan, and the United States.

Negotiations were held between ITJ and IDC decided to embark on business separately, and Prime Minister Nakasone announced at the Venice Economic Summit that both ITJ and IDC would be given licenses as international Type 1 carriers. IDC began private leased line service in May 1989, covering the United States, Singapore, Hong Kong, Canada, the United Kingdom, West Germany and South Korea.

IDC is using Intelsat (IDC owns ground stations) and participates in joint ownership of trans-Pacific Ocean cables, like KDD and ITJ. However, it is now constructing the Northern Pacific Submarine Cable (NPC) for its own exclusive use. This is an optical fiber cable with an equivalent capacity of 17,000 telephone circuits, directly connecting the United States and Japan. Many experts worry that this cable will create excess capacity on the route. Those who support the traditional international framework argue that price reduction, caused by the oversupply or severe competition may benefit Japanese and U.S. industries, but be harmful to a world system like Intelsat.

In 1985 KDD started lowering its tariffs for direct-dial and private leased circuits. Still, ITJ and IDC entered the market on October 1, 1989, offering rates some 20 percent below KDD's for international calls. By November they reportedly had 28 percent of the traffic to the United States and Hong Kong—even as KDD's absolute volume increased from a year earlier. KDD quickly reduced the rate differential to less than 7 percent and by 1990 it was less than 2 percent. Similar narrowing took place for leased lines. By 1990 KDD rates were less than half what they had been five years earlier. The result was flat revenue and plunging profits for KDD from 1989 to 1990.

KDD has done more than cut prices to meet competition. It has established "telehouses" in the United States and United Kingdom that enable companies (mostly Japanese multinationals) to locate computer and communications systems at shared, KDD-managed facilities.

22.5.2 International Type 2

Under the Telecommunications Business Law, Type 2 business refers to companies providing service by leasing transmission facilities from Type 1 carriers.

Early providers offered administrative or data-base services. This involved no information exchange among customers, so it did not contradict international system regulations. A problem arose, however, in the information exchange between customers using electronic mail systems and interconnection between computers. This international VAN service has become a controversial issue. Such service is permitted, in a narrow sense, if it is provided through public data communication networks such as KDD's VENUS-P. However, public circuits are costly and inflexible for large users. Therefore, VAN service providers want to use private leased lines.

CCITT's recommendation D 1 prohibits leasing international private circuits to anyone who plans resale. On the other hand, CCITT D 6 permits exceptions to D 1. Worldwide telecom service organizations such as the Societe International de Telecommunications Aeronautiques (SITA) and SWIFT were made possible by D 6. D 6 also permits exceptions for organizations designed as a RPOA.

Measures considered to bridge the gap between the Japanese domestic system and international systems were:

1. Revisions to CCITT D 1 so that resale of international private leased circuits is permitted.
2. Revisions to CCITT D 6 so that international VAN companies are regarded as exceptions like SITA and SWIFT.
3. Bilateral or multilateral agreements, independent of ITU rules, so that the concerned countries can permit operation of international VAN services using private leased circuits.
4. Granting international VAN companies RPOA status.

Instead of adopting any of these measures, MPT revised the Telecommunications Business Law in 1987. Type 1 carriers were permitted to lease circuits to Type 2s under conditions different from those available to other users. The Japanese term for them translates as a "non–tariff-based circuit." They are not as flexible as private circuits, but they are billed at a monthly fixed price. The idea is to treat international VANs as carriers in order to avoid D 1 problems. (Because the United States, in particular, continued to campaign against D 1 restrictions in various international forums, in it was reduced by CCITT to applying to "simple resale," in March 1991.)

By the early 1990s fax accounted for half the communications flow between the United States and Japan. In March 1990 the United States and Japan negotiated an agreement covering value added resale. Among other things, this allowed the delivery of a fax at a designated time and to multiple receivers beginning in May 1990. A similar arrangement was made with the UK that became effective in August 1991. By May 1991 there were five firms offering value added fax service. Because they used non-tariff lines their rates were about 20% below those of the common carriers.

There were 16 other providers of international VAN service operating in Japan in May 1991. Many are joint ventures of major multinationals (AT&T, NTT, IBM, General Electric, Fujitsu and so on). Among the more specialized VANs are Intech (of Japan), GTE Telenet, and Tymnet (both US firms). In addition to the United States and the United Kingdom, international VANs connect Japan to Hong Kong, Singapore, Canada, France, Germany, the Netherlands, and Switzerland.

22.6 Factors in Policy Change

Telecommunications policy is the mass of administrative, judicial, and legislative measures taken to achieve social goals through the electronic media. Therefore, policy is basically determined by two questions: "What are the social goals to be achieved?" and "What are the means to achieve these goals?" Because of political considerations, shifts in social values, and technological innovations, the answers change; therefore, policy must also change. Change in what means are best occurs mainly through technological innovation. In Japan, changes in social goals due to value changes or political reasons have not been important, except to the extent technological change has expanded the range of achievable social and political goals.

22.6.1 Goal Changes

Until after World War II military secrecy was an important reason for keeping common carriers under strict government control. Today, however, the military has developed separate networks. Similarly, expansion in terms of range and availability had long been a major goal, to bring Japan up to the level of advanced countries. This has been largely achieved. Without these arguments, support for a strictly controlled government monopoly is weakened.

Another long-time major goal has been to provide facilities to meet social needs for communication. These needs traditionally have been defined simply in terms of quantity—the ratio of telephones to population, and so on. In the 1980s, however, this took on other meanings—an expansion of "needs" that includes fax, videotext, data telephones, data communications, and information services. Quality—diversity—has become an element.

Generally speaking, monopoly is suited for large-scale, simple, and fixed services; it is less well-suited for diversified, flexible services. Laws and regulations governing monopolistic corporations are one of the major reasons for this; however, mere size can also be a serious obstacle to providing diversified, flexible services. This was an argument of those promoting privatization of NTT.

The idea of fostering information-related industries as a strategic industry was not common before the mid-1960s. It has since been included in the policies of most industrialized countries.

In Japan in the mid-1960s, faced with serious industrial pollution inside Japan and tough competition from newly industrialized countries such as Korea, Taiwan, and Singapore, Japanese economic policymakers began to feel that the industrial structure as a whole should become specialized in more pollution-free, knowledge- and technology-intensive industries. Foremost among these were MITI and its advisory groups, such as the Sangyo Kozo Shingikai (Industrial Structure Council). In the late 1960s and early 1970s policies were implemented to strengthen Japanese information and communication industries.

The Japanese information and communications industries along with MITI had long insisted on deregulation of communication circuits and enhanced services as a way of fostering industrial expansion. Sometimes MPT and NTT were cautious about deregulation, professing concern about maintenance of safe, stable public service. However, MPT could not deny the importance of strengthening the Japanese information and communications industries. By the mid-1980s science and technology policies such as national development plans for rockets were seen by many as related to telecommunications policies, particularly concerning satellite communication.

22.6.2 Changed Means

Because telecommunications laws and regulations have historically tended to be very specific to the available technology, and thus very narrow and restrictive, new laws or amendments to old laws ultimately have resulted each time a new service has appeared. In Japan, the Telegraph Law of 1900, Wireless Telegraph Law of 1915, Radio Law and Broadcast Law of 1950, Wire Telecommunications Law and Public Telecommunications Law of 1953, and Cable Television Broadcast Law of 1972 were all enacted to regulate new services.

Further, these laws have been amended to cope with changes caused by technological innovations. For example, the Public Telecommunications Law was amended in 1971 to enable connection of computers, terminal equipment, and fax to public networks. It was amended in 1982 to permit private leased circuits for information and communication services (''value-added'' or ''enhanced'' services). Home shopping and home banking are examples of services that require new legislation or amendments of existing laws.

Cable television started as a supplementary broadcasting system and has become a two-way telecommunications system. According to Baughcum (1986, p. 91), development of computer and microelectronics technologies has blurred both the distinction between telecommunications and data processing as well as the lines between the traditional basic grouping in telecom equipment: central office switching, transmission, and customer premises equipment. Legal adjustments are needed to cope with these changes.

Many regulations were developed on the assumption resources for telecommunications are scarce. Networks have been limited in most countries by the enormous amount of money required for construction. However, ways have been found to expand the usable radio spectrum and bring down the per circuit cost of other types of networks (see, e.g., Gilder 1991).

As the scarcity is reduced, groups wishing to take advantage of these circumstances work for legislation to allow them to do so. In Japan strong pressure emerged from three sources. First, private business circles demanded changes to permit use of telecommunication resources for profit making. Second, consumers demanded better, cheaper, more varied, and flexible services. Third, there has been mounting public criticisms of the bureaucratic inefficiency and red tape of mammoth, monopolistic government organizations. These factors combined to bring about the drastic legal reform of 1984.

22.6.3 Cost Changes

Costs and, therefore, prices usually drop as a result of technological innovation. Demand increases for products and services with falling prices. Telecommunications is a typical example. Long distance used to be expensive, but prices have fallen so much that more people telephone than write letters. Many business transactions are done on the telephone, which has required legal adjustments to validate verbal commitments. Basic technologies for facsimile have been known since the 1930s, but the cost was so high that manufacturers did not develop equipment for the general public. Fax machines can now be found in many homes. In Japan, citizens can receive most documents published by public offices via fax.

Telecommunication has long been considered subject to economies of scale. This has provided one of the strongest arguments against permitting competition in basic services. Network expansion, however, has almost reached a saturation point in advanced countries, and the means of providing telecom services have become diversified. In this environment economies of scale are not as great. Moreover, the value of competition has increased because diversified—differentiated—services are more effectively provided by independent companies competing with each other than by a monopolistic public corporation. These are major reasons Japan decided to introduce competition.

Technological innovation affects policies through its impact on social goals and through development of better means to achieve given social goals. Despite cultural and situational differences, countries face similar (if not exactly the same) socioeconomic problems from technological innovation. Therefore, by exchanging information on how to cope with these problems, every country can learn how to make the best use of technology to achieve its social goals.

Bibliography

Araki, Isao. 1984. "Changing Ways of Communication with a New Medium—The Case of Hi-OVIS in Japan." In Georgette Wang and Wimal Dissanayake, eds.,

Continuity and Change in Communication Systmes: An Asian Perspective. Norwood, NJ: Ablex.

Baughcum, Alan. 1986. ''Deregulation, Divestiture, and Competition in U.S. Telecommunications: Lessons for Other Countries.'' In Marcellus S Snow, editor, *Marketplace for Telecommunications: Regulation & Deregulation in Industrialized Democracies.* New York and London: Longman.

Gilder, George. 1991. ''What Spectrum Shortage?'' *Forbes,* May 27, p. 324–32.

Ikeda, Ken'ichi. 1985. '' 'Hi-OVIS' Seen by Users: Use and Valuation of Japanese Interactive CATV.'' *Studies of Broadcasting,* 21:95–120 (Mar.).

Ito, Youichi. 1981. ''The 'Johoka Shakai' Approach to the Study of Communication in Japan.'' In G. Cleveland Wilhoit and Harold de Bock, eds., *Mass Communication Review Yearbook,* vol. 2, p. 671–98. Beverly Hills, CA: Sage.

Ito, Youichi, and Yutaka Oishi. 1987. ''Social Impacts of the New Utopias.'' In William H. Dutton, Jay G. Blumler, and Kenneth L. Kraemer, eds., *Wired Cities: Shaping the Future of Communications.* Boston: CK Hall.

Kokusai Denshin Denwa (KDD), 1979. *KDD Kabushiki-gaisha 25-nenshi (25-year History of KDD).* Tokyo: KDD.

Kanto Denki Tsushinkyoku, ed. 1968. *Kanto Denshin Denwa Hyakunenshi (100-year History of Telegraph and Telephone in the Kanto Area).* Tokyo: Denki Tsushin Kyokai.

NTT (Nippon Telephone and Telegraph) = Nippon Denshin Denwa Kosha. 1977. *Nippon Denshin Denwa Kosha 25-Nenshi (Twenty-five Year History of NTT).* Tokyo: Denki Tsushin Kyokai.

———. 1986. *Shinsei NDD eno Kiseki (Step to the New NTT).* Joho Tsushin Sogo Kenkyujo.

Oniki, Hajime. 1991. ''Construction of Telecommunications Infrastructure in Japan: 1950–1980.'' In G. Russell Pipe, ed., *Eastern Europe: Information and Communication Technology Challenges.* Budapest: Statiqum, for TIDE 2000 Club. (TIDE = Telecommunications, Information and InterDependent Economies in the Twenty-first Century.)

23

Japan: Technology and Domestic Deregulation

SUSUMU NAGAI

A shift in emphasis from quantity to quality, liberalization of private leased circuits, institutional reform, and, of course, rapid technological innovation characterize Japanese telecommunications since the late 1970s. The most salient point associated with the Public Telecommunication Law in effect from 1953 to 1984 is that there were two monopolistic service providers—one domestic and one international. Government units—including Japan National Railways, the Ministry of Construction, and the Self Defense Forces—were permitted to build private networks. The law, however, prohibited all third parties from providing message switching services to the public, outlawed shared use of leased lines, and prevented interconnection of privately operated facilities to the public network.

During the monopoly period, subscriber needs for services advanced and diversified. Nippon Telegraph and Telephone (NTT), as a public corporation and the domestic service provider, had two major goals from its inception: construction of a nationwide automatic dialing system and dissolution of a long waiting list for telephone installation. By 1977, when the number of subscribers had reached 35 million, the second goal had been accomplished. In 1978 the first goal was also attained. NTT and its regulator, the Ministry of Post and Telecommunications (MPT), recognized that they had to redirect policy from a focus on rapid installation to increased usage. By the early 1980s NTT was the largest business corporation in Japan. Annual sales in 1983 were 4,499 billion yen (about U.S.$35 billion) and there were 318,000 employees.

A wave of new technologies hit telecommunications in the 1970s. Very large scale integrated chips (VLSICs) were developed, making it possible to develop and manufacture electronic and digital switching equipment. This spurred development of highly sophisticated new communication systems based on fiber-optics, microwaves, and satellites. If NTT offered all the emerging services, the largest corporation in Japan would only expand. If NTT did not monopolize the emerging technologies, services, and equipment, three alternatives were conceivable.

1. NTT would become a monopoly telecommunications facilities holding company leasing facilities to private service companies. As such, it would no longer provide service directly to the general public.
2. NTT would concentrate solely on traditional basic services (i.e., transmission of messages from one end to the other without any information processing or transformation). Telegraph, telex, telephone, and facsimile belong to this category. All enhanced services, including data communication, would be left to competition among private firms.
3. NTT would cease being a monopoly. There would be new entry and competition in all areas of telecommunications.

Private business circles and the Ministry of International Trade and Industry (MITI) were in favor of the second alternative. However, neither NTT nor the MPT liked the first two ideas, partly for political reasons, and chose the last alternative. Thus, in the early 1980s MPT decided to abolish the 1952 NTT Law and the 1953 Public Telecommunications Law and replace them with a new NTT Law and the Telecommunication Business Law.

This chapter continues from the previous one—in the late 1970s—to track deregulation of the domestic industry. For the impact of technology—particularly data processing—on telecommunications, this chapter goes back to the 1960s. Chapter 22 covered international communications.

23.1 Technological Pressure

The deregulation process was affected by technological changes both directly and indirectly. The principal indirect element was how to deal with data communications: There was strong pressure for market entry from the computer industry and others because of the fusion of computers and communications. This was manifest when a private company sought to construct its own on-line information processing system—using circuits leased from NTT—to rationalize its internal office work and communicate with affiliated subsidiaries.

Data transmission service was first introduced in 1964 by Japan National Railways and Japan Air Lines for their seat reservation systems. NTT's opening of data transmission service had an impact on the technology of information processing as well as on telecommunications technology in general. Pressure grew as information processing network systems were needed in both single companies and among diverse businesses. In 1968 NTT started a project called DIPS (for Dendenkosha Information Processing Systems), which entailed research with several electronics companies—including NEC, Fujitsu, and Hitachi—and by 1973 had begun to utilize the result. This kind of joint research promoted the technology capacity of Japanese electronics companies and encouraged the confluence of computer and communication technology, as well as new business stimulation.

NTT's data transmission and on-line information processing network system services increased very rapidly; NTT had installed 200 such systems by the early 1970s. NTT provided private leased circuits to the data processing industry and also owned data processing services connecting users to NTT's computers, especially at DIPS.

The data processing industry pressed MPT to open the system further, and in 1971 MPT proposed an amendment to the Public Telecommunication Law to allow public circuits to be used for data transmission. Shared use of private leased lines, called specific data circuits, was permitted when users shared a close, long-term relationship (such as that between a firm and its wholesalers and banks). However, private line users were prohibited from offering to transmit third-party communications. Only NTT could do this.

Other technological innovation provided opportunities to offer new services. In 1978 NTT offered a circuit switching service, called DDX, over its digital data exchanges. In 1980 packet switching service, DDX data networks, and carphone service were made available. NTT began providing a facsimile network service in 1981, debit cards for pay phones in 1982, and a videotex service called Character and Pattern Telephone Access Information Network (CAPTAIN) and television conference service in 1984. In September 1984 NTT started market tests on its Information Network Service (INS), which is commonly referred to as ISDN. INS was seen as the next evolutionary step in the advancement of communications systems providing for the increased and more complex information needs of individuals and businesses.

The Japanese government, through MPT as well as NTT, recognized the possibility of integrating technological innovation with demands for advanced services and proceeded to adopt a more active policy. In September 1980 MPT upgraded its Division of Telecommunications Policy into a bureau, which ranks with departments as the highest subdivision of, a ministry. In October MPT created the Telecommunications Policy into a bureau, which ranks with departments as the highest subdivision of a ministry. In October MPT created the Telecommunications Policy Conference, which issued a report in August 1981 titled "A Vision of Telecommunications Policy for the 1980s" (DTS 1981). The report urged re-examination of telecommunications administration and laws and the establishment of an integrated plan for the industry. The Conference also suggested a liberalization of data communications.

Following the report, MPT submitted an amendment to the Public Telecommunication Law permitting liberalized use of leased circuits. The ministry also proposed a bill regarding value added transmission operations, the VAN Law, allowing private companies to supply services to third parties. However, MITI, which had supported the information processing policy, and MPT disagreed on the proposal. MPT argued that regulation of VANs was necessary because they provided common carrier services and should assure the privacy of customers and avoid price discrimination. MITI argued that restrictions would interfere with development of the information processing industry.

The VAN Law proposal was ultimately set aside. In October 1982 amendments to the Public Telecommunications Law were enacted instead. Restrictions on third-party use of NTT leased circuits were substantially liberalized and small-enterprise VANs were approved as a temporary measure. This constituted the second deregulation of data communications. (The first, in 1971, allowed connection of computers, information receiving equipment—terminals—and facsimile equipment to NTT public telephone lines.)

Introduction of small-enterprise VANs meant management of companies became highly information-oriented, particularly in the retail, wholesale, and transportation industries. These industries gained new business opportunities through deregulation of telecommunications. MPT probably would have become the government agency heading telecommunications policy even if the VAN Law had been approved.

23.2 The Government Reform Movement

In the early 1980s anxiety about the huge government deficits had increased and there was growing concern with the scale of government involvement in the economy. Thus, in October the cabinet finished drafting a bill to establish a Second Provisional Commission for Administrative Reform (called *Rincho* for short and sometimes referred to as the Second Ad Hoc Council on Administrative Reform, an incorrect translation). This was a powerful agency to confront the government's financial crisis and to decrease its inflated size—more specifically, to head off tax increases by addressing government structure. Rincho's Fourth Division focused on privatization of public corporations such as the nearly bankrupt Japan National Railways. (A good summary of the reform movement is Lincoln 1988, pp. 116–22. The first reform commission, in the early 1960s, led to few changes; see Kumon 1984, pp. 145–47.)

NTT and JNR, as public corporations, were under heavy government control. For example, telephone tariffs and NTT's budget were controlled by the Diet. NTT's investment budget and even the number and salaries of employees were regulated by the Ministry of Finance (MOF). As a consequence, NTT was in many ways inefficient and lacked flexibility in operations. Still, it was considered to have an overall high productivity level.

In its third (July 1982) report Rincho recommended that NTT and JNR be privatized. As part of this, NTT would divest certain activities—such as repair and maintenance, CPE, and data communications—as a way of introducing competition. It was also recommended that NTT be divided into a main operating company handling trunk service and several local companies responsible for local service within five years of initial reform. This idea was derived from AT&T's divestiture. Rincho was very pessimistic about the possibility of new competitive entrants in the basic telephone service sector but very optimistic about competition among local companies. Such competition would have been indirect because each would maintain a monopoly in its own area. (JNR also was to be split into regional companies—a proposal that was implemented in 1987.)

Opponents of the plan stressed that there would have be substantial costs in separating local service areas from NTT, including deterioration of technological identification, different pricing structures between regions, and difficulty in separating long-distance carrier revenues from those of the several local telcos. Supporting Rincho's recommendations were those who believed NTT should be confined to basic services.

After publication of the Rincho's report, MPT expressed opposition to privatization of NTT and some of the other proposals. However, MPT changed its opinion and began to prepare a new law in consultation with the governing Liberal Democratic Party (LDP). Privatization—but no divestiture—of NTT was proposed, along with general introduction of competition.

NTT had good relations with Zendentsu, its very powerful trade union, even though NTT's wage levels were decided in parallel with those for employees of JNR, a then very inefficient company with a huge operating deficit. Zendentsu eagerly advocated more flexibility in wage negotiations, arguing that wages were below the average for workers with similar skills in other industries. At its annual conference in 1980, just before Rincho was set up, Zendentsu proposed a set of institutional changes at NTT. These included transforming its public corporation status to a more flexible one, such as "third-sector type companies" (government–private joint ventures), establishing self-management, and deregulation.

After Rincho's report, Zendentsu conducted a campaign against privatization of NTT, collecting about 10 million signatures on a petition to the Diet. However, it changed its strategy when the reform bill was presented in the Diet. Zendentsu negotiated with the LDP and its traditional allies in the opposition parties to amend the bill. Although stressing the public's interest in NTT's nationwide network and asking the government to maintain a balance between this and competition, the union soon recognized it was not going to stop privatization and had a good deal to gain if it involved itself positively in the process. (For more on Zendentsu and its relation to deregulation, see Yamagishi 1989.)

23.3 International Pressure

Deregulation happened simultaneously in several industrialized countries during the early 1980s. This raised international issues because deregulation meant introduction of competition into the global telecommunications market. There have been issues regarding access to Japanese markets. Three things in particular have created problems: supplying equipment to NTT, product standards generally, and mobile communications. The United States has been the major source of the foreign pressure, both because of its own open door to Japanese equipment makers and (irrelevantly) its overall trade imbalance.

23.3.1 Procurement

NTT's equipment procurement policies were being strongly protested by the United States as early as 1978. Up through the 1970s Japan used infant industry arguments to exclude most non-Japanese equipment. NTT established close relationships with six major suppliers—called the ''denden family''—for R&D and procurement. Since January 1981 NTT's procurement has been in line with the GATT Code on Government Procurement and the Japan–United States agreement on NTT Procurement.

U.S. insistence on opening NTT procurement reflected a desire for reciprocity, given the openness of the U.S. market, and resulting huge exports from Japan to the United States. Japanese exports increased sharply in the early 1980s in part because of the overvalued U.S. dollar. To facilitate buying from foreign manufacturers, NTT's overseas subsidiaries and representative offices accept tenders in English and provide English-language materials on NTT procurement activities. NTT's annual overseas purchases, mainly from the United States, increased almost tenfold in the early 1980s, of course from a very small base. Data are in Table 23.1.

23.3.2 Product Standards

Product standards has been another issue—both as an aspect of NTT procurement and in its own right. In talks called the United States–Japan Market Opening Sector Specific (MOSS) Consultation, the United States asked Japan to simplify procedures for approving such things as CPE and microwave communication equipment. It also asked that entry of U.S. firms be promoted, particularly in the microwave market where the U.S. has had superior technology, and that there be clarity and transparency of decision making in telecommunications policy.

Following the MOSS consultation, several U.S. firms entered Japan. For example, as one of the main shareholders, Motorola takes part in management of Tokyo Telemessage, a paging company competing with NTT in the Tokyo metro area. In satellites, JC Sat (Nihon Tsushin Eisei) linked with Hughes (a major shareholder in JC Sat) and another new common carrier called Satellite Japan (Uchu Tsushin) established ties with Ford Motor Co.

Table 23.1. NTT Procurement from Non-Japanese Sources*

Year	Yen	US$	Year	Yen	US$
1980	3.8	17	1986	37.1	232
1981	4.4	19	1987	37.9	275
1982	11.0	44	1988	41.4	323
1983	34.8	147	1989	50.4	352
1984	35.1	144	1990	65.6	465
1985	36.9	167			

Source: Information Communications Almanac 1991, p 168. Tokyo: Info Com Research Inc.

*Data are for years ending March 31. They are given in billion yen, million U.S.$.

23.3.3 Mobile Communications

A third serious issue involves mobile communications. As far as new entry is concerned, two NCCs (new common carriers) appeared. These are IDO (Nihon Ido Tsushin), a subsidiary of Teleway Japan, introduced NTT's technology, while a subsidiary of Daini Denden entered using Motorola technology. The MPT said it could not assign frequencies to two NCCs in one area (the country is split into ten regions for cellular service) and asked them to unify. Consultations on unification failed, however, and in 1988 both were both permitted to enter, but they were assigned different areas. IDO was assigned the hugely lucrative Tokyo metropolitan area and the corridor to Nagoya. Others were assigned much smaller areas. A Daini Denden subsidiary (Kansai Cellular) got Kansai—which includes Osaka—Japan's second largest market but nothing compared with what was awarded to IDO.

This settlement was obviously unsatisfactory to Daini Denden and Motorola. In 1989 the United States again asked Japan to allow Motorola technology. That June, Japan relented and IDO was directed to change its system to accept Motorola equipment. The government also agreed to consider the possibility of future assignments of frequencies to Motorola equipment in the Tokyo area. There is evidence the delay was motivated by a desire to give Japanese companies time to introduce equipment that more effectively competed with Motorola's, which they did. In any event, IDO began to sell Motorola equipment in October 1991. Future frequency assignments are set for 1993–1994.

23.4 A New Era for the Industry

In April 1985 the Telecommunications Business Law (TBL) and the NTT Law took effect. According to the TBL, carriers were divided into Type 1—those with independent lines providing various carrier services—and Type 2—which lease private lines from Type 1 carriers and provide mainly VAN services. (An excellent book in English explaining the process and nature of the TBL and NTT Law is Bruce, Cunard, and Director 1986. See also Aronson and Cowhey 1988, Kalba 1988, Hills 1986, Ito 1985, and Ito 1983.)

MPT gained regulatory power from the Diet with the new laws, although both MPT and the Diet gave up some authority to the market (or at least to NTT) because one point of the new law was to substantially liberalize Japanese telecommunications. Type 1 carriers are still regarded as public entities along with utilities such as electric power generation and gas companies; they are regulated in the same way. Thus, MPT regulates most Type 1 carrier rates (including those for enhanced services).

New Type 1 carriers must be approved by the MPT, which considers the overall balance between supply and demand. Exit is also controlled by the MPT. Contracted services and tariffs for Type 1 service must be approved, as must agreements on interconnection among carriers. These requirements provide the industry in general, and the MPT in particular, with wide-ranging powers to direct growth and determine market conditions.

Type 2 carriers, on the other hand, are divided between "special" and "general." Special carriers must receive registration approval from MPT, while the latter need simply notify MPT that they exist.

23.4.1 Controlling the Process

During the 1982 process of liberalizing NTT's private lines, and the 1983–1984 considerations on privatizing NTT, MPT was involved in policy and turf disputes with MITI (see, e.g., Fuchs 1984, pp. 123–41). MPT even insisted that carriers that do not own transmission and switching facilities should be regarded as public carriers because they provide services to third parties. The extension of this position is that special Type 2 carriers must be considered public carriers just like Type 1 carriers. MPT wanted special carriers to be required to obtain the same approval as Type 1 carriers and that foreign capital be excluded. However, in April 1984 MPT abandoned this proposal because of domestic and international opposition; MITI also gave up some of its proposals.

The division of Type 1 and Type 2 carriers was not intended to define differences based on basic versus enhanced services. Thus, there is a problem with this division. For example, leased circuits are provided by Type 1 carriers under a strictly controlled system, while Type 2 carriers can resell their leased private circuits at freely determined prices; therefore, it is possible to have both flexible and regulated prices in the same service market.

The same problem occurs in VAN service. NTT provides data transmission facilities, including on-line data processing and communication processing, while Type 2 carriers can offer the same services. The price and operation of NTT's services is regulated by the MPT, while those of Type 2 are unregulated. In information processing, where there have been rapid technological innovations, NTT particularly welcomed deregulation of data services so it could compete freely with Type 2 carriers.

23.4.2 Funds for the Government

Government finances have been a major beneficiary of NTT's privatization. MOF collected a temporary tax from NTT totalling some 680 billion yen between 1981 and 1984. This alone provides evidence that NTT was not being managed independently. It also shows the degeneration of the principle that tariffs should cover total costs plus reasonable returns.

The government will also have received a substantial amount from selling NTT to the public. Up to two-thirds of the government's holdings can be sold. The first sale came in February 1987: 1.95 million shares were offered at 1,197,000 yen, netting 2.3 trillion for the government. The stock rose quickly and in November 1987 another 1.95 million shares were sold at the market price of 2.55 million yen each, raising about 5 trillion yen. A third offering came in October 1988: 1.5 million shares at 1.9 million yen. Because of the subsequent decline in the overall stock market there have been no further offerings, although under the original plan they were to have continued each year. Stock market conditions will determine when sales resume. The first three sales put 34 percent of NTT in public hands and brought the government some 11 trillion yen.

23.4.3 Competition—Type 1

As far as the introduction of competition is concerned, there have been many new entrants, both in Type 2 services and, contrary to expectations of the Second Rincho, in Type 1 as well. These are summarized in Table 23.2.

Among the many NCCs, three started private line long-distance service between Tokyo and Osaka, a high-traffic corridor, in 1986, and general service in September 1987. DDI (Daini Denden) is a joint venture involving Kyocera as the principal company (with 25 percent), Sony (5 percent), and other companies using a microwave network. Japan Telecommunications is a subsidiary of Japan National Railways, which built a network using the right-of-way along its tracks. Teleway Japan—a joint venture of Japan Highway Public Corporation, Toyota, and others—installed a network alongside the highways JHPC operates.

The largest new local common carrier is TTNet, part of Tokyo Denryoku (Tokyo Electric Power Generation). It is operating in the Kanto (greater Tokyo) area. Many of the expenses in creating networks are only incremental costs to the new entrants' parent companies, many of which are large public utilities that have extensive internal communications needs. They have become telecom providers to diversify because of deregulation in their own industries. Indeed, by doing so they hoped to realize better economies of scope in their operations.

23.4.4 Competition—Type 2

Since 1985 there have been so many newcomers that by October 1991 General Type 2 carriers numbered 960 and Special Type 2 carriers reached thirty-three, up from 688 and twenty-six in May 1989. Special Type 2s are mainly infor-

Table 23.2. Number of New Common Carriers (NCCs) as of August 1988 and October 1991

1988	1991	Type of Carrier
3	3	long distance
2	3	satellite communications
4	7	local networks
2	2	international telecommunications
24	36	pocket beepers
3	16	mobile communication and others

mation processing and software companies. There are no formal restrictions on the entry of foreign companies as Special Type 2s and in 1991 there were twenty-two international VAN business carriers including AT&T, IBM, GE, GTE Telenet, and Tymnet. Domestic firms include Inteck and Japan Information Service, plus electronics companies such as NEC, Fujitsu, Hitachi, and Oki. General Type 2 VANs serve transportation (e.g., Yamato System Development, a spin-off of Yamato Unyu, a major package delivery company), wholesale and retail trade, and financial institutions.

Because of the high rate of technological innovation, the previous NTT monopoly on data processing was steadily challenged during the early 1980s by competitors that quickly identified new business opportunities. Not surprisingly, there has been a big push for deregulation of this sector since 1982.

The biggest change in the Type 2 sector since 1985 happened in July 1988 when NTT Data System was made a corporation separate from NTT; however, all its stock is still held by NTT. (The company is capitalized at 100 billion yen.) Separation was one of the proposals in the Second Rincho report in July 1982. Rincho felt NTT's data processing service had an unfair advantage over leased line carriers. However, there was not much discussion of this particular problem. In the end, separation was conducted for different reasons: NTT wanted it because regulation was too severe to accommodate rapid technological innovation.

Thus the NTT Data spin-off was intended by NTT to help the company meet the competition. With 6,800 employees, first year sales were over 200 billion yen, and 345 billion in 1990. Revenue comes from the development of information processing systems. The company does no manufacturing, so it is classified as a genuine software company. NTT Data also derives benefits from its previously developed public systems, including social insurance systems, and other large-scale systems, such as nationwide banking services.

Its main competitors are big manufacturers. Although NTT Data System is the largest Type 2, its share of the total on-line information processing industry in 1988 was estimated at only around 7 percent. Its sale of transmission processing services, intrinsically a VAN service, was about 30 billion yen, about 18 percent of the total transmission processing market.

23.5 Assessing Competition

For private line and basic telephone service the NCCs initially undercut NTT tariffs by approximately 20 percent. The three NCCs operating long-distance service had 13 million subscribers and revenue of about 300 billion yen in fiscal 1990 (ended March 1991), up from 13 billion in 1987; DDI and Japan Telecommunications were profitable, Teleway was not (and some question its survival). Their combined revenues were about 5 percent of NTT's for fiscal 1990. They have about 16 percent of total long-distance volume and 49 percent of Tokyo–Osaka traffic.

DDI invented an adapter that automatically chooses the least-cost carrier from among the NCCs and NTT when a user dials, and provided it to subscribers at no charge. The others soon followed with a similar device and DDI has succeeded in reducing it to a single chip, which will allow it to compete for residential customers. The adapters encourage users to give the NCCs business. They were almost a necessity given the complexity of the rate structure when the NCCs began operation—they had different rates and rate bands from NTT and even each other, so each was the low-cost provider in at least some cases. In 1991 the three NCCs adopted identical rate structures, with the exception of where the farthest band begins.

It has become easy to interconnect with NTT. When the NCCs first began offering network service, some of the local (cross-bar type) NTT switches were so old that they could not interconnect the NCCs. NTT had to add to its ID creation function within local and trunk switching or replace older switches with new digital switches. By the end of 1988 there were almost no problems with the ID creation functions in NTT's switching system, particularly in the NCCs's main service area.

NTT paid half the installation cost for points of interface (POIs). In addition, NTT did not ask for access charges, thereby subsidizing the NCCs. The NCCs pay NTT 20 yen a call for access to the local network. About 30 percent of total NCC revenue was paid to NTT as access charges in 1990.

NTT subsequently shifted from cooperation to a more competitive posture. In August 1987 private line rates were cut 10 percent, and rates for long-distance calls over 320 km (NTT's farthest band) dropped 8.3 percent in February 1988. The latter cost NTT 70 billion yen in annual revenues. Although it has no competition, NTT also cut charges on calls to various isolated islands, giving up 10 billion yen annually.

There were a number of cuts in 1989. Rates were cut 10 percent on calls over 320 km (the farthest band) in February and, in response to a request from MPT, on those within 20 km. In November farthest-band calls were cut again. In March 1991 there was a further reduction by having the farthest band begin at 160 km. Cumulative cuts in farthest-band rates since those in effect just before privatization is 40 percent. The NCCs have responded to each NTT cut with their own reductions, although their rates are now much closer to NTT's than they were initially.

23.5.1 Policy Regarding Long Distance

There is a strong procompetition attitude regarding the long-distance market, but this means different things to different participants. There seem to be two general opinions, and discussions of policy largely revolve around them. One opinion is that there will be more competition from now on, and the NCCs will get more market share; in fact, the three NCCs have begun to broaden their service area beyond Tokyo and Osaka and are installing more POIs with NTT. Indeed, the NCCs had installed POIs in every prefecture by 1991 and so could provide nationwide interconnection service. According to this view, competition will make NTT management more efficient and may encourage price rebalancing, particularly between loss-generating monthly rates and local call charges and profit-making long distance. If this scenario proves correct, it will be necessary to change the present asymmetric regulatory system, which protects new entrants and controls NTT.

Focusing on NTT's sheer size and its control of the local network needed by the NCCs for interexchange, the second opinion is that the present state of competition in long distance is comparable to that between ants and an elephant. This seems to be the MPT view. The policy prescription is that it will be necessary to continue asymmetric regulation to protect the NCCs, or even for NTT to divest some operations, in order to place competition on a more equal footing.

From an economic point of view long distance is a partial monopoly (i.e., there is a dominant company and small fringe ones). The dominant firm is assumed to pursue profit maximization. This means that its prices are an umbrella that fringe suppliers can undercut to gain price-sensitive business. The dominant firm then only satisfies the residual demand, which is total demand minus fringe competitor supply. (For more detail, see Nagai 1990.) Three interesting results can be observed from this model.

1. The greater the elasticity of total demand, the smaller the market power of the dominant firm.
2. The greater the elasticity of supply from fringe competitors, the greater the elasticity of demand for the dominant firm, and the smaller its market power, other things equal.
3. The larger the market share of the dominant firm, the greater its market power.

The last proposition receives considerable attention, but the first two propositions have important consequences to competition. Even if the dominant firm's share is kept at a high level, the higher level of the fringe group's elasticity of supply raises the demand elasticity of the dominant firm and reduces its market power. Further, the dominant carrier generally cannot keep its price enough above its marginal cost.

NTT, the dominant carrier, is obliged to provide universal service and so must subsidize deficit generating services. This restricts NTT's ability to meet its competitors through price competition, which enables the competitors to capture cash flow with which to make further equipment investment. This allows them to capture even more share—as their growth is essentially supply constrained.

If the fringe competitors are protected by regulators—as they often are because of proposition 3—it is much easier for them to increase their capacity. This is characteristic of asymmetric regulation. Since 1989 NTT has had absolute volume losses in some long-distance call markets. In particular, traffic at exchange offices in the Tokyo central business district has decreased. This was partly because of the movement of big business users from public switched network service to private network communications, but inroads by the NCCs also have contributed.

MPT points out that the NCCs must depend on NTT's local network. NTT's local network is therefore said to be the bottleneck for NCC operations. If NTT increases the price of local calls, then the relative advantage of the NCCs is lost. This explains why, in February 1989, MPT pushed NTT to reduce the price of the closest long-distance calls. Other disadvantages NCCs have include the fact they cannot decide the location of POIs, and they do not know which switching equipment is adequate for ID creation. They basically do not have enough network information from NTT, such as how many subscribers there are in each message area and how large the traffic flows are between areas. In early 1989, therefore, MPT also pushed NTT to disclose various network information, including figures on its costs and revenues for local and trunk call services. These claims have been made by MPT to help the NCCs and to discourage NTT from predatory pricing though cross subsidization.

23.5.2 Further Deregulation of Divestiture of NTT?

Several studies by agencies such as MITI and the Fair Trade Commission were made during 1986–1987. These suggested that telecommunications—both Type 1 and Type 2—should be deregulated further. They indicated, for example, that MPT's control of new entrants to the satellite business and international communications was very discretionary and was not transparent regulation. MPT made forecasts of future demand and capacity of production, and based its decisions on them. It was argued that MPT should deregulate pricing for more services—controlling just the core, such as local calls and monthly rental charges. Above all, the reports insisted that deregulation be the general rule and regulation the exception.

In March 1988 MPT responded with a report reviewing the deregulation process up until then. There have been no problems with the deregulation process, the report stated, and the time was not yet ripe for re-examination of the regulation system. It was not necessary to reconsider the Telecommunications Business Law.

MPT instead proposed reconsideration of NTT's management system, pointing out the necessity of dividing NTT into several companies. Such a break-up had originally been suggested by the Second Rincho in 1982. The reason for reviving the proposal, MPT said, was that although competition had been introduced in 1985, it was not really substantial yet and was not occurring on an equal footing—specifically referring to NTT's network information. This is a consideration in the proposal for divestiture of NTT submitted to the Telecommunication Policy Council by MPT.

After its interim report in August 1988 the Council issued a final report in March 1990. It stated that by 1992 NTT should spin off its mobile phones operations and that by 1995 long-distance should be spun off. In other words, NTT was to be split in three. The cabinet, and MOF in particular, opposed this in part out of concern for those who had bought NTT shares. MITI also opposed breaking up NTT—although its Information Industry Advisory Committee, which had issued a report on telecommunications a few months before the

Council's, was strongly critical of the company and called for more competition. MPT indicated it would defer a decision on a split up, but would ask NTT to structure its operations into the three lines of business.

In mid-1991 MPT issued "administrative guidelines" for NTT. By April 1994 NTT was to provide a local network interface for any rivals in each of the forty-seven prefectures. By June 1993 NTT was to release cost data that would be used to create interconnection tariffs to take effect in April 1995.

23.6 Conclusion

Since privatization, NTT has tried to increase its productivity through management reorganization and a 6-percent decrease in its work force during the first three years (1985–1988). Thus, despite price reductions for long-distance and private lines, NTT continues to earn good profits—approximately 373 billion yen pretax for the year ending March 31, 1986, and 411 billion for fiscal 1990. It also increased its R&D expenditures from 136 billion yen in 1986 to 262 billion in 1990. NTT plans to fully digitalize its network system, which will requires a major commitment of capital.

It seems privatization of NTT and introduction of competition have been very successful in many ways. Indeed, according to the 1988 annual White Paper of the Economic Planning Agency, new entrants made 700 billion yen in investment during the first three years. Still, prospects remain unclear.

Counterpoised against the appearance of deregulatory success are problems concerning the structure of competition. NTT's tariff does not reflect the competitive climate. As competition proceeds, there will need to be changes in the tariff structure. Prices have to reflect costs, including opportunity costs. My expectation is that some new method, such as a system of access charges, volume discounting, or even price discrimination between low- and high-traffic routes, will be introduced.

Digitalization of the network will further promote the fusion of communications and computer technology, as well as the structure of competition. For example, a Type 1 carrier can afford to provide enhanced services just as easily as VAN service providers (Type 2 carriers, which currently lease private digital data circuits from Type 1 carriers). With competition in ISDN, the economies of integration, including economies of scale and scope, must be reconsidered.

As digitalization encourages multimedia services, the public-interest aspects of telecommunications that justify government regulation will change. This means there will be much more change in both competition and regulation in Japanese telecommunications.

Bibliography

Aronson, Johathan D., and Peter F. Cowhey. 1988. *When Countries Talk: International Trade in Telecommunications Services*. Cambridge, MA: Ballinger.

Bruce, Robert R., Jeffrey P. Cunard, and Mark D. Director. 1986. *From Telecommunications To Electronic Services*. London: Butterworths.

Denki Tsushin Seisaku Kondankai Teigen. 1981. *80-Nendai No Denki Tsushin Seisaku No Arikata*. Tokyo.

Fuchs, Peter E. 1984. "Regulatory Reform and Japan's Telecommunications Revolution." In *U.S.–Japan Relations: New Attitudes for a New Era*, 1983–1984 Annual Review of the Program on U.S.–Japan Relations. Harvard Univ., Center for International Affairs.

Hills, Jill. 1986. *Deregulating Telecoms: Competition and Control in the United States, Japan, and Britain*. London: Frances Pinter.

Ito, Youichi. 1983. "Recent Trends in Telecommunications Regulation and Markets in Japan." In *Forum 83, 4th World Telecommunication Forum* Part 3 *Legal Symposium on International Information Networks*. Geneva: International Telecommunications Union. Reprinted in *Jurimetrics* 25(1): 70–81 (1984).

———. 1985. "Implications of the Telecommunications Policy Reform in Japan." *Keio Communications Review* 6: 7–18.

———. 1986. "Telecommunications and Industrial Policies in Japan; Recent Development." In Marcellus S. Snow, ed., *Telecommunications, Regulation and Deregulation in Industrialized Democracies*. Amsterdam: North–Holland.

Kalba, Kas. 1988. "Opening Japan's Telecommunications Market." *Journal of Communication* 38(1): 96–106 (Winter).

Kumon, Shumpei. 1984. "Japan Faces Its Furture: The Political Economics of Administrative Reform." *Journal of Japanese Studies* 10:155 (Winter).

Lincoln, Edward J. 1988. *Japan—Facing Economic Maturity*. Washington, D.C.: The Brookings Institution.

Nagai, Susumu. 1987. *Seifu Kiseikanwa Bunya nitsuiteno Hyoka to Mondaiten no Bunseki (Evaluation and Analysis of Problems with Deregulated Industry, In Case of Telecommunication)*. Fair Trade Commission Monograph Series.

———. 1990. "On the Competition of Telecommunications under Regulation in Japan." *Journal of International Economic Studies* (Hosei Univ., Institute of Comparative Economic Studies), p. 4.

Nippon Telephone and Telegraph (Nippon Denshin Denwa Kosha). 1986. *Shinsei NTT eno Kiseki (Steps to the New NTT)*. Joho Tsushin Sogo Kenkyujo. Tokyo: NTT.

Yamagishi, Akira. 1989. *NTT Ni Asuwa Aruka (Is There Tomorrow for NTT?)*. Tokyo: Nihon Shoronsha.

24

The United States

ELI M. NOAM

Telecommunications in the United States began in 1836 with Samuel Morse and the electromagnetic telegraph. The first U.S. telegraph message, sent from Baltimore to Washington in 1844, was, "What hath God wrought?" The same question was being asked one and a half centuries later when there was fear U.S. telecommunications had been severely crippled by the policy of deregulation, and by the divestiture of the dominant telecommunications institution, American Telephone and Telegraph (AT&T).

24.1 History

The United States has always been at the forefront of change in telecommunications, partly due to internal and external geographic distances, partly due to a high level of technological innovation. From the beginning, the telecommunications system was never the centralized monopoly system prevalent in many other countries.

When private financing was slow to initiate telegraphy operations, the U.S. government, although with considerable reluctance, subsidized the new medium, inaugurating a tradition of alternating governmental rejection and the embrace of an active role in the telecommunications sector. Morse's 1836 invention of a simple and workable electric telegraph faced competing companies and technical systems. In 1851, several telegraph companies consolidated into the New York and Mississippi Valley Printing Telegraph Company. It and Morse dominated telegraphy until the Civil War (1861–1865) by merging with smaller firms and aggressively expanding. The New York and Mississippi Company, renamed Western Union, was successful at securing protective rights-of-way from railroads, and adding patents. It soon became the dominant carrier and enjoyed healthy profits. By 1876, the year of the telephone's introduction, Western Union had over 300,000 km of lines and 7,500 offices. However, its high prices as well as developments in technology allowed small competitors to enter niche markets, especially in international service. Western Union's dominance in basic domestic telegraph service remained until well after World War II (see Brock 1981).

In 1876, Alexander Graham Bell, a teacher of the deaf in Boston, introduced a workable telephone. The new device created a sensation, and Bell's father-in-law and other wealthy investors launched the Bell Telephone Company. Recognizing the huge task at hand, they first offered the patent rights to Western Union, but that company chose to protect its existing market rather than enter the new one. It considered the telephone a complement rather than a competitor because the telephone was then limited to local service, while the telegraph was primarily used over distances. Thus, Western Union declined to acquire the Bell patents, which were available for less than $1 million. Western Union never recovered from its imperfect foresight, and its fate has always been a scary reminder to telephone companies to remain at the forefront of services.

The Bell firm grew and prospered. Telephony expanded nationally through the franchising of independent local operations. Later, when the original patents expired in the mid-1890s, Bell Telephone positioned itself to maintain its monopoly by a variety of means: vertical integration of equipment and services; development of interexchange long-distance service; aggressive pricing strategies; acquisition of substantial competitors; acquisition of additional patents; restricting interconnection of alternative equipment; and by preventing interconnection of rival local networks to Bell local networks and to the Bell (AT&T) long-distance system (see Brooks 1975).

By 1897 there were some 500,000 telephones in service across the United States, 80 percent of them on Bell lines. Once the basic Bell patents expired, independent competitors entered those areas not serviced by Bell operations or concessionaires, especially in rural districts and areas facing particularly high prices. In some cities several systems competed side-by-side without interconnection. As the number of independents grew, they began to form regional agreements to provide service among themselves.

After a few years the independents were nearly equal in size to Bell; robust competition existed in both the provision of local service as well as in the manufacturing of switching and customer equipment. The one main difference between the two segments, however, was interconnection. While the Bell Telephone system was fully interconnected on a national level through its long-distance network, the independents operated on a fairly limited regional scale. By 1907, when the population was 87 million, the total number of telephones had grown to 6 million.

The eroding market share of Bell Telephone, reorganized and renamed American Telephone and Telegraph, led to a more aggressive policy. Theodore Vail, who was brought back by Wall Street financiers led by J. P. Morgan for a second tour as president of AT&T, devised a three-prong strategy to increase market strength. This included aggressive acquisitions of independent telephone companies, the embracing of regulation in order to avoid antitrust suits, and a major increase in R&D in order to acquire a technological edge. Backed by Morgan, AT&T was also able to acquire a majority interest in Western Union.

AT&T grew rapidly over the next years, but several independent companies brought antitrust complaints against the firm. As the number of lawsuits mounted, and as they were joined by Justice Department actions, AT&T chose in 1913 to negotiate an agreement with the U.S. government known as the Kingsbury Commitment. AT&T sold its stake in Western Union and left telegraphy. It also guaranteed existing independent telephone companies access to its long-distance network and agreed not to expand further geographically by acquiring competitors or entering their territories. This governmental action to limit AT&T from total market dominance was part of a general trend of antitrust policy. Americans had become concerned with the enormous growth of business entities in the late nineteenth century, in the decades following the Civil War. There has always been a strong populist current opposing domination by big firms. This distrust was shared by the political left, farmers, small businesses, and westerners.

The same political constellations led to the establishment of a regulatory system of utility commissions on the state level that supervised privately owned utilities, including telephone companies. This arrangement contrasts sharply with the system of state telephone administrations prevalent in most countries.

The Kingsbury Commitment did not confine AT&T's operations to markets related to telephone service. For the next twenty years, AT&T was able to enter new industries such as commercial radio and sound movie technology. During World War I the company played an important role in the military effort and was deemed to be of enough significance that it was briefly nationalized toward the end of the war.

By 1934 AT&T manufactured and owned 80 percent of all telephones in the United States and operated the only national long-distance network. It still enjoyed relative security, although its integration into equipment manufacturing was attacked by the Walker Report, authored by one of members of the new Federal Communications Commission (FCC). The FCC was created as part of the more general "New Deal" effort to establish stronger governmental controls on a depressed economy. World War II delayed any follow-up to the Walker recommendations, but once the war was over, the Justice Department filed an antitrust suit in 1949.

Intervention by the Defense Department, as well as the 1952 presidential election, stalled the case. In 1956, under a more supportive national administration, AT&T achieved a favorable consent agreement. It was not forced to divest itself of Western Electric, its manufacturing arm, but its activities were limited to telephony. Western Electric was confined to telephone-related research and manufacturing operations, and had to take a more liberal policy in the licensing of its patents. On the whole, however, AT&T had succeeded in avoiding a possibly disastrous antitrust judgment, although it had also, once again, watched its routes of expansion close.

New technologies and their innovative uses continued to emerge and their sponsors, seeking to compete, sought help from the FCC. By the 1990s, universal service penetration in the United States would be largely completed. The telephone reached most households, and an increasingly elaborate system of transfers kept residential rates low. This soon led to pressures for change.

After 1956, the FCC had begun to allow new entrants. Under pressure from the aggressive electronids industry, two key decisions in the area of terminal equipment were *Hush-A-Phone* (1956) and *Carterfone* (1968), which permitted non-AT&T equipment to be attached to the network.

In 1959, with the "Above 890" decision sought by microwave equipment firms such as Motorola, the FCC permitted large users to operate in-house microwave long-distance service. These users felt that they were increasingly subsidizing local service and small customers, and they had incentives to drop off the common system, at least partially. This soon led to major changes. By 1969 one microwave delivery company, MCI, won a court ruling against a reluctant FCC and an adamant AT&T to provide private-line service for other users as a carrier; eventually, all specialized carriers were permitted to provide private-line service. It soon offered service to large users that did not want to operate their own systems. This was soon expanded into general public switched service, with rights to interconnect with AT&T's local networks in order to reach customers. By 1975 AT&T found itself facing regular facilities-based service competition in telephony for the first time in more than fifty years.

24.2 Policy Transformation

The policy changes were partly due to a general political and economic philosophy of limiting the role of the state, which made government institutions more receptive to allowing new entrants as an offset to corporate power. This philosophy far preceded the conservative Reagan and Bush administrations. Inspired by Lockean principles of natural law, the classic American ideology of government seeks individualism, fragmentation of private power, limitation of government (with the major exception of its role in national security), and protection of property rights and contracts. As applied to telecommunications policy, this philosophy justified a governmental role that is far narrower than in most other countries: It centered on permitting competitive markets to limit the exercise of dominance by any single firm and in permitting users to choose among service providers. This view is shared by those Democrats who are distrustful of concentration of private economic power and those Republicans opposed to government interference.

In the 1970s and 1980s telecommunications continued to undergo changes of structure and policy subsumed by the terms *deregulation* and *liberalization*. These developments eventually led to the break-up of AT&T, the world's largest communications organization at the time. It was brought about by a 1974 Justice Department antitrust suit based on unfair business practices the firm allegedly employed to suppress its competitors. It resulted, after a 1982 consent decree, in the most massive reorganization in business history in 1984. The divestiture agreement put AT&T's local operating companies—approximately two-thirds of its assets and employees—into seven regional holding companies (RHCs, often called Baby Bells or Regional Bell Operating Companies, RBOCs). These provided mostly traditional telephone service, but increasingly and aggressively sought other opportunities inside and outside the communications field.

In further developments, through several so-called Computer Inquiry decisions by the FCC, AT&T and the RHCs were permitted to enter new and unregulated markets such as data processing and computer fields. By the late 1980s the FCC and some states were in the process of dropping rate-of-return regulation in favor of price caps regulation, instituting liberalized interconnection and access rules (Open Network Architecture), and introducing local service competition, starting in New York.

Thus, a centralized system of one near-monopoly telephone carrier, one dominant domestic telegraph company, and a handful of international telegraph companies was transformed within a few years into a highly differentiated system with a bewildering number of participants and institutions.

24.3 Regulatory Structures

The basic framework of government involvement in U.S. telecommunications is complex. Unlike most other countries, the public sector did not own or operate civilian services, except for a few small municipally owned cable television operations, rural telephone systems, and educational television broadcasting stations. Although almost all civilian telecommunications facilities are privately owned, their use is often—but not always—subject to licensing and regulatory oversight. These regulations are set on the federal, state, and occasionally the local level.

For all the talk of deregulation, the number of regulatory bodies, in two senses of the word, is larger in the United States than anywhere else. Federal policy emanates primarily from the FCC, a body of five commissioners, from both parties, appointed by the president and confirmed by the Senate, but thereafter independent from both, in theory and often in practice. It tends to be dominated by its chairman. The FCC, as other independent commissions, operates as a hybrid within the American constitutional order, exercising legislative powers (adoption of regulations), executive authority (enforcement of its rules), and a judicial role (adjudication of cases). It allocates frequencies and regulates all broadcasting, satellite, and other civilian uses of the electromagnetic spectrum. The FCC is in charge of *inter*state telephony (e.g., transmissions from one state to another) and everything affecting interstate communications. The FCC also has jurisdiction over cable television.

State regulatory commissions, generally known as Public Service or Public Utility Commissions (PSCs and PUCs), are independent of the FCC. They play an important role in regulating *intra*state telephony, and in some instances cable television. Commissioners are appointed by the governor in most states; in others they are popularly elected. Municipal authorities regulate cable television through their power to grant franchises.

There was no *federal* regulation for the first thirty-five years of telephony until the 1910 Mann–Elkins Act, which gave an undefined regulatory authority to the Interstate Commerce Commission (ICC). The ICC was established to oversee the railroads and showed little interest in telecommunications, which were regulated by the various state utility commissions that were created in the early part of the century. When the Communications Act of 1934 was drafted, creating a more specialized and potentially activist FCC, the states urged a statutory limitation on the new FCC's powers over intrastate wire communications. Congress responded positively. Its report on the bill stated that "some 97.5 or 98 percent of all telephone communications is intrastate, *which this bill does not affect*." This assurance to the states proved empty, however, because separating the national from the regional regulation of an integrated network is difficult.

Public policymakers were under continuous pressure to reconcile the statutory fiction of separation of intrastate and interstate network components with the reality of their integration. What emerged was a system of coregulation, based on shared goals. For several decades, the cooperative spirit was so great that the federal level permitted a system of revenue transfers to the state-regulated domains to support low local rates for which the federal government had no direct oversight responsibility. The system, however, could not last when its constituents' fundamental goals diverged. This occurred when the FCC began to embrace the economic concepts of efficiency, competition, markets, and entry, while the state commissions continued to emphasize equity and redistribution.

The split between the states and the FCC emerged first in a serious fashion in the 1960s when the FCC and federal courts opened the terminal equipment market to rivals of AT&T. Many states, on the other hand, advocated a restrictive approach, largely for fear of having the phone companies lose revenue that subsidized residential rates.

The FCC prevailed, however, in the landmark *North Carolina* v *FCC* decision (1976). The court read the state-reserved part of telecommunications very narrowly and rendered it almost meaningless. Throughout the 1980s, preemption of state regulation by the FCC moved forward, but this trend was slowed in the 1990s by a pro–state's rights majority on the Supreme Court.

On the federal executive level, the Commerce Department's National Telecommunications and Information Administration (NTIA) helps to coordinate the executive branch's overall policy. It plays a role in international communications, together with the Office of U.S. Trade Representative and the State Department, which is the lead agency in international negotiations.

In addition, the Department of Justice plays a major role through its Antitrust Division, which oversees much of the telephone industry by way of enforcing the 1982 court order that broke up AT&T. The primary authority in that case is Federal District Court Judge Harold Greene, who frequently decides whether the Bell Companies and other parties are complying with his divestiture decree, and who has thus become a major presence in telecommunications matters.

Conforming to a broader policy trend in U.S. government decisionmaking process, other federal courts—particularly the Court of Appeals for the District of Columbia—have also become a significant locus of de facto policymaking. These courts hear appeals from trial courts and administrative agencies; their decisions can be reviewed only by the Supreme Court, which hears only a small fraction of appellate cases. For example, the Court of Appeals in Washington, D.C., forced the FCC in the *Hush-A-Phone* case to allow non-AT&T equipment manufacturers to sell terminal units for connection into the local AT&T exchanges, making competition in the equipment market possible. The Justice Department and the Federal Trade Commission (FTC) also play a role in regulating industry competitive behavior and structural changes—primarily mergers and acquisitions—and by forcing divestitures as with AT&T.

The fundamental law is the Communications Act of 1934, which has rarely been amended, despite many attempts. Congress—the legislative branch—does often wield power indirectly, giving signals to the FCC through bills, resolutions, hearings, and the budgetary process.

The political parties of the United States have had at best an indirect impact on the formation and exercise of telecommunications policy. The nature of the political party in power generally did not greatly affect the direction of change in telecommunications policy, although it did sometimes affect its pace. There is a substantial amount of overlap between the two parties over telecommunications issues, and in the philosophy of rate setting, but the tone or emphasis can be slightly different. The Democratic position has been somewhat more oriented toward protecting residential users; conversely, Republicans have placed somewhat more emphasis on economic development and large users. This has translated into a greater reliance on market forces, although Democratic-dominated FCCs have been just as active in that direction, and indeed the AT&T divestiture case was initiated under liberal Democrats and was concluded under conservative Republicans.

Access rates to local exchange networks by long-distance carriers are of particular importance in the regulatory arena. In the past, complex financial accounting rules (''separations and settlements'') provided an internal contribution from AT&T's long-distance service to local exchange providers—Bell and independent. Complicated FCC tariffs also governed the access charges paid by the rival long-distance carriers. After divestiture, this system was revamped, with equal access charges for all carriers phased in.

The rates and terms of service of intrastate communication are regulated by state commissions, traditionally on the principle of rate-of-return regulation. Several states have relaxed these rules either by outright deregulation or by instituting price regulation in place of rate-of-return rules. Due to the dominance of the local exchange companies in local residential distribution, full deregulation of local charges is unlikely soon, but substantial relaxation of such regulation is taking place. One state, Nebraska, has already largely deregulated local exchange prices.

The principle of rate-of-return regulation is to permit a ''fair'' return on invested capital. Because this return is aggregated, some cross subsidies can exist from one type of service to another. Furthermore, rates tend to cover less of the costs for rural than for suburban users, and less for residential than for business users. Because price-setting is meaningless without a definition of the product, federal and state regulators also set service quality requirements.

In order to permit value added networks (VANs) and transmission carriers access and interconnection into the "public" network, American regulators embarked in the late 1980s on a set of actions to establish an open network architecture (ONA). ONA expanded the concepts of service alternatives and network fragmentation into the very core of the networks, and aimed to lower barriers to entry for rival and varied communications services. Under an ONA the various functions of the network, including the central office functions, would be accessible and separately available and subject to potential competition.

In the long run, the implications of an open network include:

- A future competition in central exchange services, including potential incursions across franchise territories by some local exchange company's (LEC's) exchange services and even facilities.
- A major enhancement in the possibilities of local transport competition and of private group networks.
- A move toward a "distributed" rather than centralized physical architecture of public central office functions, analogous to the computer industry's evolution into distributed processing.
- The establishment of systems integrators as major institutions in telecommunications.

The multiplicity of decisionmaking bodies at several levels of government can frustrate coordinated and comprehensive policy-making. This process, however, also accommodates decentralized and ad hoc decisions, many of which are responses to specific problems, rather than part of a grand design. Lack of a central dominant force thus has not prevented a fairly rapid reorientation of U.S. telecommunications policy; indeed, it has probably helped considerably.

24.4 Constituencies

Each government entity that can affect telecommunications has various constituencies to consider, which include AT&T and the RHCs, whose numerous employees and shareholders carry some weight. During the 1980s other companies also gained a voice, in particular the rival long-distance carriers, whose survival is at the heart of the pro–competitive policy. The independent telephone companies, of which the largest, GTE, is comparable in size to an RHC, have strong support especially in areas. Information and enhanced service providers (IPs and ESPs) are younger and smaller industry segments, but they are often allied with the powerful media and computer industries. Newspaper publishers are concerned about the telecommunications industry's ability to become a provider of information services. The electronics industry, with its links to the defense sector, carries much weight.

Residential users desire ubiquitous, reliable service at reasonable rates. They are represented by a variety of private nonprofit groups and governmental consumer advocacy bodies. Rural users are primarily concerned with receiving basic service at rates similar to urban and suburban ones; they enjoy wide congressional support. Large business users want innovative technology options, dynamic and reliable service, prices that are not above cost, and minimal restrictions on operating their own private networks. A variety of industry organizations speak for these large users. The Communications Workers of America (CWA) and the International Brotherhood of Electrical Workers (IBEW) are the primary labor unions operating in this sector, and they wield a voice, particularly on changes that affect employment.

The cable television industry has gone through a remarkable building period, in which it wired most of America in a very short time with a second communications link and established a system of program supply that overcame the traditional television network bottleneck. Cable is supplemented by various over-the-air broadcasters, including microwave multipoint–multichannel distribution (MMDS), also known as "wireless cable" and direct satellite program providers. Together these media have led to a considerable diversity and have offset the power of the telephone industry in some areas. Both segments depend on each other to some extent and there is turbulence in the increasing areas of overlap. In particular, the cable industry fears incursions by the telephone companies, supported by the latter's "deep pockets." In the 1990s, cable companies began to enter telecommunications services, both in the mobile field and through the acquisition of alternative local telecommunications companies (ALTs) that emerged, led by Teleport Communications and Meptropolitan Fiber Systems (MFS). These were accorded vital "comparably efficient" interconnection rights, first by the New York Public Service Commission in 1989.

24.5 The Network

There are twenty-two BOCs, some specific to a single state (e.g., the New York Telephone Company), others covering several states (e.g., Southern Bell). They are organized under seven Bell RHCs. The BOCs, together with GTE whose size is equal to an RHC, provide the bulk of local service. Additionally, more than 1,000 mostly small independent telephone companies serve approximately half of the nation's geographic area and operate 20 percent of all access lines.

Local Bell companies, but not other LECs are restricted under the AT&T divestiture decree to service within their local access and transport areas (LATAs); they may not enter interLATA long-distance or international communications. The extent of restrictions on their parent companies have been the subject of continuing legal and political battles. ALTs have begun to compete with the established local exchange companies through a number of technologies.

In long-distance ("interexchange") service AT&T in 1990 controlled about 51 percent of the market defined as including intraLATA interexchange service, and about 65 percent of the market defined as interLATA service, measured by revenues, according to the FCC. BOCs provide long distance within their own LATAs, accounting for about 20 percent of the market. The principal competitors in interLATA service are MCI, with about 11 percent of the market, and Sprint, with 8 percent. There are also hundreds of resellers.

Specialized companies—including data networks and VANs such as Telenet and Tymnet—provide packet switching and other value-added services. Satellite carriers lease transponder capacity to other carriers and private users.

Cellular telephone service in the United States operates as a duopoly. There were 6.5 million subscribers in 1991. Customers in each major service area have a choice of two licensed cellular providers, one being their local "wireline" telephone company. The other was an independent provider. There has been a major consolidation in the industry, with most independents being acquired by telephone companies from other regions. McCaw, the major independent firm left, leads the industry with 12 percent of the market. GTE and BellSouth follow with 11 and 8 percent, respectively. Revenues have been increasing at over 30 percent each year for much of the 1980s due to the growing subscriber base, but has plateaued in the 1990s. The systems are analog, but digital transmission is anticipated, as is the entry of microcellular service providers.

Packet-switched networks have existed in the United States since the early 1970s. They originated at the Pentagon, whose Defense Advanced Research Projects Agency (DARPA) had a private firm, Bolt, Beranek, and Newman (BBN), develop the "Arpanet" nationwide network to link researchers with each other. Arpanet was a major success, and it induced BBN to start Telenet, a commercial network in operation since 1975, as the precursor to packet switched networks around the world.

Telenet was eventually sold to GTE. Expansion was costly and the network broke even only after 1983. In 1986 GTE Telenet, together with GTE's long-distance carrier, Sprint, were combined into a joint venture with United Telecom and its Uninet. United Telecom eventually controlled Telenet and Sprint.

Another packet-switched network, Tymnet served computer time-sharing customers. Tymnet and its parent Tymshare were acquired by aircraft manufacturer McDonnel Douglas in 1986 and subsequently sold to British Telecom in 1988.

Common carriage provides nondiscriminatory access and usage rights to all users, including resellers that compete with a carrier. Local exchange companies must grant access to all long-distance carriers and to all telephone users. Customers indicate their "primary" carrier to which domestic and international long-distance calls are automatically routed by a local exchange. Other carriers can be accessed by dialing a prefix number. Such a system may be extended in the future to intraLATA long-distance service. Large customers also can utilize their PBXs to select a different long-distance carrier for each call according to a programmed "least-cost-routing."

The reselling domestic local and long-distance transmission is allowed and is extensive. This includes sharing bandwidth on satellite transponders, reselling local transmission, and competing coin and credit card public telephones. Resellers do not require an FCC authorization; to sell directly to the public, they need only file a notification with the FCC and some state PSCs. Where there is no general offering (i.e., one bank reselling its surplus transmission capacity to another) no filing is necessary. Private networking is prevalent for large users, usually on leased facilities under software-defined "virtual" arrangements, with equipment manufacturers often providing the integration and network management function.

24.6 International Services

The volume of international telecommunications traffic has grown much faster than international trade. Part of the impetus has been the dramatic decrease in the costs of circuits. In many countries this has not been matched by an equal drop in rates, where carriers did not face competition. Low international rates in the United States are partly the result of overcoming market segmentation. Numerous boundaries still existed in 1964, when the FCC prohibited AT&T from entering the international record market (telegraph, telex, and data transmission). Among record services, the FCC made a further distinction between *domestic* services, from which Western Union was restricted, and *international* services, which were provided by *international* record carriers (IRCs). IRCs could only operate in the United States from certain limited and approved gateways. A telegram from Cleveland to Singapore, for example, would be routed by Western Union to an IRC gateway, transmitted by an IRC to Singapore, and passed on to the Singapore PTT.

This market segmentation led to a lack of competition as well as to substantial profit margins. Partly because of the profitability, the situation became unstable and cracks began to appear. In a series of rulings in 1979 and 1980, the FCC largely removed the dichotomy of voice and record carriage. It also eliminated the rules prohibiting AT&T and the IRCs from entering each others' markets. It also removed many of the restrictions on the expansion by domestic and international record carriers to new gateway cities.

Prior to the 1980s, AT&T provided the bulk of international voice service. Other carriers such as MCI and Sprint now provide service to countries whose PTTs have allowed it. In the Pacific, Hawaiian Telephone, owned by GTE, handles a substantial portion of the international traffic.

Comsat, the U.S. signatory to Intelsat and Inmarsat, whose ownership had been shared by the government and private companies, subsequently became entirely privately held. Originally operated solely as a carrier's carrier for Intelsat service, it is now able to access users directly. For international civilian satellite communications (as distinguished from cable or microwave) Intelsat was the sole link, although this has also been opened up to new carriers. New international satellite carrier systems have been approved, with PanAmSat the furthest along in operation; similarly, rival transatlantic and transpacific cable operations emerged.

Most foreign administrations observed the changes in the United States with some misgivings, as it challenged long-established arrangements and rate structures. They also had a potential advantage in the situation. As the only address within their countries for AT&T, MCI Sprint and others, the Post, telegraph, and telephone services were in a position to force rival U.S. carriers to compete for operating agreements. To prevent such "whipsawing" the FCC established rules for uniform settlement rates for the same routes. It also embarked on a course challenging the traditional system of settlements prevailing in international telecommunications.

24.7 The Impact of Deregulation and Divestiture

The transformation of telecommunications in the United States from monopoly toward a more pluralistic system was accompanied by grave predictions of doom and gloom: residential rates would skyrocket; universal service could no longer survive; service quality would fall; productivity would suffer; research and development would decline; employment would drop; AT&T would dominate; and so on. However, most of these fears did not materialize.

For example, despite scenarios of several hundred percent in rate increases, local rates in real terms rose from 1985 to 1990 at an annual rate of 4.7 percent, while interstate long-distance rates declined by 6.0 percent annually in the same period. According to the FCC, overall telephone rates (long distance and local) rose from 1984 to 1990 by a total of about 17 percent, which is less than cumulative inflation (CPI) of 27 percent during that period. Furthermore, the telecommunications price index does not include the sometimes substantial savings from lower equipment costs. Rates did not rise as much as initially feared, in part because costs were contained through lower interest rates and taxes, higher productivity, and lower equipment prices.

Equipment prices fell as the Bell Companies gained the freedom to shop around. Central exchange equipment costs declined from $230 per line in 1983 to less than $100 in 1992. Overall, annual expenses per access line, not including reduced taxes, declined from about $38 to about $30. Revenue increased from about $82 to $95 per line.

The prediction of steep rate increases did not take into account the working of a political-regulatory system where strong commitment to social concerns protected local service rates. Furthermore, social safety nets were introduced. In New York, for example, subsidized "Lifeline" service of $1/month for basic dial tone was instituted in 1987. An estimated 1.5 million users (about 15 percent of households) are eligible—defined as membership in one of several social support programs such as welfare.

Thus, despite fears, overall telephone penetration did not decline. Rather, it slightly increased from 91 percent in November 1983 to 93 percent in 1992. For the middle class (above $30,000/year household income) penetration was 98 percent and higher, and 95 percent of all farms had telephones. Even for some of the very poor ($5,000–7,500 per year income), it rose from 83 to 84 percent.

Service quality held steady, at least for those dimensions that are not ''labor intensive.'' On the other hand, several major service breakdowns pointed to the increased vulnerability of society to any network failure.

There was a great fear about a technological decline because Bell Labs R&D might be curtailed. Actually, the opposite occurred. One study found that total R&D employment rose from 24,100 in 1981 to 33,500 in 1985. (AT&T and Bellcore, the RHC joint R&D firm, combined) (Noll, 1987). By 1988, the regional companies had added their own laboratories, and total R&D employment had risen to an estimated 35,600.

Labor productivity rose since the AT&T divestiture by about 40 percent, although at some expense of employment, which dropped from 953,000 in 1984 to 879,000 in 1990. The old system had permitted costs to drift upward, and the new environment put pressure on labor.

AT&T's long-distance market share steadily declined each term, reaching around 67 percent in 1990. The market, although flat in dollar terms, grew strongly in terms of traffic, increasing by 13% annually and doubling usage from 37 billion minutes in 1984 to 75 billion in 1990. Americans make substantially more telephone calls per capita (1,700) than users in other countries—two and three times as many as the British (800), Japanese (550), Germans (500), and French (400). Similarly, American companies are very communications-intensive, and are steadily becoming more so.

The upgrading of the network proceeded after liberalization. For example, local Bell operating companies increased their fiber use in the network by 32 percent in 1990 and 28 percent in 1989 to 2.7 million fiber miles. Urban fiber carriers deployed some 55,000 fiber miles, and the interexchange carriers increased their fiber trunk lines by 12 percent, to 2.1 million fiber miles. According to the U.S. Department of Commerce, in 1989 96 percent of all lines were electronically switched, half of them digitally.

In ISDN, the United States is several years behind the high level of activity of several European countries and of Japan. On the other hand, fully digital lines that do not correspond to the CCITT 2B+D standard (and are therefore not considered ''pure'' ISDN) have become frequent. Usage of high capacity digital lines such as T-1 and DS-3 lines is high.

24.8 The Equipment Market

The connection of terminal equipment to the interstate network is governed by the Communications Act and FCC regulations. Part 68 of the FCC's rules sets minimum technical standards equipment must meet. Vendors must register their products with the FCC before marketing them. Registration requires the disclosure of technical specifications so the FCC's staff can identify any possible system degradation. There is, however, no approval necessary.

The U.S. market for central office (local exchange) switching equipment was characterized in the past by foreclosure by AT&T, except for independent telephone companies, among whom GTE had its own equipment operations.

Although most analysts expected the BOCs to cling to AT&T as their equipment supplier after divestiture, they in fact embraced a wide variety of non-AT&T equipment quite rapidly. In central office switches, AT&T's share dropped from 70 to 46 percent in four years, mostly to the benefit of Northern Telecom, which transformed itself from a Canadian to a North American company.

Technical network standards are coordinated for the BOCs by Bellcore. There is no evidence that Bellcore is favoring AT&T or other U.S. manufacturers. Procurement of network equipment by local telcos is governed by their obligation to state regulators to pay the lowest possible prices. They are under pressure to keep rates low. The ability to compare cost trends for the LECs also forces them to seek low-cost equipment. Because of divestiture, BOCs no longer have an incentive to increase AT&T's profits, as none of those profits are returned to the BOCs.

However, in the equipment market the U.S. trade reversed from a slightly positive balance in 1983 to an over $2 billion deficit in 1989. This was partly due to the general strength of Asian countries in consumer electronics, and partly the result of the divestiture-induced severing of AT&T's vertical integration of equipment and local exchange network services that had closed most of the U.S. market to other suppliers.

24.9 The Electronics Industry

The electronics industry in the United States is characterized by large and established firms on the one hand, and smaller entrepreneurial firms fueled by an active venture capital market on the other hand.

AT&T's Bell Labs invented the transistor in 1949, launching the age of microelectronics based on semi-conductors. In subsequent years the main development was the move from discrete devices to increasingly integrated circuits. These innovations made mass production easier and facilitated substantial component integration within one chip. Young companies that were wedded neither intellectually nor financially to the older ways moved into the new technology. These firms left the traditional, vertically integrated American and European tube manufacturers far behind.

Total sales of the electronics industry increased an average 9 percent in the 1980s and measured about $300 billion by 1992. Total imports to the United States were $79 billion in 1990; exports amounted to $72 billion.

Large electronics manufacturing firms include AT&T, General Electric, Hewlett-Packard, IBM, Texas Instruments, Motorola, Digital, and Apple. AT&T's manufacturing arm used to be known as Western Electric and is now called AT&T Technologies. It operates mostly in the telecommunications industry but has been active in other areas of electronics and computers through its research arm Bell Laboratories. AT&T earned $522 million on sales of $63 billion in 1991, and had 317,100 employees, one-third of its predivestiture size of over 1 million. Due to the U.S. Justice Department's 1956 Consent Decree, AT&T was originally prevented from entering the computer and electronic component manufacturing industry, except for internal use. These restrictions were lifted in 1984 with divestiture, but its financial success has been modest in these competitive lines of business. In 1990 it acquired the large computer manufacturer NCR; perhaps the oldest of all firms operating in the electronics industry. Formed as the National Cash Register Company in 1844, it produced a variety of business information systems. It had 62,000 employees and revenues of $5 billion before it was purchased by AT&T.

IBM, founded as International Business Machines in 1924, initially manufactured Holerith punch card equipment. By the late 1950s, its primary business had become the development and manufacturing of computers. In 1991, IBM had 377,000 employees worldwide. As the largest U.S. corporation in 1990, IBM had net earnings of $6 billion, but it lost $2 billion in 1991.

IBM's market share is very large, and the firm often commanded a premium price for its products due to its reputation and ubiquity. IBM's power was at its peak in 1964, when it held 70 percent of the computer market. Its power was short-lived, however, as other companies successfully developed "plug-compatible" peripheral equipment, forcing IBM to sharply cut its prices. When it also employed non–price tactics to make compatibility more difficult, the U.S. government initiated a mammoth antitrust lawsuit (DeLamarter 1986). The government's lawsuit was dropped in 1981, partly because the market had not stood still in the meantime, and new types of equipment and new domestic and overseas entrants were challenging IBM in most markets. IBM was forced to compete in many fields: in the supercomputer market with Cray and with Japanese and European firms; in the component manufacturing field, with AT&T, Texas Instruments, Motorola, Intel, and their highly effective Japanese counterparts; in minicomputers, with Hewlett-Packard, Digital, Prime, and Data General; and in microcomputers, with a large number of small, inventive competitors such as Apple and Compaq, and with a host of Asian producers.

In telecommunications, IBM entered the competitive long-distance transmission field with Comsat and the insurance firm Aetna as partners by launching Satellite Business Systems (SBS), a venture that proved highly unsuccessful. In the PBX market, IBM acquired Rolm, but eventually sold it to Siemens.

General Electric, the third largest of all U.S. corporations, was formed in 1878 to pursue Thomas Edison's applications of electricity. It has a very broad range of activities in manufacturing, high-technology development, and service businesses. Its total revenues in 1991 were $60 billion and profits were $2.6 billion. It employed 284,000.

Texas Instruments (TI) was founded in 1938, and manufactures components and equipment. TI is pursuing semiconductor markets in the Pacific Basin area and is a major defense contractor domestically. Its sales in 1991 were $6.8 billion, but it was financially in the red. It employs 63,000 people.

Motorola, which dates back to 1928, is a leading manufacturer of equipment and components. It employs about 102,000 and had profits of $450 million on sales of $11.3 billion in 1991.

Small electronics, computers, and software start-up firms tended to cluster regionally, creating economies of agglomeration where those of scale were absent. Perhaps the best known of these is ''Silicon Valley,'' near Stanford University and San Francisco. It is the home to some 2,700 young electronics, high-technology, and engineering firms.

Hewlett-Packard Co. (HP) was the first major ''Silicon Valley'' electronics and high technology firm. It was started in 1947 by independent engineers with relatively modest funds. HP developed into a major designer and manufacturer with some 82,000 workers, and revenues of $8 billion.

Another Silicon Valley firm is Apple Computer. Founded in 1977 by two young college dropouts, Apple employs 14,000 and its reported profit for 1991 was $300 million on sales of $6 billion.

Other high-tech centers in the United States include ''Route 128'' outside Boston, home to Lotus, and Digital; the Research Triangle in North Carolina; and the suburban districts of metropolitan New York and Los Angeles. In many instances, strong universities provided the nucleus around which industries grew.

24.10 Outlook: From the Network of Networks to the System of Systems

U.S. telecommunications is coming to resemble the rest of its economic system—a complex reflection of an underlying pluralist society and economy. Being farthest along in the transformation of its telecommunications system, the United States is likely to bear the brunt of new conflicts, both domestically among the numerous interest groups and participants, and internationally as new U.S. policies affect established global arrangements.

In the United States, the day is not far off, historically speaking, when entry will be wide open; when fiber is widespread in all stages of most networks; when radio-based carriers fill in the still substantial white spots in the map of telecommunications ubiquity; and when foreign carriers operate freely domestically.

Yet diversity can lead to fragmentation, noncompatibility, and inconvenience. From the user perspective, there is a great need for the functional integration of networks. To provide such coherence, a new category of ''systems integrators,'' who create packages of equipment and services in a one-stop fashion, is emerging.

Today, systems integrators exist for large customers. They have also begun to be active in establishing group networks. In the future, however, systems integrators will also put together individualized networks for personal use, creating *personal* networks. As these personal, group, and interorganizational networks develop, they will access into each other and form a complex interconnected whole, sprawling across carriers, service providers, and national frontiers. The telecommunications environment thus evolves from the unified network to the ''network of networks,'' in which carriers interconnect, and from there to the ''system of systems,'' in which systems integrators link up with each other.

Where does such a system of customized networks leave government regulation? Regulation in the United States had been essential to the old system, partly to protect against monopoly, partly to protect the monopoly itself. In the transition to competition, what was left of regulation was seen as temporary, shrinking reciprocally with the growth of competition.

At that point, could one expect the "system of systems" to be totally self-regulating, with no role for government? There are several public policy goals underlying regulation. They include universal coverage, affordable rates, free flow of information, restriction of market power, technological progress, and so on. To assure these goals, U.S. regulators in the past instituted a variety of policies, such as rate subsidies, universal service obligation, common carriage, interconnection rules, access charges, quality standards, and limited liability for carriers. Government regulation existed to right the imbalance of power between huge monopoly suppliers on the one hand, and small and technologically unsophisticated users on the other hand. In the future environment, however, systems integrators will act as the users' representative vis-à-vis the underlying carriers. They could, for example, protect users against carriers' underperformance in quality and price, and make regulatory control over these issues unneccesary. On the other hand, some traditional policy goals are not necessarily resolved that way, such as the maintenance of low rates for low-income and rural users, or the free flow of information across carriers, or the interconnectivity among carriers. This suggests some continuing role for government.

In the 1980s, U.S. telecommunications policy was centered on open entry. In the 1990s, however, a different emphasis is likely. Now, issues of integration of the various network parts come to the forefront. Reconciling the centrifugal pressures with the needs to interoperate and intercommunicate represents the main challenge to U.S. policymakers for the next decade. This means to provide a competitive system with tools of interoperation where they are not self-generating by market forces.

The openness of the evolving network system will not stop at the national frontiers, and the notion of each country having full territorial control over electronic communications will become anachronistic. This undermines attempts to administratively set rules for prices and service conditions. No country can be truly an island anymore, not even a large nation as the United States, and the international collaboration of its carriers, users, manufacturers, and governments with those of other countries will therefore be at the center of American telecommunications evolution and policy in coming decades.

Acknowledgment

Douglas Conn's assistance is gratefully acknowledged.

Bibliography

Brock, Gerald W. 1981. *The Telecommunications Industry.* Harvard University Press.

Brooks, John. 1975. *Telephone, The First Hundred Years.* New York: Harper & Row.

Cole, Barry G., editor. 1991. *After the Breakup: Assessing the New Post-AT&T Divestiture Era.* New York: Columbia University Press.

Cowhey, Peter S and Jonathan B Aronson. 1988. *When Countries Talk: International Trade in Services.* Cambridge, MA: Ballinger Publishing.

Crandall, Robert W. 1991. *After the Breakup: U.S. Telecommunications in a More Competitive Era.* Washington, D.C.: Brookings Institution.

DeLamarter, Richard Thomas, 1986, *Big Blue.* New York: Dodd, Mead, and Company.

Garnett, Robert W. 1985. *The Telephone Enterprise.* Baltimore: The Johns Hopkins University Press.

Levin, Richard. 1982. ''The Semi-Conductor Industry.'' In Richard R. Nelson, editor, *Government and Technical Progress: A Cross Industry Analysis.* New York: Pergamon Press.

National Telecommunications and Information Administration, U.S. Department of Commerce. 1991. ''NTIA Infrastructure Report: Telecommunications in the Age of Information.'' NTIA Special Publication 91-26 (Oct.).

National Telecommunications and Information Administration, U.S. Department of Commerce. 1988. ''NTIA Telecom 2000: Charting the Course for a New Century'' NTIA Special Publication 88-21 (Oct.).

Noam, Eli M. 1992. *Telecommunications in Europe.* New York: Oxford University Press.

——. 1989. ''Network Pluralism and Regulatory Pluralism.'' in Paula R. Newberg, editor. *New Directions in Telecommunications Policy. Volume 1 Regulatory Policy: Telephony and Mass Media.* Durham: Duke University Press, pp. 66–91.

Noll, A. Michael. 1987. ''The Effects of Divestiture on Telecommunications Research.'' *Journal of Communication* 37(1); 73–80.

Temin, Peter, with Louis Galambos. 1987. *The Fall of the Bell System.* New York: Cambridge University Press.

U.S. General Accounting Office. 1983. ''FCC Needs to Monitor a Changing International Telecommunications Market.'' GAO/RCED-865-92 (Mar. 14).

25

The Important Links in Pacific Basin Telecommunications

DOUGLAS A. CONN

Although there are immense geographical distances between the countries of the Pacific Basin, economic and political forces have effectively reduced them. As the chapters of this book have repeatedly pointed out, these linkages have relied heavily on international telecommunications.

The past decades were a time of enormous vitality for the communications sector in the Pacific Basin, mirroring and supporting the rapid expansion of the economies in that region, with international telephone traffic growth rates of as much as 25 percent in some countries. In comparison, annual growth in the United States was about 18 percent and 15 percent in the European Community. The Pacific Basin is the fastest growing region for telecommunications investment and traffic growth.

The twenty-four preceding chapters have examined the myriad changes of the telecommunications sector in the Pacific Basin. This chapter first maps out the regional collaborations and interactions that have spurred liberalization. It then discusses common policy themes in the region.

25.1 International Links

The fast growing economies of the Pacific Basin are highly interdependent with the rest of world. Collaborations that facilitate these connections have become a necessity. They include:

- Trade agreements in telecommunications equipment and services
- Joint ventures and foreign investment in manufacturing and service provision
- Physical links via satellite, cables, or other transmission technologies
- Common policies towards both domestic and regional telecommunications.

National control over links, long a predominant concern, is no longer the first priority. International links have tended to raise the expectations of users and providers alike. For example, more Pacific Basin countries have opened up their international long-distance markets faster to competition than any other region in the world. This may be attributed to their need to meet the rising demand of the region to communicate and conduct business overseas.

There are many reasons telecommunications are undergoing change throughout the region. Some are due to each country's unique political and economic system, while others can be attributed to more general trends. Collaboration between countries (bilateralism), among blocks of countries (multilateralism), or among the regions of the world (interregionalism) have changed the topology of regional telecommunications networks and the agreements that both linked and separated them.

Pan-Asian bodies with working groups on telecommunications such as the Asia Pacific Economic Council (APEC), Pacific Trade and Development Conference (PAFTAD), and the South Pacific Forum (SPF), as well as the Pacific Telecommunications Council (PTC) and the Asia-Pacific Telecommunity (ATC), are attempting to forge a regional bloc that can encourage the rapid growth of advanced telecommunications, not unlike the transnational organizations of Europe (CEPT, RACE, etc.). Regional collaboration has already resulted in more open trade in equipment and services as well as, in some cases, in trade friction, both within the region and with other regions.

25.2 Trade and Foreign Investment

Trade in telecommunications equipment throughout the region has risen in step with the increasing liberalization of telecommunications markets. Large multinational manufacturers in Japan and Korea have created a trade surplus in telecommunications equipment between their countries and the rest of the world. However, many other countries in the region do not manufacture products and have noticeable trade deficits. The United States, which is a leader in equipment manufacturing has nevertheless realized a trade deficit in telecommunications equipment since 1983.

In terminal equipment, the countries of the Pacific Basin are by far the largest manufacturers in the world. The top five exporters to the United States, for example, are Japan, Canada, China, Malaysia, and Taiwan and the region overall accounts for nearly 90 percent of all U.S. customer-premises equipment (CPE) imports. In 1991, foreign imports captured some 60 percent of the U.S. market, with Japan alone accounting for some 30 percent. The United States continues to be successful in large-scale telecommunications equipment trade. In satellites, it holds almost two thirds of the world market.

Direct foreign investment in domestic markets for manufacturing and training has increased dramatically in the region over the past decade. Assembly plants for electronics and telecommunications equipment have been prevalent in low wage countries such as Malaysia, Indonesia, Thailand, and the Philippines.

In services, foreign operation, (and in some cases ownership), of domestic networks and local service is becoming a field of expansion for North American and European operators. The lack of Japanese presence in network services in foreign markets has been primarily due to de facto restrictions concerning overseas operations in domestic markets placed on NTT and KDD by the Japanese government.

25.3 Satellites and Undersea Cables

As international telecommunications networks have grown so have the minutes of international traffic. The international routes with the highest level of traffic in the Pacific region are between the United States and Canada, the United States and Japan, and, interestingly, between Hong Kong and China. More calls are made to China from Hong Kong than any other route except for the United States to Japan. This is all the more remarkable given China's relatively slow development of telecommunications.

The region holds many of the fastest growing international telecommunications carriers in the world. As Table 25.1 shows, of the top ten fastest growing international carriers of public switched service between 1986 and 1990, seven were carriers based in the Pacific Basin. The number would be eight, if Cable & Wireless (C&W) were included geographically—not an unreasonable proposition since a large share of its international traffic originates from Hong Kong, where it operates the local telephone company. The number would again be

Table 25.1. The Fastest Growing International Telecommunications Carriers*

Carrier	Cumulative Growth, 1986–1990 (%)
Sprint (U.S.)	686
MCI (U.S.)	649
Comm Auth of Thailand	570
Cable & Wireless (U.K.)	565
Embratel (Brazil)	196
Bezeq (Israel)	187
OTC (Australia)	159
Teleglobe Canada	153
China PTT	142
KDD (Japan)	139

Source: Staples 1991

*Determined by minutes of telecommunications traffic—public circuits only.

Note: Carriers not in service in 1986 are not included. Sprint and MCI traffic excludes traffic to Canada and Mexico. China PTT is for 1988–1990.

higher if Japan's two new international common carriers, International Digital Communications (IDC) and International Telecom Japan (ITJ) (which began service in 1989), were included. Regionally, about 40 percent of all outgoing international calls in the region are made within Pacific Asia, with approximately 20 percent made to North America and 10 percent to the European Community (see Staples 1991).

Satellites and underseas cable have been the primary linkage mechanisms. For satellites, the emergence of private carriers has meant that the politics of transborder flows of voice, data, and television is forever changed. For underseas cable, the multiparty ownership of individual cables has resulted in a plurality of public and private owners, users, and service offerings, previously unknown in the region, but prevalent in the Atlantic. As a result, the suboceanic cables are more and more perceived as dedicated rather than public networks.

In the past, international satellite communications were primarily the domain of Intelsat. Domestic satellite systems include JCSat and SCC of Japan, Palapa of Indonesia, and Aussat in Australia. The latter two have also offered regional service to neighboring countries' government-operated telephone companies.

The first major breach with the traditional ownership of satellites came with the launch in 1990 of Asiasat. Asiasat was literally an overnight success; its entire capacity was sold out by 1992. Owned and operated by a large Hong Kong firm, with equity stakes taken by C&W and the China Investment and Trading International Corporation, its footprint extended from the Pacific across Asia to the Middle East. Asiasat's crossborder transmissions raised cultural issues due to its English and Mandarin chinese programming and real economic issues because it represented the first private carrier with the potential to compete directly with Intelsat and other domestic satellites in Pacific Asia.

Asiasat's de facto opening of the Asian skies sent encouraging signs to other carriers to enter the pan-Asian market. In addition to Indonesia's Pasifik Satelit Nusantara (PSN) (owned by Palapa), other smaller countries, such as Thailand and Malaysia, sought the autonomy and revenues that their own satellites would offer. A number of new private international satellite carriers with substantial non-Asian ownership have also emerged. These include Celestar, a subsidiary of U.S. McCaw Cellular, Pacific Satellite, Inc. owned by TRT Communications (a subsidiary of Pacific Telecom–U.S.) and a number of regional PTTs, Orbx, a subsidiary of U.S. Alpha Lyracom, and TongaSat, whose major funding would come from an American entrepreneur and other investors.

Private suboceanic cables are also a permanent part of telecommunications in the Pacific Basin. There has been much diversity in the ownership of cables; in fact, this is true to a much greater extent than satellite ownership. Submarine cables were traditionally owned and operated by government PTTs or monopoly telephone companies. However, since the first private link was introduced in 1989, the Pacific Region has witnessed a dramatic increase in the number of entirely private cables both constructed and under consideration. In 1988, the first major fiberoptic submarine cable, TPC-3, was installed. All installed cables since then have used fiber, which offers much greater capacity, shorter response times, and lower unit costs per circuit.

Since 1989, some sixteen cables have either been constructed or proposed. There are a wide range of ownership models for many of these systems. Some are owned by as few as two or three carriers, but others are owned by many more. For example, the HJK cable, which links Hong Kong, Japan, and Korea, came on line in 1990 with some twenty-seven partners. Usually, the telephone operator in the landing country will have a stake, as will other private operators and large corporate users with a particular interest in a specific route. Large international carriers have actively participated such as C&W, France Telecom, AT&T, and Sprint.

Where these multinational cables converge is an important question. Several countries are vying to become high-tech hub sites where the region's international traffic is aggregated and then redirected, much like some countries that have established open ports of trade. The benefits that accrue to the hub site are major including substantial revenues from transiting traffic, siting advantages for multinational corporations, and more rapid technology transfer. Both Hong Kong and Singapore have positioned themselves as the primary hub sites for telecommunications passing in and out of the region. Tokyo is also bidding for the position, and with a strong market behind it—although its local costs for land, labor, and materials tend to be higher than either Hong Kong or Singapore.

As the number of international and regional satellite and underseas cables increase, they drastically lower prices, increase usage, and thus shrink the distance between and within countries. This further accelerates the policy and structural changes in the various countries. International links will push domestic liberalization and act as a transmission belt for an exportation in telecommunications liberalization across the region (Noam 1992).

25.4 Domestic and Regional Policies

The policy changes occurring in telecommunications in the Pacific Basin are common to the industrialized regions of the world. Market structure within the telecommunications sector of various countries is changing rapidly, national monopolies are being privatized, and firms are turning to global markets for growth while positioning themselves to compete on the international level.

Though the Pacific Basin is very heterogeneous, some economic, political, and social issues are common.

1. Wherever one looks, the telecommunications sector is receiving increasingly high national priority; however, this manifests itself in different ways. In some countries, the private sector is invited in as part of a rejuvenation. For example, the privatization of New Zealand Telecom was a means to stimulate economic growth. On the other hand, some countries strengthen the role of the state. In Singapore, for example, an activist government has dominated the

electronic sector and instituted various national initiatives in computers and telecommunications. It began to turn to privatization only after that mission was a success.

2. A liberalization and restructuring of the telephone monopoly is occurring simultaneously and rapidly throughout most of the region. New experiments in competition policy are often being developed outside the largest countries. For instance, Hong Kong has permitted a new entrant to provide cable television with the goal of creating competition in the delivery of all local telecommunications services. New Zealand has opened its entire market, both domestic and international, to competition.

3. Whatever the new policies, the emphasis by governments on the expansion or preservation of affordable universal service remains strong. In some countries like Indonesia, the Philippines and many of the Pacific Islands, basic telephone service penetration remains at low levels and the spread of basic telephony is therefore priority. In these countries, a primary concern is the financing of a basic network. In countries with more advanced telecommunications, such as Taiwan and South Korea, large and accelerating investments in universal service are taking place, too, as these countries reach higher stages of industrialization. In advanced countries such as the United States, Canada, and Japan, basic service is being actively protected by government policy from erosion, while the definition of universal service is being broadened beyond voice service.

4. There appears to be no significant reduction in the use of the telephone network for purposes of raising and distributing revenue for a variety of social goals. Cross subsidization remains an important factor everywhere as network development continues—including cross subsidization between operations, geographical areas, and governmental functions. In China, funds are shifted from the central Ministry of Posts and Telecommunications to provincial authorities as a means of redistributing income from economically revitalized eastern provinces to less developed western provinces. In the United States, poor people are subsidized through "lifeline" service at rates as low as $1 per month, and rural users are supported in a variety of ways.

5. Large users—domestic and foreign, private and public—have become an effective lobbying force for change in telecommunications regulation. They have applied pressure for the introduction of new service and for restructuring of telecommunications institutions. In South Korea, the introduction of DACOM, a competitive carrier, was followed by increased lobbying of larger conglomerates to establish more competition in telecommunications. In several countries, private and public sector users have gone beyond political avenues and fashioned their own facilities-based private networks. Private networks based on dedicated leased circuits grew rapidly everywhere.

6. There has also been a trend toward network pluralism in access, provision, and ownership. As deregulatory policies were introduced, some sectors of the industry experienced a rapid increase in the number of providers. In Japan, South Korea, and Taiwan, for example, new entrants provide services through value added network services (VANs). In Japan, Korea, the United

States, Canada, Australia, and New Zealand, alternative facilities-based carriers operating their own networks compete with the former monopoly carrier.

7. Resistance to the introduction of competition in the highly profitable international telecommunications services is declining. The United States, Japan, and South Korea have permitted the entry of alternative carriers in their international long-distance market. This has paved the way for a number of countries to permit these new carriers landing rights to their national systems.

8. Domestic carriers whose own markets have become more open, have begun to branch out and increase their attention to overseas markets as a means of improving prospects for growth. This has often meant strategic international alliances. The American carrier BellSouth has a stake in Optus Communications Group, an Australian long-distance and cellular carrier. In New Zealand, two other Bell Companies, Ameritech and Bell Atlantic, have purchased a majority stake in New Zealand Telecom. Bell Canada and MCI have also invested there. NYNEX, an American Bell company, has a substantial stake in a fiber ring network in Thailand. NTT has ventures with local monopoly carriers in countries such as Indonesia, the Philippines, and Thailand C&W has a long-standing presence in the region, especially in Hong Kong and the Pacific Island nations.

9. As telecommunications activities have expanded overseas, trade frictions have developed, especially between countries that have liberalized their markets and those seeking to protect their traditional networks. Foreign pressure has played a considerable role in the opening of some markets. In the cases of Japan, Korea, and Taiwan, trade negotiations with the United States had an impact on the government's equipment procurement policies. In Japan, the cellular industry was restructured, procurement policies were changed, and partial foreign ownership was permitted of VANs. In both cases, concessions were made to guarantee reciprocity in U.S. markets.

10. Political constellations are shifting. The dynamics of the telecommunications industry in the Pacific Basin go well beyond technology. They are a reflection of many political and economic changes. Political institutions that were once tied ideologically to a strong central government and nationalized services, have often consented to a dismantling of the national telecommunications monopoly. The Labour Party in New Zealand is an example. The entire telecommunications sector there has been opened to competition and the national carrier has been privatized and sold to foreign entities. At the same time, Australia's strong trade unions and its Labour government played an integral role in slowing liberalization.

These observations indicate the breadth of change occurring in the region. The forces of globalization and their impact on economic development have caused economies to be more open to change and regional cooperation. As the more-industrialized countries in the region have opened their markets, and the less-developed have followed with their own initiatives, the rest of the world has begun to take notice. The lessons learned from these new developments will be valuable to those countries undergoing or anticipating economic and technological change. Telecommunications linkages will play a crucial role in the expansion of the world's most economically vibrant region.

Acknowledgments

The author wishes to thank Eli Noam, Ken Donow, and Larry Meissner for their suggestions and useful comments.

Bibliography

Aronson, Jonathon and Peter Cowhey. 1988. *When Countries Talk: International Trade in Telecommunications Services*. Cambridge, MA: Ballinger Publishing Company.

Barber, Richard. 1989. "Pacific Telecommunications: The Role of Regional Telecommunications." *Columbia Journal of World Business* 24(1): 101–3 (Spring).

Conn, Douglas. 1992. "Domestic Telecommunications Policies in the U.S. and Japan: Their Impact on Trade Relations." In *United States–Japan Trade in Telecommunications: Conflict & Compromise,* Meheroo Jussawalla, ed. Westport, CT: Greenwood Publishing.

Conn, Douglas. 1992. *Telecommunications Equipment Trade Between Japan and the United States: The Cases of Satellites and Cellular Telephony*. Working Paper # 478. New York: Columbia Institute for Tele-Information. Columbia University.

Noam, Eli M. 1992. *Telecommunications in Europe*. New York: Oxford University Press.

Poe, Robert, and Joyce Quek. 1990. "Projects Merged." *Communications Week International,* p. 6 (Sep. 3).

Sato, Harumasa, and Rodney Stevenson, "Telecommunications in Japan: After Privatization and Liberalization," *Columbia Journal of World Business* 24(1): 31–41.

Schwartz, Adam. 1992. "A Giant Stride." *Far Eastern Economic Review,* pp. 47–49 (Jan. 23).

Staples, Gregory, ed. 1991. *The Global Telecommunications Traffic Report—1991*. Washington, D.C.: International Institute for Communications.

Westlake, Michael. 1991. "Bed-time chats." *Far Eastern Economic Review,* pp. 52–53 (Mar. 21).

Acronyms

Acronyms are generally spelled out the first time they appear in any chapter. Location is in parentheses where applicable.

ABC	Australian Broadcasting Corporation
ACMC	ASEAN Cable Management Committee
ACPL	ASEAN Cableship Private Ltd.
AEEMA	Australian Electrical and Electronic Manufacturers Association
AEIA	Australian Electrical and Electronic Manufacturers Association
AEU	Asia Electronic Union
AGT	Alberta Government Telephones
AIIA	Australia Information Industry Association
AIT	Asian Institute of Technology
AIT-RCC	AIT-Regional Computing Center
ALTs	Alternative Local Telecommunications Companies
AMTS	Automatic Mobile Telephone System
ANZCERTA	Australia-New Zealand Closer Economic Relations and Trade Agreement
AOTC	Australian and Overseas Telecommunications
APEC	Asia Pacific Economic Council
APPU	Asian Pacific Postal Union
APT	Asia Pacific Telecommunity
ASEAN	Association fo Southeast Asian Nations
ASEAN I-S	ASEAN Indonesia-Singapore
ASEAN P-S	ASEAN Philippines-Singapore
ASICs	Application-Specific Integrated Circuits
ATC	Asia-Pacific Telecommunity
ATEA	Australian Telecom Employee Association
ATEA	Australia Telecommunications Employees Association
ATM	Automatic Teller Machine
ATUG	Australian Telecommunications Users Group
ATUNET	AIT-Thailand Inter-University Network
ATUR	Automatic Telephone Using Radio

ATUR	Automatic Telephone Using Radio
BBC	British Broadcasting Corporation
BBN	Bolt, Beranek, and Newman
BC Tel	British Columbia Telephones
BNA	Billing Name and Address
BNR	Bell Northern Research
BOC	Bank of China
BT	British Telecom
C&W	Cable and Wireless
C&WHK	Cable and Wireless Hong Kong
CAAS	Civil Aviation Authority of Singapore
CAC	The Consumer's Association of Canada
CAC	Communications Advisory Council
CAD	Computer Aided Design
CAM	Computer Aided Manufacturing
Cancom	Canadian Satellite Communications
CAPTAIN	Character And Pattern Telephone Access Information Network System
CAT	Communications Authority of Thailand
CBC	Canadian Broadcasting Corporation
CBTA	Canadian Business Telecommunications Alliance
CCIR	the Consultative Committee on Radio
CCIR	International Radio Consultative Committee
CCIS	Community Cable Information Service
CCITT	The International Telegraph and Telephone Consultative Committee
CCL	Computer and Communication Research Lab
CCPA	Computer and Communication Promotion Association
CEPD	Council for Economic Planning and Development
CFCW	Canadian Federation of Communications Workers
CGRA	Chinese Government Radio Administration
CIDA	China Interdisciplinary Association
CITC	Cook Islands Telecommunications Corp
CITIC	China International Trust and Investment Corporation
CMC	Committee for National Computerization
CMTS	Cellular Mobile Telephone System
CNPAC	Packet Switched Network
CNR	Canadian National Railways
CO	Central Offices
COMET	Computer Oriented Message Switching Exchange
COTC	Canadian Overseas Telecommunications Corporation
CP	Canadian Pacific Ltd.
CP	Charoen Pokphand
CPCNs	Certificates of Public Convenience and Necessity
CPE	Customer Premises Equipment
CPI	Consumer Price Index
CRTC	Canadian Radio-television and Telecommunications Commission

CSL	Communications Services Ltd
CSM	Citra Sari Makmur
CSO	Community Service Obligation
CTHK	Cable Television Hong Kong
CTI	Citra Telekomunikasi Indonesia
CWA	Communications Workers of America
Dacom	Data Communications Corp. of Korea
DAMA	Demand Assigned Multiple Access
DARPA	Defense Advanced Research Projects Agency
DBS	Direct Broadcast Satellites
DCI	Data Communications Institute
DCI	Data Communication Institution
DDI	Daini Denden
DDS	Digital Data Service
DGP	Directorate General of Post
DGT	Directorate General of Telecommunications
DIPS	Dendenkosha Information Processing Systems
DOC	Department of Communications
Domsat	Domestic Satellite Philippines
DPT	Department of Posts and Telecommunications
DSIR	Department of Scientific and Industrial Research
DTC	Department of Transport and Communications
ECC	Economic Council of Canada
EDACS	Electronic Digital Access and Cross Connect System
EDB	Economic Development Board
EDCF	Economic Development Cooperation Fund
EDS	Electronic Data Systems
EEC	European Economic Community
EKTS	Electronic Key Telephone Systems
ELG	Electronic Leading Group
EPB	Economic Planning Board
ERSO	Electronic Research and Service Organization
ESP	Electronics Sector Plan
ESPs	Enhanced Service Providers
ETP	Energy, Telecommunications, and Post
ETPI	Eastern Telecommunications Philippines
ETRI	Korean Telectronics and Telecommunications Research Institute
ETSI	European Telecommunications Standards Institute
FCC	Federal Communications Commission
FINTEL	Fiji International Telecommunications Ltd.
FSCS	Frontline Services Computer System
FSM	Federated States of Micronesia
FSMTC	the FSM Telecommunications Corporation
FTA	Canada-U.S. Free Trade Agreement
FTC	Federal Trade Commission
FX	Foreign Exchange

GATT	General Agreement on Tariffs and Trade
GDP	Gross Domestic Product
GEISCO	General Electric Information Service Co.
GIMM	Global Information Movement and Management
GNP	Gross National Product
GNTC	Great Northern Telegraph
GOES	Geosynchronous Orbiting Environmental Satellite
GPO	General Post Office
GPT	GEC Plessey Telecom
GTA	Guam Telephone Authority
GTE	General Telephone and Electronics
HAW-4	Hawaii-U.S. mainland submarine cable
HF	High Frequency
HKCC	Hong Kong Cable Communications
HKT	Hong Kong Telecom
HP	Hewlett-Packard
IBEW	International Brotherhood of Electrical Workers
IBS	Intelsat Business Services
IBS	Integrated Business Systems Ltd
ICC	Information Culture Center
ICC	Interstate Commerce Commission
ICs	Integrated Circuits
IDAR	International Database Access and Remote Computing Service
IDC	International Digital Communications
IDD	International Direct Dialing
IDN	Integrated Digital Network
IICWTD	International Independent Commission for Telecommunication Development
Inmarsat	International Maritime Satellite Organization
INS	Information Network Service
Intelsat	International Telecommunication Satellite Organization
IOCOM	Indian Ocean–Commonwealth
IPs	Information Providers
IRCs	International Record Carriers
IRU	Indefeasible Rights of User
IRV	Indefeasible Right of User
ISD	International Subscriber Dialing
ISDN	Integrated Services Digital Network
IT	Information Technology
IT&E	Island Telecommunications and Engineering Corp
ITA	International Telecommunications Authority
ITAC	Information Technology Association of Canada
ITANZ	Information Technology Association of New Zealand
ITDC	International Telecommunications Development Corporation
ITE	International Telephone Exchange
ITJ	International Telecom Japan

ITRI	Industrial Technology Research Institute
ITT	International Telephone and Telephone
ITU	,International Telecommunications Union
IVAN	International Value Added Network
JTM	Jabatan Telekom Malaysia
KBL	New Society Movement
KBS	Korean Broadcasting System
KCCPA	Korean Computer and Communications Promotion Association
KCE	Korea Communications Engineers
KDD	Kokusai Denshin Denwa
KIIA	Korean Information Industry Association
KISDI	Korea Information Society Development Institute
KIST	Korean Institute of Science and Technology
KMT	Kuomintang Party
KMTC	Korean Mobile Telecommunications Corp
KOTIS	Korea Travel Information Service Company Ltd
KTA	Korea Telecommunications Authority
KTAI	KTA International
KTS	Key Telephone Systems
LAN	Local Area Network
LATA	Local Access and Transport Area
LATAs	Local Access and Transport Area
LDCs	Less Developed Countries
LDP	Laban ng Demokratikong Philipino
LDP	Liberal Democratic Party
LDTA	Long Distance Telecommunication Administration
LEC	Local Exchange Company
MAFF	Market Access Fact Finding
MAMPU	Malaysian Administration Modernization & Manpower Planning Unit
MAYCIS	Malaysian Circuit Switched Public Data Network
MAYPAC	Malaysian Packet Switched Public Data Network
MBC	Mun Wha Broadcasting
MFS	Metropolitan Fiber Systems
MIS	Management Information Systems
MITI	Ministry of International Trade and Industry
MLS	Multi-Line Systems
MMBEI	Ministry for Machine Building and Electronics Industry
MMDS	Multipoint–Multichannel Distribution
MOA	Ministry of Audit
MOC	Ministry of Communications
MOSS	Market Opening Sector Specific
MOTC	Ministry of Transportation and Communication
MPT	Ministry of Post and Telecommunications
MTEL	Mobile Telecommunications Technologies
MTS	Mobile Telephone Service Corp
NAPLPS	North American Presentation Level Protocol Syntax

NCB	National Computer Board
NCC	New Common Carriers
NDCs	Newly Developing Countries
NDP	New Democratic Party
NEP	New Economic Policy
NERA	National Economic Research Associates
NITP	National Information Technology Plan
NOAA	National Oceanic and Atmospheric Agency
NPC	Northern Pacific Submarine Cable
NTA	National Telecommunications Authority
NTC	National Telecommunications Commission
NTDC	National Telecommunications Development Committee
NTIA	National Telecommunications and Information Administration
NTT	Nippon Telegraph and Telephone Corporation
NUTE	National Union of Telecom Employees
NZCS	New Zealand Computer Society
OFTEL	Office of Telecommunications
OLUHO	Okinawa-Luzon-Hong Kong
ONA	Open Network Architecture
OTC	Overseas Telecommunications Commission
OTEC	Oriental Telephone and Electric Company
Otelco	Oriental Telecom
OWNI	Oceanic Wireless Network
PAFTAD	Pacific Trade and Development Conference
PAP	People's Action Party
PBX	Private Branch Exchanges
PBX	Business Office Switchboard
PCFI	Philippine Consumers Foundation
PCM	Pulse Code Modulation
PCN	Personal Communication Networks
PDP-Laban	Philipino Democratic Party and Lakas ng Bayan
PETEF	Philippine Electronics and Telecommunications Federation
Philcomsat	Philippine Communications Satellite Corporation
PIC	People in Communication
PINs	Pacific Island Nations
PLA	People's Liberation Army
PLDT	Philippine Long Distance Telephone Company
PMG	Postmaster-General's Department
PNG	Papua New Guinea
POIs	Points of Interface
POTC	Philippine Overseas Telecommunications Corporation
POTS	Basic Rates for Basic Phone Service
POTS	Plain Old Telephone Service
PPT	Provincial Post and Telecommunications
PSA	Port of Singapore Authority
PSC	Public Service Commission

PSDN	Public Switched Data Network
PSN	Pasifik Satelit Nusantara
PSTN	Public Switched Telephone Network
PT	Post and Telecommunications
PT&T	Philippine Telegram and Telephone
PTAs	Post and Telecommunications Administrations
PTC	Pacific Telecommunications Council
PTD	Post and telegraph Department
PTIC	Post and Telecommunication Corporation
PTO	Post and Telegraph Office
PTT	Postal, Telephone and Telegraph
PUB	Public Utility Board
PUC	Public Utility Commission
PURC	Public Utilities Review Commission
RACE	R&D in Advanced Telecommunications Technology in Europe
RBOCs	Regional Bell Operating Companies
RCI	Rogers Communications Inc.
RCPI	Radio Communications of the Philippines
REA	Rural Electrification Administration
RFS	Radio Frequency Service
RHCs	Regional Holding Companies
RITE	Research Institute for Telecommunications and Economics
ROCs	Regional Operating Companies
RPOAS	Recognized Private Operating Agencies
RPOA	Recognized Private Operating Agency
RTD	Rural Telecommunications Development
RTPC	Restrictive Trade Practices Commission
SBS	Seoul Broadcasting System
SBS	Satellite Business Systems
SCPC	Single Channel Per Carrier System
SDS	Samsung Data Systems
SEACOM	South East Asia Commonwealth
SEL	Standard Electrik Lorenz
SEZ	Special Economic Zone
SIP	Subscriber Investment Plan
SITA	Societé International de Telecommunications Aeronautiques
SKDP	Sambungan Komunikasi Data Paket
SOEs	State-Owned Enterprises
SPC	Stored Program Controlled
SPC	Stored Program Control
SPC	South Pacific Commission
SPEC	South Pacific Bureau for Economic Cooperation
SPF	South Pacific Forum
SPRINT	Strategic Program for Information Technology
SPTDP	South Pacific Telecommunications Development Program
SSB	Single Side-Band

SSTC	State Science and Technology Commission
STB	Singapore Telephone Board
STD	Subscriber Trunk Dialing
STD	Subscriber Toll Dialing
STI	Singapore Telecom International
STM	Syrarikat Telekom Malaysia
TA	Telecommunications Authority
TAC	Telephone Association of Canada
TAILU	Taiwan-Luzon
TAP	Terminal Attachment Program
TAS	Telecommunication Authority of Singapore
TAT-1	Trans-Atlantic Coaxial Cable 1
TBL	Telecommunications Business Law
TCA	TransCAnada Airlines
TCTS	TransCanada Telephone System
TDM	Time Division Multiplexing
TDRSS	Tracking Data Relay Satellite System
TI	Texas Instruments
TLs	Telecommunications Laboratories
TL	Telecommunication Laboratory
TNS	Telecommunication Network-based Service
TOT	Telephone Organization of Thailand
TPC-1	the Philippine-Guam Submarine Cable System
TPC-3	Trans-Pacific Coaxial Cable 3
TTA	Taiwan Telecommunications Administration
TTI	Telecommunications Training Institute
TTWU	Taiwan Telecommunication Workers Union
TUGP	Telecommunications Users Group of the Philippines
TVRO	Television Receive-Only
TWU	Canadian Telecommunications Workers' Union
UDAS	Universal Database Access Service
UMC	United Microelectronics Corporation
UNDP	United Nations Development Program
UNIDO	United Nationalist Democratic Organization
UPS	Universal Public Services
UPU	Universal Postal Union
USTR	U.S. Trade Representative
VAN	Value Added Network
VAS	Value Added Service
VENUS-P	Valuable and Efficient Network Utility Service—Packet
VLSICs	Very Large Scale Integrated Chips
VSATs	Very Small Aperture Terminals
WAN	Wide Area Network
WATTC	World Administrative Telephone and Telegraph

Index